Cave of the Winds

The Remarkable History of the Langley Full-Scale Wind Tunnel

Joseph R. Chambers

Library of Congress Cataloging-in-Publication Data

Chambers, Joseph R.
Cave of the Winds : the remarkable history of the Langley full-scale wind tunnel / by Joseph R. Chambers.
pages cm
"NASA/SP-2014-614."
Includes index.
1. Wind tunnels--History. 2. Langley Research Center--History. 3. Aeronautics--Research--United States--History. 4. Astronautics--Research--United States--History. I. Title.
TL567.W5C47 2014
629.134'52--dc23
2014006904

Table of Contents

Table of Contents

The huge Langley Full-Scale Tunnel building dominated the skyline of Langley Air Force Base for 81 years (1930–2011). The results of critical tests conducted within its massive test section contributed to many of the Nation's most important aeronautics and space programs. (NASA L-79-7344)

Preface

On October 14, 2009, over 300 employees and friends of the National Aeronautics and Space Administration (NASA) Langley Full-Scale Tunnel gathered at the H.J.E. Reid Conference Center of the Langley Research Center to reflect on the tunnel's history and celebrate its contributions to the aerospace heritage of the Nation. After 78 years of research activities, the wind tunnel had completed its last test on September 4 and was scheduled to be dismantled and demolished. On this beautiful fall afternoon, a multimedia presentation on the history and contributions of the facility was enjoyed by all, guided tours of the tunnel were held for attendees and former staff members, and old friends relived careers and memories. To those unfamiliar with this fabled structure, the first visit to its massive test section left them in awe and deeply impressed. Standing in the facility's open test section, the crowd was treated to onsite briefings on the high-priority projects and famous people who had used its testing capabilities. Combining its gigantic size and workman-like atmosphere, the facility truly earned its legendary status. Those in attendance who had participated in the tunnel's historic test programs experienced deep feelings of pride and dedication for the countless achievements that advanced the state of the art for civil and military aircraft. The afternoon's events and camaraderie provided an emotional reflection on a national treasure that will live forever in the lore of aerospace history.

The Full-Scale Tunnel was constructed by the National Advisory Committee for Aeronautics (NACA) during an era when biplanes and dirigibles dominated aviation. The huge, cathedral-like facility was the largest wind tunnel in the world when it began operations in 1931. When the press and public first viewed the unprecedented facility, they were highly impressed by its gigantic dimensions. Writers for magazines such as *Popular Science* and newspapers such as the *New York Times* referred to the tunnel as the "Cave of the Winds" from its first operations through the end of World War II. George Gray's 1947 book about the accomplishments of the NACA during the war, *Frontiers of Flight*, also discussed the "old cave of the winds" at Langley.[1] Even Abe Silverstein, who helped design the tunnel and later became head of its operations, used the name in day-to-day discussions.[2]

By providing the capability to test full-scale aircraft within the carefully controlled and instrumented conditions of a wind tunnel, the Full-Scale Tunnel produced invaluable data for aircraft designers while conveying a message to the international scientific community and the media that the NACA was a world-class research institution. The unique design of the tunnel included the first semi-elliptical open-throat test section with twin propellers for airspeed control and a floating-frame aircraft-support mechanism mounted on dial-type scales for measurements of aerodynamic data.

The legendary contributions of the Full-Scale Tunnel and its staff to the Nation's military and civil aerospace programs began in the 1930s. The initial objective was to provide fundamental aerodynamic data on full-scale aircraft, including the effects of components such as landing gear, propellers, and wing designs for the fabric-covered, fixed–landing gear airplanes of the period. This objective was met within a few years, but in an often-repeated

The Full-Scale Tunnel became the centerpiece of National Advisory Committee for Aeronautics conferences and visits by dignitaries from industry, the military, and Congress. This photograph shows attendees of the NACA Eighth Annual Aircraft Engineering Research Conference in 1933. The aircraft in the test section is a Navy XO4U-2. (NASA L-8481)

example of the timeless value of the facility, new requirements and applications arose for the tunnel far beyond those envisioned by its designers. For example, the early testing had indicated unexpectedly high performance penalties from aircraft components and fabrication techniques, such as antennae, air intakes, rivet heads, and lapped panels for metal aircraft. During World War II, applications of these previous discoveries served as a foundation for improvements in the Nation's military aircraft through drag reduction and better aerodynamic engine cooling. The tunnel operated around the clock, 7 days a week, during the war years. Prototypes and operational versions of virtually every high-performance fighter aircraft were evaluated, resulting in countless design improvements that gave American pilots a critical edge in combat. The researchers at the Full-Scale Tunnel became specialists at providing solutions to problems caused by inadequate engine cooling and deficient aerodynamic stability and control.

After the war, the mission of the facility in the 1950s reverted back to basic studies of emerging configuration variables such as swept-back wings and advanced high-lift concepts such as active boundary-layer control for wings. Although the international focus on jet

During World War II, the Full-Scale Tunnel played a key role in resolving operational issues encountered by U.S. Army and U.S. Navy pilots. Here, a Chance Vought F4U Corsair is being prepared for testing in February 1945. (NACA LMAL 42728)

aircraft in the postwar period emphasized an interest in "higher and faster" capabilities, transonic and supersonic airplanes often exhibited poor low-speed performance and stability characteristics that could be readily evaluated and improved by tests in the Full-Scale Tunnel. The facility remained an important test facility for the NACA and its successor, the National Aeronautics and Space Administration, in the space age. New capabilities were developed for unique free-flying-model tests, which were in high demand to test unconventional vehicles such as vertical takeoff and landing aircraft, parawing vehicles, supersonic transport designs, and reentry configurations. The tunnel also became a unique national facility for evaluating the dynamic stability and control characteristics of high-performance fighter aircraft at high angles of attack. From the 1960s through the 1990s, these tests for the military services included every fighter design from the F-4 to the F-22.

The Full-Scale Tunnel had an extended lifetime despite its low air speeds (maximum speed of about 100 miles per hour [mph]) because all reusable aerospace vehicles—whether in subsonic, transonic, supersonic, or hypersonic configurations—must take off, land, or transition through low-speed flight where off-design aerodynamic issues might unexpectedly occur. Although numerous high-speed wind tunnels have been developed by NASA to evaluate the capability of vehicles to perform the high-speed design mission, subsonic tunnels are still required for studies of critical issues and the ability to complete operations in a satisfactory manner. The versatility of low-speed testing in a large subsonic tunnel without a closed test section was another factor that extended the life of the tunnel. This capability attracted many civil and military organizations to request evaluations of unusual test subjects, including dirigibles, submarines, radar antennae, gliding parachutes, inflatable airplanes, ground vehicles, and designs for other wind tunnels.

As a NASA facility, the tunnel continued to function under a double-shift operation, with typical backlogs of 1 year of scheduled tests. On many occasions the demand for data was so high that the day and night shifts involved different test subjects. For example, day-shift operations might have involved free-flight tests of a new fighter configuration while night-shift testing evaluated the aerodynamic characteristics of a subscale model of an advanced supersonic transport. The variety and intensity of the tunnel test programs were invigorating to the tunnel staff and led to high morale and a team spirit of accomplishment. After decades of operation, the tunnel underwent modifications and upgrades in 1977 and 1984 to improve its mechanical status and data-acquisition systems.

Arguably, the Full-Scale Tunnel test programs involved more configuration-specific projects and problem-solving exercises than any other Langley tunnel, leading to very popular media publicity and interactions with high-priority national projects. The resulting industry, military, and academic interests in the test activities brought a continuing flow of interesting and critical work to the tunnel and its staff. In addition to attracting world-class NASA researchers, the environment stimulated university students who took advantage of NASA's cooperative education program through onsite work assignments at Langley. Word of the activities at the Full-Scale Tunnel quickly spread through incoming co-op student groups, and given a choice of assignments, the students invariably requested working at least one tour of duty at the tunnel. Many a student was motivated to pursue a career in engineering as a result of such assignments.

Attendees at the closing ceremony for the Full-Scale Tunnel in 2009 were treated to a guided tour of the decommissioned facility by former members of the Full-Scale Tunnel staff. Here, NASA retiree H. Clyde McLemore points out some of the features of the test section of the Full-Scale Tunnel. (NASA 901P0410)

By 1995, dramatic changes in the domestic and international political and technical scenarios began to overtake the future of the tunnel. After reviews of the Nation's anticipated wind tunnel testing requirements, as well as its own future aerospace programs, NASA declared that the Full-Scale Tunnel exceeded mission needs. The facility was the oldest operating wind tunnel in NASA's inventory. It was decommissioned in October 1995—but its story continued.

Within a year, the facility began operations again under an agreement with Old Dominion University (ODU). ODU operations provided a versatile nonprofit engineering research facility for studies by graduate students and contracted use by private customers in the fields of aviation, nonaerospace technology, and automotive transportation, including the NASCAR community.

The author's personal memories of the Full-Scale Tunnel and its importance to the history of U.S. aerospace technology are especially vivid, fueled by 19 years of onsite work experiences and almost 50 years of association with the facility. Arriving at the tunnel building on my first day of work at Langley in July 1962, I reached down to open the entrance door when it suddenly opened and out came astronaut Scott Carpenter—outfitted in his spacesuit—with his support crew, having just completed a session in the Project Mercury flight trainer, which was located in an office area in the building. Seeing a NASA astronaut face-to-face for the first time was a thrill, but as I entered the office hallway I was overwhelmed by a huge display of oversized wall photos of famous World War II fighter aircraft that had been tested there during the war years. After checking in and meeting the staff, I was given a tour of the building and the tunnel test section, where I was overwhelmed by the size of the gigantic 30- by 60-foot test area. Later that day, I participated in a test of a free-flying model of the M2-F2 lifting body (made famous in the *Six Million Dollar Man* television show) to determine its dynamic stability and control. I finished my day by helping with tests of a model of the emerging F-111 fighter-bomber, and I took home a stack of NACA reports detailing the results of famous Full-Scale Tunnel projects. And I was being paid to work there!

In 1974, I was honored to become the head of the Full-Scale Tunnel, and I vowed to document the remarkable story of this historic facility after my retirement. One of my early actions was to have the existing technical reports in the organization's files, which were written by the tunnel's staff through the years, bound for preservation. Many of these documents were classified when written and had been very limited in distribution, with only a few copies dispersed to the military. Luckily, these rare and valuable documents continued to be cared for after I left the facility, and as a result, they were available for research at the start of this writing task. Unfortunately, many of the reports are no longer available in the NASA library; therefore, the primary source for documents was the bound volumes now in the care of the Langley Flight Dynamics Branch.

Having 50 years of association with the tunnel, I began to formulate the contents of this book with a comfortable feeling of knowledge of the activities, people, and contributions of the facility—but as I interviewed former staff members and reviewed surviving documents and photographs, I quickly realized that many previously unknown facts and historic events had not been recorded because of national security, lack of time, or intentional withholding of information (sometimes to hide embarrassing situations or results). In addition, I was struck by the fact that past documentation of Langley's Full-Scale Tunnel did not elaborate on interesting details of the test programs, such as the stimulus and sponsor for the activity, the test objective, important results, and the subsequent impact on aeronautics. Finally, the staff and its leaders had become, in many instances, some of the most important and colorful individuals in the histories of the NACA and NASA. Brief documentation of their personalities and careers was another goal of this task.

The research required to document the history of the tunnel and its staff members has been very challenging. Many of the NACA records of tests in the Full-Scale Tunnel were given very limited distribution as classified memorandum reports for the services, a large number have been destroyed or lost, many of the tunnel's results were only transmitted verbally, and key contacts have long since passed on. The surviving tunnel test log does not include many classified studies that were conducted from World War II to 1995, and it does not list several years of important free-flight-model projects conducted in the 1960s. Such omissions required considerable research into technical reports, dating of photographs, and fitting pieces of the puzzle together. In collating material and researching facts for the preparation of this book, my own knowledge of the extremely significant aeronautical events associated with the tunnel has been broadened and enriched. Errors and omissions of material in technical discussions or other sections can be totally attributed to me and hopefully accepted with great apologies. I sincerely wish that my personal selection of topics and individuals discussed herein meet with the approval of the reading audience, who I hope will understand the difficulty in making such selections.

The material is presented by time periods in the life of the Full-Scale Tunnel, and the discussions in each chapter are organized according to the following topics:

- The environment (technical thrusts, personnel issues, etc.)
- The leaders (organizational leaders)
- The facility (modifications, improvements, rehabilitation)
- Research activities (brief summaries of important events)
- Endnotes containing references and more details for specific topics

The historical significance of the Full-Scale Tunnel was formally recognized when it was designated a National Historic Landmark in 1985 by the National Park Service. Under the outstanding leadership of Mary Gainer, Langley has developed a multimedia Web site that captures the historical memorabilia of the tunnel, including its history, photographs of technical activities, virtual tours of the interior of the tunnel, scanned documents, videos of test projects, and interviews with retired researchers. The site is available for readers desiring more detailed information at *http://crgis.ndc.nasa.gov/historic/Langley_Research_Center*.

Endnotes

1. George Gray, *Frontiers of Flight: The Story of NACA Research* (New York: Alfred A. Knopf, 1948), p. 39.
2. P. Kenneth Pierpont (member of the staff of the Full-Scale Tunnel from 1942 to 1944), interview by author, September 12, 2011.

A general NACA wind tunnel and flight investigation of aerodynamic loads on canopy enclosures included tests of a Curtiss SB2C-4E. Langley's chief test pilot Herbert Hoover experienced potentially fatal injuries as a result of an impact with a departing Helldiver canopy during an earlier test flight. (NACA LMAL 46741)

Acknowledgments

Sincere thanks to the dozens of current and retired NACA and NASA employees who shared their invaluable memories and personal files in the preparation of this book on the historic Langley Full-Scale Tunnel. The effort to chronicle and document nearly 80 years of research and development activities associated with the facility was only possible with the assistance and encouragement of these individuals. Thanks to their participation in this activity, the rich history of the projects, people, and accomplishments associated with the tunnel will now be preserved for current and future generations. The unique and remarkable story of the Full-Scale Tunnel is an indelible part of the advancement of aeronautical technology by the United States and demonstrates the value of investing in research.

Special thanks go to Anthony M. Springer of NASA Headquarters for providing the encouragement and mechanism for this undertaking, and to Gail S. Langevin of the Langley Research Center for coordinating the preparation and administration of the effort. Gail's provision of access to the Langley historical archives was especially valuable. The personal efforts of Langley's Sue B. Grafton to preserve many historic documents and photographs of activities at the Full-Scale Tunnel over the years was a key element in the success of this undertaking and deserves special recognition. Mary E. Gainer and her staff of interns provided thousands of scanned photographs, digitized films, and documents that were within their responsibilities for preserving Langley's culture and history. The contributions of interns Stacey Jones, James Baldwin, Matthew Robinson, and George Sydnor are especially appreciated. Thanks also to Caroline A. Diehl for her support and encouragement. H. Garland Gouger and the staff of the Langley Technical Library provided invaluable documents and information, especially for the early NACA days. Langley's photo archivist, Teresa L. Hornbuckle, contributed personal research efforts and access to photographs in the Langley collection. Sandra M. Gibbs, Paul R. Bagby, George H. Homich, and Earl E. Williams contributed their photographic talents during the demolition of the tunnel. Graphics specialist Gerald Lee Pollard of Langley supplied his outstanding talent and materials.

Thanks are also extended to historian Robert S. Arrighi of the NASA Glenn Research Center, who provided material on the career and personal life of Abe Silverstein. Bob's award-winning book on the history of the Altitude Wind Tunnel served as an inspiration to me during this project.[1] Noted historians John W. "Jack" Boyd and Glenn E. Bugos of the NASA Ames Research Center contributed information on the careers of Russell Robinson and Smith J. DeFrance. Dr. Deborah G. Douglas of the MIT Museum also provided material for the manuscript.

Dr. Ernest J. Cross, Drew Landman, Dr. Robert L. Ash, and Dr. Colin P. Britcher provided material on tunnel operations under Old Dominion University. Thanks also go to John N. Ralston and Edward G. Dickes of Bihrle Applied Research, Inc., for their inputs on aircraft testing in the Full-Scale Tunnel during its management by ODU, and to Robin D. Imbar of Naval Air Systems Command for material on U.S. Navy testing in those years.

Brian Nicklas and Dr. John Anderson of the Smithsonian Institution were of considerable help in searching for rare documents concerning activities in the Full-Scale Tunnel.

The contributions of active and retired Langley personnel provided the foundation for the technical discussion and personal anecdotes presented in this work. Sincere thanks to Langley retirees Roy V. Harris, Jr., John P. Becker, P. Kenneth Pierpont, Robert J. Huston, Joseph L. Johnson, Jr., George M. Ware, Sue B. Grafton, H. Clyde McLemore, William J. Block, Joseph S. Denn, Philip F. Walker, William I. Scallion, Donald D. Davis, Long P. Yip, Dana L. Dunham, Frank L. Jordan, Jr., Donald R. Riley, Norma K. Campbell, Gloria R. Champine, Wilmer H. Reed, Charles T. Schrum, Norman L. Crabill, George W. Brooks, and Thomas B. Sellers.

Contributors from within the current Langley staff included Luat T. Nguyen, Bruce Owens, Jay Brandon, Mark A. Croom, Daniel G. Murri, Charles M. Fremaux, David E. Hahne, Gautam Shah, Dan D. Vicroy, Kim F. "Skip" Schroeder, Cheryl L. Allen, Michael D. Mastaler, John A. Cline, Thomas M. Walker, and Laura L. Eure.

E. Richard White of ViGYAN provided information on many projects he participated in at the tunnel. Vic L. Johnston of the U.S. Air Force Air Combat Command at Langley Air Force Base is also acknowledged for his help in clearing photographs for release to the public.

Special thanks to those who reviewed the draft for accuracy, content, and readability. Roy V. Harris, Bob Arrighi, Mary Gainer, John Anderson, Dan Murri, and Glenn Bugos provided superb comments and suggestions for changes that resulted in a vastly improved product. I would also like to recognize Chris Yates, Kurt von Tish, and Ben Weinstein at NASA Headquarters for their review of the manuscript, photos, and layout.

By far the most significant memory of my association with the tunnel is the honor of working with the brilliant, dedicated staff of the tunnel and those who supported us, including hundreds of engineers, technicians, data-reduction specialists, and administrative support personnel. For those who worked in the Full-Scale Tunnel, the hope is that this book rekindles your memories of the tunnel. For those unfamiliar with the facility, the hope is that you gain a perspective of the truly remarkable history of the facility and the projects and people that worked there.

Endnotes

1. Robert S. Arrighi, Revolutionary Atmosphere: *The Story of the Altitude Wind Tunnel and the Space Power Chambers* (Washington, DC: NASA SP-2010-4319, 2010).

The National Advisory Committee for Aeronautics (NACA) began flight research projects at Langley Field in 1919 using airplanes borrowed from the military. A Curtiss JNS-1 "Jenny" is shown in flight with a trailing pitot-static tube for airspeed calibration in August 1922. (NASA EL-2001-00375)

CHAPTER 1
The Awakening
1914–1928

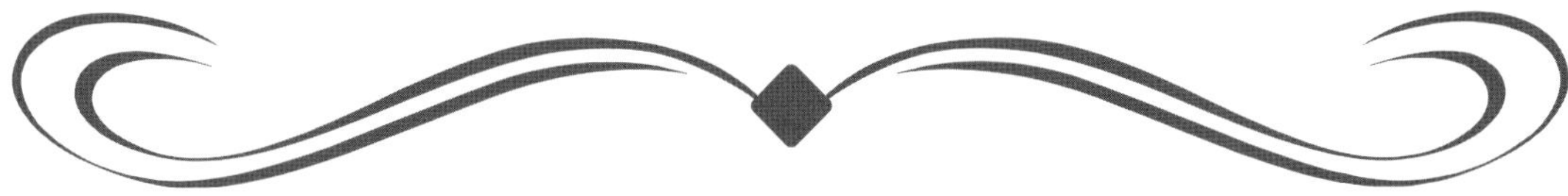

Self-Assessment

The interplay of factors that led to the construction of the Langley Full-Scale Tunnel began during a period in which the United States found itself significantly lagging in aviation technology and the development of the airplane. Following the first heavier-than-air controlled flight by the Wright brothers in 1903, the world's attention on aviation rapidly intensified, with continual demonstrations of the increasing capability of aircraft by visionary individuals in Europe. However, little or no appreciation of the potential applications of this new technology for civil or military missions was apparent in America.[1] As a result of the Nation's lack of interest in aviation, several thousand military aircraft existed in European nations at the beginning of World War I while the United States had only 23. In 1913, the new Secretary of the Smithsonian Institution, Charles D. Walcott, attempted to stimulate Congress into creating a new agency for aeronautical research and oversight. President Woodrow Wilson did not favor the proposal, but thanks to the untiring efforts of Walcott and a small group of scientists and military officers, Congress unceremoniously created a rider to the 1915 Naval Appropriations Act that created a new aeronautical advisory committee to organize and direct aeronautical research and development for the Nation.[2]

Language in the rider also included the potential future establishment of an aeronautical research laboratory. By law, the advisory committee reported directly to the President, who personally appointed its members with no salaries. Passed without debate or fanfare, the legislation gave birth to a new agency initially named the Advisory Committee for Aeronautics, mimicking an established advisory group in England. The committee members were to number 12, with 2 members from the U.S. Army, 2 from the U.S. Navy, and 1 each from the Smithsonian Institute, the National Bureau of Standards, and the Weather Bureau. An additional five members were selected from the engineering and scientific communities.[3] At its first meeting, the group proclaimed itself the National Advisory Committee for Aeronautics (NACA).

The Langley Memorial Aeronautical Laboratory in 1924. The laboratory building in the foreground housed administration and services. The building at the upper right contained Wind Tunnel 1 and the building immediately to its left was the Variable Density Tunnel. The two temporary buildings in the upper left corner were the Dynamometer Labs that later played a role in the development of the Full-Scale Tunnel. Note the railroad tracks that served Langley Field from nearby Hampton. (NACA 1375)

The Langley Memorial Aeronautical Laboratory

In 1916, Walcott consulted Army and Navy leaders on a proposal to establish a joint Army-Navy-NACA experimental field and proving ground for aircraft. Extended studies and debates among the committee members followed regarding 15 potential sites for the research laboratory. Their deliberations focused on political as well as technical issues. The Army alone had funding for the acquisition of property. Finally, the War Department procured 1,650 acres of land in 1917 near Hampton, VA, for the combined use of the services and the NACA in aeronautical operations and research.[4] The site was named Langley Field in honor of early flight pioneer Samuel Pierpont Langley. The NACA began construction of the first civilian research laboratory in 1917. The establishment, known as the Langley Memorial Aeronautical Laboratory (LMAL), was within designated areas of today's Langley Air Force Base.

The construction process proved to be a tremendous challenge because of the isolated location, swarms of mosquitoes, outbreaks of influenza, inadequate housing, and poor relations between the Army and the NACA. Delays in construction of facilities and bickering between the intended organizations led to the Army transferring its research organization to McCook Field in Ohio while retaining an aviation training mission at Langley Field. The Navy also abandoned its role at the new site and moved its testing of seaplanes to Norfolk, VA. Buildings and wind tunnels created on the property assigned to the NACA laboratory (initially Plot 16) would remain the focus of U.S. civilian aeronautical research through 1939, when the pressures and demands of World War II would result in an expansion of NACA property across the airfield's main runways. This expanded area of operations became known as the West Area, whereas the original NACA property was known as the East Area.[5]

Initial construction efforts by the NACA at the LMAL began on July 17, 1917, with excavation efforts for a laboratory building. The building was completed in 1918 and construction of the first NACA wind tunnel began. A few years later, the NACA buildings included an administration building, an atmospheric wind tunnel, a dynamometer lab, and a small warehouse. The LMAL officially dedicated its laboratory in conjunction with the completion of its wind tunnel on June 11, 1920. At that occasion, NACA Chairman W.F. Durand stated that the station should be named in honor of Samuel P. Langley, and Gen. William "Billy" Mitchell led a 25-plane flyover in salute to the NACA.[6] No one in attendance that day could have possibly foreseen the vital role that this embryonic research site would have in providing the Nation with the technology to lead the world in aeronautical accomplishments and space missions.

Initially, the NACA grasped for a unique role and mission in aeronautics. The Agency was involved in all aspects of aviation, including the resolution of patent and licensing disputes, development of navigational aids, assistance with military procurement problems, and participation in air mail experiments. While the Agency's parent committee in Washington formulated an approach to its mission requirement to "supervise and direct the scientific study of the problems of flight with a view toward their practical solution,"[7] the staff at Langley embarked on research involving aerodynamics, aircraft power plants, and flight operations.

The First NACA Wind Tunnels

Elementary Lessons: Wind Tunnel Number 1

A key decision was made when the NACA management focused the laboratory's primary efforts into the field of aerodynamics. Although wind tunnels had been in international use since the 1800s, only two such facilities existed in the United States in 1910: the Wright Brothers' Tunnel in Dayton, OH, and Albert Zahm's tunnel at Catholic University of America in Washington, DC. The Langley laboratory's first wind tunnel was primarily intended to educate its young, inexperienced NACA staff in the fundamentals of aerodynamics and provide initial training with wind tunnel testing while trying to catch up with the

Langley's first wind tunnel was used primarily to educate its young staff in the field of experimental aerodynamics and did not produce notable technical results. (NACA 4)

Europeans. As a first step in the process, a replica of a 10-year-old British 5-foot-diameter wind tunnel at the British National Physical Laboratory was constructed at Langley for initial aerodynamic experiments. Other than providing a foundation of aerodynamic testing methods and analysis techniques, Wind Tunnel Number 1 produced little in the way of breakthrough technology or advances in wind tunnel applications. In reality, the tunnel was obsolete when it was built. However, with the experience gained using the facility, the energetic young scientists of the Langley staff rapidly conceived, advocated, and put into operation a series of innovative new wind tunnels that dramatically leapfrogged existing capabilities elsewhere in the world. A brief review of three early Langley wind tunnels is presented here because experiences with the tunnels had significant impacts on the subsequent justification and applications of the Langley Full-Scale Tunnel.

Breakthrough: The Variable Density Tunnel

The first major contribution of Langley to wind tunnel technology involved the problem of extrapolating aerodynamic predictions from the results of small-scale model test conditions to full-scale aircraft flight conditions. It was widely known at the time that the inability to simulate a "scaled-down" atmosphere when testing scale models could result in erroneous predictions of several critical aircraft aerodynamic characteristics that influenced performance and landing speeds. In 1921, Langley's Dr. Max Munk proposed a revolutionary concept that involved placing a conventional wind tunnel inside a pressure vessel and pressurizing the air to levels as high as 20 atmospheres. In this manner, critical physical properties of the air-test medium could be changed under pressurization to more accurately simulate full-scale flight conditions. Known as the Variable Density Tunnel (VDT), Langley's Wind

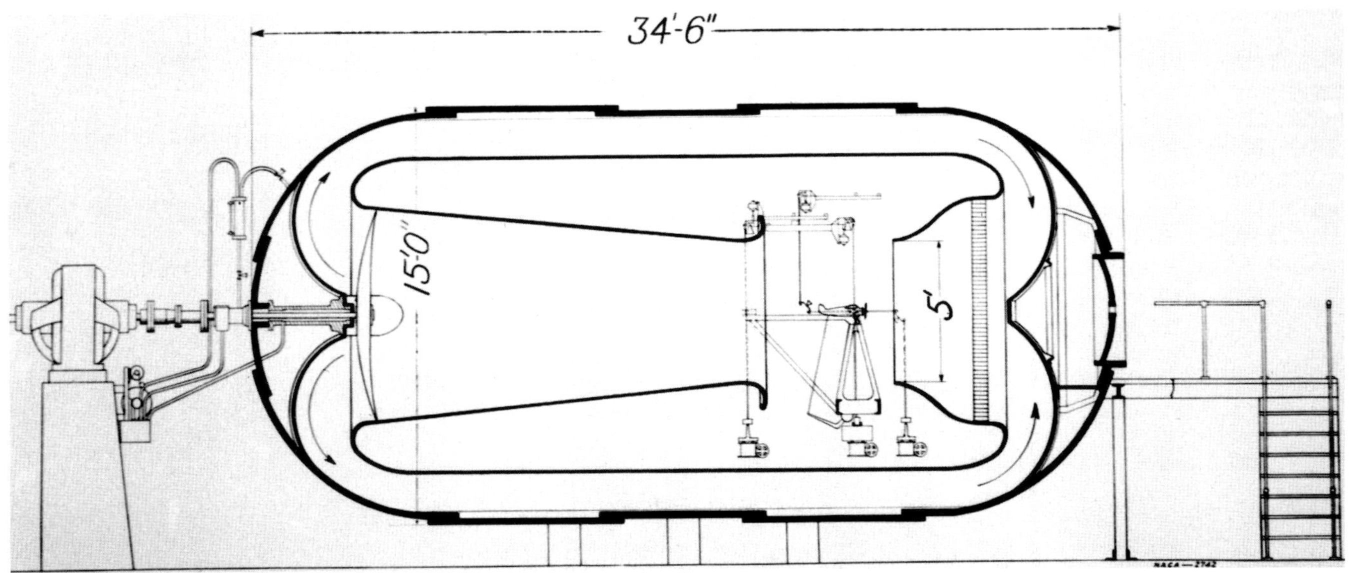

Schematic of the Langley Variable Density Tunnel showing the 5-foot wind tunnel contained in an outer pressure shell. Shown on the left are the tunnel drive motor and plumbing for pressurizing the shell. This drawing depicts the VDT after the test section was briefly changed from a closed configuration to an open-throat design in an attempt to reduce turbulence. (NASA EL-1999-00302)

Tunnel Number 2 made an immediate impact on aeronautical technology and positioned Langley in an internationally recognized leadership role in aerodynamics.

The VDT, which first operated in October 1922, was especially well suited to provide aerodynamic data on the performance of airfoils for aircraft wings, and that task became the focus of testing for over 10 years. An exhaustive effort was undertaken, including detailed studies of the effects of shape, thickness, and curvature (i.e., camber) on airfoil performance, resulting in a comprehensive collection of data on a series of 78 airfoil sections by 1933. Aircraft designers now had a reliable design tool for selecting airfoils that were the most suitable for performance objectives within design constraints such as structural weight. The legacy of the VDT was exemplified by the advances in airfoil technology and the beginning of the pressurized wind tunnels that are now routinely used on a worldwide basis.[8]

Dr. Max Munk inspects the Variable Density Tunnel on June 1, 1922. (NASA EL-1999-00258)

The staff of the Propeller Research Tunnel pauses for pictures during testing in 1928. Famous engineer, airplane designer, and head of the tunnel, Fred Weick, is on the left. Donald Wood, who would be a part of the design team for the Full-Scale Tunnel, is on the right. (NASA EL-1999-00333)

Typical test setup in the Propeller Research Tunnel in 1928 showing positions of data-acquisition personnel within the open balance frame and scale assembly, with the tunnel-speed controller to the left near the wall. All of these features would be carried into the initial design of the Full-Scale Tunnel. (NASA EL-1997-00140)

A Collier Trophy: The Propeller Research Tunnel

A critical problem emerged in the early 1920s as a result of poor agreement between data obtained on propeller performance in small-scale wind tunnel tests and in full-scale flight. The importance of this disagreement was noted by leaders in the military community, resulting in recommendations that the NACA conduct research to understand and alleviate the differences in the performance data. The NACA quickly responded to this problem in 1927 with a new large wind tunnel at Langley that would permit testing of actual full-scale propellers along with their powered engines and supporting fuselage shapes. Langley's third wind tunnel was known as the 20-Foot Propeller Research Tunnel (PRT) and was designed by Dr. Munk and Elton Miller. The tunnel was powered by a single eight-blade fan, 28 feet in diameter, with an open-throat test section and dual return passages within its exterior walls. The PRT was placed into operation in 1927 within a building that was 166 feet long, 89 feet wide, and 56 feet tall. The structure was a wood-walled, steel-framed structure, with walls on the inside of the framing so as to permit smooth flow around the tunnel circuit.

As the first large wind tunnel at Langley, the PRT provided data for first-ever analyses of full-scale aircraft components as well as measurements of engine cooling characteristics and overall propulsive efficiency. The most significant result obtained from testing in the PRT was that a streamlined enclosure (i.e., cowling) covering the exposed air-cooled engine cylinders that were in vogue at the time could significantly reduce the aerodynamic drag of the cylinders and their cooling fins, thereby reducing the fuselage drag of an aircraft by almost one third. In addition, the NACA-developed engine cowling concept provided much better engine cooling with properly designed internal baffling. Initial results of the NACA cowling were first summarized in late 1928 and were quickly disseminated to industry, where the cowling concept was immediately applied to some of the Nation's most famous civil and military aircraft, including the Douglas DC-3.

In recognition of the immediate and wide-ranging impact of this research, the NACA was awarded the Robert J. Collier Trophy for 1929 for "Development of cowling for radial air-cooled engines." The award was for "the greatest achievement in aeronautics in America, with respect to improving the performance, efficiency, and safety of air vehicles, the value of which has been thoroughly demonstrated by actual use during the preceding year."[9]

Bigger and Better

By 1928, the NACA had taken its place at the forefront of excellence in aerodynamic technology as a result of the successful programs in the VDT and the PRT; however, the science of aerodynamics was still in its infancy with regard to the level of understanding of flow phenomena and the art of wind tunnel testing. At the same time, congressional support for the NACA and its mission had increased dramatically in response to the widespread acclaim for the output coming from research at Langley and the recognition provided by the Collier Trophy event. Both scenarios set the stage for the birth of the Full-Scale Tunnel.

At first glance, the VDT could simulate the physical parameters of actual flight conditions, but it was limited in applications by two considerations. Although the VDT produced invaluable aerodynamic data for airfoils, its inherent design possessed a high level of unnatural turbulence in its airflow that produced undesirable artificial effects on test results, particularly for high-angle-of-attack conditions where maximum lift occurs for wings. This limitation of the tunnel was well known by researchers at Langley, where several approaches to the shortcoming were under consideration. In addition, the VDT could not be used to address aerodynamic issues pertinent to complete full-scale configurations. For example, the impact of the slipstream of rotating propellers and deflections of fabric-covered surfaces under air loads could not be assessed. More significantly, drag penalties due to real-world details such as surface gaps, air leaks, and engine cooling installation could not be evaluated with small subscale models. In fact, it was extremely difficult to model many of the actual physical attributes of full-scale aircraft. The PRT was capable of testing isolated components of full-scale aircraft, but it could not accommodate complete aircraft because of size limitations of its test section.

The stage was set within NACA management and in Congress for a favorable reception to a proposal for the construction of a wind tunnel capable of testing full-scale aircraft under powered conditions. In 1928, preliminary studies of the layout of a radical new full-scale tunnel began at Langley. The aerodynamic design of the massive new tunnel would prove to pose significant challenges to its designers.

Endnotes

1. J.C. Hunsaker, "Forty Years of Aeronautical Research," *Annual Report to the Board of Regents of the Smithsonian Institution for 1955*, Publication 4232 (Washington, DC: U.S. Government Printing Office, 1955), p. 243.
2. George W. Gray, *Frontiers of Flight: The Story of NACA Research* (New York: Alfred A. Knopf, 1948), pp. 10–11.
3. In 1929, Congress increased the membership to 15; in 1938, an additional amendment by Congress stated that 2 of the 15 members should be from the Commerce Department. In 1948, membership was increased to 17.
4. The NACA's initial share of the property was only 5.8 acres, but as the research facilities multiplied in ensuing years the War Department released more land, and the NACA occupied 28 acres by 1939.
5. James R. Hansen, *Engineer in Charge: A History of the Langley Aeronautical Laboratory, 1917–1958*, (Washington, DC: NASA SP-4305, 1987).
6. Ironically, Mitchell would become an outspoken critic of the NACA and railed for its disestablishment. See Michael H. Gorn, "The N.A.C.A. and its Military Patrons During the Golden Age of Aviation 1915–1939," *Air Power History* 58, no. 2 (summer 2011): p. 21.
7. Naval Appropriations Act of 1916, Public Law No. 271, 38 Stat. 930 (March 3, 1915).
8. The Langley VDT was designated a National Historic Landmark in 1985. Today, its pressure shell is preserved on the grounds of the H.J.E. Reid Conference Center at Langley.
9. National Aeronautic Association, "Collier Trophy: About The Award," *http://naa.aero/html/awards/index.cfm?cmsid=62*, accessed November 19, 2013.

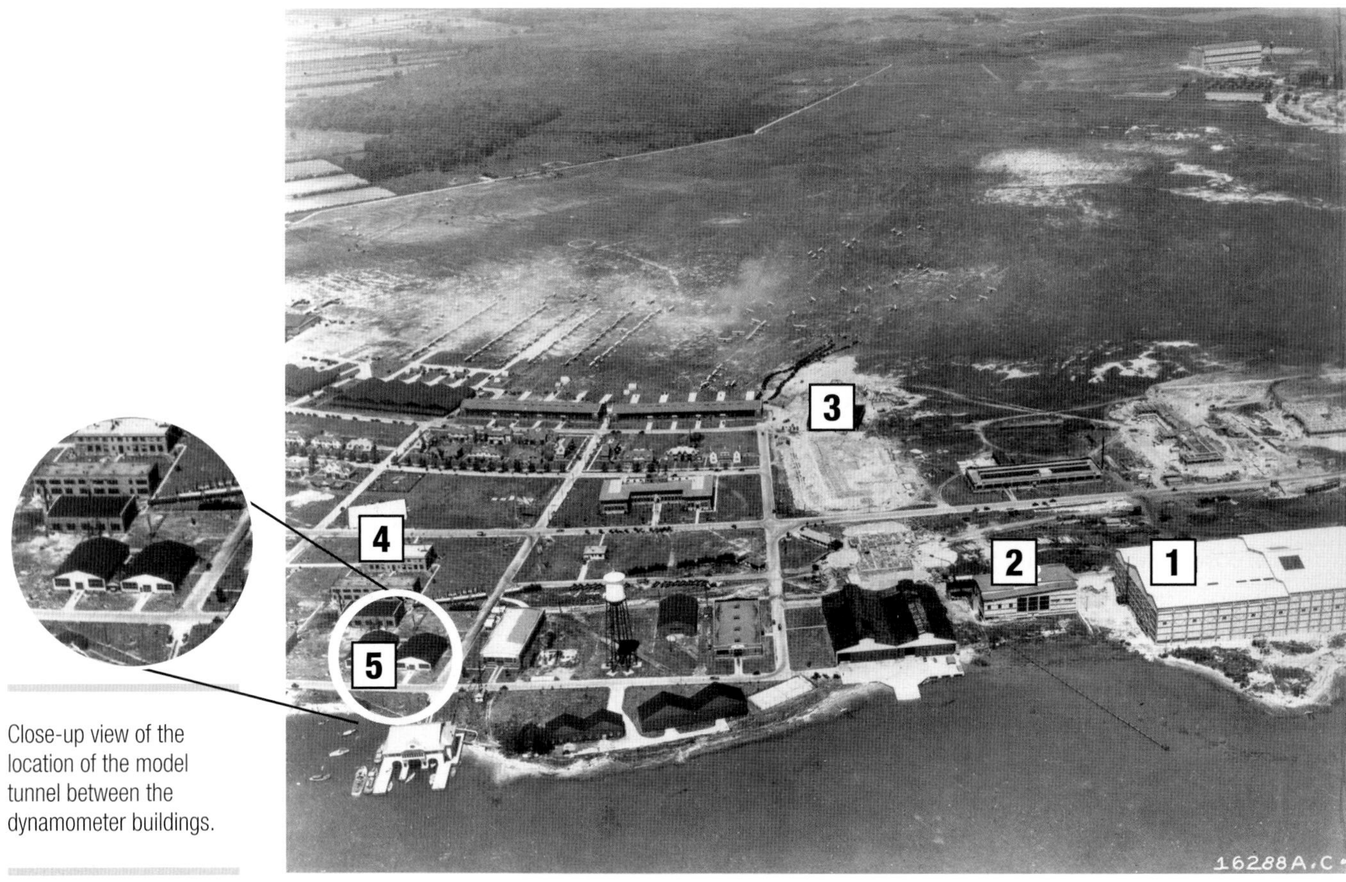

Close-up view of the location of the model tunnel between the dynamometer buildings.

This aerial view of Langley Field taken in 1931 (immediately after the Full-Scale Tunnel was constructed) shows the location of the site used in the construction and testing of the 1/15-scale model tunnel. The photograph shows the relative positions of (1) the Full-Scale Tunnel, (2) the Propeller Research Tunnel, (3) the original NACA Flight Hangars, (4) the NACA Headquarters building, and (5) the twin dynamometer buildings between which the model tunnel was located. (U.S. Air Force)

CHAPTER 2

The Birth of a Legend

1929–1931

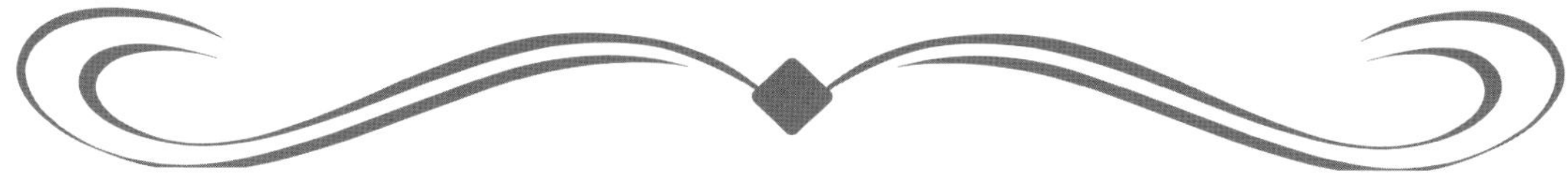

Concepts and Go-Ahead

Proposals for a "Giant Wind Tunnel" capable of testing full-scale aircraft had first surfaced at Langley during planning for the Propeller Research Tunnel.[1] George W. Lewis, who had been named Director of Aeronautical Research for the NACA in 1924, had been involved in discussions regarding the new facility and personally led the advocacy for the PRT project. Lewis had been stimulated by interactions with Dr. Max Munk, who believed that a wind tunnel with a test section of 20-foot diameter would be appropriate for the PRT in both size and costs. A meeting between Lewis and Leigh Griffith, Langley's engineer in charge at the time, was scheduled to discuss design decisions for the PRT. As a result of preliminary discussions within Langley, researcher Elliott G. Reid suggested to Griffith on April 3, 1925, that consideration be given to implementing a larger, 30-foot test section to enable testing of the same article in the wind tunnel and in flight. Reid was regarded as one of the truly outstanding aerodynamicists at Langley.

George Lewis did not respond to Reid's vision at the time, largely due to a negative review by Dr. Munk, who maintained that the 20-foot test-section dimension was adequate. Leigh Griffith tried a second time to influence the decision on the test section size for the PRT, but by then the decision to build the smaller 20-foot test section had been finalized. Dejected by the decision and other matters, Elliott Reid resigned from Langley in 1927 and later became a noted professor at Stanford University.[2]

George W. Lewis, the NACA director of aeronautical research, became the chief advocate for a full-scale wind tunnel. (NASA EL-1997-00143)

In 1928, after Reid's departure, Joseph Ames sent a letter to the Director of the U.S. Bureau of the Budget outlining the need for a new tunnel for testing complete aircraft.[3] The NACA was riding the crest of the success of the PRT, and LMAL researchers had begun homework on formulating plans for the new full-scale tunnel. The NACA submitted a request for $5,000 to the Director of the Budget for the development and design of a wind tunnel suitable for research on full-scale aircraft.[4] Congress approved the request on May 16, 1928, and in mid-1928, Henry J.E. Reid, Langley's new engineer in charge, formed a design team under the leadership of Smith J. DeFrance, with key members Abe Silverstein and Clinton H. "Clint" Dearborn, to conduct conceptual studies of possible layouts and costs of the tunnel. DeFrance and Dearborn had begun their careers at Langley conducting flight research, while Silverstein was a mechanical engineer with no training

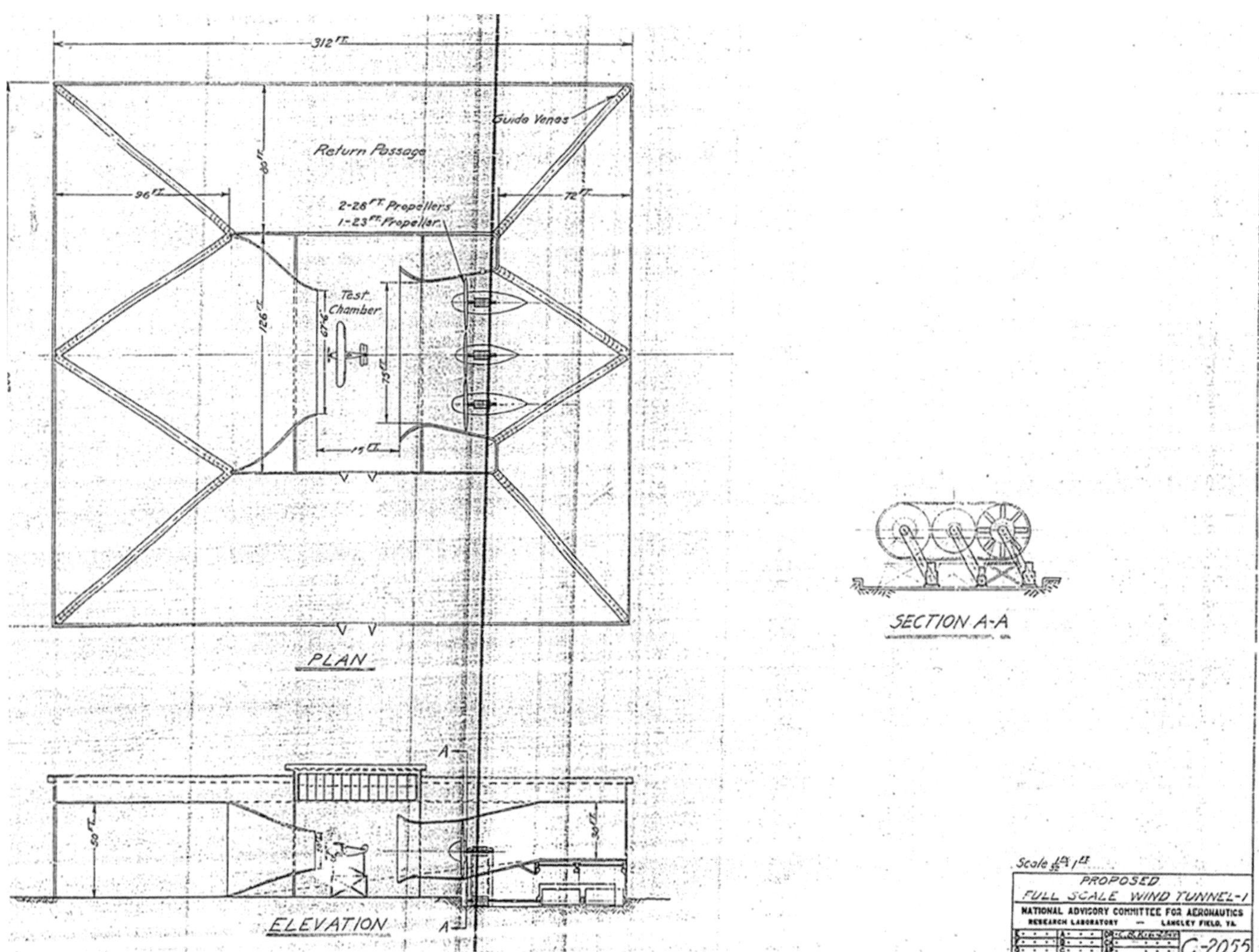

This 1928 drawing shows one of the earliest concepts considered by Smith DeFrance's team for the new tunnel. (NASA Langley Research Center Electronic Engineering Drawing 2055-1-C-1)

in aerodynamics. At the onset, the team addressed the issues of power requirements, the number of fans required, and the general arrangement of the flow path within the facility.

Engineering drawings of two early concepts for the tunnel layout in 1928 indicate a strong influence of the PRT's features on the preliminary designs.[5] The open-jet test section and airflow circuit of the new tunnel would be housed in a much larger structure, the outside walls of which would serve as the outer walls of the return passages in a layout similar to the PRT. However, estimates of the power required for a proposed 20- by 67.5-foot test section resulted in a requirement for multiple drive motors rather than the single motor used in the PRT. In one preliminary design, the team envisioned using four 2,000-horsepower motors to power 27-foot-diameter four-blade propellers on each motor. A second preliminary design proposed the use of three drive motors with an unusual propeller configuration consisting of two 28-foot propellers and one 23-foot propeller, and an abrupt change in flow path following the propeller drive section into the flow return passages. In retrospect, both layouts would have probably been unsatisfactory in view of the short-coupled flow path behind the drive propellers, which would not have provided efficient flow turning. In addition, both design concepts had dramatically oversized return passageways. A comparison of the

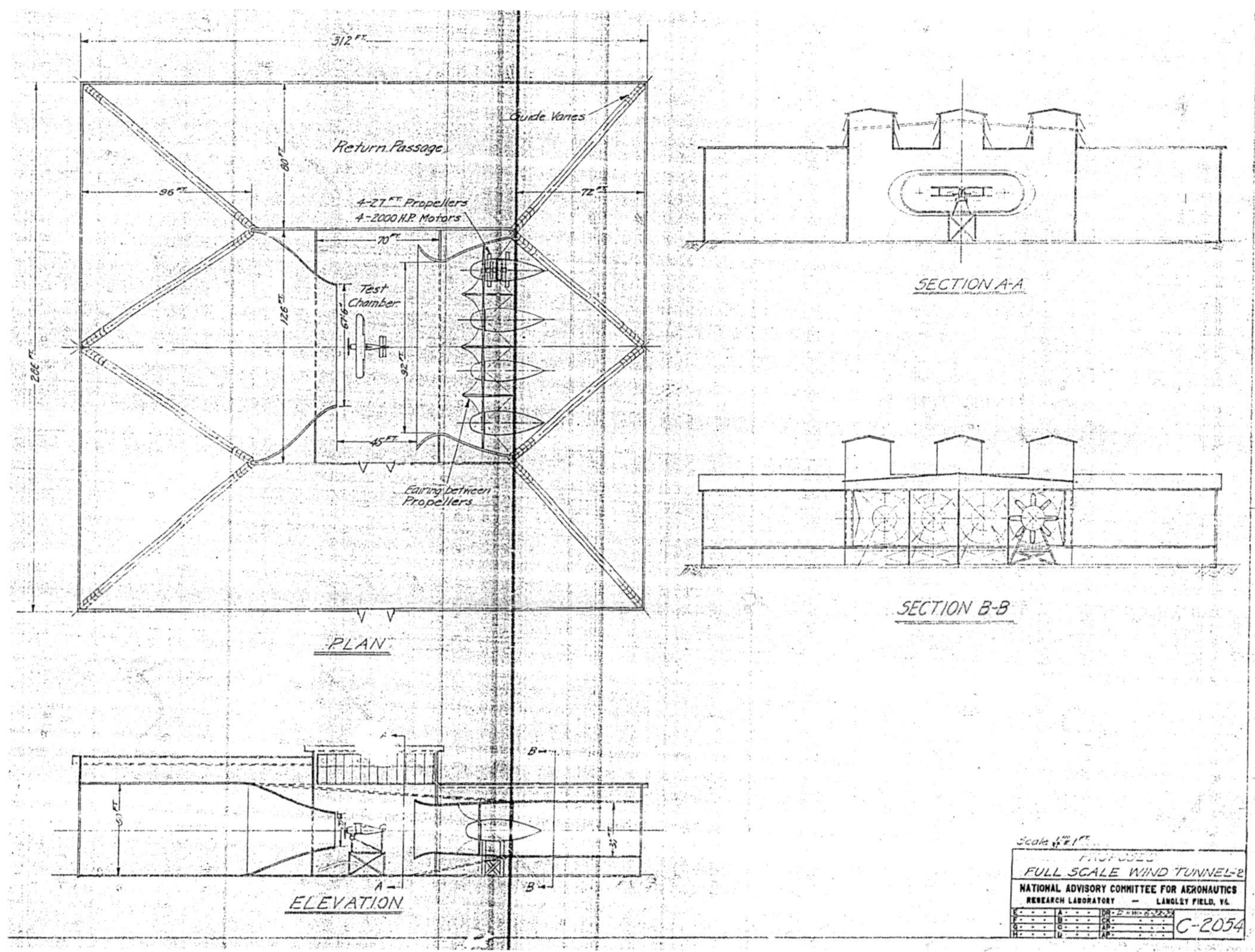

This candidate concept for a new Full-Scale Tunnel included four drive motors. The drawing was made in 1928 by Russell G. Robinson, a member of DeFrance's team. (NASA Langley Research Center Engineering Electronic Drawing 2054-1-C-1)

conceptual layouts with the final design of the Full-Scale Tunnel reveals large differences between the preliminary and final designs.

Armed with a preliminary cost estimate and layouts of the general arrangement of the new tunnel, George Lewis succeeded in obtaining NACA approval of the new tunnel, followed by congressional approval for initial funding for the construction of the Langley Full-Scale Tunnel on February 20, 1929. The next month, the Chairman of the NACA, Joseph Ames, sent a letter to the Secretary of War requesting permission to erect the new tunnel as well as a new seaplane channel (i.e., tow tank).

In March, Smith DeFrance was named head of a new Full-Scale Wind Tunnel section with responsibilities to lead the construction and operations of the new facility. Immediately thereafter, the Nation plunged into the Great Depression. Contracting prices and labor costs were, of course, very favorable during the Depression and there was no shortage of manpower. The fiscal 1930 budget approved by the 70th Congress included $525,000 of construction funds for initiation of the project, based on a total cost estimate of $900,000. The remaining funds of $375,000 were provided in fiscal year 1931. By the time the project was completed in May 1931, the total cost was just over $1 million. The expenditures amounted to a very significant investment for the NACA. This single wind tunnel would ultimately cost three times as much as all the other buildings constructed at Langley in the laboratory's first 12 years, including three laboratory buildings, Wind Tunnel Number 1, the VDT, aircraft hangars, and the PRT.[6] Preliminary work on the project officially began on April 27, 1929.

The Model Tunnel

The Full-Scale Tunnel would be the first open-throat semi-elliptical wind tunnel to be powered by two side-by-side propellers.[7] The open-throat feature offered the advantages of minimal wind tunnel wall effects compared to a closed-throat tunnel, as well as easy access for large aircraft to be mounted in the test section. However, experiences with open-throat tunnels had indicated that flow quality problems might be encountered. In particular, the design had to avoid flow pulsations caused by "organ-pipe" effects—a well-known problem peculiar to open-throat tunnels.[8] Smith DeFrance and his team recognized that a multitude of design issues needed to be addressed, including the effects on energy requirements of variables such as the transition of cross sections around the flow circuit—from the circular sections at the drive propellers through the rectangular return passages and back to the semi-elliptical entrance cone. Other primary variables were the included angle for the exit cone, the detailed design of guide vanes to ensure smooth airflow through the circuit, and the level of power required for operations. The challenge was so great that the team was concerned that a more complicated scheme, such as boundary-layer control, might be required to ensure satisfactory flow characteristics.

A request was made to George Lewis for the construction of a 1⁄15-scale model of the new tunnel for use in evaluating and modifying the performance of the full-scale version. On June 27, 1929, Lewis approved the project, which would use 2- by 4-inch framing and wooden ribs for construction at a cost (including drive motors) of approximately $3,000 and a fabrication period of 3 months. The funds for the project were allotted within the budget for the construction of the Full-Scale Tunnel. DeFrance assigned the task of directing the design, construction, and exploratory studies of the model tunnel to Clint Dearborn. Meanwhile, DeFrance and Silverstein worked with supporting groups at Langley on the design and preparation for construction of the Full-Scale Tunnel.[9]

Interestingly, management's approval of the request for construction of the model tunnel noted that it may need to be built outside beneath a roof because a suitable location was not available in any of Langley's existing buildings. The specific location of the model tunnel during its construction and operation is a prime example of the flexibility, ingenuity, and "can do attitude" of the young NACA staff. Construction of the wooden model tunnel proceeded outdoors in August 1929 in a vacant area between the two LMAL dynamometer (i.e., engine powerplant) buildings near the NACA administration building. The buildings, known as the North and South Dynamometer Labs, had been in operation since 1922.

Scenes at the construction site for the 1/15-scale model tunnel in 1929. In the photograph at the top, a workman holds a section of guide vanes for installation in the tunnel. The tarpaulin roof is seen on the right, the open door of the South Dynamometer Building is in the middle background, and the Variable Density Tunnel building can be seen behind the automobile on the left. The photographs below showing the workman in the tunnel circuit shows the drive-section end of the model tunnel. Two electric motors were mounted on the horizontal shelf outside the tunnel wall with their drive shafts penetrating the outer wall. (NASA EL-1999-00399, NASA EL-1999-00339, and EL-1999-00337)

During the ensuing construction and subsequent operations, the model-tunnel site resembled a shantytown dwelling, with exposed lumber framework and a tarpaulin-covered roof. Dearborn's group conducted a successful test program that provided vital design information on energy requirements to DeFrance and Silverstein. After the task was completed, the model tunnel was moved in the early 1930s to the new Full-Scale Tunnel building, where it was subsequently used for over 25 years as a small-scale wind tunnel facility.[10]

The actual design of the model tunnel did not duplicate the full-scale version's layout. For example, the two drive motors for the model were located outside the tunnel structure with drive shafts that penetrated the exterior wall to drive two fans in the flow circuit. The model also incorporated return passages that were geometrically different than the full-scale arrangement. Unfortunately, the aerodynamic flow quality predictions of the Full-Scale Tunnel based on the model tests and concurrent design analysis by Silverstein and DeFrance were in complete disagreement with the initial characteristics of the full-scale version of the tunnel, as will be discussed.

The Final Design: Layout and Building

With the results of preliminary design studies and data from the model tunnel in hand, DeFrance and his team submitted recommendations in November 1929 for the general design of the Full-Scale Tunnel. DeFrance described the construction components: "It is proposed that invitations be sent out for bids on five groups of items. The first would be for one contract on the complete structure; the second the same as first, including the erection of the cones but not the fabrication, since this would be more of a shipyard job; the third would cover structural steel, cover, sash and doors, but not cones or foundation; the fourth, foundations; and the fifth, fabrication of cones."[11] The main construction contract with the J.A. Jones Company of Charlotte, NC, was signed on February 12, 1930, for $400,459, and work started on March 8. The Jones Company completed the job on December 19 of the same year. Other contracts included one of $145,239 to the General Electric Company (GE) of Schenectady, NY, on May 5, 1930, for electrical equipment. GE started the job on June 23 and completed it on February 24, 1931. The Toledo Scale Company of Toledo, OH, was awarded a contract of $16,700 for the balance scale assembly on November 29, 1930, which was completed on April 3, 1931.[12]

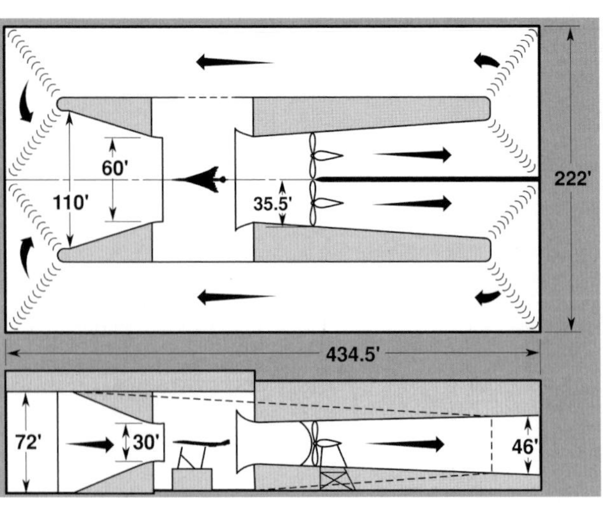

Cross-section of the Langley Full-Scale Tunnel showing the general layout and dimensions of the facility. (NASA graphic by G. Lee Pollard)

The site selected for construction of the tunnel was within the relatively small NACA property allocation near the shore of the Back River adjacent to the PRT. To the observer, the building was a peculiar structure, as the building's steel framework was visible on the outside of the exterior walls that served as the outer walls of the return passages. The overall length of the huge building was 434

The gigantic airship hangar at Langley Field that was operated by the Army. Its construction materials influenced the NACA decisions for the Full-Scale Tunnel. (U.S. Air Force)

feet and 6 inches, the width 222 feet, and the maximum height was 97 feet. The ground area covered by the building was 2.4 acres. The length of the building and the tunnel's test section were oriented in a north-south direction, with office spaces located at the south end of the building. The Full-Scale Tunnel building easily dwarfed the PRT building beside it and became a landmark at Langley Field, second only to an enormous hangar that housed Army dirigibles.

The concrete floors of the facility were 5 inches thick. The building framework was of structural steel, and the walls and roof were constructed of corrugated cement-asbestos sheets (5⁄16 inches thick, 42 inches wide, and 62 inches long). The asbestos sheets were known as Careystone Corrugated Asbestos Siding, and they were manufactured by the Philip Carey Company of Cincinnati, OH. The NACA's choice of Careystone had been based on several factors. First and foremost was its low cost, but NACA engineers were also greatly influenced by the durability, low maintenance, and fireproof qualities of the concrete-asbestos covering of the gigantic airship hangar at Langley Field that had been in operation since 1922.[13] In addition, the material had been used to cover the return passages of the PRT. Tests by the

This advertisement by the Philip Carey Company in the June 7, 1930, issue of the *Saturday Evening Post* used the upcoming construction of the Langley Full-Scale Tunnel to spread interest in the company's Careystone product. (NASA)

NACA of wet and dry Careystone test panels in February 1930 showed the material to be 3.8 times stronger than required (the maximum load the material was expected to withstand was 52 pounds per square foot; the breaking load was 196 pounds per square foot). The results of these tests were supplied to the manufacturer, but with the condition that the information be treated as confidential. The Philip Carey Company very much wanted to publicize the NACA test results (they had underbid the project in hopes of getting a strong return through an advertising campaign), but the company's request was rejected out of hand as a violation of Government policy. Nonetheless, the company placed a full-page advertisement in the June 7, 1930, issue of the *Saturday Evening Post* disclosing its upcoming role in the construction of the new tunnel.[14] The headline for the ad read, "They wanted a building material to hold 110-mile hurricanes—Daily!"[15]

The final design of the wind tunnel housed within the building was an open-throat, double-return concept with a semi-elliptical entrance cone, with a test-section width of 60 feet and a height of 30 feet. The cross-sectional area of the new tunnel was five times that of the PRT. A fairing within the exit cone transformed the single semi-elliptical flow from the test section into circular flow paths at the location of two 35-foot, 5-inch propellers. The entrance and exit cones were constructed of 2-inch wood planking attached to a steel frame and covered on the inside with galvanized sheet metal for protection against fire. Guide vanes to turn the flow around the corners of the return passages consisted of curved airfoils that were carefully adjusted to provide acceptable velocity distributions in the test section. The return passages were 50 feet wide, with the height varying between 46 and 72 feet. The research offices, shops, electrical equipment, and storage spaces occupied areas beneath the tunnel.

The Final Design: Drive System

In addition to geometric details, various power concepts had been considered in 1928, including a diesel powerplant arrangement that was stimulated by the previous use of two 1,000-horsepower submarine diesel engines to power the PRT. The Virginia Public Service Company had not been able to supply adequate electricity for the wind tunnels at Langley when the PRT had been built, and the diesel-power concept was adopted for that tunnel. In the case of the Full-Scale Tunnel, each of the two propellers would be powered by two diesel engines rated at 3,000 and 1,000 horsepower, respectively, with directly connected generators. The team also evaluated and rejected a power scheme using 30 Liberty engines driving 600-horsepower direct-current (DC) generator units. In mid-December 1929, however, Virginia Public Service agreed to supply service to the north end of the King Street

A worker inspects the original aluminum blades of the west drive motor in February 1931. The metal blades were replaced by Sitka spruce blades in late 1938. (NASA EL-1999-00409)

Bridge, which connected Hampton and Langley Field over the Back River, and the choice of electric drive motors was made.[16]

Each of the tunnel's two propellers was powered by a 4,000-horsepower General Electric motor, which could circulate airflow through the test section at speeds between 25 and 118 mph. The original propellers consisted of four cast-aluminum alloy blades screwed into

The original streetcar-type speed control handle for the Full-Scale Tunnel was presented to Joseph Walker, the legendary head of the tunnel's technician staff, on the occasion of his retirement from NASA in 1965. Today, the control is displayed in the office of his grandson at the Langley Research Center. (Thomas M. Walker)

a cast-steel hub. However, the original aluminum propellers were replaced in 1938 with wooden blades that served for the remaining lifetime of the tunnel—over 70 years.

DeFrance and his design team conducted detailed studies of the relative cost of powering the tunnel with conventional direct-current motors and control systems normally used in wind tunnels of that era. However, they found that alternating-current slip-ring induction motors together with their control equipment could be purchased for approximately 30 percent less than the conventional direct-current equipment. Thus, two 4,000-horsepower slip-ring induction motors with 24 steps of speeds between 75 and 300 rotations per minute (rpm) were installed. In order to obtain the desired range of speed, one pole change was provided. Other speed changes were obtained by varying the resistance in the rotor circuit. This drive system permitted variations in airspeed from 25 to 118 mph. The two motors were connected through an automatic switchboard to a drum-type "streetcar" controller located near the east wall in the test chamber. The control switchgear and resistor banks were located outside the flow circuit directly beneath the propellers and drive, while the control console and tunnel speed operator were located in the test chamber.

Over its lifetime, the tunnel's structural drive system—especially the motor mounts and motor fairings—experienced cracks and failures from vibratory loads. Although the problems were repaired, they persisted and concerns began to mount for daily operations at high speeds. As a result, the maximum airspeed was limited to less than 100 mph during the last 25 years of operation.

The Final Design: Test Chamber

The test section of the Full-Scale Tunnel was designed for aircraft having maximum wingspans of about 40 feet. Entrance for test airplanes to the test chamber was provided by 20- by 40-foot vertical-lift doors located in the inner and outer walls of the west return passage. The NACA flight operations and hangars were located only a few blocks away from the tunnel and, in most cases, flightworthy aircraft were towed from the flight line to the tunnel for aerodynamic tests. Tracks attached to the roof trusses supported a heavy-duty electric crane for lifting test subjects onto the balance support system. Two 30- by 40-foot skylights in the roof of the test chamber provided satisfactory lighting conditions for daytime operations, and floodlights were used for night operations. The skylights were subsequently removed during a roof upgrade following World War II.

A six-component truss balance system was used for measurements of aerodynamic forces and moments generated by the airplane under test.[17] Ball and socket fittings at the top of support struts at the front of the balance system were used for attaching landing gear or wing-spar fittings for forward support, and similar fittings were provided for the tail of the aircraft. The support struts were attached to a floating frame. The floating frame was in turn

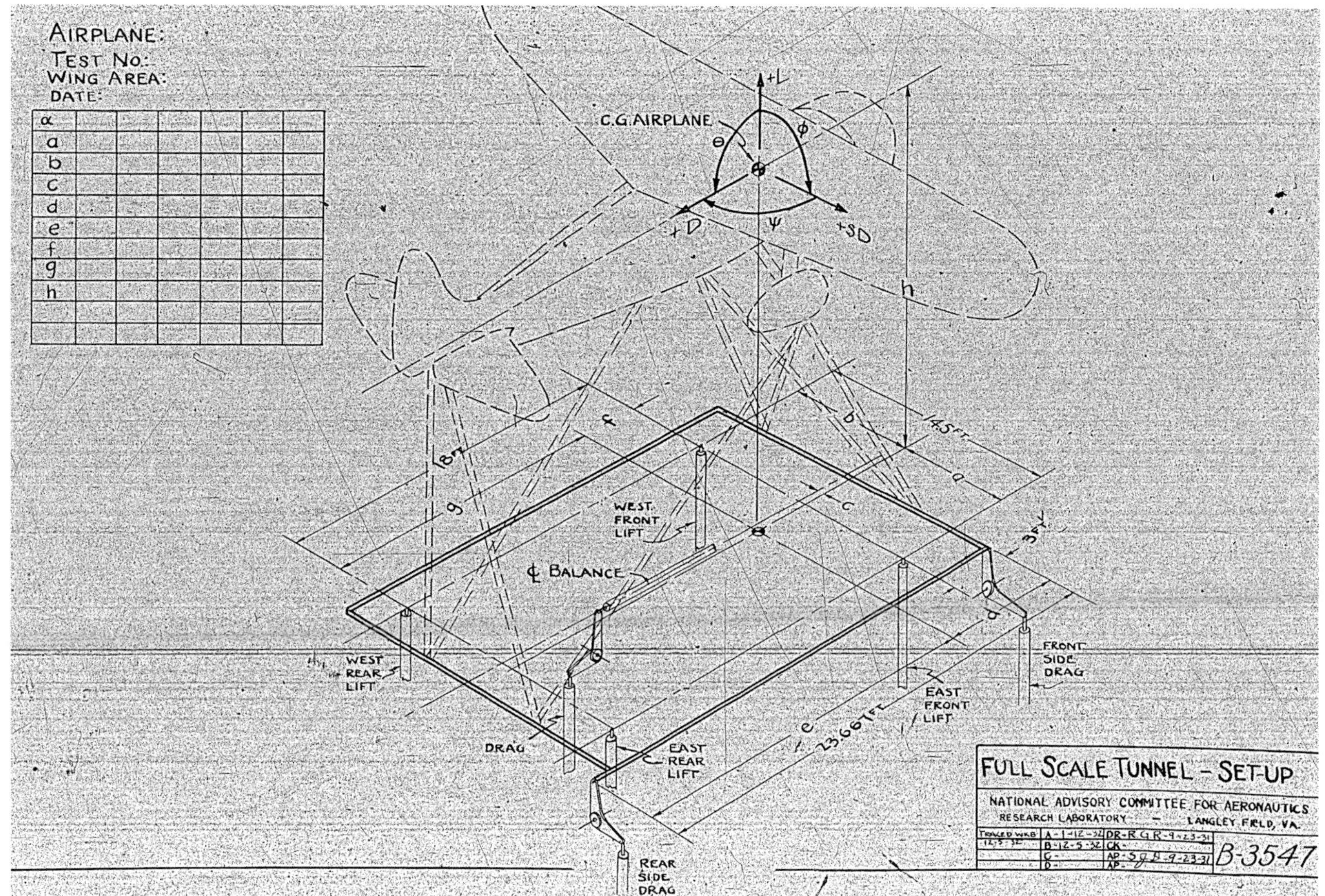

Sketch of the test chamber balance setup drawn by Russ Robinson and approved by Smith DeFrance in 1931. (NASA Langley Research Center Electronic Drawing File 3547-B-1)

mounted to a structure that transmitted lift, drag, and crosswind forces to seven dial-type Toledo scales, where they were recorded and mathematically combined to determine forces acting on the subject. By combining the magnitude of forces and known lever arms, the aerodynamic pitching, yawing, and rolling moments could be computed.

Russ Robinson's design drawing of the balance system in 1931 is an accurate rendering of the balance and floating frame installed in the tunnel. Interestingly, Smith DeFrance's report on the Full-Scale Tunnel (NACA TR-459 of 1933) erroneously indicates that a turntable was included in the original tunnel, but that was not the case. The tunnel was operated for almost 20 years without a turntable, and manually positioning the test subject in a yawed position for sideslipped conditions was a laborious, time-consuming process. In order to yaw the airplane without a turntable, the engineers and technicians had to change the relative screw-ball and socket lengths of the supporting struts such that an asymmetrical arrangement generated the yawed condition. After mathematical equations to determine the rod lengths required for each angle of yaw were solved by the data-reduction staff of the tunnel for each individual test condition for every individual airplane, the tedious task of adjusting the length of each support strut was undertaken, followed by an equally inefficient undertaking to manually align the strut shields with the oncoming wind. As will be discussed, the tunnel was finally modified in the 1950s with a turntable—to the great relief of the staff.[18]

This overhead view of a Vought OS3U-1 mounted in the tunnel during the dedication ceremony in May 1931 shows the structural detail of the floating frame and the absence of a turntable to yaw the test subject. Smith DeFrance's 1933 report erroneously notes that a turntable was installed. (NACA 5559)

Readings from the scales were provided by solenoid-operated printing devices. The early data-acquisition procedure involved oral and visual communications between the tunnel-speed operator located near the east wall at the side of the test section and test engineers and technicians near the balances. The latter crew controlled the attitude of the test airplane, its power condition, and the data-acquisition process. The tunnel-power operator sat at a table facing the rotary streetcar switch and a micro-manometer that used kerosene as a fluid to measure tunnel speed based on measurements of static pressure orifices in each return passage just ahead of the guide vanes in the entrance cone. A toggle switch on the table was used to remotely activate a light in the balance room. Within the balance room, the test director was also positioned at a table with a similar toggle switch to alert the power operator. The director could also press a switch to activate a solenoid to print the current scale readings. One or more technicians scurried around the room during tests, watching the scale loadings and changing counterweights as needed to maintain the sensitivity of the scale readings. Under steady loadings, the accuracy of the scale system was phenomenal: lift could be measured within ±2 pounds and drag measurements were accurate within ±0.5 pounds, even though the total loads were several thousand pounds.

When the test point was reached, the scale operator would activate the light at the tunnel operator's position to alert him that data were about to be taken. Upon receiving an acknowledgment signal in return, the scale operator then actuated a pushbutton switch that printed readings on all seven scales simultaneously. Typically, at least 10 readings were

Workers install the scale system and balance assembly in early 1931. (NASA EL-1999-00368)

taken at each test point. During the tunnel's first operations, the data were printed on cards, but the procedure was later modified to use paper tape for print out. The printed paper data tapes were hand-carried to the data-reduction staff, which used standard engineering parameters to compute final aerodynamic data using mechanical calculators or slide rules.

The entire floating frame and the scale assembly were first unenclosed (i.e., unshielded from the tunnel flow) in the same fashion as the PRT, but a room was constructed around the balances in the fall of 1931, as will be discussed. The support struts were shielded from direct airflow by streamlined fairings that were secured to the roof of the balance frame and were free from the balance to avoid erroneous measurements.

The overhead survey apparatus could be positioned in three-dimensional space to obtain information on the flow properties of the open jet or at specific locations over and behind aircraft. Note the research engineer on the overhead walkway. (NACA 7006)

A critical component of the tunnel's data-acquisition equipment was an overhead survey apparatus used to measure the direction and magnitude of local airflows over and behind aircraft at virtually any point of interest. Designed by Jack Parkinson, the survey rig was a car that could be rolled along the entire length of the 55-foot steel bridge attached to the roof trusses. A combined pitot (i.e., air speed), pitch, and yaw tube beneath the car could be lowered, raised, pitched, or yawed to permit alignment with the airflow at the location under study. Manometer boards were used by researchers in the survey car to align the probe with the flow by nulling readings in pitch and yaw. The survey rig was upgraded to use electronic probes in later years, but the fundamental concept served the tunnel over its entire lifetime. The ability to determine the magnitude and direction of airflow was especially important during studies of downwash behind wings and the flow around the tail surfaces of aircraft.

Construction of the new wind tunnel began in early 1930 as pilings were driven (left) and foundations were poured (right). (NASA EL-1999-00346 and EL-1999-00348)

Construction Begins

The construction work by the J.A. Jones Company began in April 1930 with foundation work at the site next to the Back River. By May 9, the area had been cleared and graded, and pilings had been driven for support of the foundation footings. The effort accelerated with simultaneous construction under way for both the Full-Scale Tunnel and the nearby NACA Tow Basin.

Russ Robinson remembered the challenging environment during the design and construction days: "Unfortunately, our design and construction shack, next to the intended site, was also next to the military's coal pile. In September, temperatures are up and humidity is about 110 percent, and sweat dripping off your nose on drafting paper covered with black coal dust that you can't keep brushed off makes for a messy completed drawing."[19]

The tasks involved in erecting and covering the Full-Scale Tunnel were inherently difficult and dangerous. The heights at which workers clung to scaffolding were very high and a misstep could be catastrophic. In October 1930, a construction worker working on the roof of the tunnel died when he stepped off the planking walkway to fetch a tool and fell through an unsupported piece of Careystone to the floor some 70 feet below.

At that time, the NACA property at Langley Field was at a premium and the shoreline of the river was within a few feet of the site, raising an obvious question about the potential impact of high water and hurricane conditions. The shoreline was later filled in and altered to avoid excessive flood threats, but floods and hurricanes still delivered damage to the site and interfered with operations numerous times through the lifetime of the tunnel.

George Lewis mandated that the Full-Scale Tunnel and the Tow Basin be dedicated together on May 27, 1931, during the Sixth Annual Aircraft Engineering Research Conference at

An aerial view of Langley Field on May 20, 1930, shows construction under way for the Langley Tow Basin (foreground) and the Full-Scale Tunnel. The Propeller Research Tunnel can be seen adjacent to the area cleared for the site of the Full-Scale Tunnel. (NASA EL-2001-00455)

The Careystone sheets were being installed in mid-August 1930 as final construction of the building was under way. Note the small construction building in the center where DeFrance's team was housed during construction. (NASA EL-1999-00355)

Aerial view of the Full-Scale Tunnel and the Seaplane Channel on November 3, 1930. (U.S. Air Force)

Langley. Work at the site therefore proceeded at a feverish pitch under great scrutiny by management. DeFrance and his men had about a year to deliver an operational tunnel. In May 1930, the foundation had been completed and the first steel framework for the tunnel structure was raised, and by August 7, workers were installing the Careystone walls and roof. The motor-mount structures and electrical power substation were constructed in September. Circuit breakers for the drive system were installed in October, and by early November the exterior of the building was completed. In February of 1931, the drive motors and the propellers were installed and the tunnel was turned over to the NACA.

When the Full-Scale Tunnel project had started, many naysayers claimed that the huge facility could never be built for the appropriated funding level. However, DeFrance and his crew got the job done in such an efficient manner that the NACA returned over $250,000 to the Government.[20] The action made a deep and favorable impression on the Appropriations Committee.

The first aircraft tested in the Langley Full-Scale Tunnel was a Vought O3U-1 "Corsair II" biplane. In this publicity picture taken on May 20, 1931, the test crew includes a data-acquisition engineer at the scales, a "pilot" in the cockpit, and a tunnel-speed controller located at a console at the bottom of the far wall. The tunnel was dedicated in a ceremony a week later. (NASA EL-2002-00594)

The Dedication Ceremony

Construction of the Full-Scale Tunnel was accelerated in the spring, and the facility was declared "operational" and ready to be formally dedicated as planned on May 27, 1931—meeting the requirement of George Lewis. The Sixth Aircraft Engineering Conference was presided over by Dr. Joseph Ames, Chairman of the NACA main committee. The dedication ceremony for the Full-Scale Tunnel was held immediately preceding the opening of afternoon sessions.[21] In preparation for the ceremony, the first test subject, a Navy Vought O3U-1 "Corsair II" biplane, was mounted to the tunnel balance struts for public display. The 200 invitees inspected the massive wind tunnel and the towing basin, after which Dr. Ames gave the tunnel's dedication address. The dedication ended with a demonstration of the operation of the tunnel; following introductory remarks by Dr. Ames, Smith DeFrance described the physical features of the tunnel and prepared the invitees for a demonstration of the tunnel operation:

> Because of the amount of power required to operate the tunnel and the small capacity of the local power plant, we are compelled to take the power on off-peak load or between midnight and 6:00 a.m. The amount of power permitted during the day is 750 kilowatts, which will give an air speed of 55 miles an hour. This afternoon we are operating at that speed. Before the tunnel is started, the pilot will climb aboard the airplane and after the air stream has been started he will start the airplane engine. Readings will be taken on the scales, and you will be notified by placards when the cards are moved and when the angle of attack and the angle of yaw of the airplane are changed. The pilot will now go aboard the

> airplane. Dr. Ames, I ask you, as Chairman of the National Advisory Committee for Aeronautics, to dedicate this full-scale wind tunnel.[22]

Dr. Ames then made a brief dedicatory statement followed by the demonstration. The "pilot" having climbed aboard the airplane, Dr. Ames pressed a button to energize the tunnel-drive motors, the drive propellers began to rotate, and the airstream began to accelerate. The pilot started the engine of the O3U-1 and readings were taken on the balances, while the members of the conference were notified by placards as to the magnitude of the airplane's angle of attack as it was varied during the demonstration.

Ready for Business

Smith DeFrance's Full-Scale Tunnel section grew to about 24 people in 1931 at a time when the entire Langley staff numbered 257. In addition to Clint Dearborn and Abe Silverstein, the group included outstanding aerodynamicists Russ Robinson and Jack Parsons, who had both come to Langley from Stanford. Together with DeFrance, they would later form the nucleus of wind tunnel operations at the NACA Ames Aeronautical Research Laboratory (today's NASA Ames Research Center). Silverstein would go on to further fame at the NACA Aircraft Engine Research Laboratory (subsequently renamed the NASA Lewis Research Center and later the NASA Glenn Research Center). Several other researchers

Smith DeFrance and Edward R. Sharp (chief clerk and property officer) pose with the original team of researchers at the Full-Scale Tunnel in 1931. Several of these individuals became legends in the legacy of NACA and NASA aeronautics and space programs. Left to right, front row: unknown, unknown, John "Jack" Parsons, Donald W. Wood, Abe Silverstein, unknown, and Russell G. "Russ" Robinson. Second row: Dale H. "Mac" McConnaha, unknown, Edward R. "Ray" Sharp, Smith J. DeFrance, and Clinton H. Dearborn. (John P. "Jack" Reeder Collection)

from DeFrance's old flight research organization at Langley also joined the section, including Dale H. "Mac" McConnaha, Manley Hood, William C. Clay, and James A. White. The group also included several technicians with special credentials for being qualified as aircraft mechanics.

At the dedication ceremony, DeFrance had addressed the application of the new tunnel:

> This is the first wind tunnel ever constructed for the purpose of testing complete full-sized airplanes, and as such it will fill a very important place in the field of aeronautics. Its principal use will be in the determination of the lift and drag characteristics of an airplane. Previously it has been necessary to do this from glide tests in flight, and sometimes the tests have been very lengthy because of inability to control test conditions. Here we will be able to control the test conditions, and to obtain the polar of an airplane in approximately one hour whereas it might take a month in flight. In this tunnel we will be able to study control, especially control at low speeds and at high angles of attack; and the drag of air-cooled engines, and of water-cooled engines with radiators, under practically the same as flight conditions.[23]

Abe Silverstein recalled that the intensity and drive of George Lewis to have his wind tunnel cathedral built had overshadowed any detailed planning for the use of the facility. In fact, Silverstein revealed that there was no specific research program in mind for the tunnel when it became operational. Although Lewis knew that building the tunnel was a good idea because it would be a vital tool for research, he had no detailed plans for the research to be conducted. In Silverstein's view, Lewis had adopted a "build it and they will come" philosophy for the new tunnel. Within the higher levels of the NACA, the vision was that industry and the military services would provide the requests and technical direction needed to effectively utilize the new facility. At the dedication ceremony, several industry and military speakers had already voiced opinions to the NACA relating to the use of the tunnel. In his dedication speech, Dr. Ames had said:

> The completion of this wind tunnel opens up a new vista of important problems, the solution of which I am confident will mean much toward increasing the safety and efficiency of aircraft. The Committee has received many suggestions for research problems from the military services and from aeronautical engineers, which will provide a research program that will keep this piece of equipment in continuous operation for a long period.[24]

Birthing Pains

Once the gigantic new wind tunnel had made its debut, DeFrance's staff raced to establish and improve the flow quality of their new tool, streamline operational procedures, and correlate lift and drag measurements obtained from the tunnel with flight data. The summer and fall of 1931 brought considerable pressure on the staff because of numerous operational issues and intense scrutiny from supporters as well as opponents of the project.

The Navy's Vought O3U-1, a two-seat observation biplane, was the first airplane tested in the Full-Scale Tunnel, at the request of the Navy's Bureau of Aeronautics. Produced at Vought's East Hartford plant in Connecticut, the aircraft could be configured as either a landplane or a seaplane. It was a distant descendent of the 1917-era Vought VE-7 (which had been the subject of many NACA flight experiments as well as tests in the PRT) and an upgraded version of the earlier Vought O2U-1 "Corsair I" that had been a workhorse for the Navy in the 1920s. The Navy named the new version the O3U-1 "Corsair II" and was elated when the prototype flew at 190 mph in February 1931.

The test program for the O3U-1 began on June 6, 1931, a week after it was displayed at the dedication ceremony. Unfortunately, the first "production" testing of the wind tunnel revealed several major operational issues. Despite extensive preliminary design studies and model-tunnel tests, the initial testing in the Full-Scale Tunnel disclosed aerodynamic problem areas that required major modifications. During the testing, massive separated airflow was experienced in the corners of both of the vertical sidewalls of the tunnel's settling chamber within the entrance cone, resulting in non-uniform, turbulent, and pulsating flow distributions across the test section. In addition, test-section flow interactions with the exposed balance components aggravated the unacceptable flow interference effects. This phenomenon came as an unexpected surprise to Smith DeFrance because the PRT had been successfully operated

After abbreviated testing of the O3U-1 aircraft revealed unacceptable airflow issues, major changes to the contours of the entrance cone were made to both the east and west walls of the tunnel. In this photograph of the east wall within the entrance cone (airflow from left to right), wooden ribs are being installed to change the local contours, and scaffolding is in place for the work. On the basis of the individual seated on the top scaffold near the ceiling, it would appear that today's occupational safety rules did not apply! (NASA EL-1999-00419)

The changes that occurred within the entrance cone during the modification are partially visible in this comparison. The photograph on top shows the original cone during December 1930, whereas the bottom photograph was taken in 1932 after the modification. (NASA EL-1999-00364 and NACA 6718)

without an enclosed balance/scale area for over 4 years. The O3U-1 was removed from the test section after only 3 days of testing (it would not return), and a major standdown was called to identify and implement solutions to the problems.[25]

Work to enclose the balance and scale areas was conducted simultaneously with the entrance-cone modifications. (NASA EL-1999-00421)

In August, a massive structural reshaping of the upper and lower side-wall lines (in particular, increasing the corner radii) of both sides of the settling chamber was implemented by inserting new wall inserts fitted on top of the original walls. At the same time, the exposed balance site was enclosed with a room-like structure to alleviate the flow interference effects as well as the impact of gusts from recirculating air and noise on the test crew. Abe Silverstein recalled that designing the contours of the tunnel contraction shape was the most difficult job of all. "We knew it had to be small here and large there, but we had no guidance on contouring the walls."[26] He also recalled that when the tunnel speed was increased to the "23rd Point" (out of a total of 24), the tunnel experienced a major power failure that validated DeFrance's claim that the power supply lines to the tunnel were too small.

Other issues were addressed during the startup operations, including significant movement of the bracing beams supporting the flow circuit when the tunnel was operating. Structural specialist Eugene E. Lundquist directed the strengthening of the structural components.[27]

Following a 3-month downtime for facility modifications, DeFrance's staff conducted extensive surveys of the flow across the test section using the overhead survey carriage to determine flow angularity, the uniformity of velocity distribution, and turbulence levels. Exhaustive manual tuning of the airfoil-shaped turning vanes in the tunnel circuit was undertaken, particularly for the vanes immediately behind the exit cone where the flow-turning process was especially critical. While the flow properties were far from perfect, DeFrance was very encouraged by the relatively low levels of turbulence, which were an order of magnitude lower than those of the Variable Density Tunnel and comparable to the PRT. Measurements in the Full-Scale Tunnel indicated turbulence levels of about 0.35 percent, whereas the level in the VDT had been measured at about 2.5 percent.[28]

A contract was awarded to the Southeastern Construction Company of Charlotte, NC, on June 17, 1931, for construction of a hangar extension to the southwest corner of the Full-Scale Tunnel building between the tunnel and the adjacent PRT building. Work for the $10,078 contract was started on June 29 and completed on September 17, 1931. The hangar proved invaluable as a staging area for upcoming projects and quick repair of active projects for the lifetime of the tunnel.

In August 1931, work was under way to construct an auxiliary hangar building to accommodate pretest components and aircraft. (NASA EL-1999-00420)

After tunnel flow surveys and fine-tuning of the guide vanes in the tunnel's flow circuit to enhance flow properties were completed, the modified tunnel was declared ready for its first projects in September 1931. The selection of projects for the test schedule was viewed as critical and was driven by two different goals. On the one hand, high-priority testing of specific military configurations for the military services (particularly for the NACA-supportive Navy) ranked high on management's list in order to maintain the support, funding, and influence of the services. But the desire of the research community to conduct more fundamental research on aerodynamic subjects also competed for positions on the schedule. This scheduling dilemma was a common factor in most wind tunnel operations at Langley and was a constant concern during the entire lifetime of the Full-Scale Tunnel.

By the time Smith DeFrance published his detailed description of the characteristics and development of the Full-Scale Tunnel in March 1934, a discussion of these birthing issues and tunnel modifications was not included.[29] The NACA had hidden its dirty laundry well.

The first test subject after the tunnel modifications was a Vought O2U-1 "Corsair I" biplane with modified tail surfaces to evaluate the control scheme patented by Robert Esnault-Pelterie. (NACA 5822)

Settling a Lawsuit: The First Test

The first project in the modified tunnel was a high-priority effort to define the aerodynamic effectiveness of a unique aircraft control-system concept conceived by Robert Esnault-Pelterie (1881–1957), a noted French aircraft designer and inventor who later became a world-famous rocket and space flight enthusiast. Esnault-Pelterie had rejected the wing-warping technique used by the Wright brothers for roll control and proceeded to invent the wing-mounted aileron control.[30] In addition, he reasoned that the multilever, body-shifting controls used by the Wrights and others such as Glenn Curtiss were unnecessarily complicated. Esnault-Pelterie invented a control concept that used one control stick, which rotated the elevator when moved fore and aft and the wing control surfaces when moved sideways. The rudder was operated with the pilot's feet.

Esnault-Pelterie's "joystick" concept for aircraft control was awarded a U.S. patent in 1914. Prior to the invention, aircraft such as the famous French Deperdussin monoplane had used a center stick for pitch control with a wheel for roll control. During WWI, many aircraft manufacturers adopted the joystick concept. After the war, Esnault-Pelterie won numerous financial settlements in French courts over infringements of his patent by French aircraft manufacturers and became a very rich man.

View of the modified "ring" tail assembly on the O2U-1 airplane. Data from the test were praised by the Department of Justice as a major contributor to the patent infringement case. (NACA 5831)

In 1924, Esnault-Pelterie filed suit against the American manufacturers Fairchild and Vought, and against the U.S. Government because Army and Navy aircraft built by the manufacturers used joysticks. Fairchild settled out of court in 1931. Vought and the Government, however, stood to lose about $2.5 million in the suit and built a defense case that the concept would never work as designed in his patent. Although using a center joystick, the patent defined a series of levers and pivots to move both the horizontal and vertical tails in a manner to provide pitch and roll control. In the Government's opinion, the concept was expected to create extreme adverse yawing moments and a single lever could not provide all control necessary. To prove its case, Vought modified a Navy Vought O2U-1 (an earlier version of the Corsair than the O3U-1 used in the tunnel dedication) with a framework truss tail exactly like that proposed by Esnault-Pelterie's patent, and scheduled tests in the Full-Scale Tunnel to obtain sufficient data to prove its hypothesis.[31]

Fourteen representatives for the defense (including the Army Air Corps, Navy Bureau of Aeronautics, Vought, the NACA, and Department of Justice) gathered at Langley to plan the test program.[32] The week-long test program began on September 16, 1931, and included tests of a "bird" tail and "ring" tail as drawn in the patent. The suit against Vought was dismissed in early 1932 in a New York Circuit Court on the grounds that the wind tunnel data showed that there were adverse aerodynamic interactions between the vertical and horizontal tails, a single lever of the type in the patent could not effect independent control in pitch and roll, and that "there was no novelty in the use of a well-known device, the lever."[33] The case was then appealed to the Circuit Court of Appeals, but once again the court ruled in Vought's favor.

George Lewis sent a memo to Langley stating that a Department of Justice letter to the NACA had been received that cited the Full-Scale Tunnel data as the key factor in favor of the Government, and that the experience highlighted the national value of the tunnel as a test facility. The department had also praised the dedication and expertise of Smith DeFrance, who served as the Government's expert witness in the case.

However, Esnault-Pelterie continued to appeal his lawsuit, taking the case to the U.S. Supreme Court. After 20 years of litigation, the Supreme Court upheld Esnault-Pelterie's suit for $1 million in 1938.[34]

Submarine Airplane: The Loening XSL-1

The second test of 1931 was a month-long study of the aerodynamic behavior of the Navy's Loening XSL-1 seaplane, which was designed to be packaged and carried on submarines for reconnaissance missions. Submarines equipped with fixed-wing aircraft for observation and attack missions had been used in WWI by several European nations with varying degrees of success. In 1931, the Loening Aircraft Engineering Corporation designed its Loening XSL-1, a small flying boat, for submarine trials aboard the Navy S-1 submarine aircraft carrier. With a wingspan of 31 feet, the airplane could be folded up and stowed in a watertight 8-foot tube on the upper deck of the submarine in 3 minutes. Its engine folded down on the fuselage for stowing in the tube, while the wings and propeller fitted into a special tank turret on the submarine.

Loening had conducted wind tunnel tests with a 1/10-scale model to determine stability and control of the configuration, but the results did not correlate with flight. In particular, during flight the aircraft exhibited ineffective rudder and elevator control at high angles of attack. Company officials contacted George Lewis requesting time in the Full-Scale Tunnel to determine stability and control of the XSL-1 for angles of sideslip from 0° to 18°, to obtain lift-drag polars, and to make "string surveys" to visualize flow on the wing.[35] Langley's chief of aerodynamics, Elton Miller, agreed to conduct the test as long as air-flow surveys and wind tunnel guide vane changes were made at the same time—a further indication of the continuing Full-Scale Tunnel airflow quality issues. Miller added that the differences in data due to configuration changes would be informative, even if the absolute values were questionable due to wind tunnel flow issues.

Unsatisfactory stability and control characteristics of the Loening XSL-1 "submarine aircraft" were analyzed in tests conducted in late 1931. (NACA 5925)

NACA test pilot Mel Gough stands next to the XSL-1 prior to his evaluation flight test at Langley. (NACA 5990)

A week after the tunnel tests were completed, the flying characteristics of the same XSL-1 were evaluated by Langley test pilot Mel Gough, who found the flight behavior to be very unsatisfactory due to an excessively aft center of gravity and extremely poor controllability, even in cruising flight. In fact, Gough could not take his hands off the control stick because of instabilities. After a few serious (non-XSL-1) accidents involving the development of submarine aircraft carriers, the concept was never accepted by the Navy's submarine service.

Beginning of a Core Expertise: Engine Cooling

In November 1931, DeFrance responded to a request from H.J.E. Reid, Langley's engineer in charge, regarding future plans for the tunnel. At that time, plans were to finalize data on the XSL-1, correct airflow in the tunnel before an upcoming test on the Navy P3M-1 engine, conduct lift-drag polars for several aircraft and include correlation with gliding flight results (for the PW-9, McDonnell "Doodlebug," Fairchild F-22, etc.), conduct tests of an autogiro, evaluate scale effects for several isolated Clark Y wings, and conduct flow surveys relevant to predicting the aerodynamic characteristics of horizontal tails.

The first year of testing in the Full-Scale Tunnel ended in December with a month-long investigation of engine cooling and drag reduction for the engine/nacelle/wing configuration used by the Navy's Martin twin-engine P3M-1 patrol airplane.[36] The test article was a full-scale Pratt & Whitney Wasp engine mounted in a P3M-1 nacelle on a 15-foot span of wing from the airplane. Previous NACA aerodynamic tests in the PRT on the effects of engine vertical placement on the leading edge of a wing using unpowered generic models had shown that the aerodynamic propulsive efficiency was low for the arrangement used by the P3M-1. The Bureau of Aeronautics requested the test in the Full-Scale Tunnel, but the Navy did not want to make any major structural alterations to the original wing and nacelle installation. The purpose of the test was to improve the cooling of the engine and to reduce the drag of the nacelle combination with several types of nacelle cowlings. Thermocouples were installed at various points on the cylinders and temperature readings were obtained for analysis. The results were transmitted to the Navy in a special report that cited the advantages of reshaping the aft end of the nacelle and of using a NACA cowling.

A section of the wing and engine assembly of a Navy P3M-1 twin-engine patrol bomber was tested to determine methods for improving engine cooling and reducing nacelle drag. (NASA EL-1999-00426)

Controversy Within Langley: The Wind Tunnel Wars

The factors that had enabled the NACA to justify the construction of the Full-Scale Tunnel included the ability to test full-scale test articles in a turbulence environment that was much lower than the Variable Density Tunnel. When the initial results from the tunnel indicated dramatically lower turbulence levels than the VDT, and that lift and drag data were in better agreement with flight tests, a major controversy ensued between the members of the Full-Scale Tunnel and the VDT.[37] Smith DeFrance was particularly outspoken regarding what he believed was the inability of the VDT to replicate flight conditions. Meanwhile, Eastman Jacobs, head of the VDT, cited the relatively expensive test articles used in the Full-Scale Tunnel, which made parametric testing of a large number of test articles impractical. Noted Langley researcher John V. Becker later recalled that staff meetings of the heads of the wind tunnels became heated debates, with DeFrance aggressively promoting his position that testing a full-scale airplane at near 100-mph conditions was the only way to obtain valid data.[38] Meanwhile, Langley theoreticians and physicists, such as Theodore Theodorsen, demanded a more theoretical emphasis in the activities of all wind tunnels.

Up and Running

The 2 years of justification, design, and operations of the Full-Scale Tunnel had given rise to a multitude of operational and programmatic issues. Most of the problems had been satisfactorily resolved, and planning for a responsive test schedule had begun to mature. Requests from industry and the military began to intensify, and the domestic and international aeronautical communities voiced their support and interest in the offspring of George Lewis's vision of the ultimate wind tunnel. The next 5 years would bring a period of concentrated efforts to address the capabilities of this unique facility.

Endnotes

1. Deborah G. Douglas, "Hurricanes Daily: Langley's Full-Scale Wind Tunnel and the Sublime," National Aerospace Conference, Dayton, OH, October 1, 1998.
2. Additional information on Elliott Reid's career is presented in the Appendix.
3. Michael David Keller, "Fifty Years of Flight Research: A Chronology of the Langley Research Center, 1917–1966," NASA TM-X-59314 (1966).
4. Joseph S. Ames and Smith J. DeFrance, "Remarks at the Dedication of the NACA Full-Scale Wind Tunnel," Langley Field, VA, May 27, 1931, Langley Historical Archives (LHA).
5. NASA, "Proposed Full-Scale Wind Tunnel-1 and Proposed Full-Scale Wind Tunnel-2," Electronic Drawing Files C-2054 and C-2055, NASA Langley Research Center.
6. Alex Roland, *Model Research: The National Advisory Committee for Aeronautics 1915–1959* (Washington, DC: NASA SP-4103, 1985), p. 108.
7. The Full-Scale Tunnel was not the first twin-propeller wind tunnel. The British 14-foot "Duplex" Wind Tunnel of the National Physical Laboratory was inspired by placing two 7-foot wind tunnels side by side, resulting in a 14- by 7-foot tunnel with a closed test section powered by two drive motors. A model of the tunnel was tested prior to the construction of the facility, and the tunnel was placed into operation in the early 1920s. See T. Lavender, "The Duplex Wind Tunnel of the National Physical Laboratory," British Reports and Memoranda Number 879 (September 1923).
8. The phenomenon is characterized by the surging of flow in the test section caused by the shedding of vortices within the flow from the periphery of the entrance cone.
9. Abe Silverstein, interview by John Mauer, March 10, 1989.
10. In March 2010, the author participated in a final walk-through inspection to acquire any historical documents that might have been left in the abandoned Full-Scale Tunnel building prior to its demolition. A northeaster had recently flooded the building with several inches of water, but the author discovered a single old blueprint a few inches above the water line in an aged filing cabinet. The blueprint was a 1929 copy entitled "1⁄15-Model of Full Size Tunnel," drawn by Clint Dearborn and approved by Smith DeFrance. The print, identified as D2510, was not found in the current Langley electronic drawing file collection and has been placed in the Langley Historic Archives.
11. Silverstein, interview by Mauer, 1989.
12. "Full-Scale Tunnel Costs," briefing chart in tunnel files provided by Donald Baals, June 30, 1940, LHA.
13. The airship hangar at Langley Field was built in the Lighter Than Air (LTA) area by the Harris Construction Company of New York from 1919 to 1922 for $490,607. It was demolished in late 1947. A parking lot and temporary Air Force quarters now occupy the site.
14. Douglas, "Hurricanes Daily."

15. Heading the advertisement was an artist's rendition of a test under way in the new tunnel. Interestingly, the sketch of the tunnel entrance cone had a semi-elliptical shape similar to the shape chosen for the Full-Scale Tunnel, but the test balance and internal walls were replicas of the Propeller Research Tunnel.
16. Douglas, "Hurricanes Daily."
17. NASA, "Full-Scale Tunnel-Set-Up," Electronic Drawing File B-3547, NASA Langley Research Center.
18. H. Clyde McLemore, interview by author, October 15, 2011; William Scallion, interview by author, September 18, 2011.
19. Russell G. Robinson, "Memoir for Three Sons, Toryn and Kra" (unpublished manuscript, 1997), Archives Reference Collection, NASA Ames History Office, p. 22.
20. Smith DeFrance, interview by Walter Bonney, September 23, 1974, LHA. Russ Robinson also mentioned this fact in his memoir.
21. James R. Hanson, ed., *The Wind and Beyond: Volume 1 the Ascent of the Airplane* (Washington, DC: NASA SP-2003-4409, 2003), pp. 631–635.
22. Ibid., p. 635.
23. Ibid., p. 634.
24. Ibid., p. 633.
25. DeFrance must have had flashbacks to a visit he had from famous aerodynamicist Theodore von Kármán, who toured the tunnel prior to the first test and told DeFrance, "Smitty, it will never work!" See interview of DeFrance by Walter Bonney, 1974, LHA.
26. Donald Baals, notes on presentation on the Full-Scale Tunnel by Abe Silverstein, Langley Research Center, June 1979, LHA, p. 23.
27. DeFrance, interview by Bonney, 1974.
28. Abe Silverstein, "Scale Effect on Clark Y Airfoil Characteristics from NACA Full-Scale Wind-Tunnel Tests," NACA Technical Report 502 (1934).
29. Smith J. DeFrance, "The NACA Full-Scale Wind Tunnel," NACA TR-459 (1934). When DeFrance was interviewed by Walter Bonney in 1974, he admitted that the flow in the test section was terrible, but once fixed the tunnel performed in a satisfactory manner.
30. A group headed by Alexander Graham Bell is also given credit for independently inventing the aileron.
31. "Aeronautics: Joy-Stick," *Time* (October 5, 1931).
32. Research Authorization (RA) Number 373, correspondence in the LHA.
33. "Aeronautics: Everybody's Joystick," *Time* (March 14, 1932).
34. *Flight* (March 10, 1938).
35. RA Number 374, correspondence in the LHA.
36. RA Number 375, correspondence in the LHA.
37. James R. Hanson, *Engineer in Charge* (Washington, DC: NASA SP-4305, 1987), pp. 101–105.
38. John V. Becker, interview by author, November 5, 2010.

The first Fairchild F-22 research aircraft procured by the NACA was tested to determine its lift and drag characteristics in 1932. Note the early NACA badge emblem on the vertical tail indicating NACA ownership of the aircraft. (NACA 6276)

CHAPTER 3

Mission Accomplished?

1932–1937

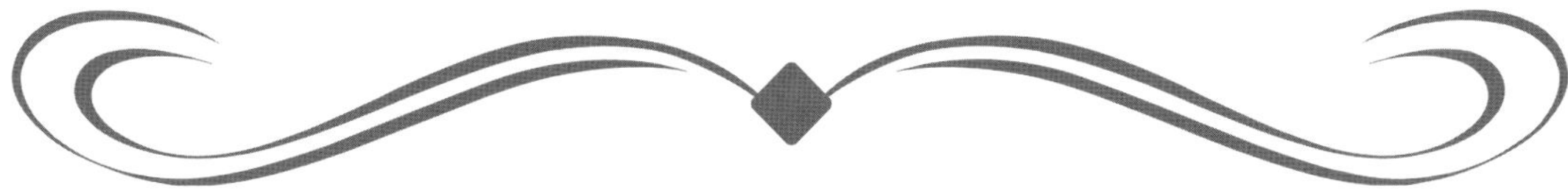

Establishing Credibility

On March 2, 1932, Smith DeFrance reported to the main NACA committee that the Full-Scale Tunnel was operating better than expected and research was quickly proceeding to establish correlation between tunnel data and flight results. Now that the visual impact of the facility had been absorbed by the aeronautical community and general public, the task at hand shifted to providing bottom-line answers regarding the accuracy of predictions based on the tunnel tests. A key part of that process was the task of correlating results obtained in flight and in the tunnel for the same aircraft. Parametric testing of general research models broadened the design database for future aircraft, and problem solving for the military was successfully accomplished. By the end of the 5-year period, the primary mission of the Full-Scale Tunnel had apparently been addressed and some began to question whether the facility had reached the end of its useful scientific life.

Prewar Research Projects

Improving Safety for Landing: The Fairchild F-22

The NACA had been conducting flight research using borrowed military aircraft at Langley Field since the early 1920s, but constraints on flight schedules and desirable modifications to aircraft soon limited the scope of the laboratory's activities. George Lewis successfully advocated to Congress for funds to acquire NACA's own research aircraft, and new aircraft began to flow into Langley by the late 1920s. Several of these NACA research aircraft became workhorse test beds for studies in the Full-Scale Tunnel. For example, Langley procured three Fairchild F-22 high-wing monoplanes that were used in very successful coordinated flight and wind tunnel studies. The first of these aircraft underwent testing in the Full-Scale Tunnel in March 1932 with the specific purpose of establishing correlation between flight results and wind tunnel data. The comparison of lift and drag data was exceptionally good and served as the earliest basis for validation of the new wind tunnel.[1]

Langley's three Fairchild F-22 aircraft also played key roles in the laboratory's focus on operating problems for flight—especially the critical landing condition. A great deal of wind tunnel and flight aerodynamic research at Langley in the early 1930s was devoted to studies of concepts to enhance safety by reducing landing speeds while maintaining satisfactory stability and control characteristics and handling behavior. New concepts for control surfaces, high-lift devices, and new wing shapes came from leaders such as Fred Weick, who had so masterfully led the Langley efforts on engine cowling technology at the Propeller Research Tunnel. Weick was promoted to oversee operations in the 7- by 10-foot Atmospheric Wind Tunnel (AWT), and his intense personal interest in promoting safety for low-speed flight stimulated his contributions while maintaining close collaboration between researchers at the Full-Scale Tunnel, the AWT, and the Flight Research Section.

In the Full-Scale Tunnel, individual test entries were conducted with the F-22 to evaluate candidate high-lift and stability and control ideas that surfaced in the Langley program. The scope of testing was quite broad and included modifying the aircraft with an auxiliary wing attached to the leading edge of the main wing, new airfoils, a "special" wing equipped with an array of spoilers and ailerons, split flaps, slot-lip ailerons, external airfoil flaps, Fowler flaps, and stall-control concepts.[2] In one test program, the high wing of the aircraft was moved to a low-wing position to study the effects of flaps on stall behavior. Although most of the investigations were primarily planned as flight projects, it was deemed advisable to check out the concepts in the wind tunnel under controlled conditions prior to flight.

In addition to the Fairchild F-22 high-angle-of-attack and stall studies, Langley researchers successfully advocated for tests of the Boeing YP-29A, an aircraft intended to be a modernized and upgraded version of the classic P-26 "Peashooter." Early flight tests had disclosed that the aircraft had vicious stalling characteristics and would be an appropriate candidate for NACA's research interests. Accordingly, the YP-29A (aircraft 34-24) entered a month-long test program in the Full-Scale Tunnel in 1935, followed by flight tests at Langley.[3]

Results gathered in the integrated Langley program on high-lift concepts and low-speed, high-angle-of-attack stability and control provided parametric variations for use by aircraft designers. Although the NACA did not invent the wing flap, it conducted extensive studies on optimizing wing configurations for high lift and lateral controls to prevent inadvertent loss of control. The results of the program are widely regarded as some of the most valuable research conducted by the NACA.

Other test objectives were also pursued with the F-22, including tests to determine the effect of wing surface finish on lift and drag. In these tests, the painted wing was replaced by a new, highly polished wing for measurements of lift-drag polars; however, the results showed negligible effects.[4] Another test to evaluate the dynamic stability of an elevator balance concept was piggybacked during one of the major test programs. The study was prompted by flight studies of the F-22 in which bob weights had been added to the longitudinal control system to enhance handling qualities. The tunnel study involved measuring the inherent damping of elevator oscillations following release of the control by the pilot. Famous Langley researcher W. Hewitt Phillips served as the wind tunnel "pilot" for the measurements, and the experience was vividly recalled by him many years later.[5] He stated, "It was a bitterly cold day in November when the tests were made with the temperature in the tunnel near freezing. I climbed up a balloon ladder into the open cockpit bundled up in an overcoat,

Coordinated flight and wind tunnel studies were conducted at Langley to evaluate the effects of high lift and stability and control concepts. Here, an auxiliary wing has been mounted to the leading edge of the main wing on the Fairchild F-22 aircraft during tests in 1932. (NACA 6650)

In this 1937 test, the wing of a Fairchild F-22 aircraft was moved to a low-wing position for evaluations of a "stall-control" concept. The modified aircraft was known as the J-2 airplane. (NACA 13844)

A Boeing PW-9 pursuit plane had been the focus of an extremely timely and valuable NACA flight study on the magnitude of structural loads encountered during maneuvers in the late 1920s. The aircraft was later tested in the Full-Scale Tunnel in 1932 to determine potential tunnel wall effects and scale effects for data previously obtained in smaller wind tunnels. (NASA EL-1999-00432)

which provided little protection when the air speed was increased to 100 miles per hour. I served as the pilot to deflect the control stick abruptly and record the motions of the elevator. The results were published as an Advance Restricted Report."[6]

Maneuvering Loads: The Boeing PW-9

Other production aircraft were applied to generic research issues. For example, in the 1920s, the LMAL had conducted high-priority flight and wind tunnel studies to address an epidemic of tail structural failures during maneuvers by military aircraft. The laboratory had acquired a Boeing PW-9 in 1926 and heavily instrumented it for measurement of aerodynamic pressures and accelerations during violent maneuvering, including dives, loops, barrel rolls, and pullups. Coupled with wind tunnel tests, this pioneering research program provided critical data and design methods for the analysis of maneuvering loads.[7] During the research project, issues were raised relative to the potential impact of wind tunnel wall effects on aerodynamic data gathered in the program. In order to address the potential wall interference effects, a brief series of tests were conducted of the Langley PW-9 in the Full-Scale Tunnel in 1932.

Generic Research

The ongoing test schedule of the Full-Scale Tunnel also included generic research projects of interest to the aeronautics industry. For example, a detailed study was made to determine the aerodynamic effects of aircraft construction techniques and components. Extensive tests were conducted to determine scale effects for a family of airfoil-wing models of various sizes

Several tunnel test entries were made in the 1930s with simple airfoil models of varying sizes to determine scale effects, wind tunnel wall effects, and the impact of surface finish and rivets on aerodynamic performance. This three-dimensional model of a Clark Y airfoil had a chord length of 6 feet and a span of 36 feet. (NACA 7000)

and to determine potential correction factors for data measured in the tunnel.[8] As early as 1932, evaluations were made of the impact of rivets on the aerodynamic performance of representative configurations, such as a 6- by 36-foot Clark Y airfoil.[9] An investigation of ground effect on the aerodynamics of a Douglas-type twin-engine transport was conducted in 1936 by testing a large, powered model above the ground board.[10]

The large airfoil models were also used for other general tests, such as a study to determine the aerodynamic drag of wing-mounted landing lights, and for experimental and theoretical studies of the flow field and wake behavior behind wings at various angles of attack.[11] Isolated horizontal-tail surfaces were also tested to develop and validate theoretical methods for predicting their effectiveness in providing stability to complete aircraft configurations.[12] Another test focused on the effects of simulated ice formations on the aerodynamic characteristics of an airfoil.[13]

Rotary-Wing Research

In addition to Jack Parsons and Russ Robinson at the Full-Scale Tunnel, their Stanford classmate John Wheatley joined the Langley Memorial Aeronautical Laboratory and pursued a career in flight testing, specializing in rotary-wing aircraft. Autogiros were of primary interest at the time, and it would be years before helicopters would be included in the research program. Langley had for several years been making intensive studies of such aircraft because of the advantages shown over conventional aircraft in safety and low-speed control. The NACA had acquired a Pitcairn PCA-2 autogiro in 1931 for research purposes. The PCA-2 had a two-place airplane-like fuselage, a tractor propeller, and a low-wing monoplane arrangement with extreme dihedral in its wingtips. The Pitcairn design was the first rotary-wing aircraft certified in the United States and is considered a forerunner of today's helicopters.

Flight tests of the Pitcairn PCA-2 autogiro had been made by Langley to determine gliding performance, loads on the fixed wing of the configuration, and motions of the rotor blades during flight. Unfortunately, evaluation of the aerodynamics of the rotor had been unsatisfactory because the drag of the rotor could not be separated from interference effects of the fuselage, nor otherwise identified in flight. In late 1933, Wheatley joined his Stanford friends at the Full-Scale Tunnel and led an investigation of the isolated autogiro rotor to determine its characteristics.[14]

Wheatley's test of the isolated PCA-2 rotor was among the first to measure high-quality quantitative information on the effects of rotor-blade pitch setting and the effect of rotor protuberances. In addition, airflow surveys were made in a plane near the blade tips. The resulting data provided fundamental information on the lift-drag ratio of the rotor and the effects of configuration variables. In coming years, the test section of the Full-Scale Tunnel would again be visited by autogiro configurations and helicopters as part of general research and specific military requests.

The NACA purchased the Pitcairn PCA-2 autogiro shown in the left-hand photo in 1931 for its initial research on rotary-wing aircraft. The isolated rotor of the aircraft was tested in the full-scale tunnel to determine its individual aerodynamic characteristics. (NASA EL-2000-00353 and NACA 9360)

Army Projects

The introduction of the Full-Scale Tunnel into the NACA wind tunnel stable of assets provided the military with an unprecedented opportunity to evaluate full-scale aircraft before flight and to resolve problems in the early stages of aircraft development programs. The military had been a key participant in the advocacy for constructing the Full-Scale Tunnel, and both the Army and the Navy were well prepared with specific requests for NACA assistance. These projects included virtually every type of air vehicle at the time, including observation aircraft, fighters, bombers, and dirigibles. The requests also extended to generic issues of a broad nature, such as the drag of landing gear and engine cooling.

A gull-winged Douglas YO-31A aircraft undergoes testing to define its performance, stability, and control characteristics in May 1932. Note the makeshift vertical-tail extension added to mitigate a problem with low directional stability. The photograph also shows early modifications to the Full-Scale Tunnel, including the balcony and walkway added to the east wall, the tunnel speed control room at the bottom of the east wall, and an air-lock door for access to the test section directly below the exit cone. (NACA 6490)

Douglas YO-31A

During the 1930s, the Army Air Corps continued acquisition of observation (i.e., scout) aircraft while aircraft configurations migrated from biplanes to monoplanes. The Douglas YO-31A evolved as a graceful gull-winged two-place monoplane for the observation mission. The design had its wing at the top of the fuselage, with gull-shaped inner wing sections faired into the fuselage. The aircraft used wire bracing and a four-strut pylon mounted atop the fuselage center section to support the wires. When early flight tests revealed deficient directional stability and other issues, the Army requested tests in the Full-Scale Tunnel to analyze and resolve the problems of the YO-31A. The configuration became the subject of many wind tunnel and flight-test activities at Langley in 1932 and 1933. The studies began with a week-long investigation of one of the prototype YO-31A aircraft in the Full-Scale Tunnel during May 1932.

A general study of the aircraft's performance, stability, and control verified the extremely low levels of directional stability experienced in flight. Analysis indicated that the problem was caused by adverse airflow shed by the pylon/gull-wing areas. The poor flow severely reduced the effectiveness of the vertical tail and rudder—particularly at low-speed, high-angle-of-attack conditions. With no possibility of modifying the unique wing-pylon arrangement, the NACA staff conducted a cut-and-try assessment of the benefits of increasing the aspect ratio of the vertical tail, using an auxiliary panel.[15] As expected, additional area at the top of the vertical tail helped mitigate the problem, and Douglas engineers designed a new tail based on the NACA's results. Subsequent models of the YO-31 incorporated a larger, higher-aspect-ratio tail.

A close-up view of the radical vertical-tail extension conceived in the Full-Scale Tunnel tests for the YO-31A. The aircraft is parked following the wind tunnel tests near the entrance doors of the recently completed NACA flight hangar in 1932. (NACA 6613)

Douglas and the Army pursued further variants of the YO-31 configuration with help from the Full-Scale Tunnel staff. Its unique gull-winged design presented a multitude of aerodynamic design issues regarding stability, performance, pilot visibility, and the aerodynamic drag of bracing wires and components such as the exposed radiator. Provided with YO-31 components and resources by the Army, the NACA proceeded to construct a full-scale powered mockup of the YO-31A for parametric configuration studies in the Full-Scale Tunnel.

The mockup, built in Langley's new flight hangar in 1933, was extremely sophisticated and included a 600-horsepower liquid-cooled engine and Prestone and oil radiators.

The primary test objective for the mockup was to evaluate the aerodynamic characteristics of four different configurations of the gull wing and wing positions in combination with a new vertical tail. The gull shape and wire-strut assembly were a major problem that led eventually to the decision to adopt a new wing design in which the gull-shaped intersection was replaced by a continuous wing center section raised several inches above the fuselage. Results of Full-Scale Tunnel tests showed that eliminating the gull intersection and moving the wing upward alleviated the directional stability problem. The raised-wing follow-on derivative of the YO-31A, known as the Y1O-43, reverted back to the original vertical tail with satisfactory results.

View of the Langley flight hangar in January 1933 with the construction of the YO-31A mockup under way. Note the engine installation and tubing entering the wings for pressure measurements. Other aircraft in the picture include a Pitcairn PCA-2 autogiro and the NACA PW-9. The NACA badge appears on the tail of the PW-9. (NACA 8106)

The modified YO-31 mockup was tested with three higher vertical wing positions that eliminated the gull-shaped intersection with the fuselage. The photo shows the highest wing position tested. Moving the wing above the fuselage eliminated the directional stability issue. (NACA 10628)

During the test program for the YO-31A mockup, the Army also requested an analysis of the aerodynamic drag associated with the radiator and oil cooler of the configuration. Smith DeFrance personally reported that the study of radiator drag revealed that, even for an aircraft as clean as the YO-31A, the radiator accounted for as much as 12 percent of the minimum drag of the airplane.[16]

Analysis of the aerodynamic drag contributed by the extended landing gear of the Lockheed Y1C-123 highlighted the magnitude of the penalty during tests in the Full-Scale Tunnel in 1932. (NACA 6714)

Lockheed Y1C-23

In response to the interests of Charles Lindbergh, Lockheed pursued the development of a fixed-gear civil aircraft for surveying oceanic flight routes. The primary aircraft developed for Lindbergh was the Lockheed Sirius, which had a fixed landing gear. Lockheed also designed a second variant with retractable undercarriage that was purchased by the Army Air Corps and given the designation Y1C-23. The single Y1C-23 was based at Bolling Field at Washington, DC, and was used to transport high-level military dignitaries. When tested at Langley, it carried the flag insignia of the Assistant Secretary of War.

In 1932, the Army requested that the NACA conduct tests of the Y1C-23 in the Langley Full-Scale Tunnel to study the effects of retractable landing gear wheel wells on takeoff performance. Both Lockheed and the Army Air Corps had suspected that the lower-surface wing openings were significantly degrading aerodynamic performance. In fact, the aircraft had gained notoriety as requiring an excessive takeoff distance. The airplane's landing gear was completely housed in the wing when retracted, but when it was extended the resulting openings in the lower wing had an area equal to the side area of the struts and wheels. During the Full-Scale Tunnel tests, lift and drag characteristics were evaluated at various angles of attack with landing gear up and down. Performance assessments were made with the landing gear up, landing gear extended and wheel wells open, and with landing gear extended and the wheel wells covered with sheet metal.[17]

During the brief, 2-week tunnel test program, it was found that covering the wheel openings with sheet metal while the landing gear was down reduced drag by only 2 percent at takeoff condition, resulting in little or no impact on the aircraft's performance during takeoff. Langley researchers suggested that the design of the landing gear could be aerodynamically refined to reduce drag; however, a more significant finding was that, with the landing gear in the retracted position, the minimum drag of the aircraft was reduced by 50 percent. Coupled with research previously undertaken in the Langley PRT, these NACA data demonstrated the advantages of retractable landing gear.

Bell XFM-1

During the mid-1930s, the Army awarded a contract to the Bell Aircraft Corporation for its innovative concept of a multiplace, long-range "bomber destroyer" to combat potential enemy long-range bombers attacking the U.S. mainland or other strategic locations, such as the Panama Canal. Bell's aircraft was designated the XFM-1 Airacuda and had been designed by Robert J. Woods, who had been a NACA researcher at Langley for a year in 1928 before

Tunnel tests of the radical Bell XFM-1 "bomber destroyer" identified major sources of aerodynamic drag and provided analyses of the performance of the engine cooling ducts. (NACA 12732)

moving to industry. Woods later became a legendary designer at Bell, leading the company's efforts in programs like the P-39 Airacobra and the X-1 supersonic research aircraft. The XFM-1 was a radical twin-engine pusher aircraft design with a wingspan larger than many contemporary bombers of the day, such as the B-25. This advanced design was crewed by two pilots and three gunners.

At the request of the Army Air Corps, a large ½-scale powered model of the XFM-1 constructed by Bell underwent aerodynamic testing in the Langley Full-Scale Tunnel under the leadership of Abe Silverstein from December 1936 to May 1937 to determine the aerodynamic characteristics of the design and the relative effectiveness of various cooling system designs for its Prestone radiators and oil coolers. Design features of the aircraft, including the mutual interference effects of the wing/nacelle junctures, flaps, and landing gear, were evaluated. In addition, the performance of aileron control surfaces and aerodynamic drag associated with the supercharger air cooler design were analyzed. Tests were conducted for both power-on and power-off conditions, and flow-visualization observations were made with wool tufts.[18] Robert Woods took a deep personal interest in the Langley test and visited the tunnel on several occasions to provide concepts for cooling and aerodynamic improvements.

The Full-Scale Tunnel staff quickly discovered that aerodynamic interference effects at the junctures of the wing and nacelles resulted in aerodynamic flow separation that seriously degraded aerodynamic performance, particularly during climb conditions. In addition, the flow separation produced unacceptable tail buffeting. Fillets were subsequently tested and incorporated into the full-scale airplane. Aerodynamic drag of the Airacuda model was assessed, along with an analysis of airflow through the supercharger air coolers, over the landing gear door covers (the gear was in the retracted position), and over a new supercharger fairing. The results of the testing provided Bell with several options for cooling inlets and ducts either on top of the wing nacelles or in the leading edge of the wing.

When the Airacuda made its maiden flight in September 1937, it sparked the interest of the public due to its radical configuration and massive size. The event provided Bell with some badly needed publicity in its role as an upstart aviation company in a very competitive industry. Unfortunately, the XFM-1 had a large number of complex systems and operational issues that led to a loss of military interest in the aircraft, and it never saw production.

Boeing P-26A

The Boeing P-26A "Peashooter" was the first Army fighter to be constructed entirely of metal and to employ the low-wing monoplane configuration. The prototype P-26 exhibited potentially dangerous landing characteristics, including a very high approach speed and the possibility of flipping on its back upon ground impact with fatal consequences for the pilot of the open-cockpit aircraft. Deliveries to Army combat squadrons had begun in 1933.

Protection for the pilot during landing mishaps was provided by an armored headrest, but the Army was interested in the development of wing flaps to reduce the landing speed and mitigate the problem. In response to an Army request, a 2-week tunnel entry in the Full-Scale Tunnel was scheduled to measure the beneficial effects of several wing-flap concepts using an aircraft from Bolling Field.[19]

The objective of tests of the Boeing P-26 aircraft in the Full-Scale Tunnel in May 1934 was to define the benefits of wing flaps on reducing the landing speed to enhance safety of flight. (NASA EL-1999-00623)

The results of the test program included aerodynamic data for several flap segment configurations across the trailing edge of the wing, including a carry-through section beneath the fuselage. Three sets of flaps were evaluated, each of which had a chord of 20 percent of the wing chord. One set extended from the wing root to the inboard edge of the ailerons; the second set was approximately 3 feet longer and would require shorter ailerons; and the third was a curved segment under the fuselage. With the longer flap, the reduction in landing speed was 8.4 miles per hour. Following the NACA wind tunnel test program, later models of the P-26 were designed with wing flaps.

Shown on the ground floor of the Full-Scale Tunnel, the P-26 exhibits a split-flap concept evaluated during the test program. Rear view shows the wing flaps as well as the curved-flap segment beneath the fuselage. (NACA 9922 and NACA 9923)

Boeing XBLR-1 (XB-15)

The Army Air Corps initiated a top-secret long-range bomber program in 1935 with the designation BLR (Bomber, Long Range) consisting of three competitive designs: the Boeing BLR-1, the Douglas BLR-2, and the Sikorsky BLR-3. The Boeing design was a four-engine configuration, redesignated the XB-15 in 1936; the Douglas design was also a four-engine aircraft, redesignated the XB-19 in 1936; and the Sikorsky design was abandoned at an early stage of the program. The XBLR-1 was designed by Boeing for a range of 5,000 miles and was the largest aircraft in America at the time, having a fuselage length of 87 feet, 7 inches and a wingspan of 149 feet. It featured enormous wings capable of carrying large amounts of fuel. The interior wing structure was large enough to permit repair crews access to the engines. The airplane also featured several modern amenities for the crew, including heated crew compartments complete with beds, a cooking area, and a bathroom. Designed for liquid-cooled engines, it instead had to use air-cooled engines of much lower power ratings.

The connection of the XBLR-1 and the Langley Full-Scale Tunnel includes an amusing story of security—or the lack thereof.[20] In his autobiography, Fred Weick states that in 1936 he was questioned by Army officers at a technical meeting at Wright Field regarding the maximum size model that could be tested in the Full-Scale Tunnel because the Army intended to have the NACA conduct power-on and power-off tests of a top-secret four-engine bomber in the tunnel. He was also told that security was very tight regarding the project and that only a handful of Langley people were briefed on the plan. After Weick returned to Langley he found that only a few people had been briefed into the program, including George Lewis, H.J.E. Reid, Elton Miller, and Smith DeFrance. Weick himself was not considered a need-to-know person, and all other Langley employees were told to avoid the Full-Scale Tunnel while the "special" tests were being done.

Later, Weick was invited to join in a luncheon at the Langley Officer's Club with Reid, who was hosting a distinguished female visitor from Nazi Germany who was supposedly in the United States to sell aircraft to the American airlines. Unofficially, the word was that she was a spy and was to be treated with caution. When the luncheon was over, the group observed two flatcars on nearby railroad tracks that served Langley Field from Hampton at the time. Each of the cars had a large wooden box on it. The German guest exclaimed, "Oh, there is the model of the four-engine Boeing bomber that you people are going to test in the Full-Scale Tunnel!"[21]

The huge ¼-scale model of the XBLR-1 had a wingspan of 37.3 feet and was supplied to the NACA by the Materiel Division of the Air Corps. Weick's memory of its arrival at Langley was off by a year as it actually arrived in late 1935, and tests began in the Full-Scale Tunnel with Jack Parsons as lead engineer in December of 1935. An extended test program was conducted to address the capabilities of the design through mid-1936 in an intensely classified environment, with results transmitted directly to the Air Corps in secret memos.[22] The scope of data gathered was extremely broad and included effects of power, flaps, asymmetric power conditions, control and trim-tab effectiveness, and many other variables.

Progress in the XBLR-1 airplane program was outpaced by Boeing's development of the smaller, higher-performing B-17 bomber, and the company had already received orders for

The huge ¼-scale model of the XB-15 was used for assessments of the aerodynamic characteristics of the airplane as well as for generic studies of engine installations by the NACA. In this 1936 photograph, the model is mounted for tests of the configuration in yaw. (NACA 12088)

The XBLR-1 model was loaned by the Army to the NACA and served as a test bed for many projects. Here the model has been modified as a twin-engine configuration with air-cooled engines in February 1940. (NACA 19406)

the B-17 before the newly designated XB-15 made its first flight in 1937. Because it was dramatically underpowered and lacked performance comparable to aircraft of the day, the XB-15 was passed over in growing Air Corps excitement for the B-17, and only one XB-15 was built.[23] The XB-15 model was loaned to Langley by the Air Corps for NACA research projects following the evaluation of the design and served as a major asset for generic powered-model tests in the Full-Scale Tunnel through 1940.[24] The model can be seen stored on the wall of the tunnel test area outside the test section in many photographs of the period.

Tests of a Kellett YG-1 autogiro for the Army in 1937 were disappointing and involved a major nonfatal accident. The project began on October 21, 1937 (upper photo), and ended abruptly with a major failure of the vehicle 4 days later. (NACA 12955 and NACA 13900)

Kellett YG-1 Autogiro

By 1937, the direct-control type of autogiro had reached a relatively advanced stage with little or no aerodynamic information available concerning its most important component, the rotor. Tests of subscale model rotors had indicated significant scale effects, and the determination of rotor characteristics in flight was impractical. In October 1937, the Army Air Corps requested that Langley test an Army Kellett YG-1 autogiro in the Full-Scale Tunnel and include lift, drag, and control force measurements for the complete vehicle, the rotor alone, and the vehicle with the rotor removed over the full range of air speeds available in the tunnel.[25]

Unfortunately, the wind tunnel measurements were in poor agreement with flight results. In particular, the maximum lift-drag ratio for the vehicle was considerably higher in flight than in the wind tunnel. In addition to these unexpected results, catastrophic dynamic resonance with the mounting system in the tunnel occurred and the project was terminated by a failure of the rotor system and destruction of the autogiro.

Navy Projects

The Navy had made its interest in the Full-Scale Tunnel known long before the tunnel became operational. The office of the NACA was initially co-located in the Navy Building in Washington, and as a result the Navy and NACA staffers became personal, day-to-day acquaintances with full knowledge of each other's technical programs and plans. Navy Lt. Cmdr. Walter S. Diehl, who headed the Navy Bureau of Aeronautics' liaison office, had a special interest in using the NACA as a research arm of the Navy. He was a close friend of George Lewis of the NACA, and their close proximity in the same building kept Diehl informed on the latest NACA projects and plans. Diehl was a brilliant technical expert in aeronautics and frequently visited Langley for briefings on its research undertakings. During the 1930s and 1940s, he authored over 30 NACA reports while a member of the Bureau of Aeronautics.[26] Not surprisingly, testing requests from the Navy dominated the early test programs at the Langley laboratory and had a great influence on the test schedule at the Full-Scale Tunnel.

The Navy's interest in the tunnel was stimulated by two factors. First, the Navy was still using dirigibles and airships in carrying out its mission, and obtaining reliable aerodynamic data required the use of large models in wind tunnels like the Full-Scale Tunnel. Second, the service was beginning the transition from slow biplanes to larger monoplanes with higher landing speeds. This greatly complicated the challenging process of landing aboard aircraft carriers; therefore, research on low-speed high-lift devices was of special interest to the Navy. In addition to requesting tests for specific military aircraft, they also requested generic investigations, such as tunnel and flight studies of the modified Fairchild F-22 discussed earlier.

Airships

The Navy's use of airships led to unique aerodynamic tests of operational issues in the Variable Density Tunnel, the Propeller Research Tunnel, and the Full-Scale Tunnel.[27] These issues were well known to the staff and included in-flight evaluations of performance, stability, and control; assessments of ground-handling problems; and aerodynamic load issues for airship hangars. Smith DeFrance and Floyd L. Thompson had both personally participated in flight experiments conducted by the NACA to measure pressures on the U.S. airship Los Angeles in the 1920s, and the staffs of all the Langley tunnels had maintained communications with the Navy regarding operational airship issues.

In 1935, the Navy's Bureau of Aeronautics requested Langley's support for Full-Scale Tunnel evaluations of the ground-handling forces for a 1⁄40-scale model of the U.S. airship Akron.[28] Actual handling experiences with large airships had experienced encounters with extremely large forces and moments that could endanger the ground crew and the airship. Small-scale models could not be reliably used for projection of such data due to scale effects; therefore, the large-model testing ability of the Full-Scale Tunnel was of great interest for measurements of aerodynamic data.

Navy-requested airship testing in the 1930s included assessments of ground-handling loads for the airship Akron (upper photo) and aerodynamic loads on the Lakehurst airship hangar (lower). The model of the Akron is being tested for rearward (flow from right to left) wind conditions. (NACA 11163 and NACA 11123)

The test program for the 20-foot-long Akron model included a modification to the tunnel to allow simulation of a velocity gradient above a new tunnel ground board representative of gradients encountered over a landing field. A screen was used across the tunnel about 24 feet upstream from the end of the entrance cone with varying height and screen density distribution to replicate an actual velocity gradient measured in tests at Langley Field. The model was supported by struts projecting through the ground board and attached directly to the balance frame. Tests were made with the model at six angles of yaw from 0° to 180° and several heights above the ground board. Smoke was used to visualize the flow over the model. As would be expected, the results of the test program showed extremely large magnitudes of lift and drag with variations in yaw angle. The date proved to be invaluable for analysis of loads during actual operations and for guidance in ground-handling equipment.

A second Navy request was for the distribution of wind loads on a 1/40-scale model of the Navy's Lakehurst airship hangar.[29] Using the same ground-board setup used for the Akron test, pressures from 70 pressure orifices on the model were measured for yaw angles of 0°, 30°, 60°, and 90° to the wind.[30]

XO4U-2

The Navy's Bureau of Aeronautics requested tests in the Full-Scale Tunnel in 1933 after the facility's entrance cone and flow conditions had been modified following the embarrassing results of the attempted first test of the O3U-1 in 1931. The test request was for the Vought XO4U-2 biplane, the next version of the Corsair series, which had fabric-covered metal wings and a metal monocoque fuselage. Only one XO4U-2 was built for evaluations of the metal airframe structure, and it first flew in 1932.

The tests of the XO4U-2 in the Full-Scale Tunnel were focused on how to cool a two-row radial engine under conditions simulating a full-throttle climb.[31] Although many cooling investigations had previously been conducted for single-row engine installations, the problem of cooling two-row engines under such severe conditions required more research. Tests were made for two-blade and three-blade propellers and with the airplane at different angles of attack, and the tunnel was run at a maximum speed of 120 mph. Data obtained in the program provided detailed cooling information for many test parameters.

Tests of the Vought XO4U-2 biplane in the Full-Scale Tunnel in April 1933 were designed to investigate cooling characteristics of two-row radial engines. (NACA 8501)

XBM-1

The Martin XBM-1 biplane was built for the NACA for use in cooperative research investigations. The aircraft was procured by the Navy and loaned to the NACA in the spirit of mutual interest in aeronautical research and development. Only one aircraft was built, and it was equipped with special pressure tubes and instruments to be used in a broad research program at Langley, including studies of load factors, pressure distributions, handling qualities, and icing. In 1933 and 1934, the Navy requested flight studies followed by wind tunnel tests of the XBM-1 in the Full-Scale Tunnel to obtain data on the wing load distribution, structural deformation and stresses, and tail loads in typical maneuvers.[32] The data were used by the Navy as a check on the structural design requirements established for this type of aircraft.

The Martin XBM-1 research aircraft mounted for tests in 1933. The aircraft was procured by the Navy for use by the NACA in many joint research projects. Note the pressure tubing that was routed from pressure orifices within the tail into the rear cockpit for pressure measurements using manometers. Wing pressures were also measured, and structural deformations were photographed by the cameras mounted in the rear cockpit. (NACA 8667)

One of the two Vought XF3U-1 aircraft was tested to evaluate its aerodynamic characteristics in late 1933. (NACA 9321)

XF3U-1

Vought built two XF3U-1 aircraft, continuing to use the name Corsair. One of the aircraft was used by Pratt & Whitney as a flying test bed, and the second was built for the Navy. Initially, the Navy envisioned the XF3U-1 as a fighter, but it decided to modify it into a scout/bomber that later became known as the XSBU-1. The XF3U-1 was the first Navy airplane to feature an enclosed canopy and began a long succession of tests of Navy aircraft in the Full-Scale Tunnel on the potential problem of carbon monoxide seepage into the cockpit during powered operations.

XFT-1

The Northrop XFT-1 was an all-metal low-wing monoplane designed for the Navy as a high-performance fighter featuring a fixed "trousered" landing gear. A single prototype was built for flight evaluations at Anacostia Naval Air Station in January 1934. Although the XFT-1 was the fastest fighter tested by the Navy, it was found to have poor handling qualities for the carrier mission and to be prone to dangerous spinning behavior. Navy test pilots hated the airplane. After being returned to Northrop, it was the subject of a month-long test in the Full-Scale Tunnel in June to measure its stability and control characteristics. It was then modified and resubmitted as the XFT-2, but its characteristics were even worse than before and it was rejected for production.

The Northrop XFT-1 was a disappointment to the Navy because of poor handling qualities. Tests were conducted to measure its stability and control characteristics, and a brief study of methods to reduce the drag of the huge landing gear trousers was completed. (NACA 10122)

This SU-2 Navy scout plane was flown to Langley from the Anacostia Naval Air Station for tests in the Full-Scale Tunnel in 1934. (NASA EL-2003-00019)

SU-2

The Vought SU-2 was a refined derivative of the O3U flown by the Navy and Marines designed specifically for the scouting mission. At the Navy's request, tests were made in August 1934 to document the lift and drag of the configuration.[33]

XBFB-1

The Boeing XBFB-1 was the last fixed-gear biplane built by Boeing for the Navy. Only one aircraft, originally intended to be a fighter design, was built because it did not have adequate maneuverability for a fighter. It was redesignated for more appropriate potential bomber applications, although several concepts were applied to try to improve its fighter-related performance. In the Full-Scale Tunnel tests, for example, the benefits of streamlining the fixed landing gear struts were examined.[34]

The Boeing XBFB-1 prototype was evaluated in 1934. Note the extreme fairings for the landing gear struts used in some tests to improve performance. (NASA EL-2003-00020)

Tests of the prototype of the famous Curtiss SBC Helldiver in 1936 included extensive measurements of aerodynamic characteristics. Note the dummy bomb shape attached to the lower wing. (NACA 12365)

XSBC-3

The Curtiss XSBC-3 Helldiver was the last biplane acquired by the Navy. It had initially been designed as a high-wing monoplane designated the XSBC-1, which crashed during company dive-bombing evaluations and was redesigned as a biplane. Equipped with an upgraded engine as the XSBC-3, it was placed into production and served with carrier-based scout squadrons beginning in 1938. Testing in the Full-Scale Tunnel was completed in 1936, involving performance and power-effect tests.[35]

In addition to obtaining aerodynamic performance data, tests of the SOC-1 in 1936 examined the potential for seepage of carbon monoxide into the enclosed cockpit. (NACA 12550)

SOC-1

The Curtiss SOC-1 scout observation biplane was designed for land or sea operations. Equipped with floats, it served with distinction on battleships and cruisers in the early years of World War II, being launched by catapult and recovered after a sea landing. The SOC-1 also operated as a land plane with conventional landing gear. The aircraft first entered service in 1935. The Navy requested tests of a production SOC-1 aircraft in the Full-Scale Tunnel in 1936 to evaluate performance and engine cooling. The staff also measured power effects on the tail surfaces, which contributed to the growing database on power effects.[36]

Incidents of carbon monoxide in the cockpit of the SOC-1 had been encountered in service, and the test program in the Full-Scale Tunnel included a segment to examine the possibility of backflow or seepage of the deadly gas into the cockpit.

SB2U-1

The last Navy-requested job in the Full-Scale Tunnel in the period from 1932 to 1937 was for the Vought SB2U-1 Vindicator dive bomber. The study was very limited, consisting of only 2 days in the wind tunnel in September 1937 to measure its aerodynamic behavior at stall. The Navy had noted that the airplane had exhibited an unusual "secondary stall" at a lower speed than the initial stall. Langley test pilot Mel Gough flew several stall maneuvers with the test airplane a week before the wind tunnel tests and reported it had outstanding stall behavior—much better than had been experienced with any other low-wing monoplane of the day. Gough was very positive regarding the handling of the SB2U-1.[37]

This Vought Vindicator, tested in the tunnel in late 1937, was the first production SB2U-1 aircraft. Its tunnel test program was abbreviated and consisted of measuring aerodynamic behavior at stall for correlation with flight tests. (NASA EL-2003-00283)

The tunnel tests of the SB2U-1 followed the flight tests and covered an angle-of-attack range from 10° to 25°. For all tests, the airplane was configured as shown in the accompanying photograph: with landing gear extended, cowl flaps closed, cockpit hoods open, wing flaps down, ailerons drooped, and the propeller locked in a vertical position. The results of the tests revealed that the lift data for the airplane displayed two lift peaks, and that a strong nose-down diving moment occurred at the stall, which enhanced stall recovery.[38]

Requests by the Department of Commerce

In addition to military requests for testing, the Full-Scale Tunnel became the focal point for important studies sponsored by the Department of Commerce in the mid-1930s. Commerce was extremely interested in evaluating aircraft design concepts that demonstrated inherent safety advantages for civil aircraft. Three of the most important airplane designs of the day were tested in the tunnel.

McDonnell Doodlebug

In the winter of 1929, philanthropist Daniel Guggenheim sponsored the Daniel Guggenheim International Safe Aircraft Competition, with an interest in advancing safety by improving the art of airplane design and fabrication. The competition included several domestic and international designs, with a prize of $100,000 for the winner. The 10 competitors included the winning Curtiss Tanager biplane and a unique, fabric-covered monoplane design conceived by famous aviation pioneer and McDonnell-Douglas Corporation founder James S. McDonnell.

McDonnell's two-place monoplane, known as the Doodlebug, incorporated several revolutionary features, including an automatic leading-edge slot across the entire span of the wing and a slotted trailing-edge flap across 70 percent of the wing, a wide-span landing gear for safer landings, and streamlined wing-support struts to enhance structural integrity with minimal drag. The trailing-edge flap and an adjustable stabilizer were interconnected by a single lever to minimize trim changes when the flaps were lowered or raised. The airplane also used a NACA cowling to reduce drag and promote improved aerodynamic performance. The aircraft had a top speed of 110 mph and a stall speed of 35 mph, with a landing ground roll of only 150 feet.

Following his unsuccessful Guggenheim Competition bid, McDonnell flew the aircraft in demonstrations for about a year and attempted to market the Doodlebug to commercial buyers, but this effort generated no "takers." The onset of the Great Depression also hindered any last hopes of being able to find a financial backer for the airplane. The NACA, however, took a keen interest in the technologies embodied in the Doodlebug for improved aviation safety, and decided to procure it for research at Langley in 1931 with the aircraft designation NACA 42.

Langley flight-test evaluations of the Doodlebug in 1931 revealed that, although the performance of the slots and flap were impressive (total increase in lift of 94 percent), the

The McDonnell Doodlebug test in the Full-Scale Tunnel in 1933 demonstrated the effectiveness of wing-root fillets to alleviate tail buffeting. In this photo, a NACA cowling has been added to the configuration. (NASA EL-2002-00593)

This front view of the Doodlebug shows the aircraft without an engine cowling and with one of several wing-root fillets conceived by the Langley staff. (NACA 8145)

airplane experienced severe tail buffeting. The Doodlebug was one of the earliest monoplanes, and worldwide concern had begun to arise over the cause and cures for adverse buffeting effects experienced by monoplane designs at the time. An earlier accident in Europe on another aircraft that disintegrated in flight had caused speculation that a very fundamental flow problem was occurring.

In preparation for the tunnel testing, the airplane was modified for evaluations of a new cowling and wing fillets. In the tunnel test setup, vertical movements of the tip of the horizontal tail were recorded by instrumentation to measure the intensity of the buffeting, and the direction and speed of the airflow of the tail were also measured. Wing-fuselage fillets were designed by the Langley staff to control the rate at which airflow in the critical wing-fuselage juncture diverged and thereby prevent flow separation. The airflow separation problem for the basic aircraft was readily apparent in flow-visualization studies, which showed that the wing-fuselage interference caused a premature stalling of the wing at the root. The problem was caused by several factors, including the tapering of the fuselage toward the rear and bottom, the additional friction drag caused by the fuselage, and the presence of the large, exposed engine cylinders.[39]

The effectiveness of the fillets and the NACA cowling in preventing flow breakdown was quite apparent both in flow-visualization studies and in overall forces and moments measured on the airplane. The tail buffeting levels were significantly reduced to amplitudes small enough to be considered unobjectionable throughout the range of angles of attack tested. The data generated by the study were quickly absorbed by the aviation industry and resulted in dramatic improvements in the performance and safety of U.S. aircraft in World War II and the general aviation fleet following the war.

Weick W-1 and W-1A

After playing a key role in the development of the NACA low-drag cowling, Fred Weick became deeply devoted to establishing the practicality of personal-owner aircraft, with ease of operation and safety being his main interests. As assistant chief of the Aerodynamics Division, Weick scheduled "brainstorming" sessions with his subordinates at the Langley 7- by 10-foot Atmospheric Wind Tunnel to review ideas and concepts for the "ideal" general aviation aircraft. The participants built free-flying models of several candidate concepts and test flew them in one of the return passages of the Langley Full-Scale Tunnel in 1931. The group agreed that the design that demonstrated the best performance and stability would be selected for full-scale production and eventual flight testing.

A pusher-propeller twin-boom design known as the W-1 won the competition. Equipped with a fixed auxiliary airfoil ahead of the leading edge of the wing and no trailing-edge flap, the two-place configuration exhibited a 30-percent increase in maximum lift and a 50-percent increase in maximum usable angle of attack compared to designs that used conventional wings. Weick was particularly concerned about the dangers of ground loops that were being experienced by contemporary aircraft during landings, and his NACA team devoted a substantial effort to the landing gear design. After considerable study, Weick arrived at a configuration he named a "tricycle landing gear," in one of the most noteworthy

The Weick W-1 airplane incorporated several safety-enhancing concepts, including a tricycle landing gear. Note the fixed auxiliary airfoil ahead of the wing leading edge. (NACA 9554)

contributions in aviation history. The breakthrough in this landing gear arrangement was a steerable nose wheel, which was a radical departure from other three-wheeled airplane designs of the time. Weick and his team began assembly of the W-1 in his garage during weekend work sessions, finally finishing the aircraft in late 1933.

The Bureau of Air Commerce had initiated a major effort to expand personal-owner flight operations in the United States during the early 1930s, and they were briefed on the W-1 design by Weick during a factfinding mission to Langley. Impressed by the potential benefits of the configuration, the bureau purchased the aircraft and requested that the NACA conduct wind tunnel and flight tests to evaluate the characteristics of the W-1.

The brief 2-week test program, which was conducted in the Full-Scale Tunnel in March 1934, measured lift, drag, and stability; the only negative result was a finding that directional stability was low.[40]

After the tunnel test, NACA test pilots conducted a brief flight evaluation of the W-1 in April, finding that it was very resistant to ground looping but confirming the tunnel prediction of very low directional stability. In addition, adverse yaw from the ailerons resulted in unsatisfactory roll control. Consequently, Weick and his team made modifications to the design, including increasing the size of the vertical tails.

A more advanced version of the W-1, known as the W-1A, was completed and sold by Weick to the Bureau of Air Commerce in 1935. The W-1A design abandoned the fixed auxiliary airfoil of the W-1, and instead included glide-control flaps that could be deflected to a maximum of 87° in an effort to provide for safer and more precise landings. The flaps

Following modifications, the original Weick design was known as the W-1A. Note the increased vertical-tail size. The glide-control flap is not deflected in this photo. (NACA 10834)

extended from the tail booms outward to the wingtips. Lateral control was provided by unique "slot-lip" ailerons, which were spoilers with a slot that permitted the air to flow through the wing to the undersurface.

In response to a formal request by the Bureau of Air Commerce, the W-1A was first tested in the Full-Scale Tunnel and later in flight at Langley.[41] In the tunnel tests, the hinge location of the flap was found to be too forward (20-percent chord), resulting in high stick forces, and directional stability of the aircraft was judged to still be low but adequate. In addition, it was determined that balance of the aircraft would be hard to maintain during flight for the planned center-of-gravity location due to large diving moments produced when the flap was deflected.

Although the W-1 and W-1A were one-of-a-kind aircraft and never made it into production, the designs were clearly ahead of their time. The steerable tricycle landing gear arrangement was later used on military aircraft such as the Lockheed P-38 Lightning and Bell P-39 Airacobra, and on Weick's famous Ercoupe general aviation aircraft. Today, it is a common feature found on most military, civil transport, and general aviation aircraft.

Nature's Wrath: The 1933 Hurricane

The close proximity of the Full-Scale Tunnel and other NACA facilities at Langley Field to the shores of the Back River and waters of the Chesapeake Bay resulted in frequent flooding during tropical storms and hurricanes. In the case of the Full-Scale Tunnel, virtually every decade of its 80-year history experienced a major flood that resulted in damage to equipment, destruction of office documents, and interruption of tunnel operations. Somewhat surprisingly, the huge Careystone and steel structure of the tunnel building withstood the storms and avoided catastrophic damage.

The first major hurricane to pound the Full-Scale Tunnel building with its fury was the unnamed Chesapeake-Potomac Hurricane of 1933. The year 1933 was a very active one for tropical storms and hurricanes with 21 storms, of which 10 became hurricanes. The great hurricane of 1933 had reached Category 3 strength at one point before weakening to Category 1 and striking on August 23, 1933, causing 30 deaths and $461 million (equivalent 2011 dollars) in damage. The center of circulation passed directly over Norfolk, and the storm surge of 9.8 feet remains a record. Nine of the eleven NACA buildings were flooded with salt water, and the water level during the flooding in the test section of the Full-Scale Tunnel building reached a depth of 5 feet.

The 1933 hurricane caused widespread damage at Langley and the local areas. As shown on the left, most of Langley Field was flooded with several feet of water. Damage to the Careystone sheeting and windows of the Full-Scale Tunnel (above) was extensive, as water flooded the office and test section areas. (U.S. Air Force and NACA 9126)

Russ Robinson later recalled the event:

> We were established in the Full-Scale Tunnel in the ground-floor space, really under the tunnel, looking out on the Back River. There was no advance warning. The siding was corrugated asbestos sheeting, and wind and water pushed right through. When we got in, it was apparent the average water level had been exactly desk high, 30 inches. Mud was left all over by receding water; fishing and pleasure boats were left all over Langley Field. An irony of the situation was that methodical people who put everything away in their desks at the end of the day lost everything; the less-disciplined people who let papers and books accumulate in a pile on their desk saved something. I won't mention names![42]

Even as the staff cleaned up the offices and tunnel, the Mid-Atlantic states were hit by another hurricane almost exactly a month later, when a Category 3 storm emerged from a disturbance in the Bahamas and came up the coast to make landfall at Cape Lookout, NC. Fortunately, the storm ended up causing only a fraction of the damage that resulted from the Chesapeake-Potomac storm.

Annual Engineering Conferences

In 1926, the NACA began a tradition of hosting an Annual Aircraft Engineering Research Conference at its Langley laboratory.[43] The invitation-only tour included briefings on ongoing NACA research programs and feedback from attendees on the ongoing research as well as new opportunities and problem areas for research. Over 300 people typically attended the sessions, and the Full-Scale Tunnel provided an appropriately awe-inspiring location for staging the group's photograph. Attendees at the inspections were technical leaders, decision makers, and very influential individuals within industry, the military, academia, and other NACA stakeholders.

Perhaps the most famous group photograph taken during the conferences was made during the NACA Ninth Annual Aircraft Engineering Research Conference in May 1934. This photo, which can be seen on page 84, documents one of the most impressive gatherings of individuals in the history of aviation, including such legendary figures as Orville Wright, Adm. Ernest King, and Charles Lindbergh (all members of the NACA committee at the time); as well as Howard Hughes, Leroy Grumman, John Northrop, Alexander de Seversky, Harold Pitcairn, Lloyd Stearman, Henry Berliner, Giuseppe Bellanca, Theodore Wright, Sherman Fairchild, James Doolittle, Elmer Sperry, Clarence Taylor, and Grover Loening.

Eight of the twelve members of the NACA posed for this photograph at the 1934 conference. Left to right: Brig. Gen. Charles A. Lindbergh, V. Adm. Arthur B. Cook, Charles G. Abbot, Dr. Joseph S. Ames (committee chairman), Orville Wright, Edward P. Warner, Fl. Adm. Ernest J. King, and Eugene L. Vidal. (NACA 9846)

Facility Modifications

The confidence in results from the Full-Scale Tunnel continued to grow during the 1930s. A detailed measurement of turbulence factors in the NACA tunnels was conducted with spheres of various sizes in 1937, and comparative results showed the following:[44]

Facility	Turbulence Factor[45]
Free air	1.0
Full-Scale Tunnel	1.1
PRT	1.2
Model of FST	1.2
AWT	1.4
Free-Spinning Tunnel	1.8
VDT	2.5

These results boosted DeFrance's claims regarding the low level of turbulence in the tunnel and greatly increased the confidence in aerodynamic results obtained from the facility.

During the early 1930s, several physical features were changed or added to the building housing the Full-Scale Tunnel. Within the test section area, walkways and a balcony on the east sidewall were added in 1932 to enhance remote visibility during test programs. Also in 1932, the ⅟15-scale model of the Full-Scale Tunnel was installed out of the airstream directly beneath the exit cone. It would operate as a low-cost, rapid-response test facility in that location for over 30 years. It was then given to the nation of Portugal as part of NASA's relationships in the Advisory Group for Aerospace Research and Development (AGARD) organization of the North Atlantic Treaty Organization (NATO).

The First Langley Gust Tunnel

The building that housed the Full-Scale Tunnel was also the site of many other historic research projects. For example, in the 1930s and 1940s, the facility was the site of an innovative NACA testing technique to acquire data for the structural design of aircraft. A daunting engineering design challenge in the early days of aviation was the development and validation of methods to predict air loads experienced during flight in gusts and turbulence. This capability was extremely important, not only from a safety-of-flight perspective, but also to prevent an overdesign of aircraft structures that would result in unnecessarily large weight penalties. Early research in the 1920s and 1930s at the NACA Langley laboratory had included theoretical studies of loads generated in specific gust fields, but flight data to substantiate the predictions were extremely difficult to obtain at that time.

In order to experimentally investigate gust loads under controlled conditions, the NACA designed and constructed a pilot "gust tunnel" in 1936 within an area of the Full-Scale

Tunnel building beneath the tunnel flow circuit. Its location was under and to the right of the exit-cone structure and to the right of the model tunnel near the east return passage wall.

The pilot gust tunnel testing technique consisted of launching dynamically scaled, free-flight airplane models through gusts of known shapes and intensities. During the flight, measurements were made of the accelerations and reactions of the models due to the gusts. The test facility consisted of a gust generator, a catapult for launching the models, and two screens used to decelerate and catch the airplane model at the end of the flight. The gust generator was a large squirrel-cage blower that supplied air to an expanding rectangular channel discharging a current of air upward. The vertical jet of air was 6 feet wide and 8 feet long, and its airspeed profile was shaped by a combination of screen meshes designed to produce the desired gust shape. After the catapulted airplane model completed its flight through the vertical-gust field, it impacted a barrier of vertical rubber strands that decelerated the model. After deceleration, the model nose (shaped like a barbed hook) engaged a burlap screen that stopped the model and held it until the model was removed by the tunnel operator. The propelling catapult was powered by a dropping weight, and the maximum model flight speed was adjustable by changing the amount of weight. The facility was capable of testing scaled airplane models with wingspans of about 3 feet at speeds up to about 50 mph.[46]

This first gust tunnel facility operated in the Full-Scale Tunnel building under the leadership of Philip Donely for almost a decade. The facility produced very valuable information on gust loads as affected by primary aircraft design variables, and results from the facility were used to justify reducing the structural design criteria that had led to overdesign of wing structures for certain types of aircraft configurations. In 1945, the pilot gust tunnel was replaced by the new Langley Gust Tunnel, which was similar in operational concept but capable of testing larger 6-foot-span models at speeds up to 100 mph. The facility was housed in a new building (Building 1218) in the Langley West Area. The blower from the

The 1/15-scale model of the Full-Scale Tunnel was moved to the facility building in 1932 and installed under the tunnel's exit cone. During operations, the staff of the model tunnel was located above the tunnel, which was accessed by stairs and a seating platform. In this photograph, the operator's site and the tunnel's test balance are in the foreground atop the model tunnel, and the Full-Scale Tunnel's exit cone is shown in the background. (NACA 6338)

pilot tunnel was retained and briefly used at the Full-Scale Tunnel for studies of flow through blades of helicopter models in hovering flight.[47]

Changing of the Guard

By mid-1933, the tunnel was up and running and the original staff of the Full-Scale Tunnel began to be reassigned to positions of increasing importance. As a result of his success in management of the design and operations of the Full-Scale Tunnel, Smith DeFrance had established himself as a premier designer and operational manager of wind tunnel facilities at Langley. He was thereafter promoted by H.J.E. Reid to oversee the design and operations of new tunnels as well as the management of existing assets. Russ Robinson, who had assisted DeFrance during the design of the Full-Scale Tunnel, was selected to design a new 500-mph high-speed wind tunnel to be located next to the Full-Scale Tunnel. This wind tunnel was envisioned to be an 8-foot high-speed companion to the Full-Scale Tunnel and, in fact, the first name selected for the tunnel was the "Full-Speed Tunnel."[48] Robinson subsequently became head of the new 8-Foot High-Speed Tunnel in 1935, and Manley Hood also left the Full-Scale Tunnel Section to become his assistant. The tunnel began research operations in 1936.

In 1935, DeFrance was appointed assistant to Elton Miller in the Aerodynamics Division, where he was assigned managerial responsibilities for the 8-Foot High-Speed Tunnel, the PRT, and the Full-Scale Tunnel. He also was given a new responsibility to design a new large tunnel to further address the issue of scale effects on aerodynamic data. This would be the large pressurized wind tunnel that Max Munk had advocated for in 1929, rather than the Full-Scale Tunnel. Conceived as a pressurized version of the PRT, the new tunnel would be known as the 19-foot Pressure Tunnel and would be built across the street from the Full-Scale Tunnel. In accomplishing this task, DeFrance assigned Jack Parsons of the Full-Scale Tunnel staff to be the chief designer. The new facility was authorized in 1936 and became operational in 1939.

Meanwhile, Clint Dearborn became the head of the Full-Scale Tunnel. Notable new engineers entered the scene at the tunnel in the mid-1930s, including Samuel Katzoff, Harry J. Goett, John P. "Jack" Reeder, and Herbert A. Wilson, Jr., who would all become legendary figures in the NACA and NASA during their careers. As will be discussed in the next chapter, the end of the 1930 era would lead to a further depletion of experienced engineers at the tunnel as career opportunities arose at new NACA laboratories.

End of the Line?

After 5 years of research focused on the correlation with flight-test results and analysis of scale effects, the staff of the Full-Scale Tunnel had built a firm foundation regarding the accuracy of tunnel tests to predict airplane aerodynamic characteristics for aircraft of the day. The

The first gust tunnel operations at Langley were conducted in an open area in the Full-Scale Tunnel building. Operations shown here include A) illustration of catapulted model about to enter the vertical gust; B) loading a recording accelerometer into the free-flying model; C) balancing the model before flight; D) preparing to launch the model on the catapult; E) modeling flight over the vertical airstream; and F) model after impacting retrieving apparatus at end of the flight. (NACA)

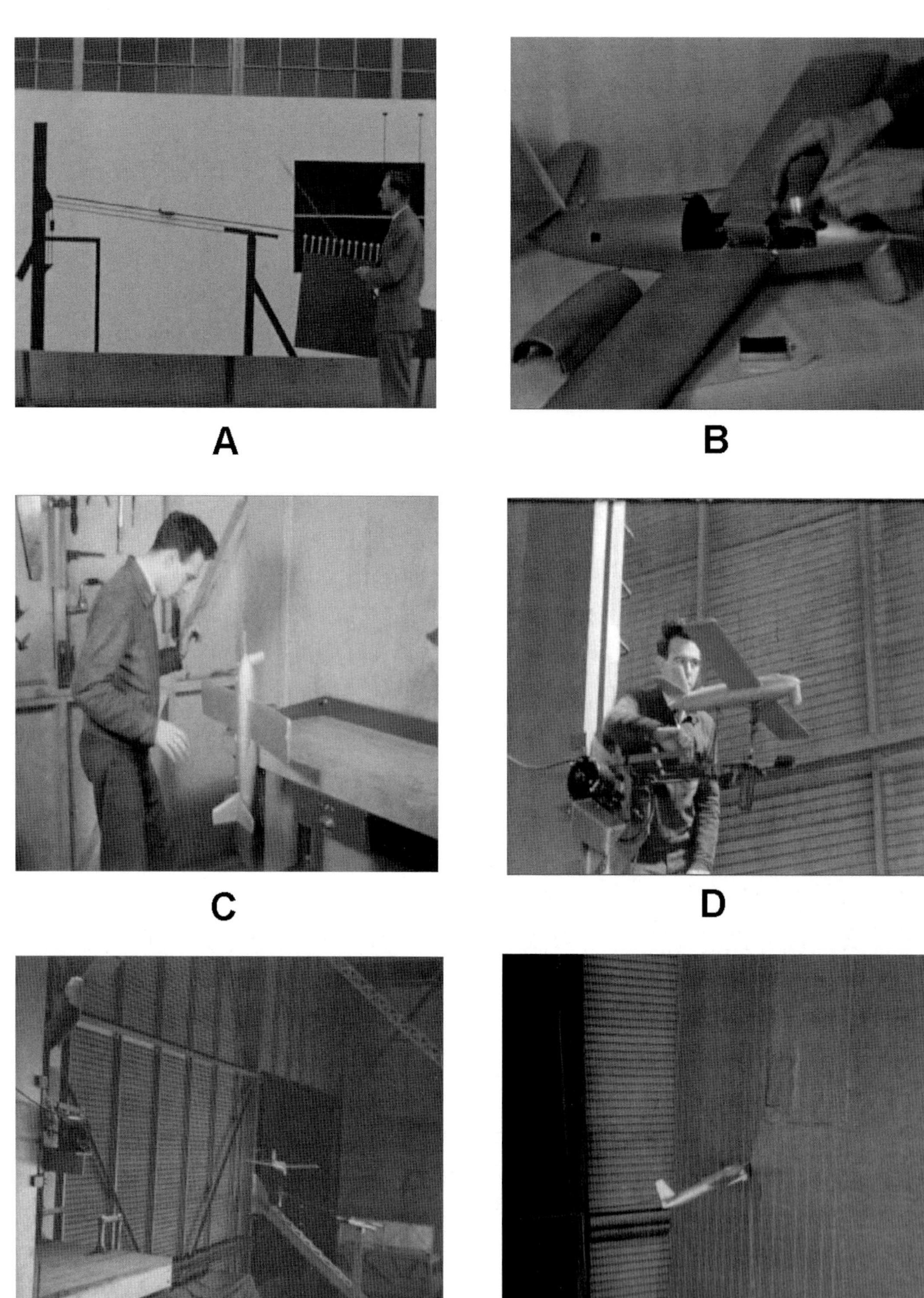

argumentative relationship that had built between Eastman Jacobs's group at the Variable Density Tunnel and Smith DeFrance's staff at the Full-Scale Tunnel regarding the effects of wind tunnel turbulence on the quality of data had eased as Jacobs conceived a method to extrapolate data from the VDT (or any other tunnel) based on the concept of a turbulence factor derived from actual turbulence measurements in the tunnel. This approach was successfully applied by many researchers to analyze wind tunnel data. Parametric testing had been accomplished on high-lift devices, engine placement, the drag of protuberances such as landing lights, and other important components.

In 1937, it appeared that many of the questions that had formed the basis of advocacy for the construction of the Full-Scale Tunnel had been answered, and a few supporters of the NACA began to wonder if the tunnel had served its purpose and reached the end of its productive life. However, the rapid deterioration of world stability as war clouds appeared brought an even more intense mission to the tunnel that would ensure its fame in aviation history. As would happen many times during its lifetime, the Nation reaped the great rewards of having invested in this legendary facility when unexpected new challenges and opportunities faced the aeronautical community. The prewar experiences on problems of drag reduction and engine cooling would soon become one of the facility's major assets as it participated in the upcoming global conflict.

Langley engineers occupy the data-gathering positions in the balance house of the Full-Scale Tunnel during a typical test in April 1936. The Toledo scales can be seen in the background, and the floating-balance structure is at the top of the photograph. Clint Dearborn, head of the tunnel, is at the left foreground, and James White is at the right. The individual in knickers is unidentified. (NACA 11968)

Endnotes

1. Smith J. DeFrance, "The NACA Full-Scale Wind Tunnel," NACA TR-459 (1934); Abe Silverstein, Samuel Katzoff, and James A. Hootman, "Comparative Flight and Full-Scale Wind-Tunnel Measurements of the Maximum Lift of an Airplane," NACA TR-618 (1938).
2. Clinton H. Dearborn and Hartley A. Soule, "Full-Scale Tunnel and Flight Tests of a Fairchild 22 Airplane Equipped with a Fowler Flap," NACA TN-578 (1936); Rudolf N. Wallace, "The Effect of Split Trailing-Edge Wing Flaps on the Aerodynamic Characteristics of a Parasol Monoplane," NACA TR-539 (1936); Clinton H. Dearborn and Hartley A. Soule, "Full-Scale Tunnel and Flight Tests of a Fairchild 22 Airplane Equipped with a Zap Flap and Zap Ailerons," NACA TN-596 (1937).
3. Apparently, no report was written on the Full-Scale Tunnel tests of the YP-29A.
4. Smith J. DeFrance, "Effect of the Surface Condition of the Wing on the Aerodynamic Characteristics of an Airplane," NACA TN 495 (1934).
5. W. Hewitt Phillips, *Journey in Aeronautical Research: A Career at NASA Langley Research Center*, NASA Monographs in Aerospace History Number 12 (Washington, DC: NASA History Office, 1998), p. 28.
6. Ibid., p. 28.
7. Michael H. Gorn, *Expanding the Envelope: Flight Research at NACA and NASA* (Lexington, KY: The University Press of Kentucky, 2001), pp. 65–88.
8. Abe Silverstein, "Scale Effect on Clark Y Airfoil Characteristics from NACA Full-Scale Wind-Tunnel Tests," NACA TR-502 (1935).
9. Clinton H. Dearborn, "The Effect of Rivet Heads on the Characteristics of a 6- by 36-Foot Clark Y Metal Airfoil," NACA TN-461 (1933).
10. Video of the flow-visualization tests of ground effects in the Full-Scale Tunnel is available at *http://www.youtube.com/watch?v=l-Q2G8w11C0*, accessed March 28, 2012.
11. C.H. Dearborn, "Full-Scale Drag Tests of Landing Lamps," NACA TN-497 (1934).
12. Abe Silverstein and Samuel Katzoff, "Design Charts for Predicting Downwash Angles and Wake Characteristics Behind Plain and Flapped Wings," NACA TR-648 (1939). See also Abe Silverstein and Samuel Katzoff, "Aerodynamic Characteristics of Horizontal Tail Surfaces," NACA TR-688 (1940).
13. Beverly G. Gulick, "Effects of Simulated Ice Formation on the Aerodynamic Characteristics of an Airfoil," NACA Wartime Report L-292 (1938). Originally issued as an Advance Confidential Report.
14. John B. Wheatley and Manley J. Hood, "Full-Scale Wind-Tunnel Tests of a PCA-2 Autogiro Rotor," NACA TR 515 (1935).
15. Peter W. Westburg and Peter M. Bowers, "The Parasols of Santa Monica," *Wings* 4, no. 2 (April 1974).

16. Smith J. DeFrance, "Drag of Prestone and Oil Radiators on the YO-31A Airplane," NACA TN 549 (1935). Many sources cite the subsequent testing of the Brewster Buffalo in 1938 as the beginning of drag cleanup tests in the Full-Scale Tunnel. NACA documentation, however, shows that the fundamentals for those studies had been laid a decade before, during the tunnel's earlier tests.
17. Smith J. DeFrance, "The Aerodynamic Effect of a Retractable Landing Gear," NACA TN-456 (1933).
18. The results of the XFM-1 tests were summarized in Abe Silverstein, "Full-Scale Wind-Tunnel Investigation of the ½-Scale Model of the XFM-1 Airplane. I. Aerodynamic Characteristics and Air Flow with Cooling Ducts in the Wing-Preliminary" (1937). This report is no longer available in the literature, but a copy was found in the files of the NASA-Langley Flight Dynamics Branch. Other reports include James A. White, "Full-Scale Wind-Tunnel Investigation of the ½-Scale Model of the XFM-1 Airplane. II. Tests Conducted Before the Installation of the Cooling System" (1937); and Arthur B. Freeman, "Full-Scale Wind-Tunnel Investigation of the ½-Scale Model of the XFM-1 Airplane. III. Trim Characteristics and Scale Effect Power Off" (1937).
19. Smith DeFrance, "Aerodynamic Characteristics of a P-26A Airplane With and Without Split Trailing-Edge Flaps," NACA Administrative Files AA248-1 (389) (1934). See also NACA, *NACA Twentieth Annual Report* (1934), p. 6.
20. Fred E. Weick and James R. Hansen, *From the Ground Up: The Autobiography of an Aeronautical Engineer* (Washington, DC: Smithsonian Institute Press, 1988).
21. Ibid., pp. 162–163.
22. Four separate secret memorandum reports (all previously classified works cited have since been declassified) were transmitted to the Air Corps. See, for example, John F. Parsons, "Preliminary Report of Power-Off Tests on ¼-Scale Model of the XBLR-1 Airplane-IV, Yaw Results," NACA Secret Memorandum Report for Army Air Corps (1936).
23. In 1938, flight research studies of the XB-15 were conducted by the NACA at Langley. The flight tests yielded valuable design information on gust loads and their effects on the structural components of large aluminum aircraft. Additional flights were devoted to handling quality assessments.
24. For example, see Herbert A. Wilson, Jr., and Robert R. Lehr, "Drag and Propulsive Characteristics of Air-Cooled Engine-Nacelle Installations for Two-Engine Airplanes," NACA Wartime Report ACR L-428 (1940).
25. John B. Wheatley and William C. Clay, "Full-Scale Wind Tunnel and Flight Tests of a YG-1 Autogiro," NACA Memorandum Report for the Army Air Corps (1937).
26. In the early days of the NACA, technical reports were often authored by non-NACA personnel. For example, see Walter S. Diehl, "The Mean Aerodynamic Chord and the Aerodynamic Center of a Tapered Wing," NACA TR-751 (1942).

27. See Ira H. Abbott, "Airship Model Testing in the Variable Density Wind Tunnel," NACA TR-394 (1932); Smith DeFrance, "Flight Test on U.S.S. Los Angeles. Part One: Full-Scale Pressure Distribution Investigation," NACA TR-324 (1930); Hugh B. Freeman, "Force Measurements on a 1/40-Scale Model of the US Airship Akron," NACA TR-432 (1933).
28. Abe Silverstein and B.G. Gulick, "Ground-Handling Forces on a 1/40-Scale Model of the U.S. Airship Akron," NACA TR-566 (1936).
29. James A. White, "Wind-Tunnel Tests of Wind Pressures on a 1/40-Scale Model of the Lakehurst Airship Hangar," NACA Memorandum Report for Bureau of Aeronautics (1935).
30. Digitized videos of flow-visualization tests of the models of the Akron and the Lakehurst hangar in the Full-Scale Tunnel may be viewed at *http://www.youtube.com/watch?v=5gqAyEwCmcA,* accessed January 4, 2012.
31. Oscar W. Shey and Vern G. Rollin, "Cooling Characteristics of a 2-Row Radial Engine," NACA TR 550 (1934).
32. Philip Donely and Henry A. Pearson, "Flight and Wind-Tunnel Test of an XBM-1 Dive Bomber," NACA TN 644 (1938).
33. Smith DeFrance, "Performance Tests on SU-2 and XBFB-1 Airplanes in the Full-Scale Tunnel," NACA Memo Report for Navy, NACA Administrative File AS320-1 (458) (1934).
34. Ibid.
35. Abe Silverstein, "Full-Scale Wind-Tunnel Tests on XSBC-3 Airplane," NACA Memorandum Report for Navy, NACA Administrative File AS320-1 (544) (1936); Samuel Katzoff, "Longitudinal Stability and Control with Special Reference to Slipstream Effects," NACA TR 690 (1939).
36. Beverly G. Gulick, "Full-Scale Wind-Tunnel Tests of the SOC-1 Airplane," NACA Memorandum Report for Navy, files of Flight Dynamics Branch, NASA Langley Research Center (1935).
37. B.G. Gulick and M.N. Gough, "Full-Scale Wind-Tunnel Tests of the SB2U-1 Airplane With Notes on Flight Tests," NACA Confidential Memorandum Report for Bureau of Aeronautics, Navy Department, files of the Flight Dynamics Branch, NASA Langley Research Center (1937).
38. J.S. Davidsen and W.H. Harries, "Full-Scale Wind-Tunnel Tests of the Navy SB2U-1 Airplane with Dive Flaps," NACA Memorandum Report for Navy (1938).
39. Manley J. Hood and James A. White, "Full-Scale Wind-Tunnel Research on Tail Buffeting and Wing-Fuselage Interference of a Low-Wing Monoplane," NACA TN-460 (1933).
40. Smith DeFrance, "Tests of W-1 Airplane in the Full-Scale Tunnel," NACA Administrative File AA248-1 (389) (1934).

41. Clinton H. Dearborn, "Full-Scale Wind-Tunnel tests of W-1A Airplane," Confidential Memorandum Report for Bureau of Air Commerce, files of Flight Dynamics Branch, NASA Langley Research Center (1935).
42. Russell G. Robinson, "Memoir for Three Sons, Toryn and Kra" (unpublished manuscript, 1997), Archives Reference Collection, NASA Ames History Office, p. 26.
43. James R. Hanson, *Engineer in Charge*, NASA SP-4305 (1987), pp. 148–158.
44. Robert C. Platt, "Turbulence Factors of NACA Wind Tunnels As Determined by Sphere Tests," NACA TR-558 (1937).
45. Turbulence factor is defined as the ratio of the critical Reynolds number of a sphere in a nonturbulent air stream to the critical Reynolds number in the tunnel.
46. Video scenes of a test in the first Gust Tunnel are available at *http://www.youtube.com/watch?v=KIakupITXK4,* accessed on January 5, 2012.
47. Philip Donely, "Summary of Information Relating to Gust Loads on Airplanes," NACA TR-997 (1950).
48. Manley Hood, interview by Walter Bonney, September 24, 1974, LHA.

NINTH ANNUAL AIRCRAFT ENGINEERING RESEARCH CONFERENCE
EXECUTIVES AND ENGINEERS OF AIRCRAFT INDUSTRY AND GOVERNMENT OFFICIALS
FULL SCALE WIND TUNNEL
NATIONAL ADVISORY COMMITTEE FOR AERONAUTICS
LANGLEY FIELD, VA MAY 23, 1934
VIEW SHOWS ARMY PURSUIT P26A MOUNTED FOR AERODYNAMIC TESTS
ARMY

Attendees at the NACA Ninth Annual Aircraft Engineering Research Conference in May 1934 pose under a Boeing P-26A mounted for tests in the Full-Scale Tunnel for one of aviation's most historic photographs. This gathering of leaders of the aeronautical community was arguably the most notable of the NACA conferences. The following attendees have been identified: 1, Howard Hughes (Fairchild Aviation Corp.); 2, Clarence G. Taylor (Taylor Aircraft Company); 3, Grover Loening (Grover Loening Aircraft Company); 4, John F. Victory (NACA Secretary); 5, Albert F. Zahm (Library of Congress); 6, Thomas Carroll (Management and Research, Inc.); 7, Temple N. Joyce (North American Aviation, Inc.); 8, Charles J. McCarthy (Chance Vought Corp.); 9, Reuben H. Fleet (Consolidated Aircraft Corp.); 10, Col. Charles A. Lindbergh (Pan Am, NACA Member); 11, Charles G. Abbot (Smithsonian Institution, NACA Member); 12, Dr. Orville Wright (NACA Member); 13, Dr. Joseph S. Ames (Johns Hopkins University, NACA Chairman); 14, Rear Adm. Ernest J. King, (U.S. Navy, NACA Member); 15, Hon. Eugene L. Vidal (Dept. of Commerce, NACA Member); 16, Alexander P. de Seversky (Seversky Aircraft Corp.); 17, Dr. Theodore Theodorsen (NACA LMAL); 18, Prof. Alexander Klemin (American Society of Aeronautical Engineers), 19, F.A. Louden (U.S. Navy); 20, Charles Lawrence (Wright Whirlwind); 21, Henry Berliner (Engineering and Research Corp.); 22, Harold F. Pitcairn (Autogiro Company of America); 23, I. Macklin Ladden (Consolidated Aircraft Corp.); 24, Giuseppe M. Bellanca (Bellanca Aircraft Corp.); 25, Luis DeFlorez (Air Associates, Inc.); 26, Edward H. Chamberlin (NACA Assistant Secretary); 27, James F. Ray (Autogiro Company of America); 28, Elmer A. Sperry, Jr. (Sperry Products); 29, George W. Lewis (NACA Director of Aeronautical Research); 30, T.P. Wright (Curtiss Aeroplane and Motor Company); 31, Leroy R. Grumman (Grumman Aircraft Engineering Corp.); 32, Charles Ward Hall (Hall-Aluminum Aircraft Corp.); 33, Watson Davis (Science Service); 34, G.L. "Albert" Mooney (Bellanca Aircraft Corp.); 35, Dr. Jerome C. Hunsaker (Massachusetts Institute of Technology); 36, Igor Sikorsky (Sikorsky Aviation Corp.); 37, Lt. Cmdr. Walter S. Diehl (U.S. Navy); 38, Maj. J.H. "Jimmy" Doolittle (Shell Petroleum Corp.); 39, John K. "Jack" Northrop (The Northrop Corp.); 40, Hon. Edward P. Warner (NACA Member); 41, Capt. Arthur B. Cook (U.S. Navy); 42, Prof. J.S. Newell (Massachusetts Institute of Technology); 43, Vincent J. Burnelli (Uppercu-Burnelli Corp.); 44, Brig. Gen. Oscar Westover (U.S. Army); 45, Sherman M. Fairchild (Fairchild Aviation Corp.); 46, Dr. Hugh L. Dryden (Bureau of Standards); 47, Kenneth Bullivant (NACA LMAL); 48, Philip Donely (NACA LMAL); 49, John F. "Jack" Parsons (NACA LMAL); 50, James White (NACA LMAL); 51, Melvin Gough (NACA LMAL); 52, Eastman N. Jacobs (NACA LMAL); 53, Eugene Lundquist (NACA LMAL); 54, H. Julian Allen (NACA LMAL); 55, Fred Weick (NACA LMAL). (NASA EL-1996-00157)

This aerial view of Langley Field in 1939 shows the relative locations of the Full-Scale Tunnel (bottom of photo) and the NACA flight hangar (top right). During WWII, aircraft would be towed from the hangar down the street to the tunnel for tests. The checkerboard and NACA markings on the hangars would soon disappear when the field was camouflaged in 1943. (U.S. Air Force)

CHAPTER 4

Serving the Greatest Generation

1938–1945

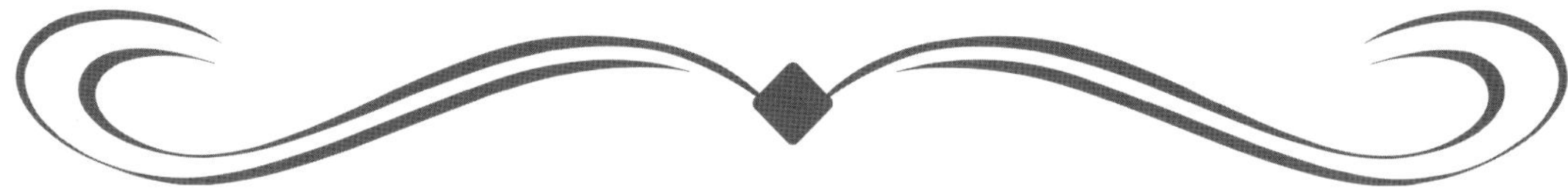

Preparing for the Storm

As war clouds began to threaten Europe with the rise of Nazi Germany, the Nation began to awaken and recognize its second-class stature among producers of military aircraft. In the late 1930s, the military services asked industry for revolutionary aircraft with unprecedented capabilities. The limited funding of the NACA had constrained its advanced research programs, and a multitude of issues would surface for industry's new aircraft designs. The NACA would soon find itself focusing totally on the war effort, and plans for generic research would be overridden by critical testing for specific military aircraft.

Plans for general research projects in the Full-Scale Tunnel during the war were necessarily placed on the back burner while problem-solving and assessments of specific military aircraft were moved to the forefront of the tunnel schedule. The facility was run at a frantic pace 24 hours a day, 7 days a week during the war, as hundreds of war planes were flown in to Langley for flight-test evaluations and, in numerous instances, tests in the Full-Scale Tunnel. The tunnel's staff adhered to the same 48-hour work-week schedule as that used by personnel at most Langley tunnels—6 days a week without holidays. The identification badge system that had been originated in 1925 for enhanced security was updated to include personal pictures in 1941, and only official visitors were allowed at the laboratory after 1938.

Before reviewing some of the most important Full-Scale Tunnel tests of the period, it would be helpful to briefly review some of the personnel, facility, and cultural changes that took place there during the war.

Navy Domination and the Army Response

It became obvious to leaders in the Army Air Corps that the buildup of military-related work at Langley was decidedly being dominated by the Navy. In response to the situation, Gen. H.H. "Hap" Arnold appointed Jean A. Roche to serve as a liaison between the Army Air Corps Materiel Division at Wright Field and Langley.[1] Roche had earned his technical credentials as the designer of the famous Aeronca aircraft and numerous other airplanes. By 1939, the value of Roche's actions in the role of liaison became evident to the Army, and

Maj. Carl F. Greene was sent to Langley as Roche's military superior to further ensure that the Army's interests were given fair consideration by the NACA and to increase the communications between Wright Field and Langley. Greene had been the leader of the Army team that won the Collier Trophy in 1937 for the first successful pressure-cabin aircraft, the XC-35. At Langley, he took a special interest in structures research and stimulated many joint programs with the NACA, such as the ditching of a B-24 bomber in the local James River in 1943.[2] Meanwhile, Roche was particularly interested in the laboratory's work in aerodynamics and was especially impressed by the expertise of the staff at the Full-Scale Tunnel. He became close friends with the staff of the tunnel and acquired many of the tunnel's instruments and test methods used for pressure measurements for the Army researchers at Wright Field.

In the face of the Army presence, the Navy continued its requests for support, particularly from the Full-Scale Tunnel. The Navy's Walter Diehl later remembered, "Over half of Langley's work was for the Navy. At the beginning of war, the Navy was providing so much equipment to Langley that Jean Roche of the Army Liaison Office expressed his anger with the Navy for dominating the NACA laboratory by calling Langley the 'Naval Advisory Committee for Aeronautics.'"[3]

DeFrance Goes West

By 1938, Smith DeFrance had become inundated by new wind tunnel and management responsibilities in his position as assistant chief of aerodynamics. In addition to leading the design of the new 19-Foot Pressure Tunnel and his supervisory responsibility over the Propeller Research Tunnel, the Full-Scale Tunnel, and the 8-Foot High-Speed Tunnel, he was selected to be the leader of the construction and operations of a new NACA laboratory at Moffett Field, CA.[4] During 1939 and 1940, DeFrance stayed at Langley while designing facilities for the new lab, and on February 15, 1940, he was relieved of his responsibilities for supervising Langley tunnels to work full time on the construction of Moffett Field facilities. Meanwhile, Russ Robinson, who had transferred to NACA Headquarters from the 8-Foot High-Speed Tunnel at the request of George Lewis in 1940, served at the California site as the NACA liaison official to the West Coast industries with the job of enhancing the efficiency of coast-to-coast communications. Edward R. Sharp was sent to California in early 1940 to head administrative matters, and he returned to Langley in the fall to plan for the second new NACA lab at Cleveland, where he would later become director. In April 1940, the new California lab was formally named the NACA Ames Aeronautical Research Laboratory in honor of Dr. Joseph Ames, who had recently retired from NACA. DeFrance arrived at Ames as engineer in charge in August, assisted by Jack Parsons from his Langley staff.[5]

The task of staffing the new Ames laboratory fell directly on Langley. With an average age of only 26 during the 1930s, many members of the young Langley staff were intrigued by the opportunities presented in California. Over half of the original Ames staff of 51 research and support people in August 1940 was from Langley. In addition to DeFrance, transfers were arranged for Jack Parsons, Harry Goett, Manley Hood, F.R. Nickle, W. Kenneth Bullivant, A.B. Freeman, and J.A. White—all of whom had been affiliated with the Full-Scale Tunnel. Other notable Langley researchers who transferred included H. Julian "Harvey" Allen from the VDT and Donald Wood from the PRT. Both men became legendary leaders at Ames.

Dearborn Moves Up

Clint Dearborn had become head of the Full-Scale Tunnel Section in 1935, assisted by Abe Silverstein. In late 1941, Dearborn was assigned to head a new propulsion installation group known as the Cooling and Cowling Group, which worked closely with industry on this critical topic in a collocated site on the third floor of the East Shop across from the Full-Scale Tunnel.[6] The operation became known as "The Cooling College," as remembered by Jean Roche: "In order to educate industry's airplane and engine designers in the shortest possible time, Col. Greene used his prestige to start a school in the loft of the model shop with Dearborn, Silverstein, Rupert, and others as instructors. Design engineers were invited from the entire airplane and engine industries. Each industry representative brought the problems of his own company, engines and cowlings were provided by the contractors, instrumented by the NACA, and tests were made in the Langley tunnels. After returning to their companies the students taught others and thus our entire industry was indoctrinated in the science of engine cooling."[7]

Dearborn was subsequently promoted to the position of chief of the new Full-Scale Research Division in July 1943, with responsibilities for many of the Langley wind tunnels including the Full-Scale Tunnel, the 19-Foot Pressure Tunnel, the Propeller Research Tunnel, and the Cooling and Cowling Group.

Silverstein Moves North

Abe Silverstein was named head of the Full-Scale Tunnel Section on January 28, 1942. His prowess at engine-airframe cooling and drag reduction was widely recognized. He simultaneously served as head of the Full-Scale Tunnel, as director of engine-cooling tests in the Full-Scale Tunnel and the PRT, and as a special consultant to the Dearborn's Cooling and Cowling Group.

In addition to staffing the Ames Aeronautical Research Laboratory, Langley provided personnel to staff a new NACA engine research laboratory at Cleveland, OH, in the early 1940s. Langley's Edward "Ray" Sharp oversaw construction of the new laboratory, known initially as the NACA Aircraft Engine Research Laboratory (AERL).[8] Sharp was named manager of the lab in 1942 and director in 1947. In October 1943, Abe Silverstein transferred to Cleveland to become chief of the new NACA Altitude Wind Tunnel at the request of Dr. George Lewis. Silverstein went on to become one of the most famous researchers and managers of the NACA and NASA. He started as a staff member and head of the Full-Scale

Tunnel. At Cleveland, he led the Wind Tunnels and Flight Division for 5 years before becoming associate director. He transferred to NASA Headquarters in 1958 to assist in the establishment of NASA and to manage efforts leading to Project Mercury and the Apollo Program. Silverstein returned to Cleveland in late 1961 to serve as the Center director for the next 8 years.

Of interest to the subject matter of this book is the fact that, in addition to Silverstein, Dale H. "Mac" McConnaha of the original Full-Scale Tunnel staff also transferred to the AERL in the early 1940s. McConnaha had started his career at Langley in 1923 in the Dynamometer Lab as an engine mechanic, followed by an assignment to the flight research section, where he participated in flight activities with the early NACA aircraft. He then joined DeFrance's tunnel staff in 1930 and was the first head of mechanics at the Full-Scale Tunnel.

Following Silverstein's departure, Herbert A. "Hack" Wilson, Jr., was named head of the Full-Scale Tunnel Section in December 1943, with responsibilities for the experimental studies conducted in the tunnel. A second research group, known as the Full-Scale Analysis Section, was also formed under Samuel Katzoff within the previous Full-Scale Tunnel organization, with responsibilities to provide an analytical understanding of the experimental data.

The careers of the original and early members of the Full-Scale Tunnel staff were truly remarkable. No less than three became directors of NACA laboratories or NASA Centers: Smith DeFrance (NACA Ames Aeronautical Research Laboratory/NASA Ames Research Center, 1940–1965), Abe Silverstein (NASA Lewis Research Center, 1961–1969), and Harry Goett (NASA Goddard Space Flight Center, 1959–1965). Many others were widely recognized for their technical and managerial skills at Langley and other laboratories and field Centers of NASA.

New Tenants at Full-Scale

Immediately after the Full-Scale Tunnel went into operation, other research organizations at Langley recognized the availability of the considerable space beneath the flow circuit return passages at the south end of the building. As previously discussed, the first Langley gust tunnel was installed adjacent to the inner wall of the east return passage and was operated by the structures and loads organization during the mid-1930s. Langley also formed a small physical research organization under the noted Langley physicist Theodore Theodorsen, with research goals directed toward fundamental and theoretical matters such as airfoil theory, propeller flutter, and noise. This group established a physical research laboratory at the southeast end of the building beneath the east return passage in 1934, adjacent to the engineering offices of the Full-Scale Tunnel staff.[9] The physical research laboratory included a sound recording room, sound generation equipment, and a shop. The complex existed until 1945, when Theodorsen's Physical Research Division moved into a new building in the NACA West Area.

Langley woodworkers fabricate new wooden blades for the Full-Scale Tunnel in the flight hangar in May 1938. Made of Sitka spruce, the blades were installed a few months later and remained operational through the life of the tunnel. (NASA EL-2000-00357)

Metal to Wooden Propellers

In 1937, an event happened that would change a primary feature of the Full-Scale Tunnel for the rest of its operational life. The tunnel's next-door neighbor, the 8-Foot High-Speed Tunnel, had been built in 1936 with cast-aluminum alloy propeller blades similar to those of the Full-Scale Tunnel and the PRT. During a night shift being run in the 8-Foot Tunnel by John Becker on October 8, 1937, a major accident occurred. As Becker recalled, "A terrible explosion came from the tunnel, smoke belched from the tunnel and I pushed the emergency stop button, bringing the drive motor to a stop, causing the electrical power in the city of Hampton to go off line. The aluminum blades had failed as a result of fatigue caused by periodic disturbances during their passage through the wakes of the support struts. The tunnel was down for repairs for six months."[10] According to Becker, after the accident Langley management decided to build and install new wooden blades in the 8-Foot Tunnel, with fabrication accomplished in-house through Langley support organizations.

As a result of concern over a similar experience occurring at the Full-Scale Tunnel, the decision was made to replace its existing aluminum blades with wooden blades made of Sitka spruce.[11] Once again, it was decided that the blades be built in-house. The Langley woodworking shop personnel under Percy Keffer took on the task of laminating the spruce and hand carving the blades in the Langley aircraft flight hangar. The blades were installed in mid-1938 with minimal impact on the tunnel schedule—during a period in which the historic drag cleanup work was beginning to ramp up. Incredibly, the wooden blades of the Full-Scale Tunnel continued to operate with only minor repair and maintenance for 71 years until the closing of the facility in 2009.

The most serious damage to the wooden propeller blades in the Full-Scale Tunnel occurred when an auxiliary air cooling intake assembly atop the East motor failed during tests of a B-24 engine nacelle in 1942. The intake pivoted forward into the trailing edges of the propeller. Note the intake debris behind the horizontal blade on the right. (NACA 28358)

No other wind tunnel blades in Langley's history came close to the operational robustness of the 71-year-old blades. The ingestion of model parts and blade failures in high-speed tunnels such as the Langley 16-Foot Transonic Tunnel and the Langley High-Speed 7- by 10-Foot Tunnel caused severe damage to their wooden blades, requiring complete replacement of the blade sets and significant impacts on tunnel test schedules. The most serious damage inflicted on the wooden blades of the Full-Scale Tunnel occurred in May 1942 during tests of a B-24 engine and nacelle, when a tear-drop shaped auxiliary cooling air intake assembly on the top of the East tunnel motor failed. The rear section of the streamlined duct failed and the entire assembly pivoted forward into the propeller blades with the tunnel running at 100 mph. However, the damage to all four blades was repaired in place and the impact was minimal.[12]

Aerial photograph shows camouflaged NACA buildings during September 1943. The Full-Scale Tunnel is at the middle-left area of the picture. (NACA LMAL 34517)

Langley Field Camouflaged

The Tidewater area of Virginia, with its concentration of Army and Navy bases, was a prime target of extensive foreign spies in the early years of World War II. German submarine attacks on U.S. shipping off the East Coast peaked in 1942, and stringent blackout procedures were put in place to minimize visual access to activities in the Hampton area.

In 1943, the Army Air Forces spent over $550,000 to camouflage Langley Field. Hangars, warehouses, and many NACA buildings along the Back River received a coat of olive drab paint. The runways were not painted, but most of the streets were also camouflaged.[13] The Full-Scale Tunnel, 20-Foot Spin Tunnel, 8-Foot High-Speed Tunnel, and other NACA assets were painted. The Army sandblasted the camouflage paint from the buildings for several months after the war and restored the prewar appearance of the buildings.[14]

Photographs of the 8-Foot High-Speed Tunnel (foreground), Propeller Research Tunnel (background), and Full-Scale Tunnel (right) show appearances of the buildings in 1936 (left image) and after being camouflaged in 1943 (right image). (NACA)

Skylights Close

In June 1945, the Full-Scale Tunnel motors and test section underwent a major maintenance activity and a new roof was installed over a 2-month period. During the process, the original large 30- by 40-foot skylights, which appear in many photographs prior to that date, were removed and replaced with a wooden, metal-finished roof structure. The diminished interior lighting was enhanced by the installation of overhead lamps above the test section. The change came as a welcome modification according to several retirees, because the heat of the Tidewater summers was amplified by the skylights, making life miserable during many test programs.

Other Large Tunnels: Foreign and Domestic

The international aeronautical community had taken notice of the unique capabilities provided by the Full-Scale Tunnel, and several nations quickly committed to building their own full-scale test facilities.[15] The French had proposed a new full-scale tunnel in 1929, inspired by an awareness of the NACA's plans for the Full-Scale Tunnel.[16] France built the second large open-throat tunnel capable of testing full-scale aircraft at Chalais-Meudon near Paris from 1932 to 1934. It became operational in 1935. Initially known as La Grande Soufflerie ("the Great Wind Tunnel"), it was later renamed the ONERA (Office National d'Etudes et de Recherches Aerospatiales) S1Ch tunnel. The tunnel was powered by six 10-blade, 29-foot-diameter fans to a maximum speed of about 112 mph; had a test section of 52 feet by 26 feet; and could accommodate powered aircraft with wingspans of about 40 feet. Interestingly, it was an atmospheric open circuit tunnel—that is, the tunnel did not have return passages, and the test flow was discharged to the atmosphere. During the German occupation in the 1940s, it was operated in support of Axis war efforts. Smith DeFrance reported that the flow in the French tunnel was very poor and that the Germans had used it unsuccessfully to test a captured P-51 Mustang in an attempt to discover the airplane's secrets.[17] The tunnel has continued to operate to this time.

Russia built an even larger low-speed tunnel known as the T-101 at its Central Aerohydrodynamic Institute (TsAGI) in 1939. Equipped with an elliptical open-throat test section measuring 39.4 feet by 78.7 feet, and driven by two fans to speeds up to 170 mph, the tunnel layout is very similar to the Full-Scale Tunnel closed-return arrangement. The T-101 tunnel continues to operate into the 21st century.

Other "full-scale" tunnels put into operation in the late 1930s included facilities in England (24-foot diameter, 17 by 35 feet).

When Smith DeFrance began design of the wind tunnel facilities for the Ames laboratory, a high-speed 16-foot transonic tunnel and two 7- by 10-foot tunnels were at the top of the NACA priority list.[18] While the tunnels were being constructed, DeFrance and his staff started the design for an updated West Coast version of the Langley Full-Scale Tunnel.[19] The result was a monstrous 40- by 80-foot closed-throat, single-return tunnel with a speed capability up to 350 mph. It was started in 1942 and completed in 1944 as the largest wind tunnel in the world. The 40- by 80-Foot Wind Tunnel was later augmented with an 80- by 120-foot test section in the 1980s in a new facility known as the NASA National Full-Scale Aerodynamics Complex (NFAC). The NFAC was decommissioned by NASA in 2003 and is now operated by the U.S. Air Force as a satellite facility of the Arnold Engineering Development Center (AEDC).

The War Years: A Dependable Workhorse

Detailed discussions of the large number of tests conducted in the Full-Scale Tunnel at the beginning and during World War II cannot possibly be covered within the scope and detail intended herein. The following discussion is mainly directed at the role the tunnel played in the development of famous U.S. aircraft during the war. Although a considerable number of generic tests (especially tradeoff studies of cooling and drag for propeller/nacelle installations, high-lift flap systems, and downwash-flow studies behind wings) were also conducted, this review is limited to only a few important generic tests. The material is presented in chronological order so that it can best portray the role the tunnel played in advances made in famous U.S. military aircraft during the period.

Cooling Wing

As previously discussed, Abe Silverstein had no formal training in aerodynamics prior to becoming a member of the staff at the Full-Scale Tunnel. However, he was an outstanding mechanical engineer and he quickly became an expert on the important issue of providing satisfactory engine cooling and the associated technologies of engine baffling, radiators, and cooling-air ducting with minimal drag penalties. While adapting to on-the-job training in aerodynamics, he initiated a series of cooling-related tests that would greatly influence the applications of the tunnel during the war years.

Silverstein quickly became the tunnel's leader on the topic of cooling methodology for air-cooled and liquid-cooled engines. He strongly opposed the conventional approaches being used for liquid-cooled engines, which were typically equipped with cowled, underslung radiators because the high-speed drag of such installations could be as much as 15 to 20 percent of the drag of the entire airplane. Instead, he promoted the concept of using expanding ducts within or partially within the wing to provide sufficient cooling air for all flight conditions and reduce cooling drag at high speeds, which accomplishes both goals without significant adverse effects on the wing's maximum lift.

In 1938, Silverstein initiated the first of a series of tests of over 100 different duct-radiator combinations.[20] The testing examined a multitude of design variables, including the spanwise length of the duct, the leading-edge radius of the duct inlet, various outlet flaps to control flow rate through the radiator, and the quantification of results in relation to practical design. The staff of the Langley 7- by 10-Foot AWT also collaborated closely with the Full-Scale Tunnel program and conducted their own investigation of wing ducts. The potential benefits of Silverstein's wing installations for cooling ducts permeated the mindset of the

Abe Silverstein directed a series of investigations of wing ducts for cooling radiators of liquid-cooled engines using this 45-foot-span, tapered airfoil. A typical duct inlet is shown in the center section of the wing. The photograph was taken in March 1938, before the tunnel propellers were changed to wood. (NACA 14920)

Full-Scale Tunnel staff during drag cleanup tests during the war, and the concept usually became the standard to beat for other cooling concepts. Unfortunately, many of the aircraft had progressed too far in internal structural design to be modified to any extent by the time they were tested in the tunnel.

Brewster F2A Buffalo

One of the most important events in the history of the Full-Scale tunnel occurred in April 1938, when the Navy let it be known that it was extremely unhappy with the performance of the prototype of its first monoplane fighter, the Brewster XF2A-1 Buffalo. Although it was the winner of a Navy competition to build a 300-mph fighter, in preliminary flight tests the aircraft had exhibited a maximum speed of only 277 mph at an altitude of 15,000 feet—significantly below expectations. An urgent Navy request from Cmdr. Walter S. Diehl was received for tests in the Full-Scale Tunnel to determine whether the aerodynamic drag of the aircraft could be reduced, and the airplane was flown to Langley on April 21 from Anacostia Naval Air Station for installation in the tunnel.

The response to the Navy request was especially challenging for the team at the Full-Scale Tunnel, which was led by Clint Dearborn. The frenzy over the deplorable performance of the airplane resulted in the XF2A-1 only being available for a brief 3-day test period that began on Saturday, April 30. The test program was so abbreviated that the propeller was removed and all testing was in a power-off condition.[21] By Monday afternoon, the tests had identified major sources of drag. The most unsatisfactory flow characteristics were noted around the engine cowl, intakes, and forward fuselage, with large regions of separated and reversed flow noted around the carburetor air scoop and blast tubes for guns.[22]

An urgent Navy request to examine the possibilities for increasing the top speed of the Brewster Buffalo in 1938 resulted in a parade of prototype and production military aircraft to the Full-Scale Tunnel for drag cleanup. This test was the last to use the tunnel's metal propeller blades before the change to wooden blades. (NACA 15336)

The maximum speed for the aircraft based on NACA predictions using results from the power-off tunnel tests was 248 mph, much less than the 277 mph demonstrated in the Navy preliminary flight tests. The difference was attributed to a combination of power-on effects in flight, extrapolation of data to flight conditions, and higher propulsion efficiency than used in the calculations. In any event, the tunnel results indicated a potential boost in top speed to 291 mph (an increase of 43 mph) if all the changes tested in the tunnel—many of which could not be made—were accepted.

The results of the test highlighted the fact that modifications to certain configuration details would permit a substantial increase in the airplane's top speed. Subsequent modifications to the XF2A-1 based on lessons learned from the Full-Scale Tunnel test resulted in a top speed of 304 mph, ensuring that the aircraft met the Navy's 300 mph requirement, and the aircraft was put into production.[23] The military services were extremely impressed by the ability of Langley to provide such valuable information in such a short time, and many prototypes and operational aircraft were run through the tunnel for drag cleanup during the period.

In addition to the drag reduction tests for the XF2A-1, the tunnel entry provided requested information to the Navy relevant to inadvertent carbon monoxide ingestion in the cockpit. For the unpowered tunnel tests, wool tufts were placed around the canopy edges to indicate unwanted inflow, an observer in the cockpit used a tuft on a wand to examine for inflow, and the passage of airflow from the wing-root aileron-control cutout and tail-wheel opening were also examined. Carbon monoxide ingestion was a problem for production F2A aircraft in service.

Following the wind tunnel test, the XF2A-1 was flown back to Brewster, where it was modified with a new engine and several of the drag-reducing recommendations from the tunnel test results. Now known as the XF2A-2, the aircraft became the subject of a little-known joint Navy-NACA investigation of compressibility effects during high-speed dives in 1940.[24] This work had commenced following several airframe component failures for different aircraft during high-speed dives years before the role of compressibility became well known as a result of the YP-38 crash in 1941. The activity had been initiated by mutual interactions between Navy personnel and Langley's Richard Rhode, who had led the previously discussed loads work on the maneuver loads for the PW-9. John Stack and John Becker were interested witnesses to the XF2A-2 dive tests, which took place on May 7, 1940, over Grandview Beach in Hampton.[25] Their interest was piqued by comparing pressure data over the wing during the dive tests with wind tunnel data from the 8-Foot High-Speed Tunnel. Terminal velocity dives from 30,000 feet produced a maximum speed of 560 mph and supercritical flow over the wing, followed by an 8-g pullup. At the conclusion of the dive tests, the aircraft was returned to Brewster in preparation for final Navy testing at Anacostia.

In late October 1940, the XF2A-2 returned to Langley in preparation for wind tunnel testing at the request of the Navy. The December 1940 tests in the Full-Scale Tunnel were to conduct a more refined assessment of drag-producing protuberances, inlets, and surface finish. In addition, extensive pressure instrumentation was used to provide estimates of potential critical compressibility locations on the aircraft.[26] Once again, all tests were power-off, and results indicated that if all of the drag-reduction modifications suggested by the NACA were carried out, an increase in top speed of 44 mph was predicted. Many suggestions were carried out for production aircraft, with significant performance benefits noted.

After the tunnel tests of 1938, the modified XF2A-1 was designated the XF2A-2 and tested in 1940 for further drag cleanup data. Note the bulbous propeller spinner and revised cowling intakes from the earlier version of the aircraft. (NACA 22532)

The XF2A-2 remained at Langley and was again tested in the Full-Scale Tunnel a few months later in April 1941 at the request of the Navy, to determine the effects of wing-mounted machine-gun and cannon installation configurations on maximum lift and drag.[27] Results showed that, for minimal impact on lift, the ends of the machine guns should be flush with the leading edge of the wing rather than protruding and that cannons carried in underwing gondolas caused excessive drag.

The Brewster Buffalo design made its last appearance in the Full-Scale Tunnel in August 1942, when a production F2A-2 airplane equipped with a new wing featuring a full-span slotted flap was tested. Interest in NACA research on full-span flaps had intensified, and the Navy requested tests to determine the maximum lift, aileron forces and effectiveness, elevator effectiveness, and flap installation effects on high-speed drag.[28] The project also included flight tests at Langley and tests in the 7- by 10-Foot AWT.[29] As was the case for all Buffalo tests in the tunnel, the propeller was removed and the project was for power-off conditions. The airplane used conventional ailerons for low flap-deflection angles and slot-lip spoiler-type ailerons for large flap deflections. No modifications to the Buffalo fleet were made as a result of the studies (production had already terminated), but the data were useful for future designs.

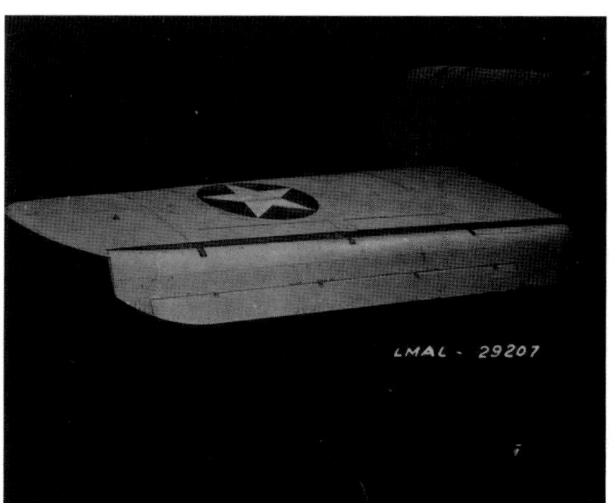

Views of the F2A-2 in the Full-Scale Tunnel with various deflections of the full-span slotted flap. Top photos show the flap deflected (left) and undeflected (right), while lower views show slot-lip ailerons deflected (left) and undeflected (right). (NACA 29206 to NACA 29209)

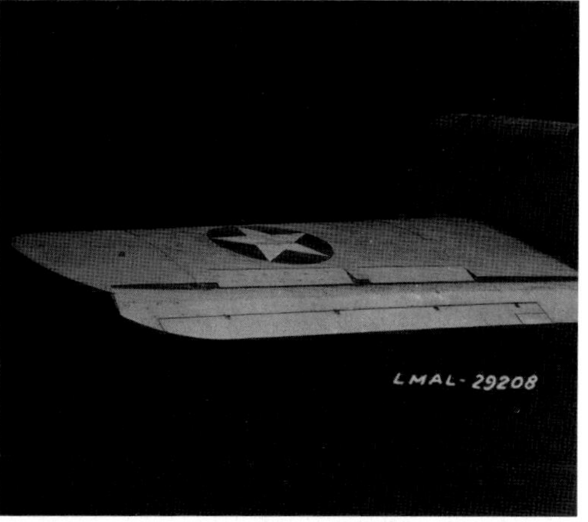

Drag Cleanup Procedure

As discussed in the previous chapter, the staff at the Full-Scale Tunnel had already begun fundamental and applied investigations of the drag produced by aircraft components prior to the Buffalo tests, but the startling results of the XF2A-1 test program impressed the military with the relatively large magnitude of drag reduction—and the increase in top speed—that could be produced by careful attention to details in aircraft construction and design. The military quickly requested more "drag cleanup" tests of a similar nature for their emerging new aircraft.

The procedure developed for such tests consisted of first conducting wool-tuft flow-visualization tests for the test subject in the "service" condition (representative of operational aircraft conditions), with the airplane at a cruising attitude at several tunnel speeds. Results quickly indicated areas of flow disturbances due to turbulent separated flows. In addition, the tests at several speeds indicated whether any significant scale effect (i.e., Reynolds number effect) was present. In most cases, the test was conducted at maximum speeds of about 100 mph. The majority of tests were conducted with the propeller removed, although in some instances a few power-on tests were made to determine the effects of the propeller slipstream on drag and local flow on the fuselage nose. Following the tuft studies, the service aircraft was tested over a range on angles of attack up to and including wing stall, and measurements were made of lift, drag, and pitching moment.

After results were obtained for the service configuration, the areas indicated by tuft studies to be suspected sources of drag-producing turbulence were faired with putty, covered, or removed. Cracks and cutouts such as gun blast tubes, intakes, canopy panels, and wheel wells were covered with aluminum sheets or tape. In some cases for radial-engine aircraft, the nose area was modified with a pointed wooden insert to create a smooth, streamlined, bullet-like shape. The "faired and streamlined" configuration was then subjected to the same test conditions to establish the level of drag that might be possible by modifications. This result represented the goal to strive for in the cleanup tests.

As the test program proceeded, various components were exposed or added to the faired configuration to establish the increment of drag produced by the component or condition. Surveys of the momentum in the wing wake were also taken at several spanwise stations to determine the local wing-profile drag for the wing in its service and faired conditions. In many cases, pressure measurements were made over certain parts of the aircraft to estimate the speeds at which compressibility effects would first begin, and to suggest modifications to delay the onset of compressibility. Typical of such locations were canopy peaks, cowl upper surfaces, and inlets. An important lesson learned in this test procedure was that, although most airframe items produced individual drag increments of only a few percent, the total increments added up to an impressive total increase in aircraft drag.

It is important to note that the drag cleanup tests were inherently limited in application but nonetheless provided very valuable guidance for manufacturers and the military. Some of the limitations included the relatively low top speed of the tunnel compared to the maximum speeds of propeller-driven fighters of the day (100 mph versus over 300 mph). As a result of this speed difference, certain key aerodynamic parameters (such as Reynolds number) were not matched between flight and tunnel tests conditions, requiring a mixture of science and art based on experience to extrapolate the tunnel data to flight. Quantitative estimates of the top speed of test subjects based on power-off results at the low tunnel speed were typically low; that is, the aircraft exhibited a higher maximum speed in flight. This result was attributed to possible scale or compressibility effects as well as propulsive effects and errors in propulsion efficiencies assumed in the calculations. In any event, the incremental contributions of components to drag proved to be representative of those experienced in subsequent flight tests of modified configurations. In many cases, the aircraft modifications identified as being most significant for drag reduction could not be carried to production (such as wheel covers, etc.), and the demand for operational aircraft in the war zones could not be delayed, except for extremely significant improvements. Nonetheless, the drag cleanup process became a legendary activity at the Full-Scale Tunnel.

Unfortunately, some confusion has arisen regarding the aircraft actually tested in the Full-Scale Tunnel for drag cleanup.[30] Langley had received formal Research Authorizations (RAs) to proceed with requests for drag-reduction efforts for a large number of aircraft, but some of the RAs were directed to other facilities. For example, the Curtiss XP-37, Curtiss YP-37, XF4U-1, and Consolidated XB-32 were not tested in the tunnel. The following table indicates aircraft that were tested, including actual full-scale aircraft and large models. The tests are further defined by "drag reduction," in which models or aircraft shapes were modified, and by "cleanup," in which the classical "service" and "faired and taped" tests were conducted.

Aircraft	Date	Type of Test
Boeing XBLR-1	December 1935	Drag Reduction
Bell XFM-1	December 1936	Drag Reduction
Vought SB2U-1	September 1937	Cleanup
Brewster XF2A-1	April 1938	Cleanup
Grumman XF4F-2	June 1938	Cleanup
Grumman F3F-2	July 1938	Drag Reduction
Grumman XF5F-1	December 1938	Drag Reduction
Northrop XBT-2	February 1939	Cleanup
Brewster XSBA-1	February 1939	Cleanup
Curtiss XP-40	April 1939	Cleanup
Martin XPB2M-1	May 1939	Drag Reduction
Seversky XP-41	June 1939	Cleanup
Bell XP-39	July 1939	Cleanup
Grumman XF4F-3	September 1939	Cleanup
Curtiss XP-46	October 1939	Drag Reduction
Bell XP-39B	March 1940	Cleanup
Curtiss XP-42	September 1940	Cleanup
Curtiss XSO3C-1	October 1940	Cleanup
Brewster XF2A-2	December 1940	Cleanup

Aircraft	Date	Type of Test
Curtiss P-36	December 1940	Cleanup
Douglas A-20A	January 1941	Cleanup
Republic XP-47	May 1941	Drag Reduction
Lockheed YP-38	December 1941	Cleanup
Consolidated B-24	February 1942	Drag Reduction
Grumman XTBF-1	June 1942	Cleanup
Republic XP-69	August 1942	Drag Reduction
Vought F4U-1	November 1942	Cleanup
Curtiss SB2C-1	January 1943	Cleanup
Grumman F6F-3	March 1943	Cleanup
Bell XP-77	June 1943	Drag Reduction
Bell P-63	August 1943	Cleanup
North American P-51B	September 1943	Cleanup
Bell YP-59A	March 1944	Cleanup
Curtiss SC-1	November 1944	Cleanup
Grumman XF8F-1	December 1944	Cleanup
Total Drag Reduction: 10	**Total Drag Cleanup: 25**	

The results of drag cleanup tests came so quickly during the war that Langley published a summary report of data from the Full-Scale Tunnel for rapid dissemination to industry in 1945.[31] During the 1970s, NASA republished relevant drag cleanup results on 23 aircraft for use by the general aviation community.[32]

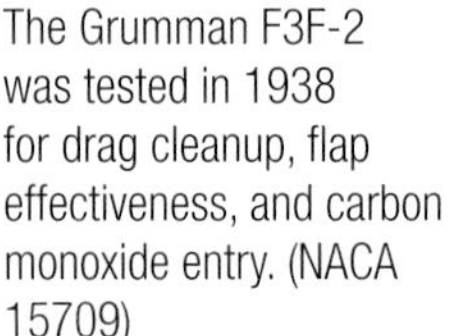

The Grumman F3F-2 was tested in 1938 for drag cleanup, flap effectiveness, and carbon monoxide entry. (NACA 15709)

Grumman F3F-2

The Grumman F3F was the last biplane fighter flown by the U.S. Navy, entering service in 1936. In July 1938, the Navy requested a 2-week test program in the Full-Scale Tunnel to determine methods to reduce the drag of the aircraft, the impact on maximum lift of adding flaps on the upper wing, probable points of entry of carbon monoxide into the cockpit, and the effectiveness of methods to reduce the entry of the potentially deadly gas.[33]

Results of the tunnel test showed that the upper fuselage of the F3F experienced massively turbulent flow all the way back to the cockpit enclosure. The problem was caused by flow separation around the machine-gun blast tube openings in the forward cowl area. In addition, a violently disturbed flow existed immediately behind the exhaust stacks. The cockpit was subjected to potential carbon monoxide entry via flow through the machine-gun blast tubes and ejector chute openings, the rear interior, and overlap of the cockpit canopy.

Grumman F4F Wildcat

Grumman had entered the Navy fighter competition in 1936 that was ultimately won by the Brewster Buffalo. Grumman's entry was initially a biplane design designated the XF4F-1, which was superficially similar to Grumman's earlier single-place F3F-2 airplane. When performance predictions by Grumman showed the XF4F-1 biplane would be outclassed by the other entrants in the competition, the company (with Navy approval) changed the design to a monoplane with the designation XF4F-2. The Navy reacted by showing its interest in having the aircraft as a backup to the inexperienced Brewster team's winning F2A fighter. The XF4F-2 made a forced landing in April 1938 and suffered minor damage, after which the Navy requested tests in the Full-Scale Tunnel in June to determine approaches to increasing its maximum speed and decreasing the stalling speed, evaluate why the engine would not respond rapidly near stall speed, and determine places of possible entry of carbon monoxide into the cockpit.[34]

Following a brief analysis of wool-tuft flow studies, the Full-Scale Tunnel staff identified several sources of incremental drag, including an underwing oil-cooler intake, the exposed wheel wells and retracted tires, and the carburetor-intake scoop on the engine cowl. Results indicated that if the wheel wells were faired flush, an underslung radiator used for cooling, the carburetor intake and cowling refaired, antenna wires removed, and wing roughness eliminated, the maximum speed could be increased from 267 to 288 mph. An extended, squared-off wingtip and extended flap segment were also found to decrease the power-on stall speed from 77 mph for the as-received round-tip wing to 71 mph. The carbon monoxide–related study found inflow into the cockpit from all openings in the firewall (especially around the fuel tank), aileron-control tubes, and the rear of the aircraft.

After the tunnel tests, the XF4F-2 was rebuilt with the extended wing with square tips, several drag-reducing modifications, and a new Pratt & Whitney Twin Wasp engine. Known as the XF4F-3, it demonstrated much improved performance, and the Navy ordered 54 production aircraft as the F4F-3 in August 1939. The XF4F-3 began a month-long test program in the Full-Scale Tunnel in September 1939.[35]

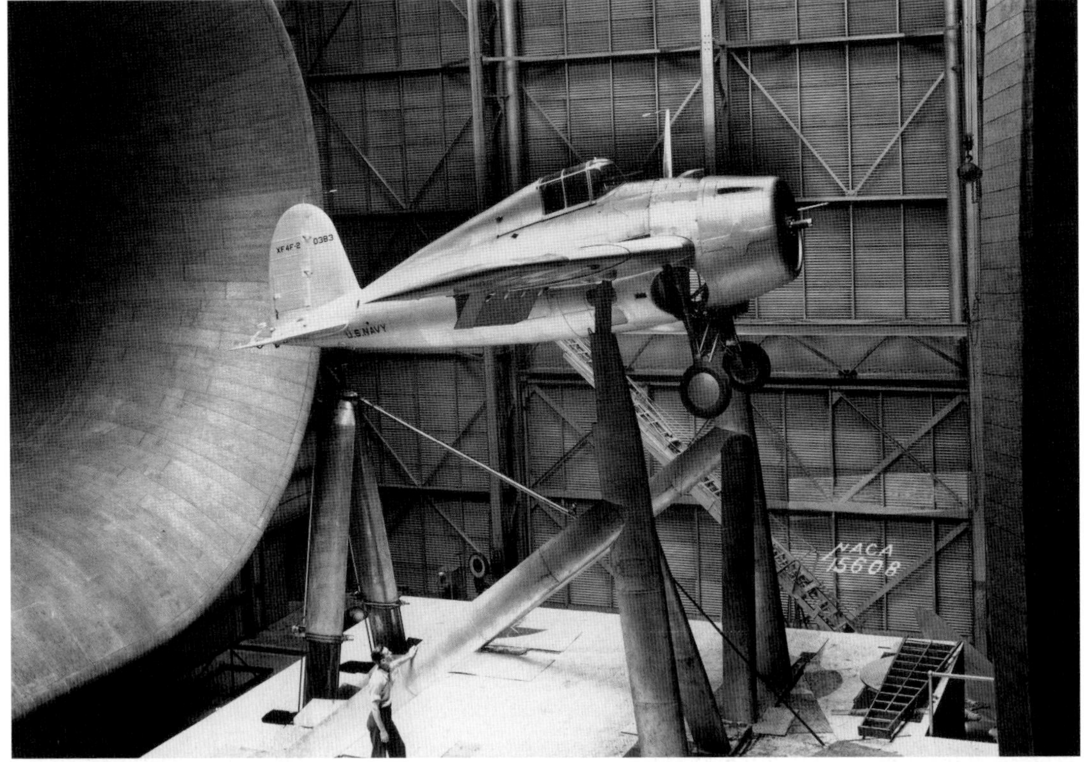

The portly Grumman XF4F-2 during tests in June 1938. The photo on the bottom shows details of the hand-cranked retractable landing gear and drag-inducing protuberances, including the underwing oil cooler and engine exhaust stub. (NACA 15608 and NACA 15600)

The XF4F-3 featured squared-off wing and tail surfaces in September 1939. In the upper photograph, the airplane is in the "as-received condition," while the lower photo shows the "faired" condition with drag-producing seams covered, protuberances removed, and a streamlined nose block installed for measurements of minimum drag. In later production versions, the vertical tail was unswept and the horizontal tail moved from the fuselage to the vertical tail. (NACA 18522 and NACA 18539)

Previous testing had shown that operating the XF4F-3 engine at the low speeds of the tunnel resulted in rough operations and insufficient accuracy, so the engine was replaced with an alternating current (AC) electric motor. The objectives of the test were to determine more refinements for drag reduction and to evaluate propeller efficiency with propeller spinners and cuffs.[36]

Results of the tests indicated that the maximum speed of the XF4F-3 could be increased by about 15 mph if certain gaps were sealed and a smaller antenna adapted. The maximum lift with the new wing was significantly increased over the earlier XF4F-2. The XF4F-3 with final modifications was ordered into production as the F4F-3 Wildcat. Although the aircraft was inferior to the Japanese Zero, it carried the fight in the Pacific for the United States during the early war years of 1941 and 1942 at the battles of Midway and Guadalcanal, and it earned a reputation as a rugged opponent.

The Vought SB2U-1 was tested in July 1938 to assess the impact of wing-mounted dive flaps and define further possibilities for drag cleanup. Note the extended dive flaps and the large oil-cooler scoop on the engine cowl. (NACA 15756)

Vought SB2U Vindicator

The same Vought Vindicator dive bomber that had been tested in 1937 returned to the Full-Scale Tunnel a year later for Navy-requested tests to evaluate the effectiveness of wing-mounted dive flaps, to determine if the maximum speed of the aircraft could be increased, and to measure maximum lift for varying spans of landing flaps.[37] Originally, Vought had intended to use a reversible propeller to act as a dive brake, but the concept was unsatisfactory. As an alternative technique, Vought constructed a dive flap that consisted of three segments of flaps on the upper- and lower-wing surfaces. The flaps were actuated by finger-like spars that, during normal flight, were flush with the wing surface but during a dive could be extended at large angles to the wing surface to slow the aircraft. The pivot point for the actuators was the wing spar.

The tunnel tests revealed that severe aileron buffeting occurred when the flaps were extended despite several different flap configurations. An alternative trailing-edge split flap (similar to the one used by the Douglas Dauntless dive bomber) was also evaluated. Langley chief test pilot Bill McAvoy sat in the cockpit, evaluated the level of aileron buffeting felt through the control stick, and declared the buffeting feedback to be unacceptable. The Navy ultimately adopted a shallower dive angle rather than implementing the dive flaps. The cleanup tests focused on the huge oil cooler installation and the landing gear. The effects of re-fairing the oil cooler scoop and refining the landing gear fairings were to provide about a 15-mph increase in speed.

The radical Grumman XF5F-1 adopted several modifications based on Full-Scale Tunnel tests of a full-scale mockup in 1938. For example, the wing-fuselage fillet fairing shown here was implemented. (NACA 16576)

Grumman XF5F-1 Skyrocket

In 1938, the Navy awarded a contract to Grumman for a single prototype of a radical new twin-engine, twin-tail fighter designated the XF5F-1. The award was surprising in view of the fact that the Navy was just beginning to transition from biplanes to relatively conservative single-engine monoplanes. In December 1938, a full-scale model of the Skyrocket built by Grumman underwent a 2-month series of detailed aerodynamic tests in the Langley Full-Scale Tunnel at the request of the Bureau of Aeronautics.[38] The test objective was to evaluate the effect of configuration variables such as nacelle shape and location on the wing, propulsion effects, and fuselage shape on aerodynamic performance and stability.

First flight of the XF5F-1 occurred on April 1, 1940, and its impressive climb performance quickly earned it the nickname Skyrocket. Several changes inspired by the recommendations produced by the NACA tunnel tests were made, but operational problems—especially cooling of the engines—resulted in serious delays in the development program. By 1942, Grumman and the Navy had become more interested in developing the more powerful new twin-engine XF7F-1 Tigercat aircraft, but the single XF5F-1 prototype was used in over 200 test flights until it experienced a belly landing forced by a landing gear failure in December 1944, after which it was retired.

The XBT-2 offered many opportunities for drag cleanup, especially in the wing, which exhibited many irregularities in construction and a perforated split-surface trailing-edge flap for speed control during dive-bombing runs. Later, the wing would be outfitted with leading-edge slots for better low-speed behavior. (NACA 16853 and NACA 16844)

Northrop XBT-2

The Northrop XBT-2 dive bomber first flew on April 25, 1938, and after successful testing the Navy placed an order for 144 aircraft. The aircraft had been designed in 1934 by John Northrop, whose company was acquired by the Douglas Company in 1937. The Northrop Corporation became the El Segundo Division of Douglas. The XBT-2 aircraft became the prototype of the famous Douglas SBD Dauntless dive bomber. The Navy requested testing of the XBT-2 at Langley in February 1939 to determine recommendations for drag cleanup and for studies of wing-stall progression and maximum lift.[39] All tests were conducted with the propeller removed and the control surfaces locked in the neutral position. After conducting a traditional wool-tuft investigation of the flow on the surfaces of the aircraft, the staff of the Full-Scale Tunnel conducted force and moment measurements over an angle-of-attack range including that for maximum lift. Extensive sealing and fairing of the airframe was accomplished, including sealing the perforations in the wing flaps.

The tuft studies revealed many areas of turbulent flows, including behind the exhaust stacks and around the wheel wells, the gun trough in the engine cowling, and the bomb bay. The most productive area for drag reduction, however, was the construction and geometry of the wing. The irregularities and flap perforations produced extremely high drag for the cruise configuration and led to a recommendation to consider a flap configuration that would present a smooth and airtight surface in the closed position. Recommendations for drag cleanup to the fuselage and wing accounted for a total possible drag reduction of 26 percent—which would extrapolate to an increase in maximum speed of 31 mph.

Later, in 1939, the designation of the aircraft was changed to the Douglas SBD-1, and further modifications led to the legendary Dauntless series of aircraft that helped provide the United States with its first major Pacific victory at the Battle of Midway 3 years later.

Brewster XSBA-1

In 1934, Brewster had won a Navy competition for a scout bomber, and one prototype, designated the XSBA-1, was ordered on October 15, 1934, for evaluations that first occurred in 1936. Because of the strain of producing and developing the Brewster Buffalo, the company was unable to produce the XSBA-1 and the Navy acquired a license to produce the aircraft at the Naval Aircraft Factory at Philadelphia, PA. In September 1938, the Navy placed an order for 30 production aircraft. In February 1939, the XSBA-1 entered the Full-Scale Tunnel for a month-long test at the request of the Navy.[40] The objectives of the tests were to investigate means of increasing the maximum speed of the airplane, to determine the change in maximum lift when the dive flap perforations were covered, and to measure rudder-pedal forces that were found in flight to be excessive for large rudder deflections. The rudder force characteristics were of further interest because the aircraft had exhibited an objectionable oscillation in yaw in flight.

The drag cleanup results identified the usual protuberance drag levels due to cooling-air drag, effects of the flap perforations, exhaust stacks, and wheel wells. In addition, sources of drag included joints in the cockpit enclosure, the gunsight, rivets, etc. The total speed increase for all recommended modifications was projected to be a gain of 37 mph. The

The Brewster/Naval Aircraft Factory XSBA-1 mounted for drag cleanup tests, maximum lift studies, and analysis of high rudder forces in early 1939. Note the increased vertical-tail area recommended by the NACA. (NACA 17132)

airplane was also found to have excellent stall characteristics and, for the highest angle of attack tested, the wing leading edge, the wingtips, and a portion of the ailerons were still not stalled. The analysis of rudder forces indicated a poor tail mechanical design, and a floating rudder tab was recommended along with an increase in size of the vertical tail.

The first production aircraft (designated the SBN) were not delivered to the fleet until 1941—when they were already obsolete. The aircraft were withdrawn from service in 1942.

Curtiss XP-40

In 1937, Curtiss saw the end of development opportunities for its P-36 fighter and obtained permission from the Army to modify a production, radial-engine P-36A airframe to a liquid-cooled Allison engine with reduced frontal area and enhanced performance. Curtiss estimated the maximum speed of the aircraft to be 350 mph. The task of creating the new design, which the Army designated the XP-40, involved much more than simply mating the new engine to the old airframe. For example, the engine radiator was initially located at the wing trailing edge under the aft fuselage, vaguely similar to the later P-51 Mustang.

The first flights of the XP-40 in October 1938 were disappointing, as the aircraft was only capable of about 300 mph. When, in December, the underslung radiator was relocated to the nose by Curtiss (some sources say this move was driven by the Curtiss sales representatives) and other modifications were made, the maximum speed increased to 342 mph. In early 1939, the XP-40 exhibited a top speed that was 40 mph faster than several competitors (including the XP-38 and XP-39) in an Army fighter fly-off competition at Wright Field and was ordered into production. The Army requested an entry in the Full-Scale Tunnel

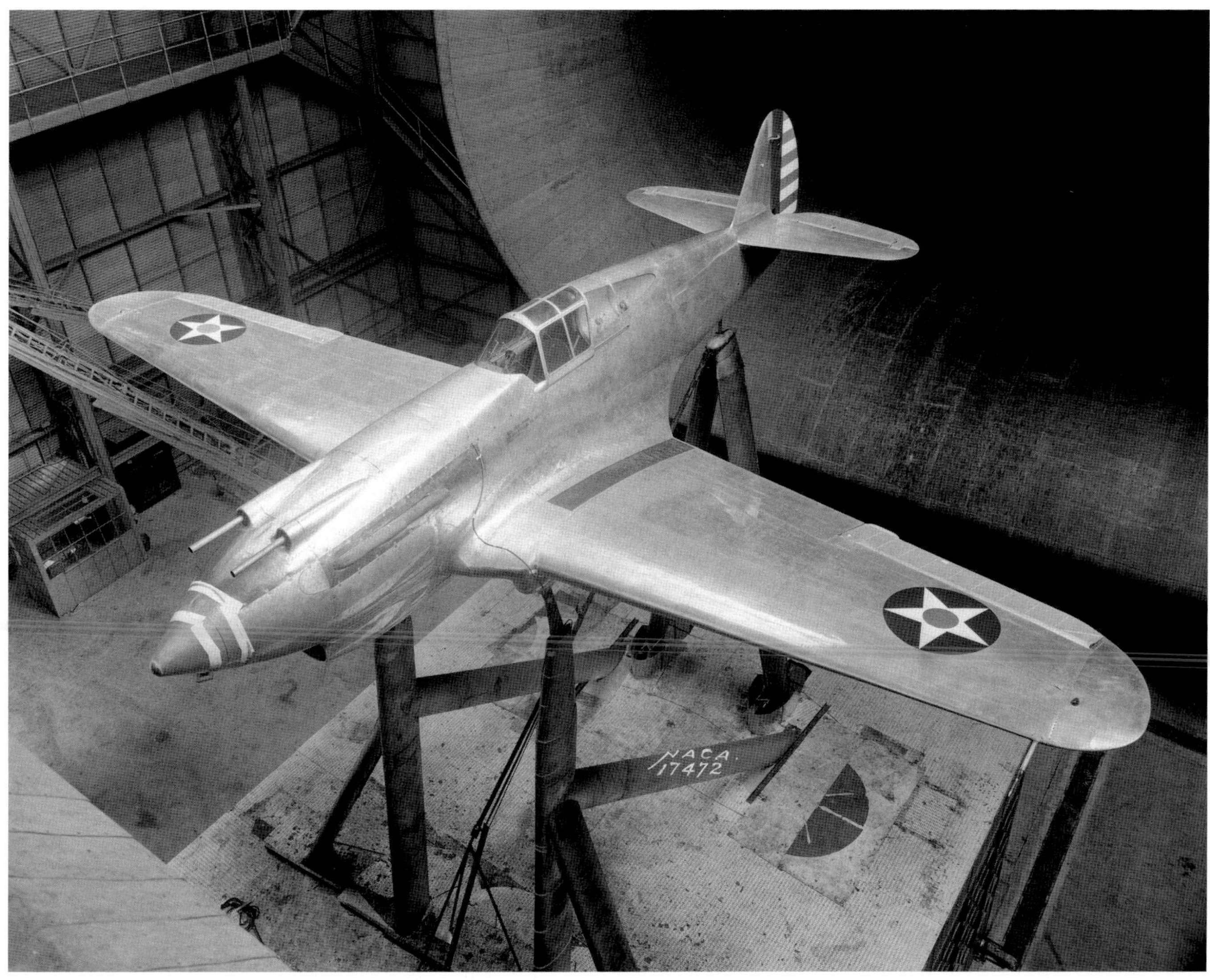

The Curtiss XP-40 mounted for tests in April 1939. The prototype had been modified prior to delivery for testing, and it would be modified again after the tests before production. Note the enclosed room for the tunnel speed operator at the bottom of the far (east) wall. (NACA 17472)

for 2 weeks beginning in April 1939. The XP-40 contract was to be the largest U.S. contract for a fighter and it therefore commanded high priority.

Management at Langley recognized the critical nature of the XP-40 program and assigned established "first team" engineers Clint Dearborn and Abe Silverstein to the job, along with newcomer John P. "Jack" Reeder. The objective of the Army request was straightforward and to the point: investigate methods to increase the maximum speed of the aircraft. The scope of the test program followed the drag cleanup process, but the focus of attention during the tests was the drag created by the radiator. The results reported to the Army stated:

> Based on the test results it is estimated that modifications to the airplane that are immediately practicable such as sealing slots, utilizing trailing antenna, closing spinner holes, fairing landing gear, and modifying the radiator installation would increase the top speed by about 23 miles per hour. Incorporating the further refinements of completely retracting the landing gear, increasing the size

> of the radiator and providing an optimum radiator duct, smoothing the wing, redesigning the carburetor inlet, redesigning the oil-cooler system so as to obtain a higher duct efficiency, and improving the wing fillets could result in a total increase in maximum speed of about 42 miles per hour.[41]

The Army subsequently accepted several of the NACA recommendations, and the XP-40 was modified. The top speed of the airplane subsequently increased to about 360 mph.[42]

A Curtiss P-40K fighter is shown during tests in the Full-Scale Tunnel in April 1943 to obtain data for analyses of structural tail failures that occurred in military operations. (NACA LMAL 32792)

The Curtiss P-40 series returned to the Full-Scale Tunnel in April 1943, when a production P-40K was tested to provide data in the analysis of several tail structural failures that had been experienced with the configuration.[43] The tail failures had occurred primarily in high-speed dives, and it was suspected that one of the factors contributing to the failures was an asymmetric tail loading that occurred as a result of rotation of the propeller slipstream and yaw of the aircraft. Under certain operational conditions, the asymmetric tail loading may have caused bending moments on the horizontal tail in excess of those predicted by design methods. The test program consisted of pressure measurements over the horizontal tail of the P-40K and flow surveys in front of the tail, with power for combinations of angles of attack and sideslip. A few force tests were also made to determine the variation of lift with angle of attack.

When requesting the tests, the Army recognized the low-speed nature of the investigation and the lack of effects of Mach number and elastic deformation; however, the criticality of the situation demanded any and all data that the NACA could provide. The results of the test showed that the magnitude of the tail-load asymmetry for unyawed flight was negligible. However, for yawed flight at high power settings (as might be encountered in a pullup from a high-speed dive), the asymmetrical loads were significant. As will be discussed in a following section of this chapter, other U.S. military aircraft—especially the P-47 Thunderbolt—also suffered tail failures, and similar tests for the P-47 had been conducted the previous year in the Full-Scale Tunnel.[44]

The Seversky XP-41 began a drag cleanup test program in the Full-Scale Tunnel in June 1939. In this photograph, the airplane is in the "sealed and faired" condition and at a cruise attitude. (NACA 17822)

Seversky XP-41

The Seversky Aircraft Corporation (which later became the Republic Aviation Corp.) had designed and produced the P-35 pursuit plane for the Army in the mid-1930s. The P-35 was the first all-metal, retractable landing gear–equipped, enclosed cockpit airplane for the Air Corps. In 1938, the company modified the final production P-35 into the XP-41 prototype by using a turbo-supercharged engine that provided considerable additional horsepower. The XP-41, which flew for the first time in March 1939, was capable of a maximum speed of 323 mph at 15,000 feet. The Army Air Corps Materiel Division requested drag cleanup tests and documentation of stalling characteristics of the design in the Full-Scale Tunnel.[45]

The XP-41 only existed as a single prototype and never made it into production, but it served as an important example of the potential benefits of drag cleanup, and the results of the tunnel studies conducted in June 1939 are now textbook examples on the topic.

The Langley tests of the XP-41 were especially thorough. During the tests, incremental drag was measured for the engine and accessory cooling systems, the gun installation, radio antenna, engine cowling, canopy, and production abnormalities in the wings. The drag increments caused by the power-plant installation (i.e., cowling and cooling airflow, carburetor air scoop, accessory cooling, exhaust stacks, intercooler, and oil cooler) increased the drag of the sealed and faired condition by over 45 percent. Additional drag caused by the gaps in cowling flaps, antenna, walkways, landing gear doors, and gun blast tubes was also very large, and the combined drag of the power-plant items and these items increased the drag of the XP-41 by a very impressive 65 percent.

After analyzing the magnitude and causes of the incremental drag created by aircraft components, the NACA researchers suggested improvements that could significantly increase the performance of the XP-41. For example, the powerplant drag could have been reduced from 45 percent to 27 percent, and the roughness and leakage drag could have been reduced from 20 percent to only 2.5 percent.

Bell P-39 Airacobra

Arguably, the most controversial test ever made in the Langley Full-Scale Tunnel occurred during August 1939, when the Bell XP-39 was tested at the request of the Army Air Corps to investigate methods for increasing its maximum speed and providing adequate engine cooling. The results of the test had a profound impact on the mission effectiveness of the production version of the P-39, and the role played by the NACA in subsequent modifications to the original design has been hotly debated by historians and aviation enthusiasts to the present day. Some in the aviation community still regard the NACA drag cleanup effort for the XP-39 as one of the gravest mistakes of the war years and vilify the NACA for critical decisions in the development of the aircraft.

The XP-39 was the result of Bell's design for a new fighter specification issued by the Army in 1937.[46] Details of the specification called for a high-altitude capability, a single liquid-cooled engine with a turbo-supercharger, a top speed of at least 360 mph, and a tricycle landing gear. The Bell design was built around a 37-millimeter cannon firing through the propeller spinner and included placement of the engine behind the cockpit, with the propeller driven via a shaft under the cockpit floor.

The sleek airplane first flew in April 1939 with a turbo-supercharged engine, achieving a speed of 390 mph at 20,000 feet, which it reached in an impressive 5 minutes. The airplane, however, did not carry armament and weighed about 5,550 pounds—much less than a production-armed version. Some references state that a production P-39 would have weighed a ton more, resulting in a decrease in top speed of 50 mph.[47] The NACA report on results of the tunnel testing, however, stated that the weight of the production version would have been 6,150 pounds.[48]

The Bell XP-39 test in 1939 was the most controversial test ever conducted in the tunnel. The upper photo shows the airplane in the faired condition without the oil-cooler intake on the right rear fuselage. The bottom photo shows the airplane with the canopy removed. The person on the ground board is Abe Silverstein. (NACA 18281 and NACA 18212)

The scope of testing in the entry for the Full-Scale Tunnel test included drag cleanup, the location of boundary-layer transition on the wing, pressure measurements to determine the location and speed for critical compressibility on the airplane, and aerodynamic characteristics with flaps up and down. The Navy maintained an awareness of the tests because a naval version of the aircraft (the XFL-1 Airabonita) was under design by Bell.[49] The test program for the XP-39 was one of the most thorough ever conducted in the tunnel. An area of particular interest during the tests was the drag increments produced by the required cooling intakes for the Prestone radiator and intercooler installations for the turbo-supercharger. The Prestone radiator was installed in a duct in the left wing, and the test program investigated the effects of changing the duct openings. The oil cooler was installed in a sharp-edged rectangular scoop on the right side of the fuselage. The original intercooler was installed in a duct on the left side of the fuselage with a sharp-edged inlet. During the tests, the intercooler was also modified into a "NACA intercooler" with more streamlined and recessed lines.

The turbo-supercharger of the XP-39 was located under the fuselage beneath the engine. Four vertical exhaust stacks, which carried engine exhaust gases to the turbo-supercharger, protruded several inches beyond the fuselage lower surface and were an immediate suspect for drag contributions.[50]

Other features of interest for drag reduction included the tricycle landing gear, which did not completely retract the wheels into the wing and left several inches of wheel protruding into the airstream, and the cockpit, which extended 15½ inches above the fuselage.

The results of tuft-flow observations showed that the oil cooler, the turbo-supercharger intercooler, the carburetor intake, and the radiator duct all exhibited disturbed flow in the wing root area. The left wing stalled early during high-lift tests because of poor design of the radiator duct intake.

A three-quarter view of the left side of the XP-39 and the intercooler duct that exhibited disturbed flow during testing. (NACA 18465)

The turbo-supercharger of the XP-39 and its cooling system created a large drag penalty. On the left, a tuft-flow study photo shows the disturbed airflow in the area (note the carburetor intake at the wing root); on the right, a view beneath the aircraft looking to the rear shows the turbo-supercharger and the pipes extending below the fuselage (note the protruding wheels). (NACA 18477 and NACA 18483E)

Views of the oil cooler duct on the right side of the XP-39 fuselage. The photo on the left shows the intake, and the photo on the right shows the exit of the duct. (NACA 18464 and NACA 18467)

The recommendations in the test report by Silverstein included enclosing the supercharger within the fuselage, which was not possible because of the relatively small size of the XP-39. The NACA team modified the intercooler with a new configuration designed to reduce drag, but the concept was not pursued because of lack of information on cooling requirements for the turbo-supercharger. The staff also recommended relocating the oil cooler and the carburetor intake in the right wing in a duct similar to that used on the left wing for the radiator. Another recommendation was to lower the cockpit enclosure to

The radiator was mounted within the left wing with a wing leading-edge intake and an upper-surface exit. These photos of the intake (upper photo) and exit (lower photo) in the left wing show the internal structural blockage to cooling flow. (NACA 18429 and NACA 18430)

The XP-39 in the faired condition. Note the wheel covers and sealed gaps in the nose-wheel doors. (NACA 18421)

reduce the drag contribution by 50 percent. Finally, it was recommended that the landing gear design be modified to produce a flush wing and fuselage surface when retracted.

A particularly critical statement appears at the end of the list of recommendations in the report:

> A further modification possible to the airplane might be the use of an Allison altitude blower engine. With this engine it is possible to eliminate the supercharger and intercooler installation making possible a material reduction in airplane drag.[51]

After the test in the Full-Scale Tunnel, the Army decided that the turbo-supercharger of the XP-39 would be removed. With the supercharger removed, the XP-39 performed well at lower altitudes, but it was not effective at high altitudes. The follow-on production P-39 aircraft were nicknamed "Iron Dogs" because of their inferior performance at typical air-combat altitudes.

The long-lived controversy over how the decision to remove the turbo-supercharger from the XP-39 was made has continued for many years.[52] Some say the NACA "ordered" the removal; others say that both the Army and Bell were very concerned over the relative immaturity and complexity of the supercharger design and were relieved when the NACA report gave them additional ammunition to terminate the concept. Whatever the truth, the P-39 was a relatively poor performer as a high-altitude fighter when it was reluctantly injected into the Pacific in the early stages of the war. Most production aircraft were sent to Russia, where they were very effective as low-level anti-tank weapons.

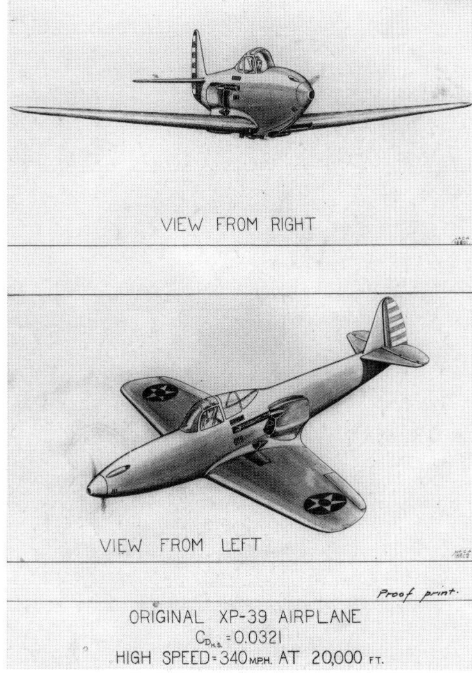

NACA projections of top speed for the XP-39 based on results from the tunnel tests. (NACA 18600, 18603 and NACA 18601, 18602)

After the XP-39 returned to Bell following the Full-Scale Tunnel tests, the turbo-supercharger was removed, the cockpit canopy was lowered, the carburetor intake was moved, and full retraction of the landing gear into the wing and fuselage was accommodated. The modified aircraft was renamed the XP-39B, and early flight tests showed its performance was lower than expected. An Army request for NACA assistance initially called for flight testing by NACA pilots at Langley, but the project was directed to the Full-Scale Tunnel for tests in March 1940. By that time, the scope of potential new fixes to the XP-39B to improve performance was severely restricted by the urgency of the tests and a stipulation that precluded any major structural modifications.[53] Most of the study was directed at the drag induced by the wing-inlet ducts, the carburetor intake and wheel-well fairing, and defining the benefits of special propeller cuffs that were designed and fabricated by Langley based on positive results obtained during tunnel testing of a XP-39 propeller in the PRT.

The results of the XP-39B tests in the Full-Scale Tunnel indicated that the high speed of the aircraft might be increased by 18 mph by drag cleanup items, and that a further increase of 4 mph could probably be obtained by the addition of new propeller cuffs. By this time, however, the die had been cast, and the production versions of the P-39 were limited to relatively low-altitude operations.

After extensive modifications and removal of its turbo-supercharger, the XP-39 returned for testing in the tunnel as the XP-39B in 1940. (NACA 19516)

The full-scale model of the Curtiss XP-46 mounted for tests in November 1939. The model was fabricated from wood with an external finish of glazing clay polished to a mirror-like smoothness. (NACA 18957)

Curtiss XP-46

In September 1939, the Army Air Corps requested Curtiss-Wright to build two prototypes of a new fighter aircraft design known as the XP-46, which was intended to replace the Curtiss P-40 with better performance. Following a request from the Army's Materiel Division, a full-scale model of the XP-46 was prepared for tests in the Full-Scale Tunnel. The scope of the test program, conducted for 2 months at the end of 1939, included an assessment to determine the optimum configuration for maximum speed, identify adequate engine-cooling concepts, determine aerodynamic stalling characteristics, and evaluate aileron and elevator effectiveness. Critical compressibility locations were also determined by surface-pressure surveys.[54]

The test results indicated that the best aerodynamic performance was obtained with the addition of a large fuselage-wing fillet and a new wing leading edge. Researchers also found that a long-fuselage-nose configuration showed better maximum lift capability than a short-nose version and that the radiator on the underside of the aircraft degraded maximum lift with inlet scoop and outlet flaps open.[55]

By the time the first XP-46 prototype flew on February 15, 1941, the aircraft's performance was inferior to the advanced version of the P-40 at the time, and the XP-46 program was cancelled.

Martin XPB2M-1 Mars

The Martin XPB2M-1 Mars was created in response to a Navy contract in 1938 for a long-range, four-engine ocean patrol seaplane. One prototype was built. A request for Full-Scale Tunnel testing to improve the top speed of the configuration was quickly forthcoming from the Navy. Tests were conducted in May 1939 with a ⅕-scale model powered by electric motors. The tests included the isolated wing as well as the complete configuration. Initial results were disappointing and the model was returned to Martin for fairing and modifications, followed quickly by follow-on testing that same year. The model returned in early 1940 for additional testing.[56]

A ⅕-scale powered model of the Martin XPB2M-1 flying boat was tested for drag cleanup during 1939 and 1940. (NACA 17702)

Preliminary tuft-flow-visualization tests revealed that the flow at the wing-fuselage juncture of the model was very turbulent. Leading-edge fairings minimized the problem and significantly reduced drag. The drag due to the tail turret and the "keel step" peculiar to seaplanes were also determined and improved by fairings. Drag increments due to the wingtip floats and underwing bomb racks were also measured.

The twin-tail configuration and mission of the XPB2M-1 would change in the next few years. It first flew in July 1942 as a patrol plane, but the mission was considered obsolete by 1943, and the design was changed into the JRM-1 transport with a single vertical tail. The Navy purchased five of the JRM-1 design, but the end of WWII terminated Navy interest in more production versions. Four surviving aircraft were mothballed by the Navy, but they were purchased for civilian use in 1959 and converted into water bombers for firefighting.

Curtiss XP-42

The Curtiss XP-42 was a 1939 modification of a production radial-engine P-36A in an attempt to incorporate a more streamlined engine cowling to reduce drag and improve performance for fighters powered by radial engines. The XP-42 used a Pratt & Whitney engine with an extended propeller shaft that placed the propeller about 20 inches farther forward than the standard position. The forward extension of the propeller enabled the use of nose shapes with a higher fineness ratio than shapes used in the standard short-nose configuration. Unfortunately, flight tests showed that the top speed of the airplane was no better than the short-nose P-36 airplane.

The original Curtiss design used a pointed fuselage nose with sharp-edge air scoops at the top and bottom of the cowling for cooling. NACA analysis and preliminary in-flight measurements indicated that the cowling scoops and cooling-flow characteristics caused excess drag. In particular, it was found that the engine-cooling air from the lower air scoop was only traveling at about half the airplane's flight speed, and that the energy of this flow was rapidly dissipated by a sharp change in the airflow direction at the rear of the scoop to a large area of the engine. This high internal energy loss due to cooling flow led to Full-Scale Tunnel studies to identify the sources of internal and external drag and suggest improvements for the design. The Langley projects for the XP-42 included a broad variety of experiments, including an extensive multiyear flight and wind tunnel evaluation of experimental engine cowling shapes to improve aerodynamic performance and cooling, and research on enhancing pitch control during high-speed dives.

Tunnel tests of the XP-42 were conducted in September 1940 to investigate the cooling and aerodynamic performance of several engine-cowling concepts.[57] The goals of the test program were to reduce external drag and increase the critical Mach number over the nose by reducing the negative pressure peak associated with high-speed flight, and to reduce cooling drag by increasing the cooling-air pressure recovery. Four modified cowling and cooling designs were evaluated, and it was demonstrated that the scoop-type inlet the original configuration used resulted in unsatisfactory aerodynamic behavior. A new annular inlet cowling proved to be very efficient. Langley designed a promising new cowling shape referred to as the NACA "D" cowling that used an annular inlet and a diffuser section for the engine-cooling air. The Full-Scale Tunnel tests were followed by an extensive series of flight tests at Langley from May 1941 to December 1942 to evaluate the effectiveness of several experimental cowling designs.

The design of the cowlings and engine installations was a project of the Air-Cooled Engine Installation Group stationed at the Langley Laboratory under Clint Dearborn's direction. The organizations associated with this project included Curtiss-Wright, Republic Aviation, Wright Aeronautical, and Pratt & Whitney. The Army Air Forces sponsored the investigation and supplied the XP-42 airplane. Curtiss-Wright handled the construction as well as the structural and detailed design of the cowlings and also supplied personnel to assist in the servicing and maintenance of the airplane and cowlings during the tests. Pratt & Whitney prepared the engine and torque meters for the tests and assisted in the operation and servicing of the engine. The propellers, cuffs, and spinners were supplied by Curtiss-Wright.

The XP-42 in its service condition (top) and with an NACA-designed annular inlet configuration (bottom). Note the sharp-edged upper and lower cowl inlets for the original design and the annular NACA design with a large spinner. The annular design was used in Germany for the Focke-Wulf FW-190 fighter. (NACA 22001 and NACA 22128)

The XP-42 proved to be slower than the Curtiss XP-40 and it therefore never entered production. Nevertheless, the important technologies advanced by the NACA's research on this aircraft provided aircraft designers with options for cowling concepts. The airplane stayed at Langley until 1947 and was used in many projects, including the first development of an all-movable horizontal stabilizer to alleviate high control forces at high speeds. All-moving horizontal tails are now a standard feature on all modern high-speed fighter aircraft designs.

A Curtiss P-36C was tested in December 1940 as part of the NACA engine cowling program. (NACA 22128)

Curtiss P-36

The interests of the Air-Cooled Engine Installation Group extended to additional testing of the Curtiss P-36 configuration for comparison with results from the XP-42 tests. After the XP-42 studies in September and October of 1940, an operational P-36 underwent testing in the Full-Scale Tunnel. A brief drag cleanup activity was undertaken, but the focus of the program was spinner and cowling studies similar to those conducted for the XP-42.[58]

Curtiss XSO3C-1 Seamew

The Curtiss XSO3C-1 Seamew was developed in 1937 as a land- or sea-based replacement for the SOC Seagull, the Navy's standard floatplane scout. In acceptance flight tests by the Navy in 1939, the airplane exhibited cylinder barrel temperatures that were 25 degrees higher than the maximum allowable for climb conditions. Pilots also reported that the airplane showed unstable characteristics and undesirable longitudinal-trim changes with rudder inputs during carrier approaches for landings.

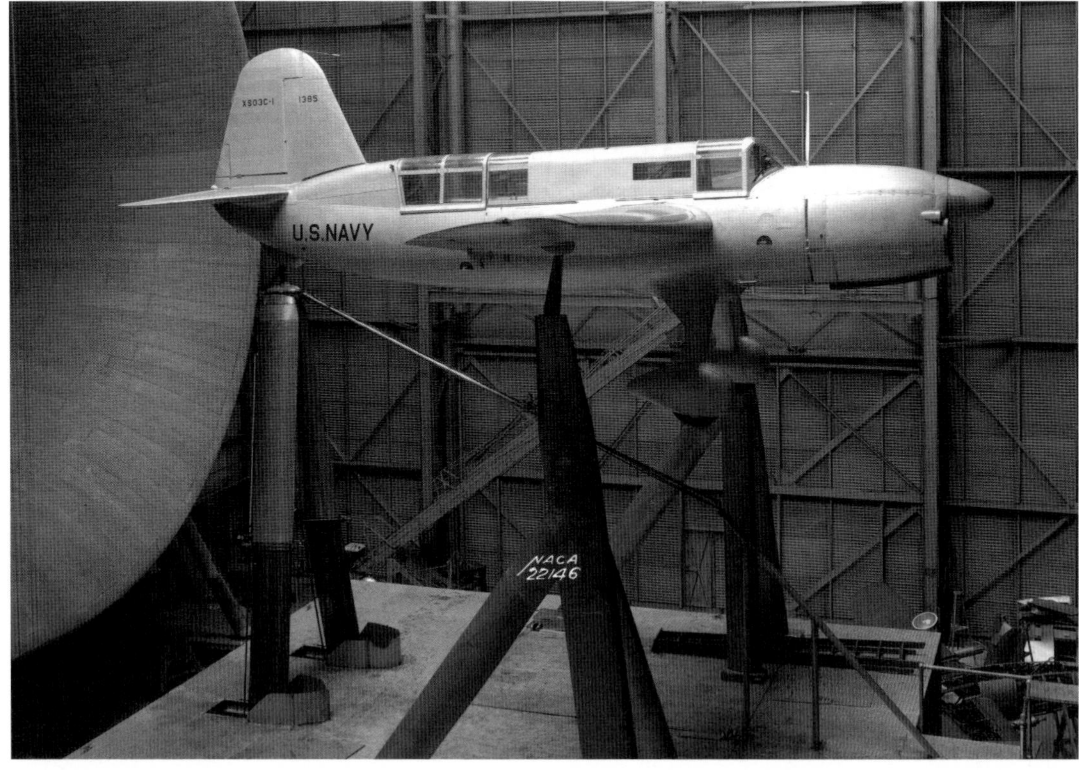

The Curtiss Seamew could be configured as a land-based or sea-based scout. It was studied as a land-based plane during tests in the Full-Scale Tunnel in 1940. (NACA 22146)

At the Navy's request, an investigation of the cooling and stability problems of the XSO3C-1 was made in the Full-Scale Tunnel in October 1940.[59] The tests included extensive power-on and power-off thermocouple measurements of the cooling behavior both in the baseline condition as well as with an improved cowling inlet, cylinder baffles, and cowling outlets. At the end of the program, a few traditional drag cleanup tests were conducted.

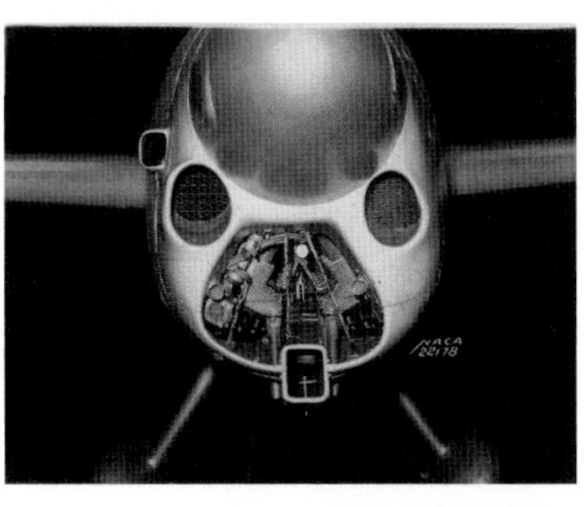

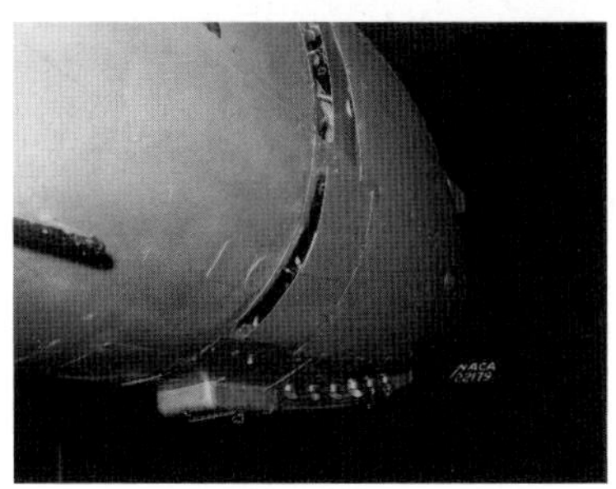

The Curtiss XSO3C-1 suffered poor cooling and stability and control characteristics during its test program in 1940. A view of the relatively small inlet and its blockage (top right) and small exit (bottom right) visibly accentuate a potential cooling problem. (NACA 22178 and NACA 22179)

The anticipated cooling problems were quickly encountered in the tunnel test program. The original cooling system was shown to have unacceptably high losses due to restrictions in inlet and outlet openings. Tests to modify the cooling characteristics with baffling or revised inlet contours within the constraints of the existing engine properties were relatively unsuccessful (only about 13 degrees decrease). The stability and control issues included changes in trim when the rudder was deflected, with different results for right or left deflections. In addition, stick forces were very high. No tests were made with sideslip, so the lateral and directional stability of the airplane were not measured. Finally, the limited drag cleanup study (of antennas, fairings, etc.) produced very mediocre results, with a projected increase in speed of only 5 mph.

The relatively negative tunnel results for the XSO3C-1 were followed by equally disappointing experiences with production versions in the fleet. Although the vertical tail was enlarged and the wingtips were angled upward to add dihedral effect for enhanced lateral directional stability, the aircraft's stability and control issues persisted and engine cooling continued to be a problem. The SO3C was finally withdrawn from first line fleet units by 1944 and exchanged with the older SOC biplanes it was intended to replace.

Douglas A-20A Havoc

In the fall of 1937, the Douglas Aircraft Company responded to an Army Air Corps specification for a new attack aircraft with a twin-engine light bomber design it named the Model 7B. Although the Army did not order the design, it attracted the attention of the French military, which placed an order for 270 of the aircraft before the war started. Ultimately, after improvements were made to the design, the Army ordered two versions: the A-20 with a supercharged engine for high-altitude missions, and the A-20A for lower-altitude operations. However, the A-20 had significant engine development problems that caused the Army to question the value of a high-altitude light-attack bomber, and it ultimately halted production after only one A-20 airplane had been built.

In initial flight tests in 1939, the A-20A drew very positive comments from pilots because of its maneuverability, but it suffered from engine-cooling problems. Specifically, the head temperatures for the top engine cylinders were above the allowable limits for the military-power climb condition. Overheating also occurred at the cruising condition, in which the power was low but the allowable engine temperature limit was also low. In addition, the top speed of the airplane was lower than expected because eight cooling holes had been cut by Douglas into the cowlings behind the cylinder baffles, in an attempt to cure the heating problem.

In January 1941, a 2-month test program was conducted at the request of the Army in the Full-Scale Tunnel, with objectives of determining the nature of the cooling problem, devising fixes that could be used in later versions of the airplane without a large penalty on top speed, and conducting a drag cleanup assessment of the design.[60] With a wingspan of 61.3 feet, the A-20A was too large to fit in the tunnel, so the outer wing panels of the aircraft were removed. Extensive pressure and temperature instrumentation was used throughout the propulsion system.

The test results showed that the top cylinder head temperatures were as much as 100° higher than those measured on the bottom cylinders. The eight cooling holes in the cowling of the original design caused a decrease in top speed of about 14 mph and did not solve the cooling issue for climb conditions. Detailed inspection of the nacelle contents showed severe flow blockages in oil-cooler ducts, exhaust collectors, cylinders, and outlet pipes. In addition, the upper cowl flaps could not be opened in flight because interference with flow over the wing resulted in severe tail buffeting. The study identified several satisfactory cowling arrangements, including an innovative use of blowers and ejector stacks. The suggested modifications contributed to the design of later variants of the A-20A, which ultimately became a highly effective ground attack/low-level bomber during the early to middle years of World War II.

A Douglas A-20A was mounted with its outer wings removed during cooling and drag cleanup tests. Note the faired condition for drag cleanup tests. (NACA 22841)

Republic P-47 Thunderbolt

The developmental history of the famous Republic Aviation Corporation P-47 Thunderbolt included testing in the Full-Scale Tunnel. The P-47 was a big, powerful, radial-engine fighter known for its size and ability to absorb damage. However, the first design for the P-47 was a lightweight, liquid-cooled fighter known within the Republic team as the model AP-10. Republic was stimulated to break from the traditional radial-engine arrangement favored by its predecessor, the Seversky Aircraft Company, because competitors such as Lockheed, Bell, and Curtiss were all turning to streamlined, liquid-cooled, high-performance configurations. The AP-10 proposal was ordered by the Army in 1940 with the designation of XP-47.

As part of the XP-47 developmental program, the Army requested tests in the Full-Scale Tunnel in 1941 to investigate several cooling concepts and to determine the stability and control characteristics of the design. This test program was one of the most extensive studies conducted in the tunnel during the era, occupying the tunnel for almost a half year in 1941.[61] Two full-scale models, referred to as Model "A" and Model "B," were involved in the program. Model A had a midwing configuration similar to the final P-47 production aircraft, whereas Model B had a low wing similar in planform to the wing of the Curtiss P-40.

A 10-foot-diameter propeller fitted with blade "cuffs" (airfoil-shaped coverings on the lower shank of the blade to improve efficiency) and driven by an electric motor was used on Model A to determine the effect of the slipstream. An investigation of carburetor air inlets included nine different inlets varying in shape, size, and location on the forward fuselage. Detailed engineering measurements were made of velocity distribution, ram pressure, and drag within the inlets. A separate study of stability and control for Model A included detailed

Full-scale model of the XP-47 Model A undergoes cooling tests in the Full-Scale Tunnel in May 1941. The model is in a clean configuration without a carburetor inlet. The propulsion scheme for the P-47 was changed from the liquid-cooled, inline concept shown here to the final classic radial-engine configuration. (NACA 25006)

measurements of longitudinal stability and control, including the effects of wing flaps and propeller slipstream. Stick-force characteristics and the contributions of tail components and stability are also determined.

Before construction could even begin on the XP-47 fighter, analysis of air-combat reports coming from Europe indicated that the lightweight XP-47 design would be unacceptably deficient in performance and armament. With the Army's approval, the Republic design team changed to a larger, more powerful radial-engine design with a turbo-supercharger and a weight nearly twice that of the earlier XP-47. The Army placed an order for the P-47B in September 1942, and the original XP-47 efforts were terminated. However, valuable engineering data had been generated by the tunnel investigations and a better understanding of aerodynamic factors influencing the cooling and stability of future designs was obtained.

The year 1941 saw an epidemic of tail failures in emerging high-performance military airplanes, including the P-38, P-40, and early versions of the P-47. Republic and the Army Air Forces requested that the NACA investigate the situation through studies of the tail loads encountered during high-speed maneuvers. The success of the Thunderbolt hinged on solving its tail failures. A production P-47B was flown to Langley, where it underwent extensive testing to measure its stability and control characteristics and tail loads in the Full-Scale Tunnel. Subsequent in-flight pressure measurements were made at Langley for comparison

An early Republic P-47B was tested in the Full-Scale Tunnel in July 1942 as tail-failure issues became known. Here the airplane is being tested at a high angle of attack with the propeller removed to determine if flows from the front of the P-47B caused structural resonant conditions at the tail. Note the humorous "skull and crossbones" sketched on the rudder as a reference to the fabric-covered rudder failures experienced in service. (NACA 29037)

with the wind tunnel results. Additional testing and analysis included tests of a model of the P-47B tail in the 8-Foot High-Speed Tunnel and a theoretical analysis of tail flutter.

Stability, control, and tail-load tests of the production P-47B were conducted in the Full-Scale Tunnel in early 1942.[62] Pressures were measured over the horizontal and vertical surfaces for several angles of attack and angles of yaw to provide a check on the design loads and predictive theories for estimating tail loads.[63] The elevator and rudder on the test airplane were fabric covered and fitted with control tabs. Over 400 pressure orifices were installed in the tail, and all tests were conducted with the propeller removed. The detailed data obtained during the 2-week test indicated that the pressure distributions of the tail were adequately predicted by existing theoretical and empirical methods, and that no structural resonance occurred due to the impingement of the wake of the forward aircraft on the tail surfaces.

Following an intensive review of all results from the Langley facilities (especially the 8-Foot High-Speed Tunnel), it was concluded that loads at high speeds were causing the fabric covering of the elevators and rudder to balloon, and that the structures were too light to provide adequate safety at higher speeds.[64] Based on this conclusion, the decision was made to change from fabric-covered elevators and rudder to a stronger metal-covered tail design. Equipped with the revised tail structure, the Thunderbolt entered service without further tail problems.

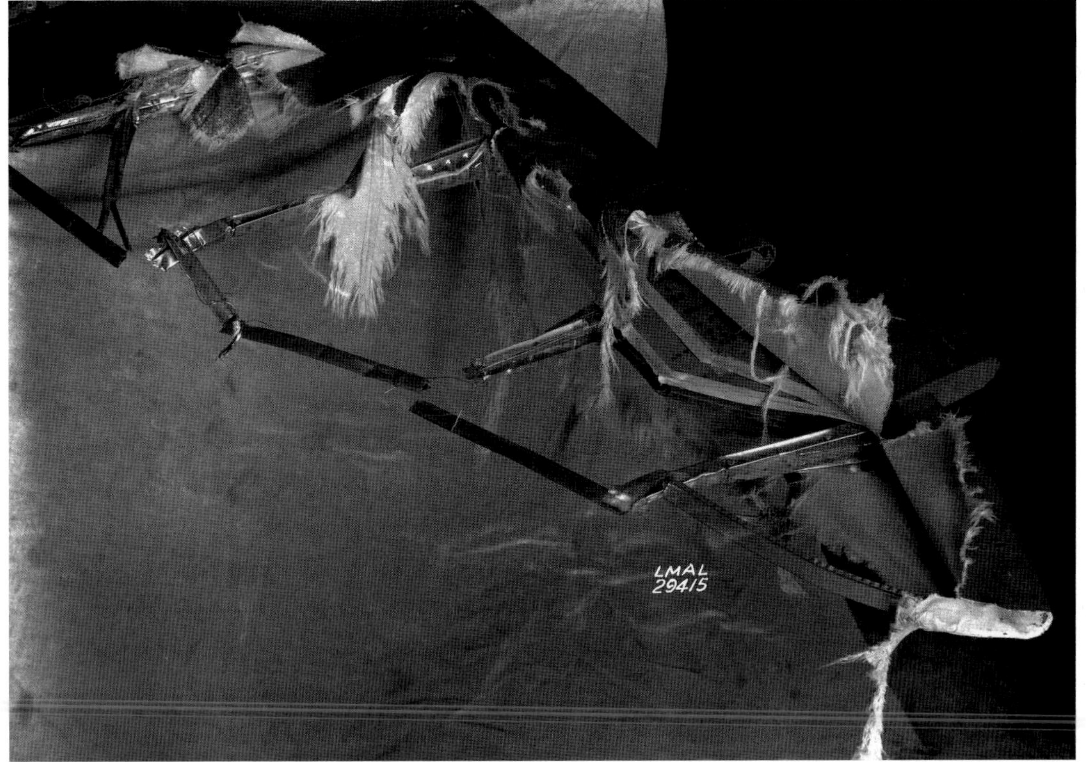

Photograph of the remains of a P-47B rudder after high-speed tests in the Langley 8-Foot High-Speed Tunnel resulted in structural failure at 460 mph. The top of the fabric-covered rudder (bottom right) failed, the fabric ripped, and the trailing edge went down the tunnel. The trailing edge segments (left) are held together with wire for the photograph. The failure occurred in less than 5 seconds. (NACA 29415)

Vought V-173 Flying Pancake

One of the most unconventional aircraft designs ever tested in the Full-Scale Tunnel was the Vought-Sikorsky V-173 Flying Pancake. Designed by Langley's Charles Zimmerman, the configuration was an attempt to provide unprecedented short takeoff and landing (STOL) capability for civil and military applications. Its low-aspect-ratio wing and large counter-rotating propellers were specifically designed for extremely low approach and landing speeds that could allow the pilot to land in almost parachute fashion without altering the flightpath or speed. Such a capability would demand a relatively low lift-to-drag ratio and a high resultant force. Zimmerman was aware that low-aspect-ratio wings might satisfy both requirements for the "parachute" effect he desired for landings.

As an entry in the NACA's 1933 competition for innovative light aircraft, Zimmerman's flying pancake design was considered the winner based on originality and innovation, but his NACA managers were not interested in the concept, declaring it to be "too advanced" and "too much of a novelty."[65] NACA programmatic support for further studies was not forthcoming. Nonetheless, Zimmerman continued to design and test his radical design concept during off-work hours at his home using small flying models.

With the approval of NACA management, Zimmerman approached United Aircraft Corporation with his novel design in 1937, then he left Langley to join the Chance Vought Aircraft Division as a project engineer to develop the aircraft. He designed the V-173 to be a flightworthy, wood-and-fabric experimental aircraft to demonstrate his STOL concept and to serve as a precursor to an advanced military fighter for the Navy, which had a definite interest in aircraft that could use short landings for operations from seaborne ships. Sometimes

The Chance Vought V-173 was a flying test bed designed to demonstrate the short takeoff and landing capability of a low-aspect ratio wing immersed in the slipstream of large propellers. The engineer shown on the tunnel ground board in the bottom photo is John "Jack" Reeder, who later transferred from the Full-Scale Tunnel to flight research and became a famous test pilot. (NACA 26369 and NACA 26368)

called the Zimmer Skimmer, the V-173 made its first flight on November 23, 1942, and flew test flights until 1947. Over 190 flights were conducted during its lifetime, including an assessment by Charles Lindbergh, who was very impressed by its STOL capability. The aircraft performed surprisingly well in flight, and in 1941, Chance Vought received a Navy contract to further develop the design under the designation XF5U-1.

One-month-long V-173 tunnel tests in the Full-Scale Tunnel were conducted in November 1941 at the request of the Navy's Bureau of Aeronautics to assess the aerodynamic performance, drag characteristics, and airflow phenomena (especially propeller-slipstream effects) of the design.[66] At the time of the tests, the relatively lightweight V-173 was flown with two 80-horsepower engines; however, the envisioned high-speed fighter version of the concept was to be powered by two 2,000-horsepower engines. The tunnel tests were intended to provide data that might be useful in the design of a higher-powered fighter (ultimately to be known as the XF5U-1).

The tunnel tests produced extensive aerodynamic data, including pressure distributions on the wing for various power-on conditions at combinations of angle of attack and angle of sideslip. The Langley engineers also investigated the effect of propeller-rotation direction on lift and drag. For these tests, the propellers (which rotated in opposite directions) rotated either up or down at the wingtips.

The results of the test program provided some very surprising fundamental information on the aerodynamic performance of the unusual configuration. The researchers found that, as had been expected, the inherently high induced drag of a low-aspect-ratio wing could be partially compensated by the favorable interaction of large-diameter propellers operating ahead of the wing. This effect was equivalent to an increase in the wingspan since it resulted in increasing the mass of air to which downward momentum was imparted.

Prior to these tests, it had been conjectured by many that the secret of Zimmerman's design was to have the propellers rotate down at the wingtip, to oppose the rotational direction of the wingtip vortices and thereby minimize the tip losses and induced drag. However, the Langley tunnel test team unexpectedly found that the reduction in induced drag due to propeller operation was only affected to a small degree by the relative direction of the rotation of the propeller at the wingtip. They also found that longitudinal stability was significantly decreased by propeller operation due to both the direct lift on the forward-located propellers when the angle of attack was increased and the effect of propeller slipstream on downwash at the fuselage tail location. Another important result showed that changing the mode of propeller rotation from down at the tip to up at the tip resulted in a large increase in longitudinal stability with only a small decrease in performance.

One-month-long tests of a powered ⅓-scale model of the XF5U-1 were conducted in the Full-Scale Tunnel at the request of the Navy in mid-1945.[67] Whereas the test program for the V-173 concentrated on low and moderate angles of attack, the XF5U-1 test program concentrated on very high angles of attack. The results of the tests with the propellers removed indicated that the engine air ducts and cockpit canopy severely reduced power-off maximum lift. For the powered XF5U-1 model, the results revealed that the propulsive efficiency was increased about 7 percent with the propellers rotating upward in the center

A 1/3-scale model of the XF5U-1 fighter was tested in 1945 and 1948 for performance, stability, and control with power on and off. In this photograph, the propellers have been removed for a power-off test in 1945. Note the wool tufts attached to the model for airflow studies. (NACA LMAL 44791)

rather than downward in the center. Although the data obtained in the tests were sufficient to determine the propulsive characteristics, effective propeller operation, and static thrust of the propellers, the test program was terminated prematurely by the failure and complete destruction of one of the model's propellers.

The Navy seriously considered using the XF5U-1, with its projected excellent STOL and high-speed characteristics, as an interceptor aboard small carriers against the Japanese kamikaze aircraft that delivered devastating blows to U.S. Naval warships in the Pacific during 1944 and 1945. However, the war ended before the aircraft could be put into service, and the emergence of the jet fighter caused the Navy to lose interest in the XF5U-1 program.

After it was repaired, the XF5U-1 model had a second postwar test entry in the Langley Full-Scale Tunnel in April 1948 to more fully test the effects of articulated propellers on the performance and stability characteristics of this STOL concept for future designs.[68] While the aircraft never made it into production, the knowledge and data obtained from the XF5U-1 program led to improvements and refinements in future vertical and/or short takeoff and landing (V/STOL) concepts developed after World War II. The testing techniques and knowledge gained from analysis of propeller-wing aerodynamic interactions was of great value when interest in powered-lift V/STOL configurations peaked in the late 1950s and early 1960s.

Lockheed YP-38 Lightning

The Lockheed P-38 Lightning was designed by Clarence "Kelly" Johnson for an Army Air Corps competition for a twin-engine high-altitude interceptor in 1937. As previously discussed, the Bell P-39 Airacobra was the winner of a simultaneous competition for a single-engine aircraft. The P-38 used a unique configuration in which the pilot and armament were located in a gondola in the center of a twin-tailed boom layout powered by liquid-cooled engines with turbo-superchargers. The XP-38 first flew in January 1939, followed by an order for 13 YP-38 prototypes.

Early YP-38 flight testing uncovered a significant problem at high speeds (approaching Mach 0.7) wherein the tail surfaces displayed severe buffeting, especially during dives. During high-speed dives, the nose of the aircraft would tuck and the pitch control was virtually immovable until lower altitudes and denser air was encountered. On November 4,

1941, Lockheed test pilot Ralph Virden was killed when the tail components of the first YP-38 failed during a high-speed dive test. The Army suspected that the crash was caused by flutter and demanded that mass balances be added to the elevator of the P-38 design, but the undesirable phenomena persisted. Meanwhile, Kelly Johnson suspected that the cause of the tail buffeting and uncontrollable high-speed dives was the onset of flow separation associated with compressibility and shock waves.

Lockheed requested that the NACA provide test time in the 8-Foot High-Speed Tunnel at Langley to analyze the problem and provide a fix, but the tunnel was occupied by critical testing of the Boeing XB-29 configuration in December 1941. In lieu of an entry in the high-speed tunnel, time was allotted for testing in the Full-Scale Tunnel. The third YP-38 was flown to Langley and installed in the tunnel for tests that coincidentally began on Pearl Harbor Day. The YP-38 tests consisted of two investigations: a classical drag cleanup study, and a study to obtain data for predicting the aircraft speed at which the onset of compressibility begin and identifying the location of critical parts of the aircraft that would first experience its effects. Airframe modifications would also be evaluated in an attempt to delay the compressibility onset to higher speeds. The staff of the Full-Scale Tunnel had successfully followed test procedures in previous tests to predict the onset of compressibility for other aircraft, even though the maximum tunnel test speed was limited to 100 mph. The critical speed values estimated by this test method had been in good agreement with flight experience.

The high-priority study to predict compressibility onset was made with extensive pressure instrumentation on the wings and fuselage of the unmodified YP-38.[69] The values of the peak negative pressures and their location obtained from the wind tunnel were used in an analytical procedure to determine the critical Mach number and critical speed for an altitude of 20,000 feet. The pressure measurements revealed that the largest negative pressures occurred in the wing-fuselage fillet, on the peak of the canopy, and on the wing between the fuselage and the booms. Using these data, it was estimated that the local speed of sound would be reached in the wing-fuselage fillet at a speed of about 404 mph at 20,000 feet. At speeds 10 to 20 mph higher, the entire region between the booms would reach the critical speed and be subjected to flow separation effects. According to Kelly Johnson, the flow separation resulted in a decrease in the wing lift, a sharp increase in wing drag, and a strong diving tendency. The flow separation created a large wake featuring oscillating motions that impinged on the tail surfaces, causing violent buffeting.

The general characteristics of the YP-38 buffeting and diving tendencies appeared to be explained by critical speeds reached over the intersection of the airplane between the booms. Therefore, modifications to the wing and fuselage in that area were made to determine if the critical speed could be increased beyond the operational envelope. The wing modification consisted of a wing leading-edge chord extension between the booms and a revision of the airfoil section to a NACA 66-series airfoil. Although chord extensions of 0.10 and 0.20 significantly increased the critical speed (by 34 mph and 64 mph, respectively), the modifications resulted in an unfavorable effect on longitudinal stability. The staff of the Full-Scale

The third Lockheed YP-38 mounted for drag cleanup studies and predictions of the onset of compressibility effects in December 1941. The photo shows the aircraft in its service condition. (NACA LMAL 26472)

Tunnel commented that the undesirable effects on stability could be mitigated if the YP-38's Prestone and oil radiator installations were moved to the extended leading edge of the wing.

Revisions of the original canopy shape were designed to eliminate a large negative pressure that occurred there, resulting in a potential increase in critical speed of about 44 mph at 20,000 feet. Finally, wing flap deflections were evaluated in an attempt to deflect the wing wake away from the tail surface.

The drag cleanup tests for the YP-38 involved modifying the Prestone radiator and intercooler installations to simulate the installations for the production P-38E airplane.[70] The results revealed that the duct efficiencies for these installations were extremely low and that a significant drag reduction could be effected by the previously discussed relocation of the radiator installations to a wing leading-edge duct. Although the change in supercharger insulation from the YP-38 to the P-38E configuration resulted in a significant reduction in drag, further reduction was possible by completely enclosing the supercharger. The top

Overhead view of YP-38 with revised cockpit geometry and extended-chord 66-series airfoil modification to inner wing. Photo was taken on Christmas Eve, 1941. (NACA LMAL 27008)

speed of the YP-38 as tested would have been increased by approximately 28 mph if all the recommended applications were adopted.

This preliminary assessment of potential compressibility details for the P-38 was a precursor to tests in the Langley 8-Foot High-Speed Tunnel in early 1942. None of the recommendations from the Full-Scale Tunnel tests were deemed acceptable by Lockheed and the Army; however, results from the 8-foot tunnel tests at high-subsonic speeds; joint discussions between NACA, Army, and Lockheed officials; and additional tests in the NACA Ames 16-Foot Transonic Tunnel ultimately led to the development and application of wing lower-surface flaps to solve the dive recovery dilemma for the P-38 as well as other aircraft such as the P-47, A-26, P-59, and P-80.[71]

Consolidated B-24 Liberator

Engine-cooling problems for the four-engine Consolidated B-24 bomber resulted in requests for support testing in the Full-Scale Tunnel in 1942. The first B-24 support test occurred over a 4-month period in the Full-Scale Tunnel beginning in February 1942.[72] The original engine installation on the XB-24B airplane had encountered serious heating problems; it was not possible to cool the engines at normal power for altitudes above 15,000 feet, and the carburetor-air temperature was as much as 23° above the allowable limit. After considerable flight testing by Consolidated, the nacelles were modified to improve cooling, after which climb and cruise operations could be continued up to 20,000 feet. By using larger carburetor jets, additional cooling was provided and the airplane exhibited marginal cooling capability to 25,000 feet.

A B-24 engine, nacelle, and wing stub were used in cooling tests in 1942. (NACA LMAL 2765)

The primary objective of the first Full-Scale Tunnel investigation was to identify the modifications necessary to fully cool the engine for cruising, normal, and military power at an altitude of 25,000 feet. In addition, the Army Air Forces requested an investigation toward obtaining satisfactory cooling at a critical altitude of 35,000 feet. The tests were made on a production B-24D engine nacelle mounted in the wind tunnel on a stub wing with a span of 40 feet and a chord of 12 feet. An 11.5-foot-diameter Hamilton Standard propeller was used for the installation.

The scope of the test included investigations of the engine cowling and cooling, the inter-cooling, the oil cooling, the turbo-supercharger installation, and the flow through the induction system. Effects on engine temperature of the cooling-air pressure drop, the fuel-air ratio, and engine power were determined. Numerous modifications of the production configuration were tested and their effects on airplane performance evaluated.

The fundamental physical principle for cooling an air-cooled engine is to provide sufficient mass flow of cooling air through the cooling fins. The flow of cooling air is dependent on the pressure difference across the engine baffles, which can be increased either by reducing the static pressures behind the engine or by increasing the total pressure in front of the engine. In the Langley test, the rear engine pressures of the B-24D nacelle were decreased by modifying the cowl outlet and increasing the chord of the cowl flaps, and the front pressures were increased by means of blowers attached to the propeller shaft. By June, the series of cooling studies had been completed and recommendations made to the Army. Perhaps the most important result of the tests was that it would be impossible to cool the engines of the B-24D while cruising at an altitude of 35,000 feet without improving the cooling-fin design on the cylinder heads.

Abe Silverstein took the opportunity to address other issues and opportunities in engine cooling while the B-24D test setup was in the tunnel. For example, specific tests were conducted to confirm that engine cooling and fuel economy at cruising power could be improved by operating at fuel-air ratios lower than those provided by conventional automatic lean carburetor settings.[73] In addition to confirming the impact of low fuel-air ratios, Silverstein's analysis of the test results emphasized the fact that the cruising range of the B-24D could be considerably extended, particularly if satisfactory cooling in cruising flight at higher fuel-air ratios required opening the nacelle cowl flaps. Silverstein had also led the application of cooling blowers for engine cooling in several investigations in the Full-Scale Tunnel, including the B-24D investigation.[74] Blowers operating at propeller speeds were found to offer substantial improvements in the pressures ahead of the cylinders in an air-cooled engine, and in many cases they provided considerably more pressure boost than was available by means of a propeller cuff. Using the B-24D data, Silverstein emphasized that blowers appeared to offer no significant structural issue while providing a method to improve the cooling of the hot engines without requiring major changes in the cowling design. These valuable contributions to the state of the art in engine-cooling technology were typical products of Silverstein's leadership and technical prowess.

The staff of the Full-Scale Tunnel responded to an urgent Army request to mitigate B-24 engine-heating problems by testing hardware outside the tunnel on the bank of the Back River. Langley engineer Bruce Esterbrook poses with the test setup. Note the oversized nacelle cowl flaps and the instrumentation read-out room behind the test article. (NACA LMAL 30227)

In the final week of October 1942, an urgent request from the Army Air Forces was received for assistance in solving a major B-24 engine-cooling problem.[75] On a Friday morning that month, an urgent call from Engineer in Charge H.J.E. Reid's office was received by Abe Silverstein demanding his immediate presence along with "about six of his best engineers" at a high-priority meeting at Reid's office. Ken Pierpont, who attended the meeting, recalled:

> The room was full of Army brass and Pratt & Whitney engineers. The meeting disclosed that the early bombing raids by heavily-loaded B-24s in the heat of North Africa had encountered unacceptable numbers of aborted B-24 missions in which the majority of the bombers had to return to base because of overheating engines and engine failures.[76] A solution to the problem was urgently needed. By Tuesday morning a completely new engine/nacelle/wing stub setup similar to that used in the earlier wind-tunnel tests was in place for static testing outside the tunnel in an area on the shore of the Back River. A new Pratt & Whitney engine was in place and the wing stub, which had been damaged during removal after the tunnel testing earlier in the year, had been rebuilt. A test house equipped with thermocouple instrumentation had been erected at the site.
>
> Within two days the NACA and Pratt & Whitney engineers had identified the problem as overheating of the upper rear row valves resulting in valve failure.[77] The team proceeded to develop an improved baffle design to solve the problem. The fix consisted of cutting new extended-length baffle fins in Langley's shop based on the experience and guidance of the Full-Scale Tunnel staff. The first three baffles from the shop worked and the critical hot cylinder-head temperatures were reduced by over 30°. The Pratt people in East Hartford immediately adopted the design and cut new baffles which were immediately sent to the units in North Africa by Saturday. No more B-24 heating problems were encountered. The work had been accomplished during a week of testing.[78]

The results of these tests, included in a NACA restricted report, were released about a year later in 1943 with a view to their general application to other cooling problems for air-cooled radial engines.[79]

Grumman XTBF-1 Avenger

In June 1942, the second Grumman XTBF-1 prototype of the famous Navy Avenger torpedo bomber entered the tunnel for drag cleanup studies at the request of the Bureau of Aeronautics. Although the Avenger production line was turning out aircraft for the fleet, they arrived too late to participate in the Battle of Midway, which ironically was waged at the same time as the wind tunnel entry at Langley.

The objective of the Full-Scale Tunnel test was to determine sources of drag that could be eliminated on the production airplane.[80] The XTBF-1 drag cleanup studies were some of the most rigorous efforts conducted by the staff and are textbook examples of drag-reduction technology. The test program consisted of the traditional cleanup process, with incremental drag contributions identified for various airplane components. Control effectiveness was measured in power-on and -off conditions, cooling-air pressures were measured, pressure distributions were measured on the canopy and turret to determine air loads, and the effects of wing-mounted radar installations and bomb mounts were evaluated. In addition to tests

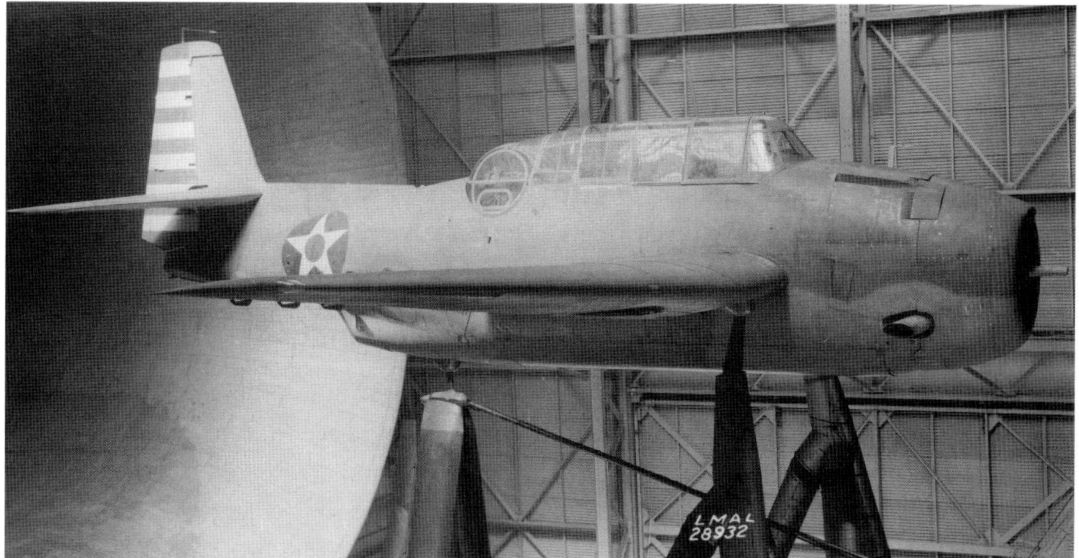

The second Grumman XTBF-1 prototype underwent drag cleanup testing in the Full-Scale Tunnel in 1942. (NACA LMAL 28932)

with the standard Grumman upper-gun turret, an evaluation was made of the effect of a special two-gun turret designed by the Martin Aircraft Company.

The results of the test program identified numerous areas around the cockpit and engine cowling that could be modified for appreciable drag reduction. Interestingly, no attempt was made to predict the incremental increase in speed that might be produced by the drag-reduction recommendations.[81]

Republic XP-69

The Republic Aviation Company designed the XP-69 in 1941 for the Army as a large fighter with contra-rotating propellers. The configuration featured an experimental 2,500-horsepower, 42-cylinder Wright R-2160 radial engine, positioned to the rear of the cockpit, driving two contra-rotating three-blade propellers via an extension shaft beneath the cockpit (similar to the P-39 Airacobra). The airplane, designed for high-altitude missions, also had a pressurized cockpit.

The Army requested that Langley conduct wind tunnel tests of a powered ¾-scale model of the XP-69 in the Langley Full-Scale Tunnel in August 1942 to assess the longitudinal and lateral directional stability and control characteristics of this unique configuration.[82] At the time, little data was available concerning the stability and control behavior of aircraft with dual-rotating propellers, and the tests were of great interest to the Army, the NACA, and industry. The large XP-69 model was built of sheet-aluminum metal outer skins that were filled and sanded to a smooth finish. Two 10-foot-diameter propellers were powered by two 25-horsepower electric motors in the fuselage. The dual-rotating propellers for the XP-69 had been previously tested in the Langley Propeller Research Tunnel.[83] Slotted Fowler-type flaps were used as high-lift devices. Aerodynamic forces and moments were measured for a range of angles of attack and sideslip for power-on and power-off conditions and with the flaps both retracted and deflected. Extensive measurements were made of the flow at the tail surfaces as affected by propeller slipstream effects.[84]

Stability and control characteristics measured in the tests revealed that the design was longitudinally stable through the full range of lift, but a significant reduction in stability occurred for power-on conditions because of the direct contribution of the propeller normal forces forward of the center of gravity. The elevator control was sufficient for all test conditions. The tests also showed that, as expected, no aileron or rudder input would be required to maintain directional trim during normal operations because of the relatively symmetrical slipstream flow (no rotational effects) behind the dual-rotating propellers. In addition, the dual rotation eliminated the asymmetric aerodynamic loading effect known as the "p-factor" experienced by single-propeller aircraft at high angles of attack.

Despite relatively promising aerodynamic test results, teething problems were encountered with the XP-69's engine and the Army Air Corps lost interest in the XP-69 program. The Army decided to encourage and pursue the development of the Republic XP-72 instead. The XP-69 program was terminated in May 1943 and never went beyond the mockup stage, but the data gathered in the wind tunnel tests remain an invaluable source of information on contra-rotating propeller configurations.

Stability and control characteristics of a powered ¾-scale model of the Republic XP-69 were assessed during tests in the Full-Scale Tunnel in 1942. Note the three-blade contra-rotating propellers of the configuration. (NACA LMAL 29654)

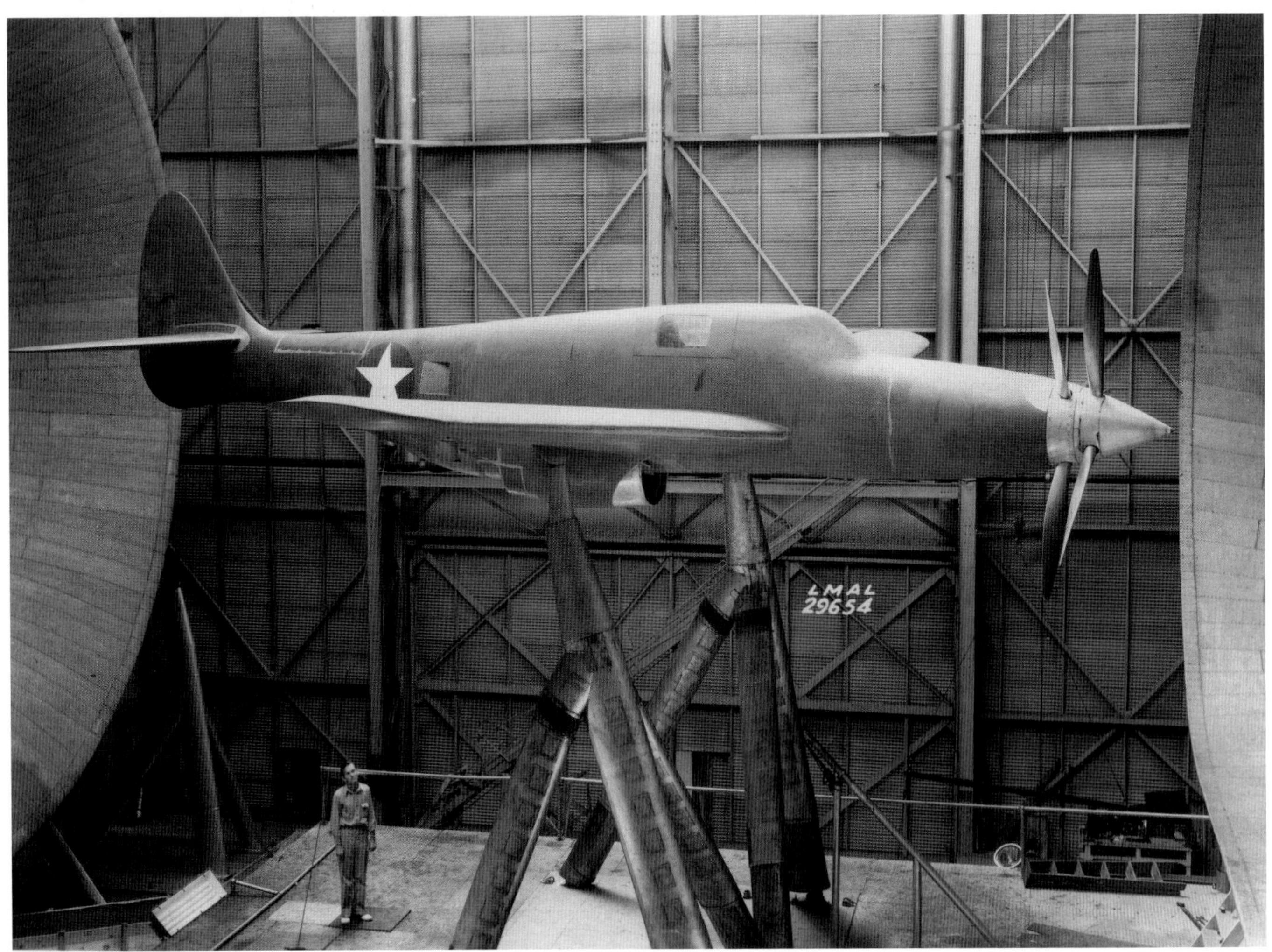

The radical canard-pusher Curtiss-Wright CW-24B flying demonstrator was tested in October 1942. Note the free-floating canard surface at the nose of the aircraft. The CW-24B was one of the first swept-wing aircraft. (NACA LMAL 30139)

Curtiss-Wright CW-24B

In 1940, the Army requested industry proposals for radical new fighter aircraft designs. The Curtiss-Wright Corporation responded with a unique free-floating canard, pusher-prop design with a swept-back wing known as the XP-55 Ascender. Curtiss-Wright received an Army contract in June 1940 that called for the construction of a powered wind tunnel model for assessments of the configuration. The contract deliverables were a wind tunnel model, some preliminary wind tunnel data, and an option for an experimental aircraft. After building and testing a large tunnel model, the Curtiss-Wright team received a less-than-enthusiastic reception from the Army based on the test results; but the company pressed on under its own funding with the design and construction of a flightworthy full-scale demonstrator aircraft known as the CW-24B. The fabric-covered, wooden-winged CW-24B was intended to be a lightweight, low-powered flying test bed for the full-scale XP-55 design. This low-cost flight demonstrator was one of the first U.S. aircraft to use a swept-back wing and encounter the aerodynamic problems exhibited by such a configuration at low speeds and high angles of attack.

Wind tunnel testing of the CW-24B configuration in 1941 included tests of a ¼-scale model at the Massachusetts Institute of Technology and the Langley 19-foot Pressure Tunnel. Both test programs identified an early concern regarding flow separation on the outer swept-wing panels at high angles of attack, which resulted in longitudinal instability (i.e., "pitch-up") near stall. Following the pitch-up at stall, the CW-24B configuration would exhibit a "deep stall" condition in which the airplane would trim at very high positive or negative angles of attack and descend in near-vertical flight in an uncontrollable condition with an almost horizontal fuselage attitude.

Initial flight tests of the CW-24B were conducted at Muroc Bombing Range (now Edwards Air Force Base), CA, in December 1941. During the tests, the underpowered CW-24B demonstrator was not capable of gaining sufficient altitude to permit spin recovery, and intentional spins were not attempted in the uneventful preliminary flight evaluation, which included over 190 flights. In July 1942, the Army Air Corps awarded a contract for three operational XP-55 aircraft.

At the end of the CW-24B flight-test program in May 1942, the Army Materiel Command procured the aircraft from Curtiss and requested that Langley undertake a Full-Scale Tunnel investigation of the aerodynamic characteristics of the aircraft.[85] Quickly approved by the NACA, the tunnel tests began in October 1942. For these tests, the CW-24B was modified to more closely represent the evolving XP-55 configuration. The canard was modified to that of the XP-55, the previously fixed landing gear of the CW-24B was removed, and the wing was resurfaced to a smooth finish. Lift, drag, pitching-moment, hinge-moment, and elevator pressure measurements were made for variations in angle of attack, wing flap deflection, elevator deflection, and elevator-tab settings.[86] All tests were made for propeller-removed conditions. Flow-separation and wing-stalling behavior were also monitored by analyzing wool tufts on the wings of the aircraft.

A key feature of the CW-24B and XP-55 designs was the longitudinal control concept, provided by the all-moveable, free-floating canard surface. The canard was directly connected to the pilot's control stick, and it also featured tabs that were controlled by a separate trim control in the cockpit.

Results of the full-scale tests correlated well with the earlier subscale model tests. Langley researchers found that the CW-24B aircraft was longitudinally unstable when the stick was fixed with the propeller removed. With the stick free (i.e., canard free-floating) and the landing flaps in a retracted position, longitudinal stability could be maintained at angles of attack below about 12°, but the aircraft exhibited pitch-up at higher angles of attack. Flow-visualization results showed that the flow on the rear of the swept wing at angles of attack approaching wing stall (17°) was in a spanwise direction and parallel to the trailing edge, promoting early flow separation and loss of lift behind the location of the center of gravity, resulting in the pitch-up tendency.[87] This result was one of the first indicators of the generic problem of longitudinal instability of high-aspect-ratio swept wings at moderate angles of attack.

The first XP-55 prototype flew for the first time on July 13, 1943. A major setback to the XP-55 program occurred on November 15, 1943, when the test pilot experienced an uncontrolled, inverted "deep stall" while conducting stall tests. The pilot had successfully conducted two stall-recovery tests with flaps up, but when he attempted to evaluate the effects of wing spoilers on stall characteristics at idling power with the wing flaps and landing gear down, the aircraft pitched rapidly in a nose-down direction, past the vertical and onto its back in an inverted deep stall similar to that predicted by earlier tunnel tests. The aircraft locked into a flat, inverted stabilized condition as altitude was lost for over 16,000 feet. The pilot finally managed to bail out of the airplane safely, but the aircraft impacted the ground in an almost horizontal attitude and was destroyed.

Even with follow-on aircraft modifications, stall characteristics of the XP-55 remained unsatisfactory. The stall occurred abruptly without warning, and rapid post-stall motions in pitch and roll were encountered. The airplane also had other disappointing characteristics, including deficient engine cooling.

A Vought-Sikorsky F4U-1 Corsair was tested in the Full-Scale Tunnel and flown by Langley test pilots in December 1942. Here, the aircraft is in the faired and sealed condition for drag cleanup tests. (NACA LMAL 30652)

Chance Vought F4U Corsair

In 1939, the Vought Aircraft Corporation united with the Sikorsky Aircraft Corporation to form Vought-Sikorsky. In 1943 the companies split, with Vought concentrating on military aircraft while Sikorsky moved to helicopter development and production. The famous Navy "bent-wing" F4U Corsair was a product of the earlier merger of 1939. The XF4U-1 first flew in May 1940, and production of the first F4U-1 production versions began in 1942. The powerful F4U-1 used a 2,000-horsepower engine with an extremely large 13.25-foot three-blade propeller that necessitated a unique, inverted gull-wing configuration for ground clearance of the propeller blades. The Corsair was the first U.S. fighter capable of exceeding 400 mph in level flight.

When the Navy requested the first of a series of tests of the F4U in the Full-Scale Tunnel in November 1942, the tests marked the return of the name Corsair to the tunnel test log, in addition to the earlier 1931 tests of the biplane Corsairs discussed in a previous chapter. The tunnel test objectives included a determination of sources of drag that might be eliminated in future production models; an evaluation of methods to reduce aileron hinge moments; measurements of the maximum lift of the airplane for various modifications; pressure distributions at the engine, oil cooler, and the supercharger intercooler; determination of the critical speeds for compressibility onset at the wing root, the wing-duct lip, and gun-blast tubes; and an assessment of the longitudinal stability of the airplane.[88] The Corsair had been designed with the wing leading-edge duct cooling inlets for the oil coolers and supercharger favored by Abe Silverstein and the staff at the Full-Scale Tunnel.

The aircraft had displayed several problems relevant to operations on aircraft carriers. The landing gear and arresting gear required revisions, and pilot vision over the long nose made carrier landings difficult for new pilots. One of the major aerodynamic problems of the early F4U versions was that the left wing would stall and drop rapidly during landing approaches at high angles of attack with power on. If power was abruptly increased, the stall and rapid roll to the left could be so severe that the aircraft would flip over onto its back.

The drag reduction analysis of the aerodynamic data gathered during the tunnel tests showed that the drag coefficient of the service aircraft was about 39 percent greater than that for the completely sealed and fared condition. After consideration of reasonable modifications for drag reduction, an increase in top speed of about 12 mph was predicted. Attempts to enhance aileron effectiveness and reduce hinge moments centered on an evaluation of ailerons with beveled trailing edges. Results indicated that the roll rate available at a representative stick force would be increased by about 59 percent over the original wooden ailerons.

The results of flow-visualization tests indicated that flow separation always began at the root section of the trailing edge of the wing for the propeller-removed condition. However, with the propeller operating, the flow began separating at the trough of the inverted gull sections. The left wing exhibited an abrupt stall for the power-on condition due to an increased upwash at the left wing's leading edge caused by the wake of the huge propeller. Flight tests of this particular airplane at Langley verified the roll-off tendency. The fix found for the problem was to attach a triangular leading-edge spoiler to the right wing outboard of the wing duct and inboard of the wing fold line. Measurements of rolling moment and flow-visualization studies in the Full-Scale Tunnel indicated that a uniform stall pattern for both wings was obtained with the asymmetric spoiler installation. The maximum lift for the aircraft with this modification was reduced about 12 percent.

In addition to detailed measurements of the pressure performance of the standard cowling, two additional cowls were tested following guidelines gathered in separate tests of a model of the F4U-1 in the Langley 8-Foot High-Speed Tunnel. A propeller-speed blower similar to that evaluated in the previously discussed B-24D engine study was also evaluated. The planned investigation to determine locations of compressibility onset on the airframe was limited because a more detailed series of F4U tests were under way in the high-speed tunnel. Critical locations revealed in the Full-Scale Tunnel tests for early onset included the wing-duct inlet lip.

In February 1945, the Corsair series returned for testing in the Full-Scale Tunnel when the Navy requested wing-operation testing of the Chance Vought F4U-1D (Vought was renamed in honor of its founder after its split from Sikorsky).[89] These tests were precipitated by reports from aircraft carriers that difficulty was being experienced in spreading and folding the wings of the F4U-1D in high winds. The study called for a determination of the airspeed and yaw angles critical for both wing spreading and wing folding, and also any simple modifications that would offer improvements in these operations. A service airplane was used for the tests, followed by later tests on another F4U-1D that incorporated modifications to improve wing operations. The scope of tests included a range of yaw angles from 30° nose right to 30° nose left, and a range of tunnel speeds up to the airspeed in which the upstream wing refused to spread. The instrumentation used included position indicators and strain gages for each wing and a hydraulic pressure gauge in the airplane hydraulic pressure indicator. Tests were made with four rocket launchers on each outer wing panel. With the rockets installed, a favorable weight moment was available for spreading the wing. Weights were also added in the wing ammunition boxes to simulate a full ammunition load. A modified hydraulic system was also tested.

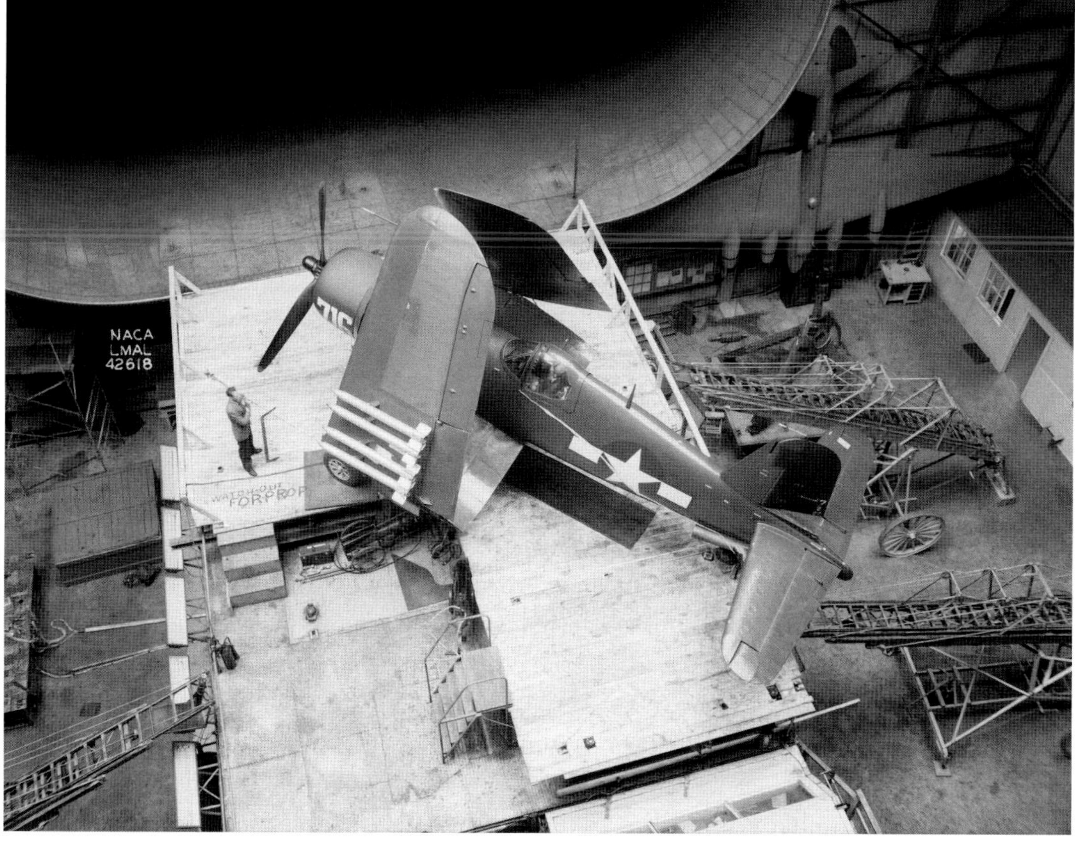

A Chance Vought F4U-1D Corsair prepares for wing-fold and wing-spread tests in 1945. Note the angle of yaw simulating a crosswind condition. Also of interest is the large Boeing B-15 model on the wall of the building. (NACA LMAL 42618)

The results of the Full-Scale Tunnel test showed that, for the range of yaw positions and airspeeds investigated, the most critical condition encountered was spreading the wings with the airplane at a 30° yaw position. Wing-folding operations in every condition tested were satisfactory. The best improvement in wing-spread operation was achieved by increasing the hydraulic pressure from the service value of 1,500 pounds per square inch (psi) to at least 1,800 psi. It was also found that operations were enhanced by operating the cowl flaps, or by moving the control stick toward the wing that was having difficulty in spreading. The project also resulted in a candidate modified hydraulic system and a method to evaluate the pressure required for wing-spread operation, including comparison with the tunnel results.

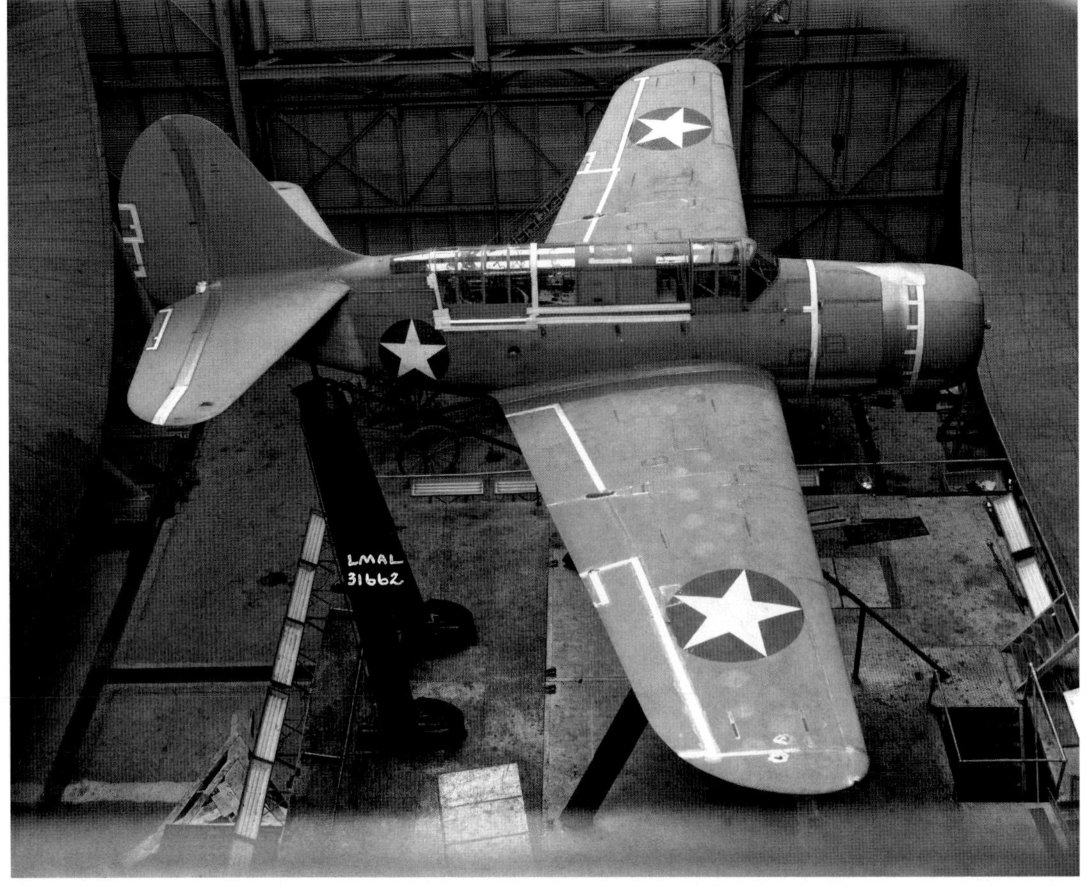

Curtiss SB2C-1 scout bomber in the sealed and faired condition during drag cleanup tests in January 1943. (NACA LMAL 31662)

Curtiss SB2C Helldiver

The Navy Curtiss SB2C Helldiver two-place scout bomber was one of the most controversial U.S. military aircraft of WWII. Intended to replace the Douglas SBD Dauntless dive bomber, the SB2C was beset by numerous problems in its early production models, requiring over 800 modifications for satisfactory mission effectiveness. The early SB2C-1 version was hated by its pilots, many of whom wanted to retain the older SBD. By the end of the war, the aircraft's reputation had finally been restored and fleet pilots showed more enthusiasm for the big Helldiver.

The Full-Scale Tunnel was used in two test programs in support of the SB2C at the request of the Navy. The first test consisted of a 2-month tunnel entry in February 1943 for the first operational version known as the SB2C-1. The scope of the investigation included a drag cleanup study; obtaining data for calculation of tail loads (prior to the tunnel tests, the horizontal tail of an SB2C-1 had failed during a high-speed dive); determining the effect of partial-span wing leading-edge slats on lift, wing stall progression, and aileron control; measurements of wing flap loads for prediction of critical loads in landing and diving flight; and the effects of a two-gun 50-caliber Emerson gun turret on aircraft performance.[90]

The drag cleanup tests of the SB2C-1 revealed that the faired and sealed configuration would have an incremental speed increase of about 30 mph compared to the service condition. Over one third of the drag difference between the two configurations was created by the large-bore stovepipe exhaust stack design due to the protruding shape, leakage around the stack, and the turbulent wake over the fuselage. Further, discharging the exhaust gases at a low velocity and at an angle to the airstream created an additional loss of thrust. Several different exhaust stack configurations were tested, including individual jet-type exhaust stacks with significantly reduced drag.

Shown are several major drag-contributing features of the SB2C-1, including the large-bore exhaust stacks and exposed wheel wells. (NACA LMAL 31980)

Additional results from the test program included in-depth visual flow analysis, which indicated that the existing wing-slat design was extremely poor and resulted in premature stalling of the wing. A slat modification was designed and tested that improved the airflow, increased the effectiveness of the aileron, and increased the maximum lift capability of the aircraft. The test also showed that the SB2C-1 was neutrally stable throughout the speed range at full power.

A second Helldiver test program was conducted in the Full-Scale Tunnel in January 1946 as part of a Navy-requested wind tunnel and flight investigation of canopy loads for the Navy's Grumman F6F-3 Hellcat, the Curtiss SB2C-4E Helldiver, and the Grumman F8F-1 Bearcat. As aircraft performance rapidly increased toward the end of the war, high-speed air-combat maneuvers and dives had resulted in numerous experiences in which canopies had departed aircraft under loads, sometimes with catastrophic results. The occurrence of canopy failures indicated that existing load requirements used in the design of canopies and their components might not be adequate. The staff of the Full-Scale Tunnel therefore initiated the project to determine critical load requirements by means of external and internal pressure measurements on aircraft having three different types of canopy installations.[91] The three canopy types of interest included single sliding enclosures (F6F-type), front and rear sliding enclosures (SB2C-type), and bubble enclosures (F8F-type).

A general NACA wind tunnel and flight investigation of aerodynamic loads on canopy enclosures included tests of a Curtiss SB2C-4E. Langley's chief test pilot Herbert Hoover experienced potentially fatal injuries as a result of an impact with a departing Helldiver canopy during an earlier test flight. (NACA LMAL 46741)

The canopy tests for the SB2C-4E were of particular interest to the Langley community because of an accident that occurred at Langley in 1943. Herbert H. Hoover, head of the Flight Operations Section and a nationally recognized test pilot, was involved in conducting pullout tests from dives in a Helldiver to address serious deficiencies exhibited by the airplane.[92] During a relatively low-speed instrumentation calibration flight, the canopy of the aircraft tore loose. As it departed the aircraft, an edge of the canopy smashed through Hoover's helmet and goggles. Although in great pain and almost blinded by blood streaming down his forehead, Hoover kept his seat in the now-open airplane and safely brought the Helldiver in for a landing.

Results of the pressure measurements indicated that, based on the maximum differential between external and internal pressures, the maximum loading for the front and rear canopies was obtained for the high-speed flight condition. The highest loads on the front canopy were in the exploding direction and occurred with both canopies closed. The highest loads on the rear canopy were in the crushing direction and occurred with the front canopy open and the rear canopy closed. The effects of propeller operation at high power and sideslip also increased the canopy loads and caused variations of exploding and crushing pressures.

Japanese Mitsubishi Zero

Arguably one of the most secret tests ever conducted in the Full-Scale Tunnel occurred in the days immediately following the first SB2C-1 tunnel entry in 1943. In an extremely secure environment, the first captured Japanese Mitsubishi Zero was tested in the tunnel during a brief stop at the LMAL.

On June 3 and 4, 1942, Japanese warplanes attacked the American military base at Dutch Harbor in Alaska's Aleutian archipelago.[93] The attack on Alaska was intended to draw part of the U.S. fleet north from Pearl Harbor, away from Midway Island, where the Japanese were setting a trap. During the attack on June 4, Japanese Zero pilot Tadayoshi Koga's aircraft was hit by ground fire, and one of the bullets severed the return oil line between the oil cooler and the engine. As the engine continued to run, it pumped oil from the broken line. Koga flew his oil-spewing airplane to a designated emergency landing site on Akutan Island, 25 miles away. A Japanese submarine was positioned nearby to pick up downed pilots. Unknown to the Japanese, the landing site was a bog with knee-high grass concealing water and mud. After Koga lowered his wheels and flaps and landed, his main wheels dug in, the Zero flipped onto its back, and Koga was killed.

The wrecked Zero lay in the bog for more than a month, unseen by U.S. patrol planes and offshore ships. However, on July 10, the crew of a U.S. Navy PBY Catalina amphibian returning from overnight patrol spotted the Zero. Inspections of the wreck by ground troops indicated that the aircraft was salvageable, resulting in intense activity to bring the priceless war prize back to the United States for analysis. Prior to this event, the Zero was regarded as a formidable foe that was virtually unbeatable in air combat, and the delivery of a flyable aircraft would provide valuable information on the characteristics of the aircraft and proper tactics for fighting it. Koga's rebuilt Zero was the first flyable aircraft of its type acquired and tested in the United States. Only 2 months after it was found, the aircraft had been shipped 2,800 miles to North Island Naval Air Station in San Diego, repaired, and was flying in simulated dogfights against frontline U.S. fighters to determine tactics to be used against the Zero in the Pacific theater. There is no evidence that the Japanese ever knew that the United States had salvaged Koga's plane.

During September and October of 1942, American pilots flew in and against the Zero in the best U.S. Army and Navy fighters at the time, and learned that the airplane had superior maneuverability only at the lower speeds used in dogfighting, with a short turning radius and excellent aileron control at very low speeds. However, immediately apparent was the fact that the ailerons froze up at speeds above 200 knots, so that rolling maneuvers at those speeds were slow and required significant force on the control stick. It rolled to the left much easier than to the right. Also, its engine would cut out under negative acceleration (as when nosing into a dive) due to its float-type carburetor. Based on these evaluations, U.S. pilots were provided with tactics on how to escape a pursuing Zero. The recommendation was to go into a vertical power dive—using negative acceleration, if possible—to open the range quickly and gain advantageous speed while the Zero's engine was stopped. At about 200 knots, the pilots were told to roll hard to the right before the Zero pilot could get his sights lined up.

Japanese Mitsubishi Zero at the Langley flight line on March 8, 1943. The aircraft had been repainted in U.S. Navy blue-grey colors and insignia but retained its Japanese serial number 4593, barely visible on the vertical tail. Note the short, stubby wooden radio mast that replaced the original mast that had been destroyed in its crash landing. Aircraft is shown after installation of NACA wingtip boom for flight tests at Anacostia Naval Air Station (NAS). (NASA EL-1997-00167)

In early 1943, the Zero was flown to the Navy's Anacostia Naval Air Station in Washington, DC, for more detailed Navy evaluation flights. The Navy requested the well-known expertise of the Langley laboratory in the field of flight instrumentation to outfit the aircraft for flight tests. The Zero was flown from Anacostia to Langley for installation of the instrumentation. Langley aircraft flight records indicate that the aircraft arrived at the NACA flight hangar in the East Area of Langley Field at about 3 p.m. on Friday, March 5, 1943. The Zero's presence at Langley is well documented in several books, and photos of the aircraft are posted on the NASA multimedia site.[94] However, the secret activities of its whereabouts during its visit were only recently revealed after 67 years during interviews with Langley retirees.[95]

As a staff member of the Full-Scale Tunnel, P. Kenneth Pierpont was invited to inspect the Zero at the flight line along with 30 high-level Army, Navy, and NACA officials. Abe Silverstein, then head of the Full-Scale Tunnel Section, and Chief of Aerodynamics Elton Miller were granted permission by the Navy to borrow the aircraft for aerodynamic tests in the Full-Scale Tunnel over the weekend under tight security and to return it to the flight line as quickly as possible. After sundown, the airplane was covered and towed a few blocks to the wind tunnel, where Friday night was spent mounting the aircraft to the struts (which necessitated top-priority fabrication of special mounting hardware in the Langley shops).

Wind-on tests commenced in the tunnel about noon on Saturday, March 6. The intense scope of testing (all in the power-off condition) included wake surveys to determine the drag of aircraft components; tunnel scale measurements of lift, drag, control effectiveness; and sideslip tests. In addition, comparative drag tests were made for the aircraft in the service condition and in a faired and streamlined configuration. Testing continued all Saturday night and Sunday, March 7, until darkness fell, when the aircraft was removed from the test section, covered, and moved back to the flight line in the same position it had occupied upon arrival.[96]

The security measures taken during the tests were remarkable. Spies were very active in the Peninsula area at that time, and the tests were conducted under strict need-to-know guidelines. Even the wind tunnel test log of the Full-Scale Tunnel was modified to ensure that the test was not acknowledged, and no known photographs were taken during the program, nor were any NACA reports written regarding the tests or their results. The data measured in the tests were retained by Abe Silverstein, and it is unknown whether he shared copies with other organizations. It is also unknown whether military personnel witnessed the tunnel tests.

On Monday, March 8, NACA photographers took pictures of the aircraft on the flight line as instrumentation was under way for the upcoming flight test at Anacostia. During its stay at Langley, the Zero was never flown by NACA test pilots.[97] On Thursday, March 11, the aircraft departed for Anacostia, where the Navy conducted simulated air combat flights with an F4F-3 and an XF4F-8 to evaluate handling qualities, including in-flight measurements of flight parameters. W. Hewitt Phillips of Langley analyzed the data and later authored two NACA reports for the Navy.[98]

The capture and analysis of the performance of Koga's Zero is regarded today by many historians as a major turning point in the war. They point out that many U.S. pilots vividly remembered briefings on recommended tactics for fighting the Zero that had been learned from the captured Zero, and several owed their lives to the information. Some of the historians believe that the capture of the Zero was as devastating to the Japanese war effort as the U.S. victory during the battle for Midway Island.

Grumman F6F-3 Hellcat

Ironically, the week after the Zero tests were completed in the Full-Scale Tunnel, the next aircraft mounted in the tunnel was its archenemy and domineering rival, the Navy's Grumman F6F-3 Hellcat.

The primary objective of the first Hellcat entry in March 1943, as requested by the Navy, was drag cleanup, but additional tests were conducted to study the stalling characteristics of the aircraft and determine the effects of several wing modifications on maximum lift; determine the longitudinal stability of the airplane and its control effectiveness; investigate characteristics of airflow in the oil-cooler and intercooler ducts; determine the critical compressibility speeds; measure the pressure distribution over the engine cowling; and investigate the nature of the airflow in the region of the tail. In addition, tests were made to determine the effects of a wing-mounted radar pod, service guns, and a detachable fuel tank on aircraft characteristics.[99]

Preliminary testing in the drag cleanup activities disclosed that an early stall occurred at the wing-fuselage juncture that resulted in low values of maximum lift, and fillets and inner wing airfoil changes were evaluated to minimize the flow separation. The major results of the tunnel entry were that the aircraft's top speed could be increased by about 13 mph if the wheel wells were sealed with full-length wheel fairings and various gaps in the upper surface of the wing were sealed. The maximum lift could be significantly increased if leakage through the wing-fold gap was eliminated and if a wing-fuselage fillet was used.

Drag cleanup tests of a Grumman F6F-3 fighter were conducted in March 1943. In this photo, the aircraft is in the service condition. The Hellcat was also used in two other tests in the Full-Scale Tunnel. (NACA LMAL 32408)

The one-of-a-kind XF6F-4 Hellcat was used for a NACA research effort on factors affecting directional stability and trim. The airplane was the subject of tunnel and flight tests at Langley. (NACA LMAL 38069)

In October 1944, another test of the F6F series occurred when the Navy Grumman XF6F-4 was used to investigate factors that affect the directional stability and trim characteristics of a typical fighter-type airplane. Separate contributions to directional stability and trim of the wing-fuselage combination, the vertical tail, and the propeller were determined for eight representative flight conditions. The XF6F-4 was a one-off variant of the Hellcat series, using a two-speed turbo-supercharged engine rather than the gear-driven supercharger used by earlier versions. The data gathered in the study were used in the development of analytical methods for predicting general trends in stability and trim.

The previously tested F6F-3 returned for canopy loads testing in November 1945 as part of the general canopy load program requested by the Navy. Canopy interior and exterior pressures were measured in a manner similar to the procedures used in the canopy loads tests for the Curtiss SB2C-4E discussed previously. As was the case for the SB2C-4E, the results indicated that net aerodynamic loads on the canopy were greatest when the canopy was closed at high speeds. Opening the canopy reduced the pressure differential of the exploding forces.

The final test of the F6F-3 was instigated by issues concerning the effect of rate of change of angle of attack on maximum lift. The Full-Scale Tunnel had conducted many tests to determine the maximum lift of aircraft before and during WWII, and the data had been summarized in a report for wide dissemination.[100] Because of its importance to the landing performance of an aircraft, an accurate prediction of maximum lift is critical in airplane design. The experience of the Full-Scale Tunnel staff indicated that good agreement could be obtained between results obtained in the tunnel and results obtained in flight tests only if the rate of change of angle of attack approaching the stall was the same for each. In order to evaluate the effect of variables, such as rate of change of angle of attack and wing surface conditions, on maximum lift, the previously tested F6F-3 was mounted on a special tail support with high-speed gearing capable of producing a continuous change in angle of attack at rates from 0° to 0.85° per second.[101] Wool tufts were used to visualize the flow over the wing during the pitching motions. The tests were conducted over a 2-month period beginning in February 1946.

The results of the study indicated that good agreement between wind tunnel and flight-test values of maximum lift can be obtained if both tests are carefully controlled so that the rate of change of angle of attack, propeller operation, Reynolds number, and wing surface roughness are reproduced, and if the airplane being tested in the wind tunnel is not too large in comparison with the size of the wind tunnel test section.

Bell XP-77

In late 1941, the Army became concerned over a possible shortage of aluminum and other critical aircraft fabrication materials.[102] In October, the Bell Aircraft Corporation responded to a request by the Army for the design of an unconventional lightweight fighter to be constructed from "non-strategic" materials. The resulting Bell XP-77 design was a single-engine, low-wing configuration constructed almost entirely of wood and magnesium alloy and featuring an NACA laminar-flow airfoil, tricycle landing gear, and a bubble canopy for enhanced pilot vision. Bell experienced considerable delays in the XP-77 program, and the Army ordered only two prototypes. The two airplanes were to have different operational missions. The first version of the airplane was a low-altitude fighter with a Ranger SGV-770C-1B engine and a design altitude of 12,000 feet, while the second version was a high-altitude fighter with a Ranger SGV-770D-4 engine and a design altitude of 27,000 feet.

The Army Air Forces requested tests to measure and analyze the stability, cooling, and air loads of the XP-77 in the Full-Scale Tunnel during two separate entries in June and October of 1943. A full-scale mockup of the aircraft was fabricated and tested in the tunnel with a propeller thrust line location similar to the low-altitude version of the airplane but using a 10.5-foot propeller intended for the high-altitude version. The landing gear was removed from the mockup for all tunnel tests.[103]

A full-size mockup of the Bell XP-77 "lightweight" fighter is prepared for tunnel tests in June 1943. Note the extremely small size of the configuration, which had a wingspan of only 27.5 feet. (NACA LMAL 33475)

The scope of the test program was to determine longitudinal and lateral directional stability, overall aircraft drag, the internal and external airflow qualities of a new NACA-designed cowling, and the air loads and critical speeds on both the cowling and the canopy. The cowling design was the product of extensive research that had been conducted with the same mockup in the Propeller Research Tunnel earlier in the year to investigate the cowling and cooling limits of the Ranger SGV-770 engine.[104]

The results of the Full-Scale Tunnel tests revealed that with power on, the cowling with a modified cooling-air exit area provided excessive engine cooling for cruise, but the cooling was found to be inadequate for a climbing attitude. As a result, Langley recommended that either the cooling-air exit area be enlarged or exit flaps be used in the cowling design to facilitate adequate engine cooling. The Ranger SGV-770C-1B engine used for the investigation failed while undergoing further tests, forcing the termination of the studies with no additional recommendations being made to enhance the aerodynamic performance of the aircraft. The results of the tunnel tests to determine stability and control characteristics did not disclose any serious problems.

Developmental problems plagued the XP-77 program (especially weight growth), and first flight of the first prototype was delayed until April 1, 1944. Initial flights exposed vibration difficulties associated with resonance of the unique wooden structure under powered conditions.[105] In addition, test pilots complained of restricted vision due to the long nose of the airplane and the rearward location of the cockpit. The aircraft, which had an unsupercharged XV-770-7 engine, was severely underpowered. Testing of the second XP-77 prototype at Eglin Field, FL, resulted in the destruction of the aircraft when the pilot unsuccessfully attempted an Immelmann maneuver that evolved into an uncontrollable inverted spin, forcing the pilot to bail out. In December 1944, the XP-77 program was officially cancelled with no production orders.

Northrop MX-334

In September 1942, Northrop began a design effort for a radical rocket-powered interceptor that would ultimately be known as the XP-79. These conceptual studies resulted in the company winning a contract for the development of three wooden single-place piloted gliders to serve as flying mockups for investigations of the handling characteristics of the unconventional design. Two of the 36-foot-wingspan plywood gliders were unpowered and became known as the MX-334, while the third glider, designated the MX-324, would be powered by an Aerojet rocket motor. The pilot lay prone during flight to eliminate aerodynamic drag created by a conventional canopy. For improved stability, Northrop later decided to add a vertical fin to the design.

The Materiel Command of the Army Air Forces requested Full-Scale Tunnel tests of the unpowered MX-334 in July 1943 to study the longitudinal and lateral stability and control characteristics of the all-wing glider.[106] The design team and Langley researchers were particularly interested in the aerodynamic performance of the design, which had neither a conventional fuselage nor vertical tails. These tests also included investigations aimed at identifying an appropriate wingtip leading-edge slat configuration capable of enhancing the glider's static longitudinal stability and maximum lift capability. In addition, the program included a drag analysis and an evaluation of directional stability with vertical fins incorporated in the design. The effectiveness of the unconventional air-operated directional control system was also evaluated. In this system, the inboard sections of the trailing-edge surfaces were operated by air bellows and provided both dive braking and directional control.

The tunnel test results showed that, with slats removed, the aircraft was longitudinally unstable and exhibited tip stall of the swept wing at high angles of attack, resulting in a serious pitch-up problem. It was also found that the original wing slats proposed by Northrop were not as effective as desired in eliminating the deficiency. Addition of large-span slats maintained attached airflow at the tips until after the center wing section had stalled, thereby

The Northrop MX-334 piloted glider was tested in the Full-Scale Tunnel in July 1943 to evaluate its stability and control characteristics. Note the cockpit enclosure wherein a prone pilot was stationed and the outboard wing leading-edge slats evaluated during the study. (NASA EL-2003-00289)

eliminating the pitch-up problem while enhancing lift. In addition, the incorporation of vertical fins at the wing center section greatly enhanced lateral directional characteristics at high angles of sideslip. The Langley researchers expressed concern that the stick forces for maximum aileron deflection at high speeds might be excessive for control operation from a prone position. The results also indicated that the unorthodox directional control of the aircraft was insufficient. Rather than using the designed air-operated bellows to operate inboard elevons for yaw control, the Full-Scale Tunnel staff recommended that the duct system be modified if the existing directional control system was employed. Further recommendations included detailed modifications to the inlets, ducting, and butterfly control valve.

Flight testing of the MX-334 began on October 2, 1943, with a vertical tail added to the configuration.[107] These tests, in which the glider was towed behind a P-38 tow plane, proved to be hazardous. On one of the flights, the MX-334 encountered the wake of the tow aircraft and became uncontrollable. The little glider entered a spin and eventually regained stability, although the aircraft was inverted and uncontrollable in an apparent inverted deep stall condition. While the pilot was able to parachute to safety, the out-of-control glider descended to ground in a series of circles and was destroyed on impact.

Flight testing of the rocket-powered MX-324 proved to be more successful. On July 4, 1944, the rocket-powered glider took to the air. Following release from the P-38 tow aircraft, the pilot performed a near flawless flight in which the glider remained in the air for a little over 4 minutes. The United States had finally demonstrated the feasibility of rocket-powered flight almost 3 years after the first flight of the German Me-163 rocket-powered interceptor.

The Bell P-63A Kingcobra was the subject of a coordinated wind tunnel and flight investigation at Langley in 1943. In this photo, the aircraft is mounted in the Full-Scale Tunnel for drag cleanup tests. (NACA LMAL 34204)

Bell P-63A Kingcobra

The Bell P-63A Kingcobra evolved from the earlier P-39 Airacobra aircraft, being larger with a four-blade propeller, a new uprated engine, and a wing that incorporated a NACA laminar-flow airfoil. In September 1942, the Army ordered the P-63A into production with deliveries beginning in October 1943. The Army Air Forces considered the P-63A to be inferior to other fighters such as the P-51 Mustang and dramatically reduced its interest in the aircraft. However, the Soviet Union was in urgent need of fighter aircraft and became the principal operator of the aircraft during the war. Over 3,000 P-63s were built, with over 70 percent delivered to the Soviet Union.

An Army Air Forces request for drag cleanup tests of the Bell P-63A in the Full-Scale Tunnel resulted in a month-long entry in August 1943.[108] The phases of the investigation included preliminary flight tests to evaluate the maximum speed of the aircraft and to make a brief investigation of the losses in the cooling-duct system; drag cleanup tests in the Full-Scale Tunnel to obtain a drag analysis and to develop fairings, seals, and other modifications that would increase the aircraft's speed; and final flight tests of the modifications developed during the Full-Scale Tunnel tests. In addition to standard drag cleanup tests, the characteristics of the airflow through the oil cooler, the Prestone cooler, and the carburetor air ducts were investigated, resulting in changes to the wing-duct inlets in the carburetor air duct.

The results of the P-63A drag cleanup tests were a strong indicator of the advances being made by American military aircraft designers as the war progressed. Designers had become impressed with the need to pay attention to details such as sealing of surfaces, surface finishes, filleting, and duct shapes. After a thorough investigation of potential drag reduction modifications to the P-63A—including sealing canopy leakage paths, installing fairings on the fuselage gun blast tubes, sealing all holes and gaps in the cooling ducts, and streamlining—the projected increase in top speed was only 10 mph.

A P-51B Mustang arrived at Langley in August 1943 for Full-Scale Tunnel tests. (NACA LMAL 34312)

North American P-51B Mustang

Widely regarded as the best propeller-driven fighter of World War II, the P-51 Mustang utilized many NACA concepts and received extensive support from the NACA laboratories. It was the first production aircraft to use an NACA laminar-flow airfoil and was provided with many wind tunnel entries in the facilities at Langley and Ames laboratories as well as the AERL at Cleveland. In 1943, the Army Air Forces requested a drag cleanup test in the Full-Scale Tunnel to determine the sources and quantity of parasite drag with the P-51B version of the Mustang in a service condition. In addition, the effect of armament installations on the aerodynamic characteristics of the airplane was also investigated. A brief 2-week test program began in late September 1943.[109]

Data obtained at a maximum tunnel speed of 100 mph predicted a top speed of the airplane in the service condition of 464 mph, which was optimistic because no account had been made of the degradation in performance due to compressibility effects. The final NACA report to the Army on the test results was unusually brief and to the point:

> The P-51B is an unusually clean airplane. A great deal of care is taken with the detailed design and there are a few minor modifications that will increase the speed. A drag increment caused by wing roughness and leakage was measured near the armament installation.[110]

The P-51B remained at Langley to participate in many NACA flight studies through January 1951. (NACA LMAL 34589)

The estimated increase in top speed of the P-51B following all observations of the drag cleanup tests was only about 3 mph.

Boeing XB-39 Spirit of Lincoln

Boeing's highly successful B-29 Superfortress heavy bomber was one of the most famous aircraft of World War II and the Korean War. However, as the airplane was being developed in the early 1940s, the Army was concerned that the pace of its development would be severely impacted if problems arose with the Wright radial engines used by the design. As a result, an experimental variant of the B-29 known as the XB-39 Spirit of Lincoln was also developed using in-line liquid-cooled engines rather than the conventional air-cooled radial engines of the B-29. The XB-39 was actually the first YB-29, and only one aircraft was produced and extensively tested in 1944.

Ground testing of the XB-39's Allison V-3420-11 liquid-cooled engine-nacelle configuration revealed engine-cooling problems that had resulted in engine modifications. The Army Air Forces then requested that Langley conduct studies of the aerodynamic and cooling characteristics of the modified XB-39 engine installation in the Full-Scale Tunnel.[111] A 3-month test program was conducted, beginning in October 1943, to verify that the design modifications to the engine powerplant had corrected the cooling problem. The tunnel tests included tests of the exhaust-shroud system and four modifications jointly proposed by Langley and the Allison Division of the General Motors Corporation. The coolant, oil, and charge-air cooling systems were studied during the tests with the propeller removed and in power-on tests. The research team also analyzed the drag of the engine nacelle and cooling flap. In addition to performing the wind tunnel tests, the staff of the Full-Scale Tunnel also

Tests of the Boeing XB-39 engine and engine-nacelle components were conducted in the Full-Scale Tunnel (upper photo) in October 1943 as well as at an outdoor ground facility (bottom photo) in 1944. The test setup and procedure was very similar to those used in B-24D engine tests of 1942. Note the data-acquisition and engine-control rooms at the outdoor site. (NACA LMAL 35052 and NACA LMAL 37951)

conducted follow-up outdoor static tests of the instrumented engine powerplant in May 1944 at a ground test site outside the tunnel that had been used for the B-24D engine tests discussed earlier.[112]

Results of the wind tunnel tests showed that the original shroud system would be inadequate for cooling purposes during a military-power climb at 35,000 feet. With the modifications to the shroud system recommended by the NACA, it was estimated that adequate cooling could be achieved at all altitudes and flight conditions. In addition, intercooling inadequacies were also identified. To correct these inadequacies, the Full-Scale Tunnel staff recommended that the intercooling air outlet be reshaped and the outlet area be enlarged. They also recommended that coolant radiators with larger frontal and exit areas be used to facilitate better engine cooling.

Development of the XB-39 proceeded in parallel with the effort for the B-29, and the first flight of the only B-39 was made in December 1944. Meanwhile, the initial B-29 fleet had entered service in June and, despite continuing B-29 engine problems, the XB-39 program was terminated with no production models ever being produced.

PV-2 Rotor

During World War II, very few efforts related to rotary-wing aircraft were conducted at the Full-Scale Tunnel because of high-priority efforts on production military aircraft. However, interest in special missions such as antisubmarine warfare and the rescue of downed pilots began to stimulate military leaders.[113] In response, the NACA began wind tunnel investigations of a small helicopter rotor known as the PV-2 rotor.

Tests of the PV-2 isolated helicopter rotor in 1944 were the first studies of this type in the Full-Scale Tunnel. Adapting the tunnel's balance system to helicopter tests proved to be a challenging task. (NACA LMAL 45748)

In 1943, Frank Piasecki became the second American (Igor Sikorsky was the first) to build and fly a successful helicopter, known as the Piasecki-Venzie PV-2. Later that year, he flew the helicopter at Washington's National Airport for a large crowd of onlookers, including military representatives. The Navy was particularly interested in the demonstration because it had been under fire from Congress since the Army had taken the lead in funding emerging helicopter developments. The helicopter was subsequently produced by the PV Engineering Forum, a company formed by Piasecki and Venzie. The Bureau of Aeronautics requested tests of an isolated PV-2 rotor in the Langley Full-Scale Tunnel in 1944.[114]

No tests of rotating-wing aircraft had been made in the Full-Scale Tunnel during the previous 6 years, and it was apparent that a considerable effort would be required to train personnel and develop adequate testing equipment.[115] While small rotors such as the PV-2 (25-foot diameter) could be handled without too much difficulty if damping and stiffening were used in the test setup, the balance and support system of the Full-Scale Tunnel proved difficult to adapt to the testing of a complete helicopter (recall that the Kellett YG-1 autogiro had previously been destroyed by resonance conditions during testing in 1937). The conventional tunnel balance system presented too much flexibility and later had to be bypassed for helicopter tests. In addition, the issue of potential interference effects from the ground board and correction procedures proved to be a source of concern.

The PV-2 articulated rotor was the first helicopter rotor tested in the Full-Scale Tunnel. The objective of the month-long project in February 1944 was to determine the performance of the rotor, along with detailed data for correlation with emerging rotor theories. The tests included blade motion photographic studies as well as measurements of rotor forces and power input to the rotor. Before the rotor tests were made, vibration surveys were conducted to determine the vibration characteristics of the wind tunnel setup. It was found that the flexibility of the supporting structure resulted in its natural frequency being too low for safe operations, and the support structure was then reinforced and several auxiliary wires were used to guy the main support and raise the natural frequency enough to permit testing up to high rotor speeds. After the successful completion of the initial testing, the Navy requested that the program be extended to include tests of a cambered rotor airfoil section (the original PV-2 rotor blades had symmetrical airfoil sections) for higher lift capability in 1945.[116] The original PV-2 blades had a tubular steel spar to which wooden ribs were attached. The forward parts of the blades were covered with plywood and the rear portions of the blades were solid wood wrapped in fabric, doped, and polished to a smooth finish. A follow-on test in 1948 was also conducted to evaluate the performance of metal PV-2 blades of improved surface condition.

Bell YP-59 Airacomet

One enduring criticism of the NACA has been its lack of leadership in the development of turbojet-powered aircraft technology during World War II. Although the noted Langley engineer Eastman Jacobs had explored the design of an axial-flow compressor in 1938, the laboratory knew nothing regarding the development of the world's first jet aircraft, the Heinkel 178, which flew for the first time in August 1939. The subsequent interests of the military and eventual NACA actions are summarized in detail in James Hansen's *Engineer in Charge*.[117] During a tour of England in 1941, Gen. Hap Arnold had discovered that the British were preparing to flight test the Whittle E 28 turbojet-powered aircraft, and he proceeded to initiate a top-secret project to develop a U.S. version of the engine by General Electric. Langley was initially kept out of the project and seriously lagged in developing technologies required for this new radical form of propulsion until 1943.

The YP-59A developmental prototype of America's first jet-powered military aircraft underwent extensive testing in the Full-Scale Tunnel in 1944. In this photograph, the aircraft is in the service condition. (NACA LMAL 37255)

Meanwhile, the Army had moved forward in 1941 with a contract to Bell Aircraft for America's first jet-powered aircraft, the P-59 Airacomet. The NACA was finally informed of the aircraft development program in 1942. The XP-59 first flew on October 1, 1942, and only a few of the Langley staff were briefed on the project until mid-1943. Once aware of the new propulsion concept, the NACA began to contribute under tight security. Abe Silverstein, who had left Langley in October 1943 to become chief of the AERL's new Altitude Wind Tunnel, led top-secret testing of the YP-59 and its GE turbojet at that facility in early 1944.

The Army procured 13 YP-59A developmental aircraft, which were delivered with GE I-16 jet engines in 1943. At the request of the Army Air Forces' Air Technical Service Command, the 10th YP-59A began a 3-month test program in the Full-Scale Tunnel in late March 1944. The goals of the activity were to determine means of improving the high-speed performance of the Airacomet and to determine the stability and control characteristics of the airplane.

The drag cleanup tests of the YP-59A were of special interest. Although the aircraft's wing used laminar-flow airfoils, its high-speed performance was disappointing and showed no advantage over advanced propeller-driven fighters near the end of the war. This lack of performance is particularly striking when comparing the P-59 with the German Messerschmitt Me-262 jet aircraft.[118] The Airacomet had a thrust-to-weight ratio that was 30 percent greater than that of the Me-262, but its top speed was 130 mph slower than the German jet. The P-59 had a much larger wing and a thicker airfoil (by 14 percent) that greatly increased aerodynamic profile drag. The large wing area was used in Bell's design for a lower wing loading to achieve acceptable low-speed performance using only small trailing-edge flaps. In comparison, the Me-262 used sophisticated wing leading-edge and trailing-edge high-lift devices to lower takeoff and landing speeds and a smaller, thinner wing for less drag at high speeds.

The scope of drag studies was exhaustive and included basic force measurements to determine the sources of drag and the effects of modifications to reduce drag; an investigation of the airflow characteristics in and around the engine nacelles and boundary-layer ducts; and an investigation of the critical compressibility speeds for the canopy, nacelle inlet lips, and cabin intercooler scoop.[119] In addition, the aerodynamic drag of seven different external wing fuel tank arrangements was measured; the effects of airplane modifications on maximum lift were studied; and the location and extent of the engine jet wake for the high-speed and climb conditions were determined. Most of the drag cleanup tests were conducted with power off, but the nacelle airflow drag investigations were made over a range of angles of attack with engines operating at several levels of thrust.

Results of the drag studies revealed that the airflow into and around the original nacelle inlet was very unstable and exhibited widespread flow separation in the fuselage boundary layer ahead of the duct inlet because of insufficient boundary-layer removal. Modified inlets were designed and fabricated, resulting in stable flow and satisfactory behavior for all test conditions. The traditional drag cleanup test procedure indicated that the top speed of the airplane could be increased by about 27 mph through relatively minor changes. Almost half of the improvement could be obtained by modifying the engine nacelle inlets and boundary-layer removal ducts. The investigation also showed that the estimated critical speeds for compressibility of the YP-59A original nacelle inlets and of the canopy-windshield configuration were less than the estimated top speed of the airplane. The revised nacelle inlets designed by the NACA increased the estimated critical speeds by over 170 mph.

At the time of the YP-59A tests, very little data were available relevant to the effect of jets on the stability of airplanes. The stability and control phase of the testing concentrated on the impact of power on the longitudinal stability characteristics of the Airacomet.[120] The results of the test indicated that most of the change in longitudinal stability due to power operation was caused by the thrust moment of the jet. Although a sufficient amount of data for a complete analysis was not obtained, tests made with power off indicated that the airplane, with landing flaps fully deflected, would be laterally unstable in flight at low speeds, in agreement with flight test experience.

This photo of the cockpit area of the YP-59A was taken on April 26, 1944, and shows the cockpit damage sustained during an engine fire under powered conditions. The damage was repaired to an acceptable state and the tests continued until July. (NACA LMAL 37434)

One of the events that occurred during testing of the YP-59A has become embedded in the historical lore of the Full-Scale Tunnel. During one of the first power-on tests on April 25, 1944, the left engine-bay area of the airplane caught fire and burned extensively, damaging the cockpit area in the process.[121] Apparently, the decision of whether to fuel the jet engines from the airplane's fuel tanks or from an external fuel source was late, and in the process a tunnel technician had begun to reroute incoming fuel lines when the decision to fuel from onboard tanks was made, resulting in an undetected fuel leak. The technician was known thereafter as "Hacksaw" Smith.

Sikorsky YR-4B Helicopter

The first military production helicopter was designed by Igor Sikorsky as a two-place three-blade configuration, and it made its first flight in 1942. Built under contract to the Army, 27 model YR-4B helicopters were built for developmental evaluations, and 7 were delivered to the Navy with the designation HNS-1.

At the request of the Army, a YR-4B was tested in the Full-Scale Tunnel for 3 months beginning in July 1944.[122] The investigation began with an evaluation of the static-thrust performance of six different rotor blade designs that differed in surface condition, pitch distribution, airfoil section, and planform geometry.[123] The results of the preliminary tests indicated that surface condition was a major factor in rotor performance. The production rotor blades of the YR-4B had a radius of 19 feet and were constructed of a tubular steel spar to which 36 wooden ribs were attached. Spruce strips were used to contour the forward airfoil shape, a wire cable formed the trailing edge, and the entire blade was fabric covered.

Since the balance and support system of the Full-Scale Tunnel could not be adapted due to flexibility, they were bypassed with braces, and three flexible six-component strain-gaged members were used at the top of the support struts. According to Frederic Gustafson, "The behavior of the mounted helicopter was not such as to soothe the nerves."[124]

Test installation for the Army YR-4B helicopter in the Full-Scale Tunnel in October 1944. The technician in the foreground is adjusting a high-speed camera system for photos of the rotor deformations during forward flight. Special strain-gage mounts were used to bypass the normal mounts and reduce vibrations. This helicopter was subsequently flown in Langley research programs by Jack Reeder, who had transferred from the Full-Scale Tunnel to become a test pilot. (NACA LMAL 40416)

Perhaps the most impressive result gathered in the study was disclosed by extensive high-speed photographs of the production rotor blades during simulated forward flight conditions.[125] The photos showed extensive fabric sagging and bulging caused by the centrifugal forces acting on the mass of air enclosed by the blade.[126]

The pioneering effort of this tunnel test cannot be overstated. Former staff member Don D. Davis recalled that "[a] visiting Navy admiral saw the YR-4B mounted in the tunnel during an annual inspection (NACA Engineering Conference) and asked me whether I thought it might ever be possible to land a helicopter on a battleship!"[127]

Curtiss SC-1 Seahawk

The Curtiss SC-1 Seahawk design was stimulated by a request for proposals for a new scout seaplane issued by the Navy's Bureau of Aeronautics. The Seahawk was equipped with main and wingtip floats and a four-blade propeller designed to absorb the engine power in the limited diameter allowed by the main float. Armed with two machine guns, it also had full-span leading-edge slats and partial-span trailing-edge flaps. The main float was outfitted with

Curtiss SC-1 Seahawk scout mounted for drag cleanup, stability and control, and cooling tests in December 1944. (NACA LMAL 41442)

a bomb bay that could also be used for additional fuel for long-range missions. The first flight of the prototype XSC-1 occurred in February 1944, followed by 577 production aircraft.

In November 1944 an SC-1 was tested in the Full-Scale Tunnel for drag cleanup and to determine its stability and control characteristics, the critical speed of the cowling lips and the canopy, and pressure losses in the cooling system. Supplementary tests were also made to determine the maximum lift and stalling characteristics of the SC-1.[128] Although the SC-1 was adaptable to land-based landing gear, the tunnel tests were made for the water-landing configuration, and the test program included unpowered and power-on test conditions. The Seahawk is widely regarded as the best U.S. floatplane of the war, although it entered service too late (October 1944) for extended action.

Results of the drag cleanup tests revealed that the top speed of the airplane could be increased by about 21 mph through relatively minor modifications to detailed design items. Sealing gaps, especially at the wing-fold joints, proved to be the most beneficial modification. The wing-fold gap was relatively large and caused premature stalling of the wing. Although the high pressures desirable for cooling were present at the inlets of the cooler ducts, the pressures were not recovered at the coolers because of large losses in the inlet ducts. The characteristics of flow in the oil-cooler duct and the intercooler system could be dramatically improved with a controllable duct-exit flap.

The powerful Grumman XF8F-1 airplane in the faired condition for drag cleanup tests in the Full-Scale Tunnel in January 1945. A dorsal fin was also tested and later added to the configuration. (NACA LMAL 41866)

Grumman XF8F-1 Bearcat

The Grumman F8F Bearcat was the last of the famous Grumman line of piston-engine carrier-based fighters. In 1943, Leroy Grumman led a company team that was invited to England to see and fly captured German aircraft.[129] Their evaluation of the Focke-Wulf FW 190 fighter was impressive and stimulated the design philosophy for a successor to the Grumman F6F Hellcat using the same engine, but it was 20 percent lighter and smaller than the F6F. The performance of this highly maneuverable fighter would provide Navy

A view of the XF8F-1 Bearcat on the Langley flight line after the tests in the Full-Scale Tunnel shows the dorsal fin added to the vertical tail. (NACA LMAL 42397)

pilots with a considerable advantage over the emerging improved Japanese fighters near the end of the war. Design efforts began immediately upon the return of the Grumman team.

Two XF8F-1 prototypes were built with first flight occurring on August 21, 1944. With its powerful engine and light weight, the Bearcat had a rate of climb that was twice that of the Hellcat. Deliveries of production aircraft began in February 1945 with a Navy contract calling for over 2,000 aircraft.

On the last day of December 1944, month-long tests of the first XF8F-1 in the Full-Scale Tunnel began with drag cleanup studies.[130] Additional testing included determinations of the maximum lift of the airplane, its aerodynamic stalling characteristics, the pressure losses in the cooling systems, the critical airframe locations for onset of compressibility effects, and the effectiveness of aileron-tab combinations.[131]

The cleanup process for the Bearcat was extremely thorough and included analyses of drag contributions from engine seals and sealing the landing gear doors, gun-compartment doors, baggage doors, and gaps at the wing-fold joint. Revisions to the elevator hinge gap, canopy shape, tail cone shape, and tail-wheel door were also evaluated. The predicted increase in top speed resulting from the modifications would be about 17 mph. The report gave a cautionary comment regarding potential effects of compressibility on top-speed estimates at the high cruise speeds of the XF8F-1.

The Bearcat was the first Navy fighter with a bubble canopy, which provided an opportunity to gather design data for canopy loads estimation procedures for this type of canopy under a general NACA-Navy research program that included testing of the canopy loads for the F6F, F8F, and SB2C aircraft. The canopy loads testing for the XF8F-1 consisted of pressure measurements conducted during a second tunnel entry in January 1946.[132]

Kaiser Cargo Wing

During World War II, when merchant shipping fleet lifelines to Britain were being ravaged by the German U-boat threat, industrialist Henry Kaiser proposed building large cargo-carrying flying wings capable of transatlantic flight.[133] The Kaiser Cargo Wing design was to feature four piston engines located in the front center section of the wing driving 15-foot-diameter four-blade propellers, with four vertical fins located behind the four engines. The cockpit was to be located atop the center section of the wing, and the cargo was to be distributed spanwise. The aircraft's wing used NACA laminar-flow airfoils and had a span of 290 feet.

A large, wooden, 1⁄7-scale powered model of the Kaiser Cargo Wing was built by Kaiser Cargo, Inc., and tested in the Full-Scale Tunnel in March 1945 at the request of the Navy's Bureau of Aeronautics.[134] Prior to the tests, a small, 1⁄60-scale model of the configuration had been flown satisfactorily in free flight in the Langley Free-Flight Tunnel, but it was deemed advisable to obtain aerodynamic data at a higher scale. Power to the four-blade propellers (which rotated in the same direction) was supplied by four electric motors.

The tests were designed to assess the general characteristics of the airplane and to predict its stability and control qualities. During the tunnel tests, the model was tested at various power-on conditions and the effects of elevator, rudder, and aileron deflections were studied as well as the stalling characteristics and wing profile drag (assessed through wake profile

Model of the radical Kaiser Cargo Wing in powered tests. (NACA LMAL 43276)

survey analysis). The data indicated that the model was longitudinally, laterally, and directionally stable for most conditions, but as expected, the rudder effectiveness and directional stability of the model were much lower than those of conventional aircraft. The aerodynamic results of the tests were used in a brief analysis of the flying qualities of the airplane.[135]

The model was then modified to permit a comparison of tailless and twin-boom tailed versions of the design. The model was modified by inverting the wing (the original tailless design had wing reflex at the trailing-edge sections), removing the vertical tails, and installing a twin-boom tailed configuration. Tested in late 1946 and early 1947 for the same power and attitude conditions in the tunnel, the twin-boom version of the model exhibited higher elevator effectiveness, as was expected.[136]

The Kaiser Cargo Wing concept eventually lost out to the famous "Spruce Goose" giant flying boat design that was developed by Howard Hughes in cooperation with Kaiser and test flown in 1947. The Kaiser Cargo Wing concept never extended beyond model tests.

End of an Era

By the end of the war, the Langley Full-Scale Tunnel was recognized for its tremendous contributions to the Nation's war efforts and as one of the most valuable investments ever made by the United States. The leadership of DeFrance, Dearborn, Silverstein, and Wilson was internationally recognized and characterized by exceptional technical expertise and dedication. Teamed with gifted test engineers and a support staff of men and women with exceptional skills, they responded to critical national needs at levels far beyond expectations. By participating in numerous studies of different airplanes, the organization had gained an immense amount of experience that was directly useful in subsequent applications. Although the fundamental research activities envisioned for the tunnel before the war were minimized by the aggressive engineering efforts required to anticipate and mitigate practical problems for the legendary military aircraft of the day, the Full-Scale Tunnel assumed an indelible position in the world of aeronautical engineering.

Over 30 aircraft of all types, including fighters, scout planes, torpedo bombers, an attack bomber, and engine/nacelle/wing combinations, had undergone drag cleanup, cooling tests, and stability and control studies. Although many of the recommendations made as a result of the testing in the tunnel could not be accommodated because of unacceptable aircraft delivery delays during those days of urgency, the information provided options for design trades and enhanced performance.

Perhaps the most notable comments on the importance of what happened at the Full-Scale Tunnel were made by Captain Walter S. Diehl, the famous leader of the Navy's Bureau of Aeronautics from 1918 until 1951:

> Most people think that the drag cleanup work at the Full-Scale Tunnel occurred during the war, but the work had started before the war with the Grumman F3F in 1938. When its flight performance didn't live up to our expectations we put it in the tunnel and they found the problem inside of 15 minutes. When they raised the carburetor cowl opening 4 inches out of the fuselage boundary layer the problem was solved. Just little things like that were learned—those little details. They can say that's not basic research, but that's another point.
>
> They saved us. We would have been lost completely if they hadn't fixed up the F4F (Wildcat). The F4F was floundering around about 280 mph when Leroy Grumman came into my office and talked about how Langley's drag cleanup work for the Brewster Buffalo provided the guidance that made the airplane the Navy's first 300-mph fighter. Grumman asked me to do something at Langley for the F4F problem. Within a week we had the airplane in the Full-Scale Tunnel and after a few weeks they had the test completed. Results indicated we could pick up as much as 45 mph. Within weeks we had the F4F up to about 320 mph and we got the airplanes out to Guadalcanal. If they hadn't done this for the Wildcat, if they hadn't done this on the F6F (Hellcat) and the F4U (Corsair) we would've been in big trouble. They paid for themselves a thousand times over.[137]

Endnotes

1. Jean A. Roche, "The Career of an Airplane Designer" (unpublished memoirs, 1969), author's collection.
2. Greene flew as copilot of the B-24 in the ditching experiment.
3. Walter Diehl, interview by Michael D. Keller, September 12, 1967, LHA.
4. Elizabeth A. Muenger, *Searching the Horizon: A History of Ames Research Center 1940–1976* (Washington, DC: NASA SP-4304, 1985).
5. When he arrived at Ames with 30 others, the first thing DeFrance did was to visit all the aircraft companies and explain, "We are here for you. This is our purpose—we aren't even going to attempt research. We're here to help you apply to military aircraft our research findings. Our research is going to have to take a back seat to that task." See Russ Robinson, interview by Walt Bonney, September 24, 1974, LHA.
6. Mark R. Nichols, interview by Walter Bonney, March 29, 1973, LHA.
7. Roche, "Career of an Airplane Designer," p. 26.
8. Virginia P. Dawson, *Engines and Innovation: Lewis Laboratory and American Propulsion Technology* (Washington, DC: NASA SP-4306, 1991). AERL was subsequently renamed the NACA Lewis Laboratory, then the NASA Lewis Research Center, and is currently the NASA Glenn Research Center.
9. A drawing of the physical research laboratory in the Full-Scale Tunnel building is available in Langley's Electronic Drawing Files as drawing 5004-1-D.
10. John V. Becker, interview by author, March 25, 2011. Although Becker was 97 years old at the time, his recollection of the 8-Foot Tunnel accident was vivid and detailed. See also John V. Becker, *The High-Speed Frontier* (Washington, DC: NASA SP 445, 1980), p. 26.
11. Formal Langley documents on changing the blade material in the Full-Scale Tunnel were not found during searches in the Langley archives; however, Becker's perspective of the events seems plausible.
12. Robert I. Curtis, John Mitchell, and Martin Copp, *Langley Field, The Early Years 1916–1946* (Langley AFB, VA: Office of History, 4500th Air Base Wing, 1977), p. 141.
13. Review of Accident at Full-Scale Tunnel. May 15, 1942. National Archives at College Park, MD. Record Group 255, Box 119 Folder 21-26 Full-Scale Tunnel.
14. Ken Peirpont, interview by author, March 15, 2012.
15. NACA, *Twenty-First Annual Report of the NACA* (November 12, 1935), p. 1.
16. ONERA, "De l'aérostation à l'aérospatial: Le Centre De Recherche De L'ONERA à Meudon" (2007).
17. Smith DeFrance, interview by Walter Bonney, September 23, 1974, LHA.
18. Muenger, *Searching the Horizon.*
19. DeFrance, interview by Bonney, September 23, 1974. DeFrance commented that the new tunnel would be a closed-throat tunnel because—based on his experiences with the Langley Full-Scale Tunnel—large, open-throat tunnels were very difficult to design satisfactorily.

20. Abe Silverstein and F.R. Nickel, "Preliminary Full-Scale Wind-Tunnel Investigation of Wing Ducts for Radiators," NACA Confidential Memorandum Report for Manufacturers, files of the Flight Dynamics Branch, NASA Langley Research Center (1938).
21. C.H. Dearborn, "Full-Scale Wind-Tunnel Test of Navy XF2A-1 Fighter Airplane," NACA Confidential Memorandum Report for the Bureau of Aeronautics, Navy Department, files of the Flight Dynamics Branch, NASA Langley Research Center (1938).
22. Video clips of airflow studies over the XF2A-1 in the Full-Scale Tunnel tests are available at *http://www.youtube.com/watch?v=wpst_iOlhzU,* accessed January 3, 2012.
23. William Green and Gordon Swanborough, "Brewster's Benighted Buffalo," *Air Enthusiast Quarterly* no. 1 (1974): pp. 66–83.
24. George R. Inger, "The Supercritical Peanut: an Unsung Pioneer in Compressible Aerodynamics," AIAA Paper 2003-0288 (2003).
25. Becker, *The High-Speed Frontier*, p. 88.
26. G. Merritt Preston and Harold H. Sweberg, "Clean-Up Tests of the XF2A-2 Airplane in the NACA Full-Scale Tunnel," NACA Confidential Memorandum Report for the Bureau of Aeronautics (1941).
27. K.R. Czarnecki and Eugene R. Guryansky, "Tests of Machine-Gun and Cannon Installations in the NACA Full-Scale Tunnel," NACA Memorandum Report for the Bureau of Aeronautics (1941).
28. John P. Reeder and Eugene Migotsky, "Full-Scale-Tunnel Tests of a Brewster F2A-2 Airplane Equipped with Full-Span Slotted Flaps," NACA Memorandum Report for the Bureau of Aeronautics (1942).
29. See John G. Lowery, "Power-Off Wind-Tunnel Tests of the ⅛-Scale Model of the Brewster F2A Airplane," NACA WR L-543 (1941); Joseph W. Wetmore and Richard H. Sawyer, "Flight Tests of F2A-2 Airplane with Full-Span Slotted Flaps and Trailing-Edge and Slot-Lip Ailerons," NACA WR 3L07 (1943).
30. James R. Hansen, *Engineer in Charge: A History of the Langley Aeronautical Laboratory, 1917–1958* (Washington, DC: NASA SP-4305, 1987), pp. 194–196.
31. Roy H. Lange, "A Summary of Drag Results From Recent Langley Full-Scale Tunnel Tests of Army and Navy Airplanes," NACA Report ACR No. L5A30 (1945).
32. Paul L. Coe, Jr., "Review of Drag Cleanup Tests in the Langley Full-Scale Tunnel (From 1935 to 1945) Applicable to Current General Aviation Airplanes," NASA TN D-8206 (1976).
33. Harry J. Goett and W. Kenneth Bullivant, "Full-Scale Wind-Tunnel Tests of Navy F3F-2 Fighter Airplane," NACA Confidential Memorandum Report for Bureau of Aeronautics (1938). Video clips of wool-tuft airflow studies over the F3F-2 in the Full-Scale Tunnel tests are available at *http://www.youtube.com/watch?v=9Y9MhshtMZ4,* accessed January 3, 2012.
34. James A. White and Beverly G. Gulick, "Full-Scale Wind-Tunnel Tests of the Navy XF4F-2 Airplane," NACA Confidential Memorandum Report for Bureau of Aeronautics (1938).

35. Harry J. Goett and W. Kenneth Bullivant, "Full-Scale Wind-Tunnel Tests of the XF4F-3 Airplane," NACA Confidential memorandum for Bureau of Aeronautics (1939).
36. Video clips of smoke-flow-visualization tests of the XF4F-3 are available at *http://www.youtube.com/watch?v=RJglrMEVfKQ,* accessed January 6, 2012.
37. J.S. Davidsen and W.H. Harries, "Full-Scale Tunnel Tests of the Navy SB2U-1 Airplane with Dive Flaps," NACA Confidential Memorandum Report for Bureau of Aeronautics (1938).
38. Carl Babberger, "Full-Scale Wind-Tunnel Tests of the XF5F-1 Full-Scale Model," NACA Confidential Memorandum Report for the Bureau of Aeronautics (1938).
39. W. Kenneth Bullivant and J.S. Davidsen, "Test of the XBT-2 Airplane In the Full-Scale Wind Tunnel," NACA Confidential Memorandum Report for Bureau of Aeronautics (1939).
40. J.S.W. Davidsen, "Measurement in the Full-Scale Wind Tunnel of Rudder Pedal Forces on the XSBA-1 Airplane," NACA Confidential Memorandum Report for the Bureau of Aeronautics (1939); see also Ferril R. Nickle, "Tests of the XSBA-1 Airplane in the Full-Scale Wind Tunnel," NACA Confidential Memorandum Report for the Bureau of Aeronautics (1939).
41. C.H. Dearborn, Abe Silverstein, and J.P. Reeder, "Tests of XP-40 Airplane in NACA Full-Scale Tunnel," NACA Confidential Memorandum Report for Army Air Corps Material Division (1939), p. 35.
42. George Gray, *Frontiers of Flight* (New York: Alfred A. Knopf, 1948); see also Francis H. Dean, *America's Hundred Thousand* (Atglen, PA: Schiffer Publishing, 1997).
43. Harold H. Sweberg and Richard C. Dingledein, "Effects of Propeller Operation and Angle of Yaw on the Distribution of the Load on the Horizontal Tail Surface of a Typical Pursuit Airplane," NACA Wartime Report ARR No. 4B10 (1944). A second report by Sweberg and Dingledein entitled "Some Effects of Propeller Operation on the Distribution of the Load on the Vertical Tail Surface of a Typical Pursuit Airplane," NACA WR No. 4C13 (1944), presented the vertical tail pressures measured in the tests.
44. A video sequence showing in-flight tuft-flow studies of the P-40K at Langley is available at *http://www.youtube.com/watch?v=4Hx_x-Hdwq8*, accessed January 3, 2012.
45. Herbert A. Wilson, Jr., "Tests of the XP-41 Airplane in the NACA Full-Scale Wind Tunnel," Confidential Memorandum Report for the Army Air Corps Material Division (1939).
46. The specification called for a single-engine pursuit plane and a twin-engine pursuit plane. The twin-engine airplane competing with the Bell XP-39 was the Lockheed XP-38.
47. Hansen, *Engineer in Charge*, pp. 198–202.
48. Abe Silverstein and F.R. Nickle, "Tests of the XP-39 Airplane in the NACA Full-Scale Wind-Tunnel," NACA Confidential Wartime Report for Army Air Corps (1939).
49. The "navalized" version XFL-1 was a "tail dragger" and did not use the tricycle landing gear. It also did not have a turbo-supercharger. The tunnel test included measurements of drag due to an arresting hook and wing-mounted cables for flotation devices.

50. Video clips of wool-tuft flow-visualization studies of the XP-39 in the Full-Scale Tunnel are available at *http://www.youtube.com/watch?v=vKxe8jc9_rI*, accessed January 5, 2012.
51. Silverstein and Nickle, "Tests of the XP-39," p. 56.
52. See Hansen, *Engineer in Charge*; and Birch Matthews, *Cobra!: Bell Aircraft Corporation 1934–1946* (Atglen, PA: Schiffer Military/Aviation History, 1996).
53. John P. Reeder, and William J. Nelson, "Tests of the XP-39B Airplane in the NACA Full-Scale Wind Tunnel," Memorandum Report for Army (March 16, 1940).
54. F.R. Nickle and W.J. Nelson, "Tests of the XP-46 Airplane in the NACA Full-Scale Tunnel," NACA Confidential Memorandum Report for the Army Air Corps Material Division (1940).
55. Video clips of flow-visualization tests of the XP-46 model in the Full-Scale Tunnel are available at *http://www.youtube.com/watch?v=IBif2U6nFec*, accessed January 5, 2012.
56. Herbert A. Wilson and R.R. Lehr, "Full-Scale Wind-Tunnel Tests of the ⅕-Scale Model of the Revised XPB2M-1 Flying Boat," Confidential Memorandum Report for the Bureau of Aeronautics (1940).
57. Abe Silverstein and Eugene R. Guryansky, "Development of Cowling for Long-Nose Air-cooled Engine in the NACA Full-Scale Tunnel," NACA Wartime Report L-241 (1941).
58. Unfortunately, technical reports on the P-36 studies could not be found.
59. Herbert A. Wilson and J.P. Reeder, "Engine Cooling and Stability Tests of the XSO3C-1 Airplane in the Full-Scale Wind Tunnel," NACA Confidential Memorandum Report for the Bureau of Aeronautics (1941).
60. Herbert A. Wilson, Jr., and John P. Reeder, "Engine Cooling and Clean-up Tests of the A-20A Airplane in the NACA Full-Scale Wind Tunnel," NACA Confidential Memorandum Report for the Army Air Corps (1941).
61. Reports on various topics for the XP-47 included W.J. Nelson and K.R. Czarnecki, "Wind-Tunnel Investigation of Carburetor-Air Inlets," NACA Advance Restricted Report (1941); and H.R. Pass, "Wind-Tunnel Study of the Effects of Propeller Operation and Flap Deflection on the Pitching Moments and Elevator Hinge Moments of a Single-Engine Pursuit-Type Airplane," NACA Memorandum for Files (1942).
62. Eugene R. Guryansky, "Stability and Control Measurements on the P-47B Airplane in the NACA Full-Scale Tunnel," NACA Memorandum Report for Files (1942).
63. Richard Dingledein, "Full-Scale Tunnel Investigation of the Pressure Distribution Over the Tail of the P-47B Airplane," NACA Wartime Report ARR No. 3E25 (1943).
64. Theodore Theodorsen and Arthur A. Reiger, "Vibration Surveys of the P-47B Rudder and Fin-Rudder Assembly," NACA Wartime Report L-653 (1943).
65. Recall that the winner of the contest was Weick's W-1 design.
66. John P. Reeder and Gerald W. Brewer, "NACA Full-Scale Wind-Tunnel Test of Vought-Sikorsky V-173 Airplane," NACA Memorandum Report for Bureau of Aeronautics (1942).
67. Roy H. Lange, Bennie W. Cocke, Jr., and Anthony J. Proterra, "Langley Full-Scale Tunnel Investigation of a ⅓-Scale Model of the Chance Vought XF5U-1 Airplane," NACA Research Memorandum No. L6119 for the Bureau of Aeronautics (1946).

68. Roy H. Lange and Huel C. McLemore, "Static Longitudinal Stability and Control of a Convertible-Type Airplane as Affected by Articulated- and Rigid-Propeller Operation," NACA TN-2014 (1950).
69. Eugene R. Guryansky and G. Merritt Preston, "Full-Scale Wind-Tunnel Investigation of Buffeting and Diving Tendencies of the YP-38 Airplane," NACA Memorandum Report for Army Air Corps (1942).
70. G. Merritt Preston and Eugene R. Guryansky, "Drag Analysis of the Lockheed YP-38 Airplane," NACA Memorandum Report for Army Air Corps (1942).
71. Hansen, *Engineer in Charge*, p. 251. Even today, controversy still exists over whether Langley, Ames, or Lockheed (or a team thereof) pioneered the dive recovery flap.
72. Robert R. Lehr, George F. Klinghorn, and Eugene Guryansky, "Cooling Investigation of a B-24D Engine-Nacelle Installation in the NACA Full-Scale Tunnel," NACA Memorandum Report for Army Air Forces (1942).
73. Abe Silverstein and Herbert A. Wilson, Jr., "Cooling in Cruising Flight with Low Fuel-Air Ratios," NACA Memorandum Report for Army Air Forces (1942).
74. Abe Silverstein, "Test of Propeller-Speed Cooling Blowers," NACA Restricted Bulletin (1942).
75. P. Kenneth Pierpont, interview by author, December 12, 2010.
76. Pierpont could not remember the specific North African mission flown, but the Battle for El Alamein and bombing of Benghazi harbor in Libya took place from July 1942 to November 1942.
77. Valve materials used at the time lacked current-day robustness at elevated temperatures.
78. Pierpont, interview by author.
79. Abe Silverstein and George F. Klinghorn, "Improved Baffle Designs for Air-Cooled Engine Cylinders," NACA Advance Restricted Report #H16 (1943); see also George F. Klinghorn and William A. Mueller, "Investigation of Methods of Reducing the Temperature Variation Among Cylinders on Air-Cooled Aircraft Engines," NACA Wartime Report L-640 (1943).
80. John P. Reeder and William J. Biebel, "Test of Grumman XTBF-1 Airplane in the NACA Full-Scale Tunnel," NACA Memorandum Report for Bureau of Aeronautics (1942).
81. Video clips of wool-tuft flow-visualization tests of the XTBF-1 in the Full-Scale Tunnel are available at *http://www.youtube.com/watch?v=XGCVHLXxyZs*, accessed January 7, 2012.
82. Harold H. Sweberg, "Stability and Control Test of a ¾-Scale Model of the XP-69 Airplane in the NACA Full-Scale Tunnel," NACA Memorandum Report for the Army Air Forces (1942).
83. David Bierman and Edwin P. Hartman, "Wind-Tunnel Test of 4- and 6-Blade Single- and Dual-Rotating Propellers," NACA Report 747 (1942).
84. Harold H. Sweberg, "Air-Flow Surveys in the Region of the Tail Surfaces of a Single-Engine Airplane Equipped with Dual-Rotating Propellers," NACA Memorandum Report for Army Air Forces, Materiel Command (1942).
85. William J. Biebel, "Full-Scale Tunnel Test of a Flying Model of the Curtiss XP-55 Airplane," NACA Memorandum Report for Army Air Forces, Materiel Command (1943).

86. Richard C. Dingledein, "Full-Scale Tunnel Measurements of the Pressures on the Elevator and Fuselage of the Curtis XP-55 Airplane," NACA Wartime Report L-630 (1943).
87. Videos of the flow-visualization results for the CW-24 showing the progression of outer wing stall with increasing angle of attack are available at *http://www.youtube.com/watch?v=tnAZwuAXwpc*, accessed May 4, 2011.
88. Eugene Guryansky, "Tests of Vought-Sikorsky F4U-1 Airplane in NACA Full-Scale Tunnel," NACA Memorandum Report for Bureau of Aeronautics (1943).
89. William R. Prince, "Wing-Operation Tests of the Chance-Vought F4U-1D Airplane in the Langley Full-Scale Tunnel," NACA Research Memorandum No. L6H26 for Bureau of Aeronautics (1946).
90. Gerald W. Brewer and Charles H. Kelley, "Test of the Curtiss SB2C-1 Airplane in the NACA Full-Scale Tunnel," NACA Memorandum Report for the Bureau of Aeronautics (1943).
91. Howard E. Dexter and Edward A. Rickey, "Investigation of the Loads on a Conventional Front and Rear Sliding Canopy," NACA Research Memorandum No. L7D04 (1947).
92. NACA Langley, "Herbert H. Hoover Receives Octave Chanute Award for 1948," *Langley Air Scoop* (internal newsletter) 28, no. 7 (July 16, 1948).
93. Jim Rearden, *Koga's Zero: The Fighter That Changed World War II* (Missoula, MT: Pictorial Histories Publishing Company, 1996).
94. For examples, see Hansen, *Engineer in Charge*; and NASA LISAR images EL-1997-00167 and EL-2000-00223, *http://lisar.larc.nasa.gov*, accessed February 14, 2012.
95. P. Kenneth Pierpont, interview by author, October 15, 2010; William J. Block, interview by author, October 1, 2010; Phil Walker, interview by author, November 3, 2010.
96. Ken Pierpont was a member of the test crew that was responsible for making wing-wake measurements behind the Zero using the overhead survey apparatus in the Full-Scale Tunnel.
97. Stefan A. Cavallo, interview by author, January 3, 2011). Cavallo was a test pilot at Langley in 1942 and remembered seeing the Zero during its stay at Langley. His most vivid memory was of seeing the rear fuselage skin of the lightly built Zero "oil canning" when a NACA technician was inside the aircraft installing instrumentation.
98. William H. Phillips and H.W. Garris, "Measurements of Characteristics of Japanese Zero-2 Airplane in Simulated Combat Maneuvers," NACA Memorandum Report for the Bureau of Aeronautics (1943); W.H. Phillips, "Preliminary Measurements of Flying Qualities of the Japanese Mitsubishi 00 Pursuit Airplane," NACA Memorandum Report for the Bureau of Aeronautics (1943). See also W. Hewitt Phillips, *Journey in Aeronautical Research: A Career at NASA Langley Research Center*, NASA Monographs in Aerospace History, no. 12 (November 1998), pp. 63–64.
99. Bennie W. Cocke, Jr., and Roy H. Lange, "Tests of the Grumman F6F-3 Airplane in the NACA Full-Scale Tunnel," NACA Memorandum Report for the Bureau of Aeronautics (1943).

100. Harold H. Sweberg and Richard C. Dingledein, "Summary of Measurements in Langley Full-Scale Tunnel of Maximum Lift Coefficients and Stalling Characteristics of Airplanes," NACA ACR No. L5C24 (1945).
101. Don D. Davis, Jr., and Harold H. Sweberg, "Investigation of Some Factors Affecting Comparisons Of Wind-Tunnel and In-Flight Measurements of Maximum Lift Coefficients for a Fighter-Type Airplane," NACA-TN-1639 (1948).
102. Joseph R. Chambers and Mark A. Chambers, *Radical Wings and Wind Tunnels* (North Branch, MN: Specialty Press, 2008).
103. Edward A. Rickey and James A. LaHatte, Jr., "An Investigation of the Cowling and the Canopy of the Bell XP-77 Airplane in the Langley Full-Scale Tunnel," NACA Memorandum Report for the Army Air Forces MR No. L4K18 (1944). See also K.R. Czarnecki and C.J. Donlan, "Lateral Stability and Control Test of the XP-77 Airplane in the NACA Full-Scale Tunnel," NACA Memorandum Report for the Army Air Forces (1944).
104. Jack N. Neilsen and Lloyd E. Schumacher, "Analysis of the High-Altitude Cooling of the Ranger SGV-770 D-4 Engine in the Bell XP-77 Airplane," NACA Memorandum Report for the Army Air Forces (1943). See also Mark R. Nichols and Arvid L. Keith, Jr., "An Investigation of the Cowling of the Bell XP-77 Airplane in the Propeller Research Tunnel," NACA Memorandum Report for the Army Air Forces (1943).
105. Matthews, *Cobra!: Bell Aircraft Corporation 1934–1946.*
106. Gerald W. Brewer, "Tests of the Northrop MX-334 Glider Airplane in the NACA Full-Scale Tunnel," NACA Memorandum Report for the Army Air Forces (1944).
107. E.T. Wooldridge, "Northrop: The War Years," *http://www.century-of-flight.net*, accessed April 28, 2011.
108. Bennie W. Cocke, Jr., and Claude B. Hart, "Tests of the Bell P-63A Airplane in the NACA Full-Scale Tunnel," NACA Memorandum Report for the Army Air Forces (1944).
109. Charles H. Kelley, "Tests of the North American P-51B Airplane in the NACA Full-Scale Tunnel," NACA Memorandum Report for the Army Air Forces (1944).
110. Ibid., p. 32.
111. Stanley Lipson, Albert Schroeder, and William K. Haggingbothom, Jr., "Wind-Tunnel Investigation of the Aerodynamic and Cooling Characteristics of XB-39 Power-Plant Installation," NACA Memorandum Report No. L4J26 for the Army Air Forces (1944).
112. Howard E. Dexter, "Correlation of the Cylinder-Head Temperatures and Investigation of the Mixture Distribution and Oil-Heat Rejection of the Allison V-3420-11 Engine in the XB-39 Nacelle," NACA Memorandum Report for the Army Air Forces (1946).
113. Frederic B. Gustafson, "History of NACA/NASA Rotating-Wing Aircraft Research, 1915–1970," *Vertiflite* limited edition reprint VF-70 (April 1971), pp. 1–27. The information appeared in a series of articles beginning in September 1970.
114. Eugene Migotsky, "Full-Scale Tunnel Performance Tests of the PV-2 Helicopter Rotor," NACA Wartime Report L-545 (1945).
115. Herbert A. Wilson, Jr., "Review of Rotating-Wing Aircraft Research at the Langley Full-Scale Tunnel-March 1945" (1945), Memorandum for Files.

116. Stanley Lipson, "Static-Thrust Investigation of Full-Scale PV-2 Helicopter Rotors Having NACA 0012.6 and 23012.6 Airfoil Sections," NACA Wartime Report L-749 (1946).
117. Hansen, *Engineer in Charge*, pp. 223–247.
118. Laurence K. Loftin, *Quest for Performance* (Washington, DC: NASA SP-468, 1985), pp. 285–286.
119. Bennie W. Cocke, Jr., and Jerome Pasamanick, "Clean-Up Tests of the Bell YP-59A Airplane in the Langley Full-Scale Tunnel," NACA Memorandum Report No. L5E14 for the Army Air Forces, Air Technical Service Command (1945). The tests in the AWT at Cleveland also focused on the boundary-layer removal ducts and nacelle.
120. Gerald W. Brewer, "Langley Full-Scale Tunnel Stability and Control Tests of the Bell YP-59A Airplane," NACA Memorandum Report No. L5A18 for the Army Air Forces, Air Technical Service Command (1945).
121. Digitized videos showing the YP-59A fire in the Full-Scale Tunnel are available at *http://www.youtube.com/watch?v=XGCVHLXxyZs*, accessed January 6, 2012.
122. Richard C. Dingeldein and Raymond F. Schaefer, "Full-Scale Investigation of the Aerodynamic Characteristics of a Typical Single-Rotor Helicopter in Forward Flight," NACA Report Number 905 (1948).
123. Richard C. Dingeldein and Raymond F. Schaefer, "Static-Thrust Tests of Six Rotor-Blade Designs on a Helicopter in the Langley Full-Scale Tunnel," NACA Wartime Report L-101 (1945).
124. Gustafson, "History of NACA/NASA Rotating-Wing Aircraft Research," p. 21.
125. Richard C. Dingeldein and Raymond F. Schaefer, "High-Speed Photographs of a YR-4B Production Rotor Blade for Simulated Flight Conditions in the Langley Full-Scale Tunnel," NACA Wartime Report L-631 (1945).
126. Scenes of Langley's helicopter research in the late 1940s at the Full-Scale Tunnel and many other Langley facilities are available at *http://www.youtube.com/watch?v=f_dB2TVDgy0*, accessed January 5, 2012.
127. Comments by Don D. Davis in "Full-Scale Tunnel Memories," NASA NP-2010-03-252-LaRC (October 14, 2009), p. 6. This was a special publication for attendees at the closing ceremony for the Full-Scale Tunnel, Langley Research Center.
128. Herbert A. Wilson, Jr., and Stanley Lipson, "Cleanup Tests of the SC-1 Airplane in the Langley Full-Scale Tunnel," NACA Memorandum Report No. L5A31a for the Bureau of Aeronautics (1945); Stanley Lipson, "Measurements of Maximum Lift, Critical Speed, and Duct Pressures of the Curtiss SC-1 Airplane in the Langley Full-Scale Tunnel," NACA Memorandum Report No. L5F26 for the Bureau of Aeronautics (1945); Anthony J. Proterra, "Stability and Control Tests of the Curtiss SC-1 Seaplane in the Langley Full-Scale Tunnel," NACA Memorandum Report No. L5E10a for the Bureau of Aeronautics (1945).
129. Corwin H. Meyer, *Corky Meyer's Flight Journal* (North Branch, MN: Specialty Press, 2006).

130. K.R. Czarnecki and Edward A. Rickey, "Drag Cleanup Tests of the Grumman XF8F-1 Airplane in the Langley Full-Scale Tunnel," NACA Memorandum Report No. L5C14 for the Bureau of Aeronautics (1945).
131. Robert D. Harrington, Bennie W. Cocke, Jr., and Jerome Pasamanick, "Investigation of the Grumman XF8F-1 Airplane in the Langley Full-Scale Tunnel," NACA Memorandum Report No. L5G05 for the Bureau of Aeronautics (1945).
132. Bennie W. Cocke, "Investigation of the Loads on a Typical Bubble-Type Canopy," NACA Memorandum Report No. L7D07 for the Bureau of Aeronautics (1947). Video clips of the canopy tests of the XF8F-1 in the Full-Scale Tunnel are available at *http://www.youtube.com/watch?v=7GW-71CJ6yI*, accessed January 5, 2012.
133. Chambers and Chambers, *Radical Wings and Wind Tunnels*.
134. G.W. Brewer and E.A. Rickey, "Tests of a 1/7-Scale Powered Model of the Kaiser Tailless Airplane in the Langley Full-Scale Tunnel," NACA Memorandum Report No. L6C13 for the Bureau of Aeronautics (1946).
135. Gerald W. Brewer, "An Estimation of the Flying Qualities of the Kaiser Fleetwings All-Wing Airplane From Tests of a 1/7-Scale Model," NACA Research Memorandum No. L6J18 for the Bureau of Aeronautics (1946).
136. Gerald W. Brewer and Ralph W. May, Jr., "Investigation of a 1/7-Scale Powered Model of a Twin-Boom Airplane and a Comparison of Its Stability, Control, and Performance With Those of a Similar All-Wing Airplane," NACA TN No. 1649 (1948).
137. Walter S. Diehl, interview by Michael Keller, September 12, 1967, LHA.

New aircraft design concepts after World War II enabled higher operational speeds but brought challenges in low-speed flight in the areas of lift, stability, and control. A large-scale 45-degree swept-wing model was tested in 1947 to evaluate the effectiveness of wing boundary-layer control on lift. (NACA LAL 59103)

CHAPTER 5
Back to Basics
1946–1957

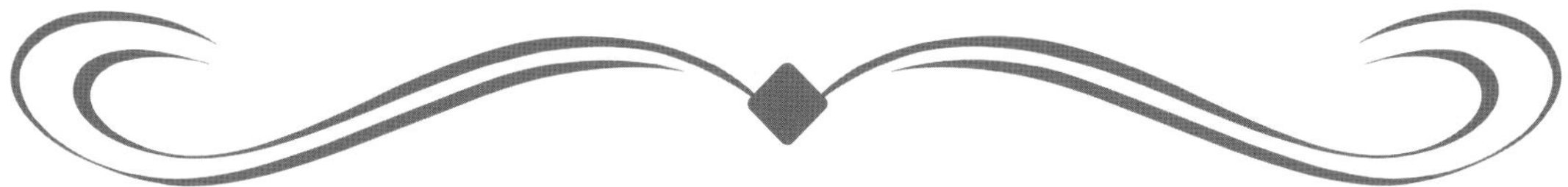

Redirection

The end of World War II dramatically changed the mission and technical activities of the NACA at Langley as well as the other NACA aeronautical laboratories that had been spawned during the war years. The concentrated efforts that had been expended on enhancing and problem solving for specific military aircraft were abruptly terminated, and management and technical organizations began to assess their priorities for more fundamental research on relevant topics of the future. The pinnacle of progress for propeller-driven high-performance aircraft had come and gone, and jet- and rocket-propulsion concepts stimulated a worldwide quest for speed and attacking the sound barrier. Virtually every organization at Langley implemented new facilities and technical programs directed at advancing the maturity level for revolutionary high-speed aircraft of the future.

In mid-1948, the name of the research laboratory was shortened from the Langley Memorial Aeronautical Laboratory to the Langley Aeronautical Laboratory.

At the Full-Scale Tunnel, the centerpiece activities of drag cleanup and engine-cooling testing that had ensured a place for the facility in the history of American aviation suddenly ended as aircraft cruising speeds entered the compressibility area far beyond the capabilities of the old tunnel. However, the emergence of aircraft configuration features necessary for transonic and supersonic flight resulted in new aerodynamic design challenges at the low subsonic speeds associated with the takeoff and landing phases of operation. The advent of swept-back and delta wings, as well as the application of wing convex airfoils with relatively sharp leading edges for efficient high-speed flight, resulted in non-optimum aerodynamic characteristics at low speeds. In particular, the relatively low maximum lift capability of the new swept wings required new concepts and devices for which very little data were available. Also, swept wings and tailless aircraft typically displayed major areas of airflow separation within the operational envelope for low-speed high-angle-of-attack conditions, often resulting in unacceptable stability and control characteristics. Many of the large-scale test activities in the Full-Scale Tunnel during this period involved investigations of concepts such as advanced high-lift devices and boundary-layer control for lift augmentation.

In addition to low-speed research for high-speed aircraft, a second major technical interest at the Full-Scale Tunnel was directed toward the opposite end of the speed spectrum—hovering flight. The emergence of the helicopter as a versatile flying machine at the end of the war had stimulated Langley's role in the evolution of rotorcraft in all aspects of technology, including aerodynamic theory, rotor-blade airfoils, structures and vibratory loads, flying qualities, and experimental aerodynamic testing in flight and in wind tunnels. As discussed in previous chapters, the Full-Scale Tunnel had been a vital component of the early autogiro and rotorcraft research activities at Langley in the 1930s and 1940s, and planning for a more aggressive research program for rotorcraft in the postwar years naturally included considerations of activities in the tunnel. Rotorcraft testing became a major element of the organization's mission, including investigations of rotor aerodynamic characteristics, behavior of single- and tandem-rotor configurations, and the development of theories to predict wind tunnel wall interference effects for rotorcraft. In addition to testing within the test section of the Full-Scale Tunnel, associated testing was conducted in the return passages of the Full-Scale Tunnel for certain studies of hovering flight, and extensive studies of the nature of airflow through rotors were conducted in the 1⁄15-scale model tunnel. By the end of this period, the first emergence of vertical takeoff and landing (VTOL) aircraft concepts stimulated an increased interest in these revolutionary vehicles.

The reduced intensity of tunnel operations and test schedules during the postwar years permitted the first opportunities for testing of unusual and radical civil and military configurations. Unconventional test subjects included airships, submarines, inflatable airplanes, and VTOL aircraft. The versatility of the Full-Scale Tunnel for testing a wide range of designs ultimately became a trademark of the facility for the remainder of its entire life.

Members of the staff also found niche specialties that included the first efforts to develop mufflers for general aviation engines, which culminated in a notable flight demonstration of a "quiet" propeller-driven airplane. In addition, research was conducted by members of the Full-Scale Tunnel staff on an alternate approach to slotted walls to mitigate wind tunnel wall interference effects at transonic speeds using concepts such as porosity.

Historically, the most critical redirection of focus within the programs conducted at the tunnel began in the early 1950s when a series of exploratory tests of dynamically scaled free-flight models proved the feasibility and advantages of conducting such tests within the cavernous tunnel test section. By the end of this era, a major reorganization took place and a new test capability was introduced that would bring the facility more fame and distinction.

New Leaders and Reorganization

As head of the Full-Scale Tunnel Section, Herbert A. "Hack" Wilson, Jr., had provided aggressive leadership for day-to-day experimental studies in the tunnel since 1943, while Samuel Katzoff had directed theoretical studies and analyses within the Full-Scale Analysis Section that focused on experimental data obtained in the tunnel. Both men had recruited and trained a new generation of researchers that conducted test programs in the immediate postwar years. Most researchers regarded Wilson and Katzoff as extremely dissimilar personalities. Wilson was a demanding, authoritative manager with a generally abrasive attitude, while Katzoff was a soft-spoken fatherly figure.

In 1948, Hack Wilson was named by Langley management to head up a Supersonic Facilities Unit for planning Langley's supersonic facilities under the new Unitary Plan Act and was relieved of his position as head of the Full-Scale Tunnel Section. After the Unitary Supersonic Tunnel was constructed and put into operation, he was named chief of the Unitary Plan Wind Tunnel Division at Langley in 1955 and remained there to the end of the laboratory's NACA years. Sam Katzoff was appointed to the position of assistant chief of the Full-Scale Research Division under Clint Dearborn in 1946 while retaining his position as head of the Full-Scale Analysis Section.

Upon Wilson's departure, staff member Gerald W. "Jerry" Brewer was named head of the Full-Scale Tunnel Section on March 1, 1948. Brewer was an "insider" who had advanced through the ranks by accumulating considerable experience in research operations at the tunnel and was well liked by the staff. Brewer's tenure as head of the tunnel came during tumultuous changes in the facility's applications and focus. Basically trained as an aeronautical engineer specializing in fixed-wing aircraft, he had suddenly inherited a program that was rapidly emphasizing rotorcraft and other nontraditional subjects. In addition to managing the operations of the Full-Scale Tunnel, Jerry Brewer was also responsible for research at the unique Langley Helicopter Test Tower, which had been built in 1947 in the West Area of the laboratory for investigations of dynamics and aerodynamics of full-scale rotors. On September 2, 1955, the Full-Scale Tunnel became part of a new organization known as the Boundary Layer and Helicopter Branch under Albert E. von Doenhoff, with Jerry Brewer as assistant branch head. The branch responsibilities included management of the Full-Scale Tunnel and Helicopter Tower.

The most significant reorganization of the Full-Scale Tunnel and its staff during the NACA years occurred in 1950, when Clint Dearborn left his position as chief of the Full-Scale Research Division and transferred to NACA Headquarters in Washington, DC, to serve as assistant to the NACA director. At Langley, Dearborn's position was taken over by his former assistant, Eugene C. Draley, who oversaw a major reorganization of the division. The Full-Scale Tunnel and its staff were included in a new Large-Scale Research Branch that also included the Langley 19-Foot Pressure Tunnel and the Langley 9- by 12-Inch Supersonic Blowdown Tunnel.

Silver Anniversary

In mid-April 1956, the Langley East Area was pounded by a destructive northeaster that resulted in major flooding in the Full-Scale Tunnel. The tides reached 7.9 feet above mean low water and were the highest experienced since September 1936. Despite the cleanup effort, planning continued for a formal event to celebrate the first quarter-century of operations of the facility.

On June 8, 1956, over 200 Langley staff members and guests attended a special dinner party celebrating the silver anniversary of the Full-Scale Tunnel and Langley's Tow Tank No. 1, both of which had been dedicated on May 27, 1931. Langley's director, Dr. H.J.E. Reid, served as the master of ceremonies, and Associate Director Floyd L. Thompson introduced guests and reflected on the history of Langley.[1] Guests included Dr. E.R. Sharp, then director of the NACA Lewis Laboratory, who had been involved in the earliest planning and construction of the tunnel. Assistant Director John Stack presented a 30-inch replica of the NACA 25-year service pin to Gene Draley and Jerry Brewer signifying the remarkable history of the tunnel and their current roles in managing its operations.

Messages were read from those unable to attend, including Smith DeFrance (then director of Ames), Abe Silverstein (then associate director at Lewis), and Clint Dearborn, who had retired as assistant to the NACA director. Dearborn, who expressed a wish that "the Full-Scale Tunnel have many more years of usefulness to the science of aeronautics,"[2] certainly could not have foreseen that the tunnel would continue operations for another 50 years.

Facility Changes

During the postwar years until the formation of NASA, the physical layout and hardware at the Full-Scale Tunnel remained relatively unchanged. The new emphasis on high-lift systems resulted in an extended ground-board installation, and the test observation room that had been built during the war for close-up monitoring and control of testing was moved to a ground-level position near the interior wall, as will be discussed later in this chapter.

Arguably, the most important facility modification that occurred during this time was the implementation of a turntable system for efficiently yawing the test subject. The turntable was a tremendous breakthrough in tunnel operations and was the brainchild of William "Bill" Scallion. After arriving at the tunnel in 1949, he was very unimpressed by the archaic and laborious method used for sideslip testing. Scallion said, "One day head mechanic Joe Walker and I got together and decided we would build a turntable. We went to the Full-Scale Tunnel 'junkyard', found a great big ring gear, and let the engineers do the design and build it. They called it 'Scallion's folly', but I feel real good about that contribution."[3] The turntable remained in the tunnel for over 50 years and permitted efficient testing, especially for semispan wings and for tests requiring large yaw angles.[4]

Former tenants in the building moved to the Langley West Area when new specialized laboratories and buildings were constructed after the war. A new Langley Gust Tunnel was built in 1945 to replace the pilot gust tunnel facility that had operated within the Full-Scale Tunnel building for over a decade, and the catapult and recovery system used there were dismantled.[5] Also in 1945, Theodore Theodorsen spearheaded the movement of his Physical Research Division from the East Area to a new building in the West Area. The office and laboratory areas that had been occupied by his staff while they engaged in noise studies within the Full-Scale Tunnel building were retained and later used by the tunnel's staff during studies of aircraft muffler systems.

Review of Test Activities

The following discussions cover some of the more important test activities conducted in the Full-Scale Tunnel and its building area after World War II and before the birth of NASA.

Lippisch DM-1 Glider

During World War II, the noted German aerodynamicist Dr. Alexander Lippisch had proposed a ramjet-powered delta-wing fighter known as the P.13a. As a first step in bringing this vision to reality, plans were made for the construction and flight evaluation of an unpowered piloted glider for assessments of the low-speed flying qualities of this advanced configuration.[6] The DM-1 fighter was constructed primarily of plywood, with the pilot's cockpit located at the nose of the delta shape. Construction work had proceeded until the DM-1 was captured by American troops at Prien in Bavaria in May 1945.[7] In August, the U.S. military proposed to finish construction and fly the glider by launching it from atop a C-47 transport, but this rash proposal was appropriately stopped. Under the auspices of Operation LUSTY (Luftwaffe Secret Technology), the DM-1 glider and other advanced German aircraft were confiscated by Allied technical intelligence personnel. The construction of the glider was completed in Germany, after which it was crated in a wooden box and transported via ship to NACA Langley in early 1946 for aerodynamic testing in the Langley Full-Scale Tunnel. Following reassembly of the glider, three separate tunnel entries took place in April, June, and November.

The initial objective for the wind tunnel test was to explore the overall low-speed aerodynamic characteristics of the new delta-wing supersonic design, but after preliminary tests were completed, interest shifted specifically to the maximum lift potential of the DM-1.[8] The results of the first tests of the DM-1 in the tunnel indicated that the maximum lift of the glider was considerably lower than values that had been measured in previous small-scale model tests in Germany and in several U.S. wind tunnels. Maximum lift for the DM-1 as received was almost 30 percent less than the maximum lift obtained for models of other delta wings with about the same aspect ratio and similar leading-edge sweep (60°). In addition, the DM-1 wing attained maximum lift at an angle of attack of about 18°, whereas the subscale models typically did not reach maximum lift until angles of attack of about 40°. As a result of the poor correlation, the NACA test program was focused on understanding why the lift was so low and how to improve the capabilities of the DM-1. The test was of special interest to industry design teams contemplating the use of delta-wing configurations for supersonic aircraft.

During the Full-Scale Tunnel tests, several modifications were made to the glider in an effort to increase maximum lift, and extensive flow-visualization tests were conducted using wool tufts. The aircraft had been designed with an airfoil section similar to the NACA 0015-64 section, with a rather bulbous nose shape. As received, the glider was equipped with a rudder for directional control and elevons for lateral and longitudinal control. However, the balance gaps on the control surfaces were relatively large. Modifications to the glider included adding a sharp leading edge to the wing semi-span, removing the vertical fin, and sealing the control-balance slots. The full-scale tests were augmented by tests of small delta-wing models having either thick or thin wing sections.[9]

The Full-Scale Tunnel tests of the DM-1 configuration revealed some of the first details of the beneficial effects of vortex flows at high angles of attack. The subscale delta-wing models had exhibited evidence of vortex flows on the upper surface at high angles of attack near maximum lift conditions but the full-scale DM-1 wing did not. This fundamental difference in aerodynamic behavior was attributed to the fact that the large leading-edge radius of the DM-1 wing promoted early wing stall and suppressed the formation of vortex flows. Dr. Sam Katzoff provided a qualitative understanding of the type of vertical flow that should be shed from wings of this planform, and Hack Wilson and J. Calvin Lovell led experimental testing to attain the flow expected.

Full-Scale Tunnel tests of the German DM-1 glider in 1946 included detailed studies of the beneficial effects of vortex flows and provided confidence that sharp-edged delta-wing supersonic configurations could have satisfactory low-speed characteristics. The top row (left to right) shows the DM-1 when captured in 1945 (LMAL 47900) and being uncrated outside the Full-Scale Tunnel. The middle row (left to right) shows the DM-1 uncrated on the shore of the Back River next to the tunnel and reassembled in the Full-Scale Tunnel hangar on January 1946 (LMAL 46853). The bottom row shows tests of the original glider (LMAL 47681) and a radical configuration modification. Note the revised thin vertical tail, faired elevons, faired sharp wing leading edge, and modified cockpit with P-80 canopy for the revised configuration. (LMAL 49146)

Observations of flow conditions on the scale models suggested that a sharp leading edge would fix flow separation at the leading edge and create powerful vortical flows for lift augmentation. When a sharp leading edge that extended halfway across the span was added to the DM-1 wing, the maximum lift coefficient was dramatically increased by about 70 percent, and the angle of attack for maximum lift increased to 31°. Additional modifications that removed the vertical fin and sealed the large control-balance gaps further increased the maximum lift coefficient of the original aircraft by 100 percent. One of the more interesting configurations consisted of a modified version of the DM-1 with a redesigned thin vertical tail, sealed elevon control-balance slots, faired semispan sharp leading edges, and a canopy from a Lockheed P-80 fighter aircraft installed.

The major result of the DM-1 test program was the conclusion that highly swept delta aircraft with wings that have the sharp leading edges or small leading-edge radii considered desirable for supersonic flight might also have acceptable low-speed high-lift characteristics. It was noted, however, that the angles of attack required to produce high lift at low speeds for delta-wing configurations would be considerably greater than those for conventional aircraft. This early Langley recognition and use of vortex flow to increase lift at high angles of attack was an important precursor to the current practice of generating vortex lift with sharp-edged auxiliary lifting surfaces. Examples of today's aircraft using vortex lift include the Lockheed Martin F-16 and the Boeing F/A-18.

The tunnel test entries also assessed the stability and control characteristics of the modified configuration.[10] Results of the stability and control investigation did not reveal major problems and indicated that delta-wing configurations with 60° leading-edge sweep and sharp leading edges could be designed to have acceptable stability characteristics at low speeds.

Results obtained from the analysis of the aerodynamics of the DM-1 glider served as the basis for the Convair XP-92 that was slated for service as a short-range interceptor by the Air Force. Although the XP-92 program was eventually cancelled, the basic delta-wing concept was later used in the experimental XF-92 program that led directly to the development of the highly successful supersonic Convair F-102 Delta Dagger and Convair F-106 Delta Dart designs, which saw service with the Air Force from the 1950s to the 1990s.

Parametric Wing Tests

The concept of using swept wings to delay the onset of compressibility effects was first discussed in 1935 by Germany's Dr. Adolf Busemann at a technical meeting in Italy. The idea was largely considered academic and was disregarded by the meeting attendees. However, the increasing speeds of jet-powered aircraft at the end of the war rekindled interest in reducing high-speed drag, including through wing modifications. In 1945, Langley researcher R.T. Jones independently identified the benefits of swept-back wings for delaying drag rise near transonic conditions. Together with captured German research data on the effects of sweep, U.S. industry quickly incorporated the concept in new aircraft. One of the first adaptions was in the design of the North American Aviation F-86 Sabre Jet.

While providing unprecedented advantages for high-speed flight, swept wings introduced new and challenging problems at low speeds. Designers quickly learned that the lack of aerodynamic data for swept wings at low-speed, high-angle-of-attack conditions was critical, and recommendations for more detailed wind tunnel studies were made by NACA advisory groups. At Langley, several wind tunnels concentrated efforts in response to the requests and the Full-Scale Tunnel began a major program to provide low-speed high-lift concepts for applications to high-speed wing shapes such as sweptback wings, delta wings, and other composite shapes. During this 9-year period, over half of the tunnel entries were devoted to this topic. The major objectives for these test programs included identification of concepts for enhancement of maximum lift while maintaining satisfactory stability and control and providing detailed design data in the form of pressures as well as direct forces and moments. The following discussions briefly review selected projects conducted in the research program during this period.

One of the early investigations in the Full-Scale Tunnel of the low-speed aerodynamic characteristics of transonic and supersonic airplane wings was a 1946 test of a large-scale 45° sweptback wing to evaluate the aerodynamic effects of applying leading-edge and trailing-edge flap configurations.[11] The wing's airfoil consisted of symmetrical circular-arc sections that resulted in a sharp leading edge. The scope of testing also included an evaluation of the effectiveness of several combinations of chordwise fences. Early test results identified the now-well-known longitudinal instability (pitch-up), caused by tip stall, that resulted from spanwise boundary-layer flow, as well as the beneficial effect of mitigating the phenomenon with chordwise fences.

This photograph of the test section of the Full-Scale Tunnel taken in preparation for a NACA conference in May 1947 provides a view of the technical research programs at the time. Boundary-layer control slots can be seen on the wings of a large, generic 45° swept-wing model at the center of the picture, while other test articles include (clockwise from upper left) the DM-1 glider, an unswept supersonic wing with raked tips, the PV-2 coaxial rotor, a small supersonic arrow-wing model, and a 45° generic swept-wing model. (NACA LMAL 53029)

A large number of test subjects were evaluated over several years in the late 1940s, including swept wings, rectangular wings, delta wings, wings with pointed tips, and cranked wings with different inboard and outboard sweep angles.[12] All the models included relatively sharp leading edges representative of supersonic wing configurations, and the test programs typically explored the effects of leading- and trailing-edge flap devices on lift and stability.

As other low-speed swept-wing wind tunnel testing in other Langley tunnels, such as the 300 mph Low-Speed 7- by 10-Foot Tunnel and the Langley Stability Tunnel, progressed, results indicated that the effect of Reynolds number on sharp-edge wings was minimal and that relevant results could be obtained from subscale testing without requiring large-scale tests in the Full-Scale Tunnel. However, the tunnel continued to be an appropriate facility for evaluations of the effectiveness of auxiliary high-lift concepts (such as boundary-layer control through blowing or suction) that could be more easily implemented with large-scale hardware.

High-lift tests of a 45° swept wing with symmetrical circular-arc sections evaluated the effect of leading- and trailing-edge flaps in 1946. (NACA LMAL 50361)

This large-scale arrow wing model with clipped wingtips was one of a series of supersonic wings tested in the Full-Scale Tunnel. (NACA LMAL 57083)

As the period came to an end, the wide-ranging parametric wing studies conducted at Langley provided designers with the detailed data on low-speed aerodynamic behavior of supersonic wings that had been so urgently requested following the war. Detailed charts for design tradeoff studies based on the tunnel results became noted examples of the value of NACA's output and perceived mission.

Boundary-Layer Control

The sudden emphasis on enhancing lift for thin, highly swept, high-speed wings in the late 1940s led to investigations of active boundary-layer control to mitigate flow separation for landing and takeoff conditions. The first major large-scale test in the Full-Scale Tunnel of boundary-layer control using suction was conducted in 1947 for a generic wing/fuselage model with a wingspan of 27.8 ft, a sweep-back angle of 47.5°, and spanwise suction slots at the 0.20-, 0.40-, and 0.70-chord stations on the outboard half of each wing panel.[13] In addition to the active-suction concept, passive concepts including full-span and semispan split flaps and partial-span leading-edge flaps were investigated. Various combinations of suction from the individual and combined ports were assessed, with results indicating that the most effective location for lift augmentation with suction was the 0.20-chord slot.

Large-scale generic wing/fuselage model with spanwise upper-surface suction slots used for investigations of boundary-layer control for enhanced lift in 1947 and 1948. (NACA LMAL 53028)

Following these initial results, suction slots were relocated near the wing leading edge, and the test program focused on the control of leading-edge separation by suction.[14] The slots were located at the 0.015- and the 0.025-chord stations, where the concept was much more effective. In the 1940s, this fundamental information was largely unknown, and the data gathered in the study helped form the foundation for evolving boundary-layer control concepts.

In addition to large-scale tests of aircraft components, the staff at the Full-Scale Tunnel conducted research on critical geometric shapes required for effective suction-slot shapes for boundary-layer control. One such investigation consisted of detailed pressure measurements and mass-flow amounts for three different cross-sectional shapes for suction slots.[15] The versatility and value of the 1⁄15-scale model of the Full-Scale Tunnel was demonstrated during the conduct of this and many other basic flow studies.

Other applications of suction control were also examined during the period, including an investigation of the use of boundary-layer control to avoid trailing-edge flow separation on relatively thick high-aspect-ratio wings. In a 1952 study coordinated between the Full-Scale Tunnel and the Langley Low Turbulence Pressure Tunnel, an aspect ratio–20 wing with full-span suction at the 60-percent chord point on the upper wing surface was used to assess the effects of suction flow rate, slot configuration, flap deflection, and wing-surface conditions on the performance of the boundary-layer control concept.[16] Results of the study demonstrated that trailing-edge separation could be controlled and that lift-drag ratios as high as 30.8 could be attained.)

The use of suction to minimize wing trailing-edge separation for thick high-aspect-ratio wings was investigated using a semispan wing model in May 1952. (NACA LAL-71710)

As boundary-layer control systems matured, hybrid suction/blowing concepts emerged as potentially more efficient active systems for lift augmentation. In 1953, tests were conducted with a large 0.4-scale, 45-foot-span powered model of the Chase C-123 transport configuration to evaluate the effectiveness of a boundary-layer control system that utilized a single pump to suck air in from inboard flaps and discharge the same air over outboard-flap segments and drooped ailerons through a blowing slot.[17] The 3-month test program included exhaustive testing of variables such as suction slot design, flap hinge positions, flap deflections, power effects, asymmetric propeller operation, and asymmetric boundary-layer control operation.

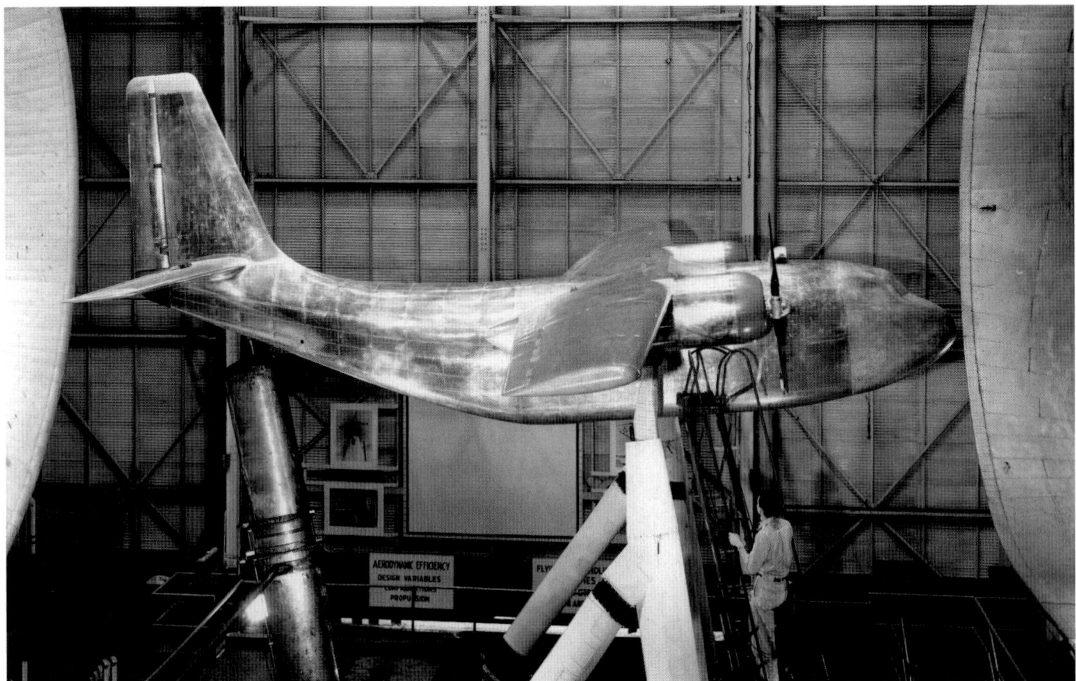

A 2/5-scale powered model of the Chase C-123B airplane was used for studies of boundary-layer control concepts in 1953. (LAL 79949)

Results of the C-123 model tests were impressive. With the boundary-layer control system operating at design flow rates, a maximum lift coefficient of 4.8 (untrimmed) was obtained for full-power propeller operating conditions. The model was stable for all conditions, the elevators were capable of trimming the model for the baseline center of gravity position, and the aileron effectiveness was adequate for control. However, extremely high values of adverse yaw were experienced for large aileron deflections, and the rudder could not provide trim for single-engine asymmetric power.

Many other investigations of boundary-layer control concepts were conducted in the Full-Scale Tunnel in the 1950s covering a wide range of suction, blowing, and hybrid systems. Applications included flapped delta wings with canards and moderately swept 45° wings. The database generated by these test programs served as fundamental information for designers of future aircraft configurations, and especially for the U.S. Navy, which used boundary-layer control to lower approach speeds for several carrier aircraft including the F-4 Phantom.

Helicopters

After successful tests of the PV-2 rotor and the Sikorsky YR-4B helicopter, helicopter-related activities became more frequent in the postwar years.[18] During these 9 years, about 25 percent of the tests conducted in the tunnel were devoted to rotorcraft aerodynamics.

The emphasis of rotor testing in the Full-Scale Tunnel shifted first to basic characteristics of multirotor configurations (coaxial and tandem arrangements) and later to basic explorations of the aerodynamics of single rotors, including blade-pressure distributions.

The PV-2 Accident

The three-blade PV-2 rotor reentered the tunnel for testing with metal blades in December 1947 and remained for 4 months until April 1948. On April 5, while undergoing testing, the rotor system failed, resulting in destruction of the rotor and drive support system. The rotor hub, with two blades still attached, departed the support assembly and embedded itself in the lower exit-cone structure of the tunnel. The third blade struck the upper exit cone, leaving a substantial puncture in the cone liner.

On April 5, 1948, a PV-2 helicopter rotor equipped with metal blades failed during forward-flight tests and the rotor and blades departed from the test assembly. Forced to the rear of the test section by the tunnel air stream, the rotor hub and two blades were embedded in the lower tunnel exit cone (lower left) while the third blade punctured the upper exit cone (upper right). Note the test control house at the rear of the tunnel test section ground plane. Installed in 1943, the house contained test personnel, engine controls, controls for model attitude and control deflections, and extensive instrumentation displays. After this accident, the observation room was removed and a new room with armored glass windows was built at ground level at the side of the tunnel. (LMAL 55849)

As might be expected, the accident prompted a thorough investigation, including recommendations for general test procedures in the Full-Scale Tunnel. Safety concerns over potential injuries to personnel and damage to instrumentation led to relocation of the test controller and observers to a new room at ground level on the east side of the test section.

Following the PV-2 accident, construction began on a new observation room at the base of the east interior wall of the Full-Scale Tunnel. Completed in early 1949, the room was protected by armored glass and retractable window shields. The photograph shows the new room with the four shields lifted. The room to the left in the photo was the office of the head technician, Joe Walker. (LAL 60116)

Rotor Configuration Studies

Rotor configuration studies included two historically significant investigations during this period.[19] A 25-foot-diameter coaxial rotor was tested in 1949 to evaluate the relative efficiency of the arrangement, including the power required for hovering and forward flight, blade motions, flow angles in the rotor wakes, and the static stability of the configuration. The experiment also served to compare experimental results with predictions from available methods using single-rotor theory. Each rotor had two blades, and the rotor system was part of an actual helicopter.

Coaxial research rotor during tests in the Full-Scale Tunnel in February 1949. Note the revised ground plane configuration with the observation room removed following the PV-2 accident the previous year. (LAL 61923)

In the second study, a general research tandem-rotor model was tested in 1951 and 1953. The arrangement had two 15-foot-diameter two-blade rotors and was designed to investigate side-by-side and tandem-rotor arrangements. The rotors could be moved toward each other to mesh the blades and could be offset vertically to cover a range of gap ratios between the rotor planes.

Tandem-rotor research model in the tunnel in March 1953. The model was designed to permit overlapping rotor configurations. (LAL 79077)

The Full-Scale Tunnel testing was augmented with innovative flow-visualization tests conducted in the 1⁄15-scale model tunnel.[20] A method was devised for visualizing the airflow through rotors by use of a balsa-dust particle technique. The method proved to be a simple means of observing flow through multiple-rotor arrangements. One of the first applications was for studies of a 1⁄15-scale model of the coaxial rotor tested in the Full-Scale Tunnel. Although smoke had been used in past studies during attempts to visualize flow through rotors, fine balsa-wood particles were found to provide the best combination of high reflectivity and low mass of any materials investigated. The technique was simple to use and required only a supply of balsa wood, a camera, and photographic lamps. Some of the grains of balsa wood were so fine that they hung almost motionless in the air during static tests. A small electric motor supplied power to the counter-rotating shafts of the coaxial-rotor model, and electric strain gages measured the rotor thrust.

Flow-visualization photographs taken during the studies were extremely informative, highlighting the general character of the wake of rotor interactions and instabilities that occurred within the wakes. The response of the rotor wake to rapid thrust increases was

The NACA balsa-dust flow-visualization technique vividly illustrates the flow through a coaxial-rotor model in the test section of the 1/15-scale model of the Full-Scale Tunnel. (LAL 66043)

observed, and the starting vortex associated with the sudden increase in thrust was easily observed as well. The balsa-dust technique has been noted as a particularly valuable contribution to rotorcraft technology.[21]

Results of the integrated tests for the coaxial and tandem-rotor configurations were of tremendous interest to the industry and the military community. The summary of results included observations that the power requirements for a coaxial rotor in static-thrust conditions could be predicted with good accuracy from available theory, although a coaxial motor required more power in level flight than an equivalent single rotor required. The tandem-rotor arrangement with rotor shafts spaced approximately one rotor diameter apart was found to have greatly improved hovering efficiency. The power requirements for the tandem arrangement in level flight could be predicted fairly well from available single-rotor theory by considering the rear rotor to be operating in the fully developed downwash of the front rotor. These early configuration studies were later augmented by more detailed experiments in the Full-Scale Tunnel and by analyses of the results during the early NASA years.[22]

Rotor in Forward Flight

Historically significant research was conducted in the Full-Scale Tunnel in the 1950s regarding detailed rotor air loads and, in particular, harmonic loads. A series of studies was designed to provide a more quantitative appreciation of rotor-blade loading, including data and correlation with predictions.[23] Detailed studies were conducted of the flow field and the aerodynamic loading on one blade of a two-blade rotor in hovering and forward flight during tunnel tests. Chordwise loading distributions at five spanwise rotor stations were determined during the in-depth study. The successful accomplishment of the test objectives is particularly noteworthy in view of the difficulty of measurements and the sophisticated instrumentation required.

Helicopter Drag Studies

Several wind tunnel entries and performance analyses were conducted by the staff of the Full-Scale Tunnel regarding approaches to reduce helicopter drag.[24] Concern over rotor-hub drag had resulted in tunnel tests of a Hiller rotor in 1946, but the subject was not pursued until the early 1950s when high-speed and long-range capabilities were of critical importance. Drawing on the experiences of efforts during drag cleanup testing in the Full-Scale Tunnel during World War II, researchers examined drag increments contributed by a typical helicopter's landing gear, rotor hub, engine-exhaust stacks, cooling losses, and air leaks through joints in the fuselage. The study concluded that a significant reduction in helicopter parasite drag was possible, but significant trades involving weight penalties for a retractable landing gear or a rotor-hub fairing required consideration.

Hughes XH-17 Helicopter

In addition to aircraft, Howard Hughes developed an interest in helicopters in the late 1940s. He acquired and funded the development of a design for a giant heavy-lift helicopter from Kellett and proceeded to construct a single prototype that made its first flight in 1952. Known as the Hughes XH-17, the helicopter had a rotor diameter of almost 130 feet and used two turbojet engines that sent hot compressed air bled from the engines up through the rotor hub, into hollow rotor blades, and out to tip jets where fuel was injected. The huge, tip-driven rotors atop a spindly-legged fuselage and landing gear gave the vehicle a bug-like appearance.[25]

The unconventional features of the XH-17 were deemed to be of general interest to designers of large helicopters, justifying the construction of a 1⁄10-scale dynamic model of the vehicle by structural and dynamic loads specialists at Langley.[26] Of particular interest to the

A 13-foot-diameter, dynamically scaled, tip-driven rotor was used to power this 1⁄10-scale model of the Hughes XH-17 for investigations of flutter characteristics. The model was tested in the west return passage of the Full-Scale Tunnel. (NACA LAL 66021)

NACA researchers were the flutter and ground resonance characteristics of the large model and its 13-foot-diameter dynamically scaled rotor. The rotor was powered by a compressed-air power supply, and blade parameters such as blade weight, chordwise center-of-gravity location, mobile rotor inertia, and stiffness were all scaled from the full-scale vehicle.

In 1954, researchers mounted the rotor and pylon suspension system of the XH-17 model on a tiltable support platform in the return passage of the Full-Scale Tunnel for flutter testing at various tip-speed-to-free-stream velocity ratios.[27] Power-on testing was conducted using the compressed-air system at the tunnel with the model located 85.5 inches above the return passage floor midway between the outer and inner return-passage walls. The upper section of the rotor support was tilted forward into the wind so that it corresponded to the nominal shaft configuration of the full-scale helicopter in cruising flight. The tunnel was powered up to a test-section speed that resulted in average speeds of about 35 feet per second at the return-passage test location.

Results of the investigation revealed that the rotor speed when flutter occurred was slightly less as the speed ratio was increased from a hovering condition, and that the nature of the flutter motion was changed from a sinusoidal-type oscillation to a random motion of comparable amplitude but without a well-defined frequency. These results were extremely valuable inputs to rotorcraft technology and design in the 1950s.

The use of the gigantic Full-Scale Tunnel return passages for various types of aeronautical and space research activities was a common occurrence during the latter days of the NACA and for the remainder of its lifetime, as will be discussed in following chapters.

Up, Up, and Away

While Langley researchers worked feverishly to mature helicopter technology and provide design information for industry, others in the public and scientific communities began to explore other approaches to achieving vertical and/or short takeoff and landing (V/STOL) capability. Beginning in the early 1950s, an astounding number of unconventional concepts emerged for vehicles offering the versatility of the helicopter at low speeds coupled with the high-speed capability of the airplane. These embryonic efforts would lead to over 20 years of concentrated research on a multitude of aircraft designs for civil and military applications. This topic will be revisited in more depth in the following chapter, which covers the early 1960s when V/STOL research reached its peak at the Langley Full-Scale Tunnel.

Controversy: The Custer Channel Wing

In April 1952, one of the most controversial tests ever conducted in the Full-Scale Tunnel occurred involving the Custer Channel Wing CCW-2 aircraft.[28] After almost 60 years, the results of this particular test and their influence on the potential applications of this novel configuration are still hotly debated within the aviation world.

As part of its interest in new concepts for hovering and low-speed flight, the NACA had begun a basic research program on the use of propeller slipstreams and blowing jets for lift augmentation at low speeds, and innovative fresh ideas were of interest. A request to test an experimental test bed of the channel-wing concept in the Full-Scale Tunnel was accepted by NACA Headquarters.

The novel channel-wing was conceived and developed by Willard R. Custer in the 1920s. The basic principle of his concept was to promote a high-energy stream of air through a semicircular channel in the wing structure to achieve low pressures on the upper surface of the channel and thereby promote lift augmentation at low speeds. Propellers mounted at the rear of the channels provided the accelerated flow through the channels. Use of this type of "powered lift" promised the advantage of increasing lift without complex high-lift devices or active systems such as suction. Although the simplicity of the principle is attractive, numerous issues needed to be addressed, including its aerodynamic performance at low and cruise speeds compared to conventional flapped wings, the profile drag of the channel surfaces, the adequacy of stability and control characteristics at low speeds, engine cross-shafting requirements, and weight penalties.

Custer had evolved his concept by using powered models and a demonstrator aircraft designated the CCW-1, which was built and flown in 1943.[29] After an evaluation by the Army Air Forces led to a negative statement of interest, Custer designed a second vehicle he named the CCW-2, which he intended to be an engineering test bed for further refinements of his concept. Using a Taylorcraft airplane fuselage and tail, the design incorporated 6-foot channels. The CCW-2 first flew in July 1948.

The Custer Channel Wing CCW-2 aircraft was tested in 1952 to evaluate the effectiveness of its unique wing design and its stability and control characteristics. (LAL 75367)

As part of gaining support for his design, in December 1951 Custer conducted a "tethered" test to demonstrate that the CCW-2 could hover like a helicopter. The test setup used a telephone-pole-like structure as an anchor, and tether lines were connected to a truss above the cockpit. Tether lines were also connected to the landing gear from ground stakes. As Custer promoted the capability of the CCW-2 to hover based on this demonstration, experts within Langley regarded the tethered test as unrealistic and a stunt. In their view, the constraints of the tether lines influenced lifting capability and mitigated a pitch-trim issue at high power.

By the time the CCW-2 arrived at Langley for testing, the investigation centered on whether the vehicle had the ability to hover. For the Full-Scale Tunnel tests, the fuselage was covered with fabric and the internal-combustion engines used to power the propellers were replaced by electric motors because the original engines were not adequately lubricated for operation at high angles of attack. In view of concerns over whether the flow characteristics through the channels might have been affected by replacing the original engines with the smaller electric motors, additional tests were made with mockup nacelles around the electric motors.

The focus of the investigation was on measurements of the magnitude and direction of total resultant force on the aircraft for various power conditions at zero airspeed. A brief investigation was also made of the static stability and control characteristics of the vehicle at zero airspeed and speeds of about 26 to 41 mph for a range of angles of attack from –2° to 46°. The static tests (i.e., at zero airspeed) were conducted with the airplane mounted on the tunnel struts out of ground effect and also with the airplane on the ground to investigate ground effect.

Measurements of the resultant force and its inclination relative to the airframe indicated that the inclination of the resultant-force vector was about 23° above the longitudinal axis of the aircraft (the propeller thrust line was only 7° above the longitudinal axis, indicating that lift augmentation had occurred). Based on this result, however, the vehicle would have to be inclined at an angle of about 67° to hover. In addition, a substantial diving pitching moment was measured, which would require a tail force downward for trim, thereby reducing the magnitude and inclination of the resultant-force vector. Flow-visualization for the static flight condition showed that the propeller slipstream was deflected well downward underneath the tail, indicating that a problem would exist in obtaining longitudinal and directional control in hovering flight.

Arguments over the scope and interpretations of the CCW-2 test data continue to this day.[30] Proponents of the channel-wing technology are critical of the NACA test's conclusion that "the increase in lift when power is applied results primarily from the component of propeller thrust in the lift direction,"[31] and of what is, in their opinion, the unjustified emphasis of the test program on hovering-flight conditions. Neutral observers of the channel-wing concept dismiss the hovering-flight arguments as unfortunate and perhaps irrelevant because the concept's major application would be for short-takeoff-and-landing (STOL) operations rather than VTOL missions. For STOL applications, opponents state a number of limitations of the channel wing compared to conventional aircraft, including higher drag because of the additional wing area of the channels; the inability to optimize the wing sections for both low-speed and cruising flight; the increased complexity of cross-shafting required for safety in event of engine failure; and adequacy of stability and control.

The channel-wing concept has reappeared in the research community in recent years, primarily as a result of the efforts of Dennis Bushnell, chief scientist of the NASA Langley Research Center. As a sponsor and stimulus for high-risk, innovative research, Bushnell sponsored a program at the Georgia Tech Research Institute to integrate channel-wing technology with another high-lift concept known as circulation control.[32]

De Lackner Aerocycle

As discussed in the previous chapter, brilliant NACA researcher Charles H. Zimmerman had left Langley in the 1930s to pursue his work in STOL aircraft with Chance Vought concerning the development of the V-173 "Flying Flapjack" and the XF5U-1 fighter. After the war, Zimmerman returned to Langley in 1948 as a senior manager in the Stability Research Division. His continued conceptualization of V/STOL vehicles greatly influenced the direction of technology at NACA and industry.

Zimmerman's stay at Vought was discouraging as progress and interest in the XF5U-1 slowed, and he began to explore a stand-on flying vehicle.[33] For several years, Zimmerman had theorized that a man in a standing position could stabilize and control a small vehicle capable of rising vertically, hovering, and translating to forward flight by his instinctive reflex responses.[34] These responses, which stabilize a person while standing and riding a bicycle, would operate in the proper sense and were referred to as kinesthetic control. The balance of the vehicle was accomplished because the lift is a force vector on which the man can maintain balance. His original proposal was for a single-place vehicle using counter-rotating propellers for lift.

In 1951, Zimmerman and Paul R. Hill of Langley conducted an exploratory study of a rudimentary stand-on flying platform. The platform was rigidly connected to a jet nozzle positioned with its thrust axis perpendicular to the platform and opposed to the pull of gravity. The study was conducted at the NACA Pilotless Aircraft Research Station (now the NASA Wallops Flight Facility) at Wallops Island, VA, because a large capacity compressed-air reservoir was available there. Results of the pioneering work were extremely impressive. Zimmerman himself was the first to fly the apparatus. It was found that a human could stabilize a jet-supported platform with little or no practice, dependent on the ability of the "pilot" to relax and permit instinctive reflexes to operate. The translational motion of the flyer and platform was easily controlled by leaning in the direction toward which motion was desired.[35]

The pioneering jet-platform experiment was followed by an evaluation of the stand-on technique for controlling a more practical rotor-powered vehicle in 1954.[36] Once again, the stability and controllability of a stand-on platform were demonstrated. However, with a teetering-rotor-supported platform, oscillations were noted, particularly in gusty air, resulting in a high level of anxiety for the flyer. The oscillations had not been experienced with the previous jet-supported platform, and further analyses indicated that the sensitivity of rotor-produced moments with airspeed was the cause of the unexpected motions.[37] Other stand-on configurations were tested in the Full-Scale Tunnel in 1954.[38]

In view of the potential military applications for a vehicle that would provide air mobility to individual troops for special operations, the U.S. military industry quickly took notice of the NACA experiments and conducted studies of stand-on flying platforms that resulted in several prototype vehicles. Benson Aircraft, Hiller Aircraft, and De Lackner Helicopters all brought forth versions of the concept.

The vehicle conceived by De Lackner was designated HZ-1 and named Aerocycle. Designed to carry its pilot and up to 120 pounds of cargo, the Aerocycle consisted of a cross-shaped frame with its small platform above a 43-horsepower outboard motor that drove 15-foot-diameter coaxial rotors. The first free flight of the HZ-1 took place in January 1955, and in 1956 the test program for a dozen vehicles was assigned by the Army to Fort Eustis, VA, for further evaluations. During the flight tests at Eustis, the vehicle proved to be more difficult to fly than had been expected, and two accidents were experienced.[39]

The De Lackner Aerocycle stand-on flying platform was tested in 1957 to determine the potential cause of an in-flight accident. The dual two-blade rotors are powered in this photo beneath the clothed dummy "flyer." The top rotor rotated clockwise as viewed from above while the lower rotor rotated counterclockwise. Note the electric-drive motor directly in front of the flyer. The flyer is standing on a six-component strain-gage balance used to measure forces and moments. (NACA LAL 57-2915)

One of the accidents occurred at a nearly maximum level-flight speed, and theories arose over whether the vehicle had become uncontrollable because of a longitudinal-stability problem or whether a collision between the blades of the coaxial rotors had occurred due to blade bending. Full-Scale Tunnel tests of an Aerocycle were requested to measure the overall aerodynamic characteristics of the powered vehicle, to determine its static stability characteristics, and to determine the clearance between the coaxial rotor tips during flight.[40] The vehicle's 40-horsepower reciprocating engine was replaced by a variable-frequency electric motor for the tunnel tests. A clothing-display dummy was installed on the vehicle to simulate aerodynamic effects of a human pilot. In addition to force and moment measurements, motor torque and rotor speeds were acquired and the clearance between the tips of the coaxial rotors was measured by means of an optical-electronic blade-tracking instrument.

The results of the test program, which was conducted during two tunnel entries in June and October 1957, concluded that the forward speed of the complete vehicle would be limited to about 17 knots because destabilizing nose-up pitching moments produced by the coaxial rotors would become greater than the trimming moments available for pilot control. Measurements showed that the tip clearance between the coaxial rotors was never less than about 5 inches for any of the test conditions. A major output of the investigation was a reasonably good correlation of predicted rotor-induced pitching moments with theory.

Versatility: Unconventional Tests

One of the most remarkable historical characteristics of the Langley Full-Scale Tunnel was its versatility for low-speed testing and the wide range of unconventional investigations that were conducted in its test section. Non-aircraft test subjects requiring large-scale aerodynamic measurements were frequent visitors to the tunnel, and unusual aircraft dominated its test schedules, particularly in the 1950s. The following examples show the vast scope of studies conducted in this particular period.

Albacore Submarine

The USS Albacore was a famous nuclear-powered research submarine whose radical teardrop shape revolutionized the design of all future submarines. The Albacore emerged from a 1949 Navy study of an efficient submarine hull for high submerged speeds and enhanced agility. Studies of generic hull forms and complete configurations had been completed at the David Taylor Model Basin and Stevens Institute as part of the Navy program, but the models used in the tests were of small scale, and it was not possible to duplicate details such as double-hull construction, flood- and vent-hole arrangement, and internal compartmenting that would exist on a full-scale submarine. The Navy's Bureau of Ships, therefore, requested tests of a ⅕-scale model of the leading candidate design in the Langley Full-Scale Tunnel to measure the drag and aerodynamic characteristics of a large-scale model that incorporated as many details as possible.[41]

A 2-month test program began in February 1950 with objectives to measure the drag, control effectiveness, and stability characteristics in pitch and yaw for a number of model configurations; obtain measurements of pressures to determine the boundary-layer conditions and flow characteristics over the rear of the submarine configuration, especially in the region near the propeller; and investigate the effects of propeller operation on aerodynamic characteristics. The 30-foot-long model included flood and vent holes placed in the skin of the external hull to replicate the full-scale design, as well as bulkheads between the external and internal hulls. The configuration consisted of the operational hull with a large bridge fairwater and an aft-located cruciform tail arrangement. A four-blade aircraft-type propeller was used for some tests on the stern of the model. The model was mounted on the six-component balance systems of the tunnel using a two-strut mounting system to minimize strut interference.

Charles Zimmerman, who had returned to Langley from Chance Vought, was brought into the study to examine the stability characteristics of the Albacore. He later recalled that the submarine would dive too deeply in response to control inputs and that it would "swoosh" out of the water when a climb command was inputted. If the dive control was commanded, reverse control had to be input quickly to avoid losing control. It was obvious to Zimmerman that the Albacore was unstable, although others claimed that the problem was caused by the fact that the submarine was too fast. The results of the tunnel test revealed that the submarine was unstable in pitch and yaw, validating Zimmerman's hypothesis. As part of his participation in the project, Zimmerman participated in actual submarine dives and evaluated its handling qualities. He then proceeded to write handling-quality requirements for submarines in cooperation with Navy personnel.[42]

Tests of a 1/5-scale model of the developmental version of the Albacore submarine were conducted in 1950. The photo on the left shows the large bridge fairwater, the cruciform tail, and the ballast-tank flow holes on the bottom of the hull. Photo on the right shows a small fairwater shape and the superstructure flood and vent holes in the outer hull. The technician is attaching wool tufts for flow-visualization studies. (LAL 64791 and LAL 64784)

The results of the study indicated that the flooding and venting holes increased the drag of the basic hull by over 60 percent and that the configuration was statically unstable in both pitch and yaw for all configurations tested. Propeller operation had little effect on stability or control effectiveness for forward-located tail surfaces, but a significant improvement in stability and control was indicated for rearward-located tails. Pressure measurements along the hull surface did not indicate any flow separation, but boundary-layer and wake surveys showed that the stern-located propeller was completely immersed in a low-energy wake region.[43]

Following the Full-Scale Tunnel tests, development of the Albacore continued, and the revolutionary warship was launched in August 1953 at the Portsmouth Naval Shipyard in Kittery, ME.[44] The submarine was used for extensive research and development efforts on systems and critical components of submarine configurations for over 20 years. The ship was decommissioned in December 1972, and it is now located at the Port of Portsmouth Maritime Museum in Portsmouth, NH. The Albacore was designated a National Historic Landmark in April 1989.

Return of the Airship

After an absence of over a quarter century, a dirigible configuration returned to the test section of the Full-Scale Tunnel in response to a Navy test request in 1954. In the early 1950s, emerging requirements for the use of airships and antisubmarine patrols required maneuver rates much higher than those stipulated in the past. The high pitch and yaw rates involved in the more strenuous maneuvers had resulted in cases of tail-surface failures during service operations and required new considerations for the structural design of tailfins for airships. The existing airship loads data available for design at the time were limited to low-aspect-ratio surfaces and relatively low angles of attack and yaw. In recognition of the absence of appropriate data, the Bureau of Aeronautics requested that a fin-loads investigation be conducted on a 1/15-scale model of the Goodyear XZP5K airship in the Full-Scale Tunnel.

Left: Evaluations of the aerodynamic efficiency of a stern-mounted propeller on a 1/20-scale model of the Goodyear ZPG3-W airship were conducted in 1960. (NASA L-60-413) Right: Measurements of fin loads for a 1/15-scale model of the Goodyear XZP5K airship were conducted in response to a Navy request following operational fin failures during severe flight maneuvers. (LAL 87171)

The scope of the tests included two types of tail surfaces representing contemporary designs. Although the primary objective of the investigation was to obtain fin-loads pressure data, total configuration force and moment data were obtained over an angle of attack range of ±20° and an angle of sideslip range of ±30° for a full range of elevator and rudder deflections. Boundary-layer surveys and wake-momentum surveys at the rear of the airship were also conducted to provide data for additional stern-propulsion design studies. Two sets of tails were used in the investigation. Both sets were inverted Y-tail arrangements with a radial spacing of 120° between fins; however, the fan shapes differed in planform and area. The NACA summary report on the investigation did not include analysis of the data and was intended to provide rapid dissemination of the information to the Navy.[45]

Although conducted later in 1960—during the early NASA years—rather than the period under consideration, a second Full-Scale Tunnel investigation of airship characteristics is noteworthy.[46] The test of a 1⁄20-scale model of the Goodyear ZPG3-W airship was precipitated by increasing requirements for a more efficient propulsive system for airships for early-warning and submarine-detection missions. For many years the idea of a stern propeller for airships had been suggested for improving propeller efficiency as well as for reducing noise and vibration for aircrew comfort. The objective of the investigation was to determine the propulsive characteristics of stern-mounted propellers designed specifically for airflow conditions existing at that location. Two propellers were designed based on best available theories, one being a four-blade configuration and the other a three-blade design. The scope of the tests included determining the propeller characteristics, aerodynamic forces, and moments for the complete model; and measuring the boundary-layer and propeller-wake characteristics on the aft hull and the surface-pressure distributions at a longitudinal station on the airship.

The results of the test indicated that a stern-mounted propeller could produce a much higher propulsive efficiency than that of a conventional-mounted installation. Operation of the stern propeller had minimal effects on the overall aerodynamic characteristics of the configuration. The increased efficiency demonstrated in the test offered significantly increased range and endurance for the vehicle.

Goodyear Inflatoplane

One of the most unusual aircraft concepts ever conceived, the Goodyear XAO-3 Inflatoplane, was tested in the Full-Scale Tunnel in 1957.[47] The Inflatoplane was an inflatable rubber aircraft built by the Goodyear Aircraft Company and powered by a two-cycle, 40-horsepower Nelson engine. The planned military application was to airdrop a packaged Inflatoplane in the vicinity of pilots who had survived being shot down in enemy territory so that they could inflate the aircraft and fly off in a self-rescue operation. The Inflatoplane could be inflated in about 5 minutes using less pressure than a car tire. All inflatable components were interconnected so that a small compressor on the engine could maintain a constant regulated pressure in the system, even with moderate leakage. The single-place aircraft had a wingspan of 22 feet and was capable of a top speed of 60 mph.

The tests in the Full-Scale Tunnel were initiated at the request of the Office of Naval Research to study the aerodynamic and structural deflection characteristics of the Inflatoplane over its operational envelope.[48] Each wing panel was restrained by two guy cables on the upper and lower surfaces, with the two upper cables anchored to the engine pylon and the two lower cables anchored to the landing gear. The tests were carried out at a variety of speeds ranging from stall speeds to speeds that resulted in wing structural failure or buckling. For testing this unique aircraft, the airplane was mounted to a special yoke so that strut retaining loads were transmitted to the fuselage through strap attachments beneath the wing quarter chord, thus leaving the wings free to deflect while being restrained only by the normal wing-fuselage and guy-cable attachments as in flight. Data measured included wing guy-cable loads, wing-distortion photographs, and aerodynamic performance and stability.

The Goodyear XAO-3 Inflatoplane had two separate test entries in the Full-Scale Tunnel. This photograph, taken during a 1957 investigation, shows the general arrangement of the inflatable aircraft. Note the wingtip ground-protection outriggers beneath the wingtips, the guy cables used to constrain the wing, and the vertical scale on the right used to measure wing deflections during tests. (NACA LAL-57-3413)

The characteristics of the configuration were also determined for a range of reduced inflation pressures simulating leakage due to battle damage or compressor malfunction. The propeller was removed for this investigation for safety considerations.

As might be expected, the results of force measurements made at various airspeeds for the Inflatoplane with nominal inflation pressure (7.0 psi) showed dramatic effects of aeroelastic deformation as speed was increased. At the lowest speed tested (40 mph), maximum lift occurred at an angle of attack of about 4°, but as tunnel speed was increased, the magnitude of maximum lift was decreased and the stall angle of attack was decreased until, at a tunnel speed of 55 mph, maximum lift occurred at an angle of attack of approximately –2°. When the tunnel speed was increased to approximately 71 mph, wing buckling occurred suddenly after about 30 seconds had elapsed at an angle of attack of about –5°. After the load was reduced, it was discovered that the aircraft did not sustain apparent damage; however, if a propeller had been installed and operating, the wing would have been destroyed. The lift produced at the buckling condition was equivalent to a load factor of about 2.0. As part of the test program, additional high-speed runs to wing buckling were conducted. On one run, the rear wing guy-cable on the lower surface of the left wing tore during a buckling cycle, and the failed wing contacted the engine and was punctured by the spark plugs and propeller shaft.[49]

In 1960, a brief 3-week test program was undertaken to obtain data for enhancing the load factor capability of the Inflatoplane and avoiding wing buckling within the operational capability of the vehicle.[50] The objective of the second test program was to determine whether load factors of the order of 4.5 to 5.0 could be obtained with appropriate deflections while avoiding wing buckling. The principal differences between the first Inflatoplane test vehicle and the second test article were that the configuration used in the second entry had a slightly larger wing, a canopy, and a more careful control of the airfoil contour. Once again, the propeller was not installed for testing. Various arrangements of guy cables were evaluated in an effort to determine a more robust configuration.[51]

Unfortunately, during a repeated high-loading test to obtain pressure distributions, the wing tore loose from the wing-fuselage bulkhead and was severely damaged. The test program was terminated before photographs could be taken of the aircraft. In spite of the untimely end of the test program, the revised guy-wire configuration significantly increased the load-factor capability of the Inflatoplane. At a normal 7.0-psi inflation pressure, load factors of about 4.5 were obtained.

In general, the Inflatoplane demonstrated handling qualities in flight that rivaled some popular general aviation aircraft designs during the late 1950s. A total of 12 Inflatoplanes were produced, and research and experimentation with the design continued well into the early 1970s. A two-seat version of the design, known as the Inflatobird, was also produced by Goodyear. By 1973, however, the Army had terminated the Inflatoplane program due to concerns over the fragility of the design and the relative ease with which the aircraft might be shot down. Surviving examples of the Inflatoplane exist at the U.S. Army Aviation Museum at Ft. Rucker, AL; the Franklin Institute in Philadelphia; and the Patuxent River Naval Air Museum at Lexington Park, MD, on loan from the Smithsonian Institute in Washington, DC.

Military Requests

During the postwar years, the intensity of military request jobs diminished at the Full-Scale Tunnel. Special military interests—such as those of the foregoing discussion on airships, submarines, helicopters, stand-on flying vehicles, and inflatable airplanes—had resulted in tunnel entries in support of the services, but developmental problems for emerging military aircraft had ended and fundamental studies became the main mission of the facility.

North American FJ-3

In the early 1950s, North American Aviation initiated an experimental flight-test program with an objective of enhancing high-speed lateral control for military fighters. An available FJ-3 airplane was selected for the program, which focused on the use of a spoiler-slot-deflector lateral control system located immediately ahead of conventional ailerons. NACA research conducted in other wind tunnels at Langley had indicated that superior control effectiveness and potentially reduced control forces could be provided by such concepts.

The Navy requested tests of the lateral-control concept in the Full-Scale Tunnel to ensure that the low-speed behavior of the modified aircraft was satisfactory before flight testing. The

A North American FJ-3 Fury Navy fighter in the Full-Scale Tunnel for assessments of a modified lateral-control system that used a spoiler-slot-deflector concept. This photograph shows the aircraft with inner-wing flaps deflected 45° and ailerons drooped 45°. The upper-wing spoilers are undeflected and the lower-wing deflectors are open. (NACA LAL 86089)

This close-up view of the right wing of the FJ-3 shows the implementation of the spoiler-slot deflector with the tunnel flow from right to left. The aileron is drooped, the upper-surface spoiler segments are deflected, and the lower-surface deflector door is open. The wing leading-edge flap is also deflected. The walls of the west return passage have been lifted, showing the street entrance to the Full-Scale Tunnel test section. (NACA LAL 86142)

configuration of interest for flight testing consisted of a drooped-aileron arrangement that resulted in a full-span flap when combined with a deflected inner-wing flap. In August 1954, a brief 3-week test was conducted to investigate the characteristics of the aircraft with the spoiler-deflector and the ailerons neutral and drooped as a flap for several spoiler-deflector configurations.[52] In addition to aerodynamic forces and moments, the spoiler-deflector hinge moments were measured.

The spoiler-slot-deflector concept consists of upper-wing spoiler segments that are interconnected to deflector doors on the lower surface of the wing. The deflector doors are hinged to open to the oncoming flow and act as "scoops" to direct the air through the aileron slot. Typically, the ratio of movement between the spoiler segments and deflector doors is about 2 to 1, and the deflection of the spoilers can vary from 0° to 70°, while the travel of the deflectors can be from 0° to 35°.

For the tests of the FJ-3, the aircraft's air intake at the nose of the airplane was faired and sealed by a metal fairing for all tests. The scope of testing included measurement of the longitudinal and lateral stability and control characteristics of the modified aircraft for angles of attack up to stall. The spoiler and deflector hinge moments were measured from strain gages mounted on special actuation rods, and although the test program was primarily devoted to studies of the high-lift configuration, data were also obtained for the cruise configuration.

Increments in maximum trimmed lift beyond that for the basic airplane were measured with the ailerons drooped. The test results revealed that the aileron required a modified nose radius and shrouds to eliminate stalled flow, which was detrimental to spoiler effectiveness. By combining spoiler-deflector combinations, an almost linear variation of rolling moments with spoiler deflection was obtained.

The spoiler-slot-deflector concept was later used on the Navy A3J Vigilante reconnaissance aircraft of the Vietnam War era.

Quiet Airplane Project

Many valuable NACA research efforts took place within the building that housed the Full-Scale Tunnel, but they did not use the facility for its intended purpose. For example, the most famous propeller-driven aircraft noise-reduction project conducted by the NACA involved members of the tunnel's staff, who developed critical engine-muffler and noise-reduction concepts for the Nation, but they never conducted a test for that purpose in the tunnel. This particularly important activity was conducted with a military liaison-type aircraft (Stinson L-5) in 1947.[53] During that time, the rapid postwar expansion of interest in personal-owner aircraft stimulated the NACA to develop methods to reduce the noise of airplanes. In the opinion of regulatory agencies, the threat of excessive noise from propeller aircraft was one of the most significant threats to the growth of civil aviation. Theories had been developed by Langley personnel under Theodore Theodorsen for the prediction of propeller noise in the late 1930s, and associated experiments had been conducted in ground facilities, but the NACA wanted to integrate these efforts with a flight-demonstration project to show how

This Stinson L-5 aircraft was modified with a five-blade propeller and muffler to dramatically reduce flyover noise. Staff members of the Full-Scale Tunnel were called upon to provide a muffler system to reduce the engine noise. (NASA EL-2000-00263)

technology might reduce aircraft noise, including noise from all sources—propeller, engine, and exhaust system. A coordinated effort among several Langley organizations to demonstrate the ability of technology to reduce the noise of a typical personal-owner aircraft was initiated and included several staff members of the Full-Scale Tunnel.

Before the project began, noisy propellers had dominated the acoustic problems for propeller aircraft, and relatively little attention had been given to engine noise. A five-blade propeller operated at low rotational speeds was designed for the experiment so as to reduce the tip speed of the propeller blades, thereby significantly reducing propeller-generated noise.[54] Working under high-priority guidelines, the propeller-noise-reduction team pushed the noise down to unprecedented levels; however, the project soon learned that the noise radiated from the engine and its exhaust had then become the dominant source of noise. Existing muffler design methods for aircraft were found to be inadequate for the task.

The program manager, A.W. Vogeley, assigned the job of developing an adequate muffler to Don D. Davis, Jr., of the Full-Scale Tunnel staff. Davis quickly learned that commercial mufflers were useless for the task, and he was appalled at the state of the art in general for muffler design. Together with a team of engineers and technicians, he set about designing an appropriate muffler in a fundamental research effort in the old physical research laboratory within the building.[55] His philosophy was to meet the noise-level requirements for the Quiet Airplane project with a new muffler design, but not necessarily one that could meet the weight requirements for practical applications. After considerable efforts beginning with the first principles of noise alleviation, the muffler design was completed and evaluated with a ground-test stand and dynamometer setup.[56] The basic engine without a muffler produced a noise level of 89 decibels at 300 feet, whereas the muffled engine-noise level was reduced to only 67 decibels, which was the same as that measured for the complete aircraft in flight.

With the Langley-developed propeller, engine, and muffler modifications, the noise pressure level of the airplane was reduced an astounding 90 percent.

The modified airplane was first flown and demonstrated as a "quiet airplane" together with flyovers of an unmodified L-5 at the Sixteenth Annual NACA Inspection at Langley in May 1947. Many in attendance did not even hear the airplane as it flew over the assembled crowd at an altitude of a few hundred feet during a lunch break at the inspection. Following the impressive demonstration, additional research flights generated detailed engineering data on noise levels for the basic and modified airplane.[57]

After the Quiet Airplane demonstration, the group led by Davis conducted an extensive series of experimental and theoretical studies to mature the science of aircraft-muffler design, including applications to the same YR-4B helicopter that had been tested in the Full-Scale Tunnel in 1944 (see previous chapter).[58] As the results of the activity were disseminated to industry, the Langley experts were sought out for presentations at offsite "schools" in locations ranging from automotive companies to appliance manufacturers. The positive reactions of the recipients of the technology included comments such as, "This is the first time the design of mufflers has had science rather than black art applied."[59]

Wall Effects in Supersonic Flow

After the highly successful muffler projects, Don Davis and his associates at the Full-Scale Tunnel tackled a problem in a completely new area—wall-reflected disturbances that would contaminate measurements made in transonic wind tunnels at low supersonic speeds. Even after Langley conceived and developed the famous slotted-wall concept for transonic wind tunnels, the presence of boundary-reflected disturbances in the low-supersonic Mach number range limited the size of test models and the collection of interference-free data.

A transonic-flow apparatus was constructed in a special room within the Full-Scale Tunnel building and used for investigations of the effect of tunnel-wall configurations at low-supersonic speeds. The 3-inch, two-dimensional test section is shown in the foreground of the photo with airflow from right to left through the closed-circuit tunnel. (NACA LAL 73103)

Beginning in 1951, Davis and his group examined the applications of several slotted– and porous–tunnel wall configurations to the problem in a 3- by 3-inch transonic flow apparatus with a two-dimensional test section and slotted walls.[60] The test apparatus was housed in a special room built near the south end of the Full-Scale Tunnel building. The 3-inch tunnel was a single-return, closed-circuit, continuously operating tunnel powered by a single-stage, variable-speed supercharger from a P-51 Mustang of World War II vintage. The supercharger was driven by an air-cooled induction motor. A separate suction source to remove the flow from the slots was provided by a second supercharger.

Extensive pressure instrumentation and a Schlieren apparatus for flow-visualization were used to evaluate numerous tunnel-wall combinations for mitigation of wall-interference effects. The Mach number range covered during testing ranged from 0.8 to 1.3, with extensive evaluations of wall configurations.[61]

Invasion of the Butterflies: Free-Flight Models

Charles H. Zimmerman's contributions to the aeronautics and space programs of the NACA and NASA were truly remarkable. Beyond his conception and development of innovative concepts such as the Vought V-173 and the flying platform, his achievements included new facilities and testing techniques that became workhorses for both Agencies. Based on the success of a British vertical free-spinning tunnel, he designed a 15-foot spin tunnel for the NACA that was put into operation in 1935.[62] Zimmerman followed that successful design

Charles Zimmerman and an unidentified engineer fly a dynamic free-flight model in the test section of a 5-foot proof-of-concept wind tunnel to determine the model's dynamic stability and control characteristics. The tunnel was mounted in a pivot-yoke arrangement that allowed the test section to be tilted downward and the unpowered model to be flown in gliding flight. The concept evolved into the Langley 12-Foot Free-Flight Tunnel. Note the large building that previously housed the Langley 15-Foot Free-Spinning Tunnel before the tunnel was replaced by the Langley 20-Foot Spin Tunnel. (NASA EL-2003-00363)

with a vision for a wind tunnel method to study the dynamic stability and control characteristics of a remotely controlled free-flying aircraft model in conventional flight. He then conceived a 5-foot-diameter proof-of-concept free-flight wind tunnel in 1937 with which two research pilots could assess the flight behavior of a dynamically scaled model.[63]

After Zimmerman departed Langley to pursue his V-173 interests at Vought, his concept for a free-flight facility was matured in 1939 into the Langley 12-Foot Free-Flight Tunnel, which was housed in a sphere next to the north end of the Full-Scale Tunnel. Free-flight tests in the facility for the next 15 years proved to be a valuable specialty area for Langley, and critical contributions were achieved for many aircraft designs, especially radical or unconventional configurations for which no databases existed.

As valuable as the Free-Flight Tunnel was, three enhancements to the free-flight technique were actively pursued by its staff under the management of John P. Campbell, who had become head of the Free-Flight Tunnel Section in 1944. The first desired improvement was the ability to build and test larger models for which more configuration details could be included, and which would allow for an accompanying increase in flying area and Reynolds number for more realistic aerodynamic behavior. The second enhancement sought by the researchers was for a free-flight facility with a larger test-section size. Out-of-control maneuvers were common in the relatively small 12-foot test section, frequently resulting in crashes into the tunnel test-section walls and major damage to models, causing significant delays in test programs. The third consideration for enhancing the free-flight technique evolved from emerging interests in NACA's research for VTOL aircraft configurations, which required the capability for models to transition from and to hovering flight.

In the early 1950s, Zimmerman (who by then had returned from Vought and simultaneously served as assistant division chief and head of the Dynamic Stability Branch), Campbell, and Marion O. McKinney of the Free-Flight Tunnel group formulated first-ever projects to assess the dynamic stability, controllability, and flying characteristics of a wide variety of innovative VTOL concepts, including tail sitters, tilt-wing configurations, and refined flying jeeps. Zimmerman had led the early work by conceiving a configuration that used a radical wing composed of segments similar to the turning vanes in a wind tunnel to deflect the slipstreams of propellers through 90° to permit vertical takeoff. The wing segments would then be rotated to a conventional wing configuration for airplane flight. The researchers called the concept the "flying venetian blind." McKinney thought the concept was much too complicated and suggested that they try simply pivoting the entire wing and propellers to a vertical attitude—the "tilt wing" concept was born.[64]

Assessments of the hovering characteristics of the VTOL models were conducted in the large three-story building that had previously housed the early 15-Foot Free-Spinning Tunnel, but the indoor technique could not permit an evaluation of the most critical phase of VTOL flight—the transition to and from hovering and conventional forward flight.

A Nose Under the Tent: Free-Flight Testing in the Full-Scale Tunnel

In 1951, John Campbell was granted test time by Jerry Brewer to pursue tests in the west return passage of the Full-Scale Tunnel to assess the takeoff, hovering, and landing

The Langley 12-Foot Free-Flight Tunnel was located in a sphere adjacent to the Full-Scale Tunnel building. In this 1985 aerial photo, facilities include (from top clockwise) the Full-Scale Tunnel, the building that housed the original 15-Foot Free-Spinning Tunnel, the 12-Foot Free-Flight Tunnel, and the 20-Foot Spin Tunnel. The Langley 16-Foot Transonic Dynamics Tunnel (TDT) is at the extreme upper right. (NASA L-85-3591)

A free-flight model of the Navy SBN-1 scout plane is supported by cables to simulate dynamic stability and control evaluations in the Langley 12-Foot Free-Flight Tunnel in 1940. Three operators were required for the test. The pilot was seated in an enclosure beneath the tunnel drive motor (out of view to the lower left of the picture) while two personnel stood next to the test section and controlled the pitch angle of the tunnel's test section and the tunnel airspeed. (NASA EL-2000-00201)

characteristics of a 0.13-scale model of the Consolidated-Vultee XFY-1 "tail sitter" VTOL design in response to a Navy request.[65] The testing even included an attempt to evaluate the model's dynamic stability and control during translational flight in the return passage with the tunnel running at low speeds; however, the flow in the return passage was too gusty at those conditions for assessments to be made. The success of those tests in late 1951 led to more hovering tests of VTOL designs in the return passage, and Campbell soon created a case for developing an upgraded free-flight testing technique to assess the flying behavior of VTOL aircraft models during the transition from hovering to forward flight in the Full-Scale Tunnel test section. As a result of an additional Navy request, the first model to be flown in the tunnel's test section was the XFY-1 model in 1952.[66] This landmark test was the initial free-flight study that was followed by over 57 years of similar flight tests in the Full-Scale Tunnel.

More free-flight testing in the test section of the Full-Scale Tunnel quickly followed. For example, in 1954, a generic delta-wing VTOL model with a vertical tail was tested.[67] The configuration had been chosen because of Langley's awareness of the emerging design of the Ryan X-13 tail-sitter VTOL aircraft at the time. In the interest of increasing the configuration details and size, the fuselage was over 6 feet long, and the model included artificial stabilization systems (activated by angular-rate gyros), a thrust-vectoring system, and instrumentation. Whereas the smaller Free-Flight Tunnel models weighed a few pounds at most, this model weighed about 45 pounds.[68]

Sequence of photographs show the transition of a free-flight model of the Convair XFY-1 flying in the Full-Scale Tunnel as it transitions from a vertical-attitude VTOL hovering condition to conventional forward flight. These tests in 1952 were the first free-flight model tests conducted in the tunnel. The numbered quadrant at the upper middle was an innovative tunnel airspeed indicator consisting of a pivoting cylinder/pointer apparatus. The pointer served as a quick "heads-up" display for awareness of airspeed by pilots. (L-87192)

The test setup for the first flight-test evaluations included a pitch pilot, a model thrust operator, and an operator of a safety-cable winch attached to the model; these staff members were located on the balcony that had been constructed on the east wall of the test section of the Full-Scale Tunnel in the early 1930s. A pilot who controlled the rolling motions of the model was seated in an observation enclosure on the tunnel ground board, and a pilot who controlled the model's yawing motions was seated at the top of the tunnel exit cone along with a camera operator.

A power and control cable consisted of plastic tubes that provided compressed air for the electro-pneumatic control actuators and electrical wires that supplied thrust (i.e., power for motors driving model propellers or direct thrust for jet configurations) and control signals from the remote pilots for the actuators. A steel safety cable was included in the flight cable

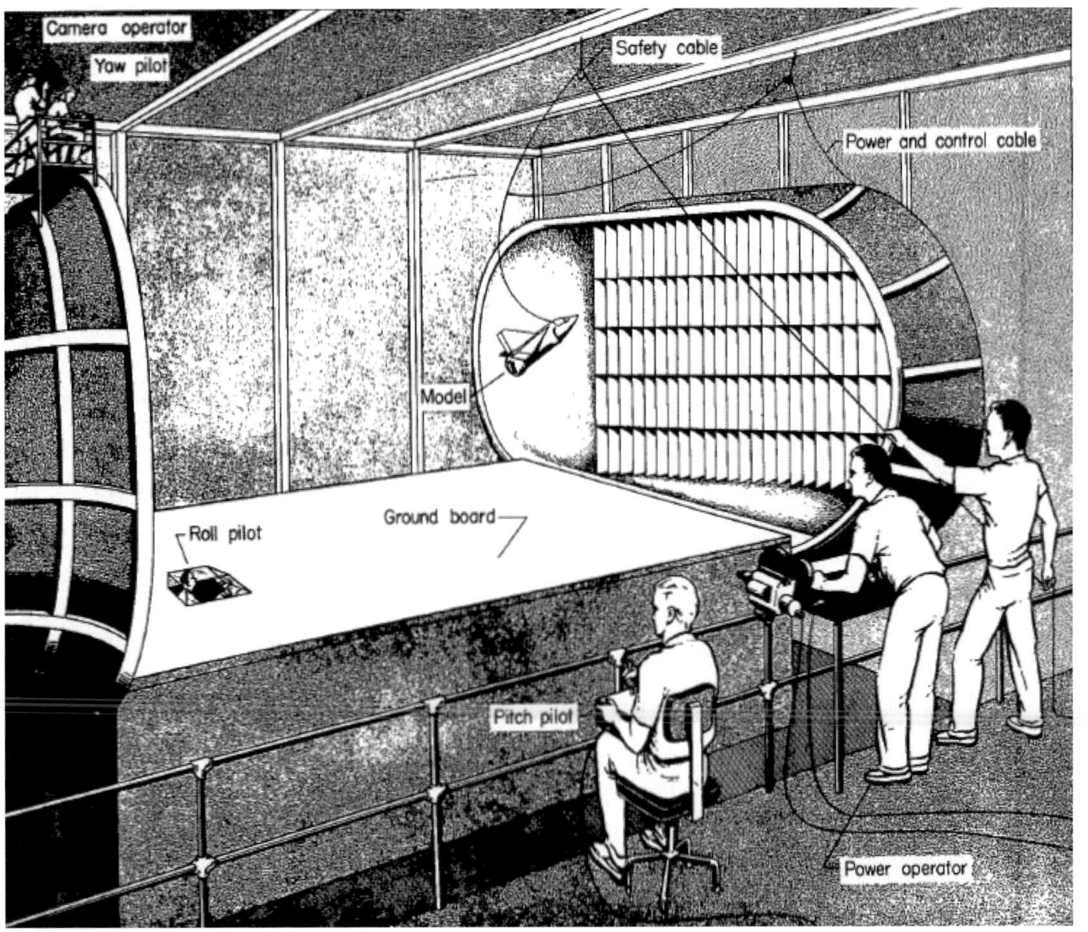

The first test setup for free-flight model testing in the Full-Scale Tunnel. The model is shown flying freely in the tunnel test section, remotely controlled by a pitch pilot in the balcony at the side of the tunnel, a roll pilot in enclosure in the tunnel ground board, and a yaw pilot at the top of the tunnel exit cone. A model power operator and safety-cable controller were also on the balcony and a camera operator participated by taking motion pictures. The relative sophistication of the testing technique and setup would change over the next 55 years, but the general principle remained the same. (NASA L-93575)

and was used for snatching the model upward out of the airflow to prevent crashes. The cable was attached to the top of the fuselage and ran over a pulley at the ceiling of the test chamber to the safety-cable operator, who adjusted the length of the cable to keep it slack during flight. During early flights in the tunnel, the safety cable operator accomplished his job by hand by paying line in and out manually.

The model used in the first evaluation was powered by electric motors turning 14-inch-diameter ducted propellers. Remotely controlled eyelids at the rear of the fuselage were used to deflect the exhaust for pitch and yaw control during hovering flight. Roll control was provided by air that was bled from the propulsion motors to jet-reaction controls at the wingtips. In conventional forward flight, the model used flap-type elevons and rudder controls.

The evaluation of the new free-flight testing in the Full-Scale Tunnel was extremely positive. The division of piloting tasks among several researchers proved to be efficient and effective, despite the rapid angular rates, because of model scaling laws.[69] The ability to vary the tunnel airspeed from hovering flight to conventional forward flight allowed transition flights to be studied in detail, including the effects of artificial stabilization. Enthusiastic about the results, Campbell, his staff, and the military began planning additional entries for other generic and specific configurations. In 1955, another VTOL configuration concept using the tilt-wing principle was tested.[70] By 1956, free-flight tests of specific configurations were being requested and conducted, including the Hiller X-18 tilt-wing transport and the Ryan X-13 VTOL tail-sitter.[71]

An Issue Arises: The Langley Control Line Facility

After free-flight tests of several VTOL models had been completed in the Full-Scale Tunnel, an issue arose regarding limitations of the facility for certain test conditions.[72] The issue arose because many of the VTOL designs exhibited severe dynamic stability problems at high angles of attack during the transition from hovering flight to conventional forward flight. Because the airspeed of the Full-Scale Tunnel could not be rapidly increased to simulate a rapid aircraft transition, it was argued that, theoretically, an unstable airplane might be able to quickly transition through an unstable angle-of-attack range before the instabilities could react and result in loss of control. Thus, the results of the free-flight tests in the tunnel might be overly pessimistic.

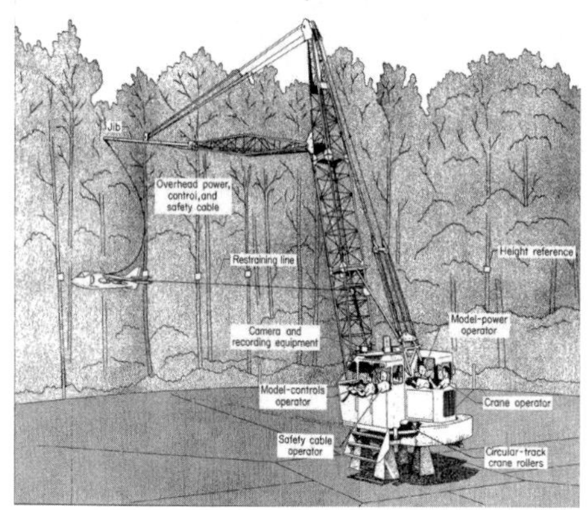

As a result of the relatively slow transition times provided in the Full-Scale Tunnel, John Campbell's staff conceived a new Langley Control Line Facility that provided rapid transition maneuvers. (SP 2009-575, p. 40)

Recognizing that this was a fundamental issue for testing VTOL aircraft, the tunnel's staff created an outdoor rapid-transition testing technique.[73] As was the case for many Langley employees, Robert O. Schade of Campbell's group of free-flight model pilots was a model airplane enthusiast, and he conceived an outdoor testing technique similar to the control-line technique used by hobbyists. Schade envisioned that such a testing method could provide the rapid-transition maneuvers that were needed. Langley subsequently acquired a large crane capable of rapid rotation and equipped it with model power systems and remote-control capability. The crane could be rotated at angular rates up to about 20 revolutions per minute and could accelerate from rest to top speed in about one quarter of a revolution. Seated in the enlarged crane cab were the model-controls operator, the safety-cable operator, the model-power operator, and the crane operator. The powered model being tested was mounted about 50 feet from the center of rotation and was restrained by wires from the model to the crane cab to oppose the centrifugal forces experienced during flight around the circle. The facility was designated the Langley Control

The Langley Control Line Facility was used for numerous rapid-transition tests of V/STOL aircraft. On the left, researchers P.M. Lovell and Joe Block pose with a model of the Convair XFY-1 during 1954. The photo on the right shows the test team for the Hawker P.1127 configuration in 1960, including (left to right) Charlie Smith, Dick Mills, Bill Veatch, Lysle Parlett, and Bob Schade. (NACA LAL 87175 and NASA L-60-7422)

Line Facility (CLF). By using the CLF, transitions could be made much more rapidly than those in the Full-Scale Tunnel and at translational rates more representative of those of full-scale VTOL airplanes.

High-Angle-of-Attack Problems

Even as the NACA was developing the free-flight testing technique in the Full-Scale Tunnel, critical issues ideally suited for applications of the technique arose. For example, during the development of the Air Force Convair B-58 Hustler jet bomber, concern arose over the dynamic stability characteristics of the configuration at high angles of attack and the flight behavior with an external pod. Free-flight model tests were made in 1956 to provide information for the design of the aircraft.[74] Free-flight tests were also requested by the Navy during the development of the Chance Vought F8U-1 fighter and were conducted in 1954.[75]

Free-flight tests of the Convair XB-58 bomber configuration provided information on the high-angle-of-attack dynamic stability control characteristics of the design. Technician Joe Block prepares the model for a flight test. (NACA L-96279)

The free-flight technique at the Full-Scale Tunnel also played a key role in the NACA's interests in high-speed aircraft and space exploration. In the case of the North American X-15 rocket-powered research airplane, a free-flight model was used in the Full-Scale Tunnel in 1956 to demonstrate the adequacy of the use of differentially deflected horizontal tail surfaces

Flight tests of this dynamically scaled model of the X-15 in the Full-Scale Tunnel helped give designers confidence that conventional wing-mounted ailerons could be eliminated from the design because differential deflection of the horizontal tail surfaces could provide satisfactory roll control. (NACA L-95227)

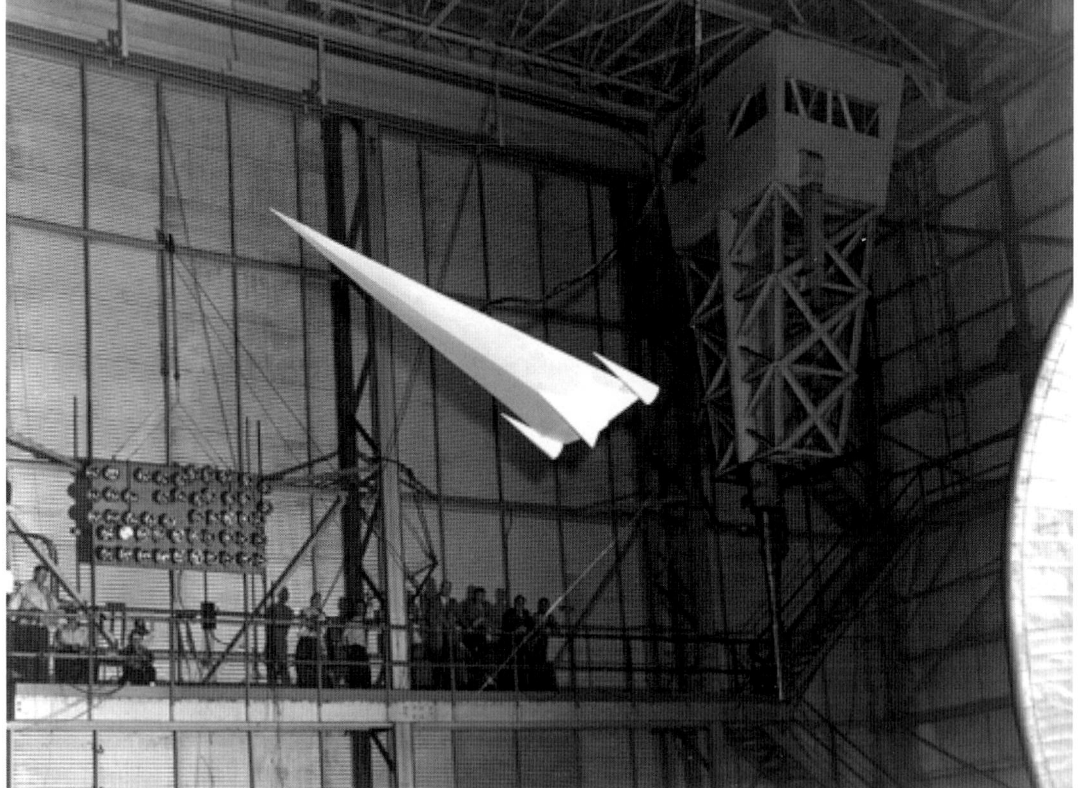

Highly swept hypersonic boost-glide reentry vehicles such as this configuration exhibit large-amplitude uncontrollable roll oscillations at moderate angles of attack. Free-flight models were used to demonstrate that artificial stabilization could mitigate the problem and result in satisfactory flight behavior. (NACA L-57-1439)

for roll control.[76] A major issue facing the program at the time was whether conventional ailerons would have to be implemented on the high-speed wing, which would add structural complexity to an already challenging wing-design problem. The very positive results of the free-flight test for the tail surfaces vividly demonstrated to management that the approach was satisfactory and that ailerons were not required.[77]

Finally, an extensive research program within the NACA in the late 1950s had been under way to evaluate the relative efficiencies and problems of hypersonic boost-glide vehicles. Many of the configurations investigated used highly swept, sharp leading-edge wings that resulted in potential issues for dynamic stability and control. In particular, many of the highly swept designs exhibited uncontrollable rolling motions at moderate angles of attack representative of those to be used during reentry maneuvers into Earth's atmosphere. Free-flight model tests in the Full-Scale Tunnel in 1957 clearly identified the problems and demonstrated that artificial stabilization could be used to mitigate the issue.[78] Additional free-flight tests of boost-glide models were conducted by the group using the Langley Control Line Facility.

The End of the NACA

By the end of 1957, major changes had occurred in the research themes and techniques used in the Langley Full-Scale Tunnel. A sweeping variety of configurations had received attention, ranging from more efficient helicopters to hypersonic-reentry vehicles. Novel testing techniques within the tunnel's test section and the building that housed the tunnel had been established. Research into space flight vehicles had begun, and the free-flight model test technique had been validated and recognized as a valuable new testing method. Hundreds of technical reports had been written, thousands of presentations had been made, and the reputation of the facility as a major national asset had been firmly established. Seeds of perspective had already been planted as to how an old wind tunnel designed for tests of biplanes in the 1930s would play a leading role as the Nation approached the Space Age.

Endnotes

1. A major article on the event with photographs of attendees is presented in the June 15, 1956, issue of Langley's internal newsletter, *The Air Scoop*.
2. Ibid., p. 3.
3. Denise Lineberry, "A Look Back with Langley's NACA Alumni: NACA's 95th Anniversary," *Langley Researcher News* (March 3, 2010).
4. Ibid.
5. Philip Donely, "Summary of Information Relating to Gust Loads on Airplanes," NACA Report 997 (1950). The original concrete steps leading up to the launch position for the gust-response models remained in the tunnel until its demolition in 2011.
6. Dr. Lippisch had led the design team for the infamous Messerschmitt Me 163 rocket-powered interceptor, but he left the project just as the prototypes were completed. He then led students from Darmstadt and Munich Universities on the P.13a design. The designation DM-1 was given to the glider in honor of the universities.
7. Joseph R. Chambers and Mark A. Chambers, *Radical Wings and Wind Tunnels* (North Branch, MN: Specialty Press, 2008).
8. Herbert A. Wilson, Jr., and J. Calvin Lovell, "Full-Scale Investigation of the Maximum Lift and Flow Characteristics of an Airplane Having Approximately Triangular Plan Form," NACA Research Memorandum (RM) L6K20 (1947).
9. Video clips of testing of the DM-1 in the Full-Scale Tunnel are available at *http://crgis.ndc.nasa.gov/crgis/images/9/9d/328.mpg*, accessed January 4, 2012, and *http://www.youtube.com/watch?v=rQNPl5DdS5I*, accessed January 6, 2012.
10. J. Calvin Lovell and Herbert A. Wilson, Jr., "Langley Full-Scale-Tunnel Investigation of Maximum Lift and Stability Characteristics of an Airplane Having Approximately Triangular Plan Form (DM-1 Glider)," NACA RM L7F16 (1947).
11. Eugene R. Guryansky and Stanley Lipson, "Effect of High-Lift Devices on the Longitudinal and Lateral Characteristics of a 45° Sweptback Wing with Symmetrical Circular-Arc Sections," NACA RM L8D06 (1948).
12. Examples include Roy H. Lange, "Langley Full-Scale Tunnel Investigation of the Maximum Lift and Stalling Characteristics of a Trapezoidal Wing of Aspect Ratio 4 with Circular-Arc Airfoil Sections," NACA RM L7H19 (1947); Roy H. Lange and Huel C. McLemore, "Low-Speed Lateral Stability and Aileron-Effectiveness Characteristics at a Reynolds Number of 3.5×10^6 of a Wing with Leading-Edge Sweepback Decreasing From 45° at the Root to 20° at the Tip," NACA RM L50D14 (1950); H. Clyde McLemore, "Low-Speed Investigation of the Effects of Wing Leading-Edge Modifications and Several Outboard Fin Arrangements on the Static Stability Characteristics of a Large-Scale Triangular Wing," NACA RM L51J05 (1952); and Roy H. Lange and Marvin P. Fink, "Effect of a Deflectable Wing-Tip Control on the Low-Speed Lateral and Longitudinal Characteristics of a Large-Scale Wing with the Leading Edge Swept Back 47.5°," NACA RM L51C07 (1951).

13. Jerome Pasamanick and Anthony J. Proterra, "The Effect of Boundary-Layer Control by Suction and Several High-Lift Devices on the Longitudinal Aerodynamic Characteristics of a 47.5° Sweptback Wing-Fuselage Combination," NACA RM L8E18 (1948).
14. Jerome Pasamanick and Thomas B. Sellers, "Full-Scale Investigation of Boundary-Layer Control by Suction Through Leading-Edge Slots on a Wing-Fuselage Configuration Having 47.5° Leading-Edge Sweep With and Without Flaps," NACA RM L50B15 (1950).
15. P. Kenneth Pierpont, "Investigation of Suction-Slot Shapes for Controlling a Turbulent Boundary Layer," NACA Technical Note 1292 (1947).
16. Bennie W. Cock, Jr., Marvin P. Fink, and Stanley M. Gottlieb, "The Aerodynamic Characteristics of an Aspect-Ratio-20 Wing Having Thick Airfoil Sections and Employing Boundary-Layer Control by Suction," NACA TN 2980 (1953).
17. Marvin P. Fink, Bennie W. Cocke, and Stanley Lipson, "A Wind-Tunnel Investigation of a 0.4-Scale Model of an Assault-Transport Airplane with Boundary-Layer Control Applied," NACA RM L55G26a (1956)
18. Frederic B. Gustafson, "History of NACA/NASA Rotating-Wing Aircraft Research, 1915–1970," *Vertiflite* limited edition reprint VF-70 (April 1971). The information appeared in a series of articles beginning in September 1970.
19. Richard C. Dingeldein, "Wind-Tunnel Studies of the Performance of Multi-Rotor Configurations," NACA TN 3236 (1954).
20. Marion K. Taylor, "A Balsa-Dust Technique for Air-Flow Visualization and Its Application to Flow through Model Helicopter Rotors in Static Thrust," NACA TN 2220 (1950).
21. Colin P. Coleman, "A Survey of Theoretical and Experimental Coaxial Rotor Aerodynamic Research," NASA Technical Paper 3675 (1997).
22. See, for example, George E. Sweet, "Hovering Measurements for Twin-Rotor Configurations With and Without Overlap," NASA TN D-534 (1960); and Robert J. Huston, "Wind-Tunnel Measurements of Performance, Blade Motions, and Blade Air Loads for Tandem-Rotor Configurations With and Without Overlap," NASA TN D-1971 (1963).
23. See Harry H. Heyson, "Preliminary Results from Flow-Field Measurements around Single and Tandem Rotors in the Langley Full-Scale Tunnel," NACA TN 3242 (1954); John P. Rabbort, Jr., and Gary B. Churchill, "Experimental Investigation of the Aerodynamic Loading on a Helicopter Rotor Blade in Forward Flight," NACA RM L56107 (1956); John P. Rabbot, Jr., "Static-Thrust Measurements of the Aerodynamic Loading on a Helicopter Rotor Blade," NACA TN 3688 (1956); and Harry H. Heyson, "Analysis and Comparison with Theory of Flow-Field Measurements Near a Lifting Rotor in the Langley Full-Scale Tunnel," NACA TN 3691 (1956).
24. Robert D. Harrington, "Reduction of Helicopter Parasite Drag," NACA TN 3234 (1954).
25. Video scenes of ground-vibration tests of the XH-17 model in the Langley Physical Research Laboratory are available at *http://www.youtube.com/watch?v=TZQpLsip3yE*, accessed January 6, 2012.

26. George W. Brooks and Maurice A. Sylvester, "Description and Investigation of a Dynamic Model of the XH-17 Two-Blade Jet-Driven Helicopter," NACA RM L50I21 (1951).
27. George W. Brooks and Maurice A. Sylvester, "The Effect of Control Stiffness and Forward Speed on the Flutter of a 1⁄10-Scale Dynamic Model of a Two-Blade Jet-Driven Helicopter Rotor," NACA TN 3376 (1955).
28. Jerome Pasamanick, "Langley Full-Scale-Tunnel Tests of the Custer Channel Wing Airplane," NACA RM L53A09 (1953).
29. Walt Boyne, "The Custer Channel Wing Story" *Airpower* 7, no. 3 (May 1977): pp. 8–19.
30. Ibid., p. 58.
31. Ibid., p. 58.
32. Robert J. Englar and Bryan A. Campbell, "Development of Pneumatic Channel Wing Powered-Lift Advanced Super-STOL Aircraft," AIAA Paper 2002-2929 (2002).
33. Charles H. Zimmerman, interview by Walter Bonney, March 30, 1973, LHA.
34. Charles H. Zimmerman, Paul R. Hill, and T.L. Kennedy, "Preliminary Experimental Investigation of the Flight of a Person Supported by a Jet Thrust Device Attached to His Feet," NACA RM L52D10 (1953).
35. Video scenes of the experiments by Zimmerman and Hill are available at *http://www.youtube.com/watch?v=vFndQulSPBQ*, accessed January 5, 2012.
36. Paul R. Hill and T.L. Kennedy, "Flight Test of a Man Standing on a Platform Supported by a Teetering Rotor," NACA RM L54B12a (1954).
37. Video scenes of a stand-on flying model being tested in the Full-Scale Tunnel are available at *http://www.youtube.com/watch?v=LPPaedMWbJk*, accessed January 6, 2012.
38. Digitized videos of tests of a stand-on generic model in the Full-Scale Tunnel are available at *http://www.youtube.com/watch?v=LPPaedMWbJk*, accessed January 4, 2012.
39. U.S. Army Transportation Museum Web site, *http://www.transchool.lee.army.mil/Museum/Transportation%20Museum/museum.htm*, accessed June 14, 2011.
40. George E. Sweet, "Static-Stability Measurements of a Stand-On Type Helicopter With Rigid Blades, Including a Comparison With Theory," NASA TN D-189 (1960).
41. Stanley Lipson, Bennie W. Cocke, and William I. Scallion, "Investigation of a 1⁄5-Scale Model of a Proposed High-Submerged-Speed Submarine in the Langley Full-Scale Tunnel," NACA RM SL50K01 for the Bureau of Ships, Department of the Navy (1950). Data from the tests are presented in Research Memorandum SL50E09a (1950) by the same authors.
42. Zimmerman interview.
43. Video scenes of the Albacore model being tested in the Full-Scale Tunnel are available at *http://www.youtube.com/watch?v=A3OBYCSPc-I*, accessed January 6, 2012.
44. Mark W. McKellar, "USS Albacore—A Revolution by Design," *http://www.hazegray.org/navhist/albacore.htm*, accessed June 16, 2011.
45. Michael D. Cannon, "Static Longitudinal and Lateral Stability and Control Data Obtained from Tests of a 1⁄15-Scale Model of the Goodyear XZP5K Airship," NACA RM SL56A11 for the Bureau of Aeronautics (1956).

46. H. Clyde McLemore, "Wind-Tunnel Tests of a 1/20-Scale Airship Model with Stern Propellers," NASA TN D-1026 (1962).
47. Chambers and Chambers, *Radical Wings and Wind Tunnels.*
48. Bennie W. Cock, Jr., "Wind-Tunnel Investigation of the Aerodynamic and Structural Deflection Characteristics of the Goodyear Inflatoplane," NACA RM L58E09 (1958).
49. Video scenes of the 1957 tests of the Inflatoplane are available at *http://www.youtube.com/watch?v=exvyt_YOe5g*, accessed January 6, 2012; video scenes of the 1960 tests of the wing buckling are available at *http://www.youtube.com/watch?v=fuHm6q9smR8*, accessed January 2, 2012.
50. H. Clyde McLemore, "Wind-Tunnel Investigation of the Aerodynamic and Structural-Deflection Characteristics of an Inflatable Airplane," NASA TM X-680 (1962).
51. Video scenes of the 1960 tests of the Inflatoplane are available at *http://www.youtube.com/watch?v=za_JOQxRlEI*, accessed January 6, 2012.
52. William I. Scallion, "Full-Scale Wind-Tunnel Tests of a North American FJ-3 Airplane with a Spoiler-Slot-Deflector Lateral Control System," NACA RM No. SL56D18 for the Bureau of Aeronautics (1956).
53. A.W. Vogeley, "Sound-Level Measurements of a Light Airplane Modified to Reduce Noise Reaching the Ground," NACA TN 1647 (1948).
54. Don D. Davis, Jr., interview by author, April 7, 2011.
55. Davis remarked that he was amazed at the lack of science involved in muffler designs at the time. He and his peers at the Full-Scale Tunnel embarked on several experiments regarded as pioneering within the muffler community.
56. K.R. Czarnecki and Don D. Davis, Jr., "Dynamometer-Stand Investigation of the Muffler Used in the Demonstration of Flight-Airplane Noise Reduction," NACA TN 1688 (1948).
57. Video scenes of flight tests of the modified airplane are available at *http://www.youtube.com/watch?v=y3-GoV5cO5U*, accessed January 3, 2012.
58. Don D. Davis, Jr., George L. Stevens, Jr., Dewey Moore, and George M. Stokes, "Theoretical and Experimental Investigation of Mufflers With Comments on Engine-Exhaust Muffler Design," NACA Technical Report 1192 (1953).
59. Davis interview.
60. Thomas B. Sellers, Don D. Davis, and George M. Stokes, "An Experimental Investigation of the Transonic-Flow-Generation and Shock-Wave-Reflection Characteristics of a Two-Dimensional Wind Tunnel with 24-Percent-Open, Deep, Multislotted Walls," NACA RM L53J28 (1953).
61. Don D. Davis, Jr., Thomas B. Sellers, and George M. Stokes, "An Experimental Investigation of the Transonic-Flow-Generation and Shock-Wave-Reflection Characteristics of a Two-Dimensional Wind Tunnel with 17-Percent-Open Perforated Walls," NACA RM L53J28 (1953).
62. The Langley 15-Foot Free-Spinning Tunnel was later replaced by a 20-Foot Spin Tunnel that became operational in 1941.
63. Joseph R. Chambers, *Modeling Flight: The Role of Dynamically Scaled Models in Support of NASA's Aerospace Programs* (Washington, DC: NASA SP 2009-575, 2010).

64. Zimmerman interview.
65. Powell M. Lovell, Jr., Charles C. Smith, Jr., and Robert H. Kirby, "Stability and Control Flight Tests of a 0.13-Scale Model of the Consolidated-Vultee XFY-1 Airplane in Take-Offs, Landings, and Hovering Flight," NACA RM SL52I26 (1952).
66. Powell M. Lovell, Jr., Robert H. Kirby, and Charles C. Smith, Jr., "Flight Investigation of the Stability and Control Characteristics of a 0.13-Scale Model of the Convair XFY-1 Vertically Rising Airplane During Constant-Altitude Transitions," NACA RM SL53E18 (1953). Scenes of free-flight tests of the XFY-1 VTOL model in the Full-Scale Tunnel can be seen at *http://www.youtube.com/watch?v=jXiq2-VUbY4*, accessed January 4, 2012.
67. Powell M. Lovell, Jr., "Flight Tests of a Delta-Wing Vertically Rising Airplane Model Powered by a Ducted Fan," NACA RM L55B17 (1955).
68. Scenes of free-flight model tests of VTOL models at the Full-Scale Tunnel and the Langley Control Line Facility can be seen at *http://www.youtube.com/watch?v=NKwc40t6YDM*, accessed January 6, 2012. Additional footage of testing at the CLF and within the 15-Foot Free-Spinning Tunnel building are at *http://www.youtube.com/watch?v=MQy2rkQhS7g*, accessed January 3, 2012.
69. Chambers, *Modeling Flight*.
70. Powell M. Lovell, Jr., and Lysle P. Parlett, "Transition-Flight Tests of a Model of a Low-Wing Transport Vertical-Take-Off Airplane with Tilting Wing and Propellers," NACA TN 3745 (1956). Other general research VTOL free-flight tests in the Full-Scale Tunnel before the establishment of NASA included Powell M. Lovell, Jr., and Lysle P. Parlett, "Flight Tests of a Model of a High-Wing Transport Vertical-Take-Off Airplane with Tilting Wing and Propellers and with Jet Controls at the Rear of the Fuselage for Pitch and Yaw Control," NACA TN 3912 (1957); Louis P. Tosti, "Transition-Flight Investigation of a Four-Engine-Transport Vertical-Take-Off Airplane Model Utilizing a Large Flap and Extensible Vanes for Redirecting the Propeller Slipstream," NACA TN 4131 (1957); Robert H. Kirby, "Flight Investigation of the Stability and Control Characteristics of a Vertically Rising Airplane Research Model with Swept or Unswept Wings and X- or ±Tails," NACA TN 3812 (1956); and Powell M. Lovell, Jr., and Lysle P. Parlett, "Effects of Wing Position and Vertical-Tail Configuration on Stability and Control Characteristics of a Jet-Powered Delta-Wing Vertically Rising Airplane Model," NACA TN 3899 (1957).
71. Charles C. Smith, Jr., "Hovering and Transition Flight Test of a ⅕-Scale Model of a Jet-Powered Vertical-Attitude VTOL Research Airplane," NASA Memorandum 10-27-58L (1958). Digital video scenes of the X-18 free-flight tests in the Full-Scale Tunnel are available at *http://www.youtube.com/watch?v=DXbb29w67SM*, accessed January 2, 2012.
72. Chambers, *Modeling Flight*.
73. Robert O. Shade, "Flight-Test Investigation on the Langley Control-Line Facility of a Model of a Propeller-Driven Tail-Sitter-Type Vertical-Take-off Airplane with Delta Wing During Rapid Transitions," NACA TN 4070 (1957).
74. John W. Paulson, "Investigation of the Low-Speed Flight Characteristics of a 1/15-Scale Model of the Convair XB-58 Airplane," NACA Research Memorandum SL57K19 for

the U.S. Air Force (1957). Digital video scenes of the XB-58 free-flight tests in the Full-Scale Tunnel are available at *http://www.youtube.com/watch?v=F85-5-HhSdk*, accessed January 3, 2012.

75. Digital video scenes of the F8U-1 free-flight tests in the Full-Scale Tunnel are available at *http://www.youtube.com/watch?v=dXBH0PUdEVc*, accessed January 6, 2012.
76. Peter C. Boisseau, "Investigation of the Low-Speed Stability and Control Characteristics of a 1/7-Scale Model of the North American X-15 Airplane," NACA RM L57D09 (1957).
77. Digital video scenes of the X-15 free-flight tests in the Full-Scale Tunnel are available at *http://www.youtube.com/watch?v=3v3KPhOX-vU*, accessed January 6, 2012, and *http://www.youtube.com/watch?v=YXMZcCUJQFY*, accessed January 6, 2012.
78. J.W. Paulson, R.E. Shanks, and J.L. Johnson, Jr., "Low-Speed Flight Characteristics of Reentry Vehicles of the Glide-Landing Type," NASA TM X-331 (1960).

NACA - Langley
September 30, 1958

NOTICE OF TRANSFER OF PERSONNEL FROM THE NATIONAL ADVISORY COMMITTEE FOR AERONAUTICS TO THE NATIONAL AERONAUTICS AND SPACE ADMINISTRATION, EFFECTIVE OCTOBER 1, 1958

Pursuant to Title III of the National Aeronautics and Space Act of 1958 (Public Law 85-568) and proclamation of the Administrator of the National Aeronautics and Space Administration published in the Federal Register for September 29, 1958, you are hereby transferred from the NACA Langley Aeronautical Laboratory, Langley Field, Virginia, to the NASA Langley Research Center, Langley Field, Virginia, effective October 1, 1958.

T. Melvin Butler

T. Melvin Butler
Personnel Officer

This announcement of the end of NACA and transfer to NASA was sent to all Langley employees. (Donald L. Loving family)

CHAPTER 6
Rebirth
1958–1968

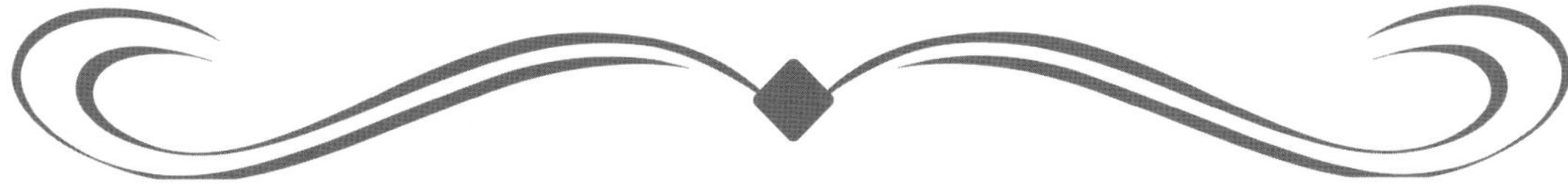

A New Agency

The launch of Sputnik by the Soviet Union in October 1957 created chaos within the Nation's scientific community. The NACA centers had already been engaged in space flight–related research, but the issue of who should lead the U.S. space program for military and civil missions became a subject of great debate. In January 1958, Dr. Hugh L. Dryden, the director of the NACA, observed, "It is the non-military aspects of spaceflight that will have the greatest impact on the thinking and future of all mankind." Dryden noted that the responsibility for military space projects had already been assigned to a new agency known as the Advanced Research Projects Agency (ARPA), but he submitted that the NACA, together with the National Science Foundation, should be responsible for management of nonmilitary space projects.[1] His views were strongly supported in several noted aviation publications, including a major editorial by Robert Hotz, editor of *Aviation Week* magazine, who stated, "If NACA gets the job our jump into space will be catapulted from a solid launching pad."[2] However, the vision of the NACA leading the space program was not to be.

On July 29, 1958, President Dwight D. Eisenhower signed HR-12575, the National Aeronautics and Space Act of 1958. Eisenhower stated, "The present NACA, with its large and competent staff and well-equipped laboratories, will provide the nucleus for the NASA." The NACA continued to exist through the summer of 1958 until October 1, when the new National Aeronautics and Space Administration officially came into existence. NASA took over all existing NACA facilities and renamed the NACA laboratories, which resulted in the Langley Aeronautical Laboratory becoming the NASA Langley Research Center.

For most employees at Langley, duties on the day following the changeover to NASA were routine. However, local lectures and educational opportunities regarding space technology that had been underway for over a year suddenly intensified, and many of the old aeronautical staff began to inquire about opportunities in the new space program.

Shakeup at the Full-Scale Tunnel

On December 10, 1958, a major reorganization of Eugene Draley's Full-Scale Research Division was announced with a direct and major impact on the staff of the Full-Scale Tunnel. Management of the tunnel was transferred from Draley's organization to the Stability Research Division under Thomas A. Harris. Albert von Doenhoff and Jerry Brewer were both relieved of their duties, respectively, as head and assistant head of the Boundary Layer and Helicopter Branch, which previously had responsibilities for the Full-Scale Tunnel. Von Doenhoff was appointed to the Compressibility Research Division, and Brewer was reassigned to special duties within the Full-Scale Research Division.[3]

The reorganization included major reassignments for the existing staff of 30 engineers, technicians, computers, and administrative personnel at the Full-Scale Tunnel. A group of 14 staff members was transferred to the Langley Flight Research Division located in the new flight hangar in the NASA West Area, and the remaining 16 personnel remained with the Full-Scale Tunnel when it was transferred to the Stability Research Division. Along with the personnel changes within the Full-Scale Tunnel staff, the responsibility for management for the Langley Helicopter Tower was transferred to the Flight Research Division.

The combined events of the new emphasis on space technology and the reorganization were unsettling and caused many of the staff members to reconsider their career paths and new opportunities. Several quickly transferred to the new NASA Space Task Group (STG) at Langley, which was preparing for Project Mercury.

Campbell's Coup: Free-Flight Moves In

The impact of the reorganization at the Full-Scale Tunnel had resulted in a significant reduction of personnel and research projects in late 1958. However, the impressive results of earlier free-flight model testing in the facility had made a significant impression on industry, the military, and NASA management. Dynamic stability issues for emerging V/STOL and reentry vehicles at subsonic speeds were an almost perfect fit for the analysis capabilities provided by the free-flight testing technique, and the Stability Research Division strongly advocated for the full-time transfer of the testing technique and associated personnel from the Langley 12-Foot Free-Flight to the Full-Scale Tunnel.

John P. Campbell and his organization moved from the Free-Flight Tunnel to new offices at the Full-Scale Tunnel in early 1959 as yet another Langley organizational change occurred.[4] On September 14, Langley combined the existing Stability Research Division and Flight Research Division into a new organization known as the Aero-Space Mechanics Division (ASMD), with Philip Donely as chief and Charles Zimmerman as associate chief. Campbell organized his Dynamic Stability Branch into two sections, known as the High-Speed Configuration Section, under John W. "Jack" Paulson, and the VTOL Section, under Marion O. "Mac" McKinney.[5]

The managers overseeing the operations of the Full-Scale Tunnel were characterized by a diverse set of qualifications and personalities. Phil Donely's background was in dynamic loads and gust response, and he was unfamiliar with the field of dynamic stability and control. A gruff and abrasive leader typical of the old NACA Division chiefs, he demanded professional work regardless of his understanding of the technical details. Campbell and Zimmerman were internationally recognized experts in the field of dynamic stability control, and both were soft-spoken, enthusiastic leaders. Paulson became the branch leader and principal participant in emerging NASA programs involving reentry vehicles, especially inter-Center programs with the NASA Flight Research Center (now the NASA Armstrong Flight Research Center) and the NASA Ames Research Center for the configurations known as lifting bodies, which built the foundation for the Space Shuttle Program. McKinney's background in dynamic stability and control had been surpassed by his intense research on V/STOL aircraft and personal collaborations with Campbell and Zimmerman. McKinney was a fiery leader with an outspoken demeanor and little regard for interpersonal relationships while accomplishing the job. Some of the older staff members at the Full-Scale Tunnel regarded him as the second coming of Hack Wilson.

New Blood

With the departure of roughly half of Jerry Brewer's staff and the arrival of an equal number of John Campbell's Free-Flight Tunnel staff, the Dynamic Stability Branch was now a mixing pot of personnel. Some of the researchers had expertise in large-scale testing in the Full-Scale Tunnel, while others who specialized in the field of dynamic stability control and free-flight model testing knew nothing about full-scale tests. The branch began operations in January 1959 with a staff of 31 engineers and support personnel.

During this era, the ebb and flow of new personnel at the Full-Scale Tunnel was significantly affected by the buildup of the new NASA space programs and in particular the unabated growth of the Apollo program. Hiring activity became frantic and intense as the Center rapidly increased its complement. For example, when the author was hired by Langley in 1962, the center hired an astounding 434 new employees—a single-year increment of new personnel that exceeded the cumulative growth of Langley during its first 20 years as a NACA laboratory.

Another major source of personnel in support of branch operations was the Langley Cooperative Education (Co-op) Program for college students, which integrated classroom studies and supervised work experiences. Initiated in 1952, the process alternated work and study in 3-month segments designed to provide the student with increasing responsibilities, resulting in enhanced academic skills and engineering experience gained from onsite assignments at Langley. Students were able to indicate their preference for work assignments while at the Center, and an assignment to the Dynamic Stability Branch was regarded as a particularly valuable learning experience. Throughout the years that the Full-Scale Tunnel was managed by NASA, the Co-op student community became acutely aware of the unique opportunities afforded by a stay at the tunnel, and it was constantly requested as a work assignment.

In addition to the influx of engineers, highly skilled technicians and graduates of the NASA apprentice program were added to the work force, resulting in a broad spectrum of support capabilities ranging from experience with full-scale aircraft systems to specialists in the areas of free-flight model systems, including instrumentation, control systems, and testing technique updates. The talents provided by the technician staff enabled tunnel operations to be particularly efficient, inexpensive, and timely. The versatile talents of the technician staff proved to be especially valuable since staff members provided capabilities to design, construct, and maintain models. These contributions benefitted the branch's ability to not only perform appropriate rapid-turnaround service for approved projects, but also to provide support for unauthorized but important "jack-leg" projects without going through the formal NASA procedures and other service organizations that frequently delayed research efforts.

Fun at the New Home

The transfer of Campbell's staff to the Full-Scale Tunnel provided impressive new surroundings in the vast areas in the building. At that time, the interpersonal and social lives of the staff retained the informal and fraternity-like atmosphere of the old NACA days. Inter-organizational athletic events were the norm, characterized by passionate competition and bragging rights. Beach parties and get-togethers were weekly affairs shared by all.

George M. Ware was hired into Campbell's group a year before it moved to the Full-Scale Tunnel, and he recalled some of his favorite memories of his 12 years at the tunnel:

> The people: The feeling of family. Everyone under the Full-Scale Tunnel roof was family. We knew all the spouses and most of the children of all our Branch members. We had free exchange between supervisors, engineers and technicians. Everyone worked together to make a test, project, or task a success. You felt that you were doing something of value. The test you are involved with and the results you published were adding to the advancement of the art. [6]

Ware also recalled his favorite anecdotes:

> Seeing who could ride a bicycle highest up the test-section entrance cone, walking around the offices with trash cans on your feet to keep dry in the frequent 10–12 inches of floodwater, watching cars float in the parking lot during floods, visits by the Center Director during the middle of night shift, birds in the return passages of the tunnel, how spooky it was during night shifts when various animals (especially raccoons) visited the tunnel, several visits by National Geographic Magazine and the ABC News team headed by Jules Bergman, the variety of flying contraptions tested, being lost in the fog and driving onto the Langley Air Force Base airstrip, the center-wide buzzer that sounded to start work, begin and end lunch, and the end of work, seeing astronauts walking around in space suits as they trained in the Mercury Procedures Simulator, the

single telephone that was provided per multi-man office with a ring code for each engineer, and…slide rules![7]

An Old Friend Retires

On June 30, 1961, the NASA Langley Research Center said goodbye to its respected Director Dr. H.J.E. Reid as he retired after over four decades of Government service.[8] Reid had started his career as a junior engineer at Langley in April 1921 and became engineer in charge of Langley in 1926. During his 34 years as the top official, Reid maintained a special interest in the operations at the Full-Scale Tunnel and was a frequent visitor. He sometimes visited during night shifts and befriended many longtime staff members at the tunnel. Although he was regarded as having limited technical prowess, Reid always supported his staff with ferocity and great effectiveness. Several staff members from the Full-Scale Tunnel were personally invited to his retirement party, including Joe Walker, the head mechanic who had maintained a long-term friendship with Reid for over 30 years.

"Mac" Takes Over

In July 1962, John Campbell was promoted and became an assistant chief of ASMD under Philip Donely. Mac McKinney was named head of the Dynamic Stability Branch with responsibilities for the operation of the Full-Scale Tunnel, and Jack Paulson was promoted to assistant branch head. Two research groups were formed: one under Robert H. Kirby that focused on V/STOL projects and reported to McKinney, and the other under Joseph L. Johnson, Jr., that conducted dynamic-stability research for non-V/STOL aerospace vehicles and reported to Paulson. In 1963, the division was reorganized as the Flight Mechanics and Technology Division, and the Dynamic Stability Branch was also assigned responsibilities for operation of the Langley 20-Foot Spin Tunnel and outdoor radio-controlled drop model flight tests in addition to the existing responsibilities for the Full-Scale Tunnel and the 12-Foot Low-Speed Tunnel.

Dynamic Stability Matures

In the mid-1960s, Mac McKinney recognized that new computer- and piloted-simulator technologies were beginning to play an increasingly important role in the analysis of dynamic stability and control characteristics of aerospace vehicles. Prior to that time, the free-flight model work conducted in the old Free-Flight Tunnel and the Full-Scale Tunnel had been primarily experimental in nature, and any accompanying analytical analyses were limited to simple linear theories used to predict the damping and period of motions or the first-order response to control inputs. However, the emergence of digital computers and simulators provided information that simply could not be obtained using remotely piloted free-flight models. The ability to simulate the flight behavior of a full-scale vehicle in real time and use a human pilot for detailed assessments of handling qualities during real-world maneuvers presented a new level of analysis capabilities. In addition to providing a more realistic piloting situation, the new analysis tools could provide representations of important nonlinear aerodynamic data and extremely sophisticated flight control systems.

In 1967, McKinney formed a Simulation and Analysis group under Joseph R. Chambers to integrate these capabilities into the Dynamic Stability Branch's repertoire. With a staff of six engineers, the group used conventional and special dynamic wind tunnel tests in the Full-Scale Tunnel and the 12-Foot Low-Speed Tunnel to gather detailed aerodynamic data for use as inputs in sophisticated analyses that included the use of Langley's piloted simulators. Many of the individuals in the group never actually conducted experimental testing in the Full-Scale Tunnel, and they were totally dependent on wind tunnel results obtained by others in the branch for realistic aerodynamic inputs required for analyses and simulation.

The general engineering approach used in these early efforts rapidly matured over the next decades of NASA's presence at the Full-Scale Tunnel. The synergistic value of combining conventional wind tunnel tests, special dynamic force tests, free-flight model tests, and piloted simulation became readily apparent—especially to the industry and the military communities that were conducting cooperative activities with the Dynamic Stability Branch in critical developmental support efforts for the Nation's first-line aircraft. As a result of the extensive experience gained by participation with military projects and feedback from full-scale flight tests, the staff was able to accumulate knowledge regarding the accuracy of their predictions of dynamic stability and control based on the correlation of results from model free-flight tests, piloted simulation, and actual full-scale aircraft behavior. More importantly, the expertise and experiences obtained by participation in numerous civil- and military-aircraft development programs resulted in a pool of experts that industry and Department of Defense (DOD) could tap for recommendations and previous lessons learned.

Facility Modifications

The ⅟15-scale model of the Full-Scale Tunnel had been used almost continuously since being moved to the Full-Scale Tunnel building in 1933. However, in the late 1950s, the model tunnel was given to Portugal by NASA Headquarters for use in that Nation's scientific programs as a good-will gesture within NATO's AGARD cooperative program. As will be discussed in a later chapter, NASA ultimately contracted for a second model tunnel during the 1980s, when flow improvement studies were conducted for the Full-Scale Tunnel.

The introduction of free-flight model testing as a full-time activity in the Full-Scale Tunnel required updates to the tunnel as well as to the hardware used during testing. During the early 1960s, the old open-air balcony that had been used for preliminary model flight investigations in the late 1950s was enclosed into a room with stations for the pilot controlling pitching motions; the model thrust operator; and other members of the test crew, including the model safety-cable operator. This latter operator's task was made considerably safer with the addition of a high-speed hydraulic rotary winch for rapidly paying out cable slack or quickly snubbing the model of the tunnel air stream in the event of loss of control.

After earlier attempts at situating the roll and yaw pilots in an enclosure on the ground board or at the top of the exit cone, a special observation and piloting room was constructed for these members of the test crew within the lower tunnel exit cone.

Continual development of propulsion systems for the free-flight models included a wide variety of concepts, including compressed-air jets or ejectors, electric motors, tip-driven fans, and ducted or free propellers. The larger models also required updated control actuators and more instrumentation.

When John Campbell's staff migrated to the Full-Scale Tunnel, free-flight testing was terminated at the old Free-Flight Tunnel. The university-class tunnel was then used for quick-response assessments of advanced aerospace configurations and concepts. With operations only requiring an engineer and technician, conventional static tests to measure aerodynamic performance, stability, and control of subscale models were efficiently conducted prior to testing in the Full-Scale Tunnel. In addition, the tunnel retained the capability for specialized dynamic force tests to measure aerodynamic phenomena related to angular motions of vehicles. The 75-year-old tunnel, renamed the Langley 12-Foot Low-Speed Tunnel in 1959, has continued to operate to this day and is currently Langley's oldest operational wind tunnel. The 12-foot tunnel is frequently requested for use in projects from other organizations at Langley.

Review of Test Activities

When test operations at the Full-Scale Tunnel began under the new organization in 1959, tunnel test time was shared between free-flight testing and large-scale helicopter and V/STOL projects. In many instances, the tunnel test section would be occupied by two different tests during a 24-hour period. For example, the day shift might be devoted to a free-flight model test, which typically required the presence of several test personnel to be available, whereas the night shift might consist of a less-complex conventional force test of a subscale model requiring only two or three personnel. By raising the free-flight model out of the test section overnight using its safety cable, the nighttime force test could be conducted with minimal interruption. This type of "double scheduling" was extremely efficient and was adhered to during the remaining 50 years of the life of the Full-Scale Tunnel—although it caused considerable confusion for outsiders regarding the tunnel schedule and projects.

The broad spectrum of testing that took place from 1958 through 1969 included both basic research and support for high-priority national programs, including space capsules, lifting bodies, V/STOL configurations, helicopters, flexible-wing concepts, general aviation aircraft, supersonic transports, and training concepts for astronauts for the first Moon landing. Over 150 technical reports were issued by NASA on the results of these tests through public dissemination and classified reports to industry and the military. Hundreds of technical presentations, onsite briefings to appropriate visitors, and status reports at NASA Headquarters were also conducted to ensure timely and pertinent research efforts and plans. The results of many of the test programs were used as justification and advocacy for new aerospace vehicles and new NASA facilities.

Fly Me to the Moon

As NASA scrambled to apply its existing facilities to the ominous Soviet space challenge, it became obvious that many of the issues relevant to the subsonic phase of reentry and the landing behavior of crewed spacecraft could be worked at the Full-Scale Tunnel. Research that had been conducted there with free-flying models of highly swept hypersonic-boost/glide vehicles during the last days of the NACA provided extremely valuable information on dynamic stability and control. The aerodynamics of crewed space capsules, lifting bodies, and concepts for enhanced "footprint" landing areas using auxiliary flexible wings were assessed. In addition, several critical capabilities used for the space program were developed in the Full-Scale Tunnel building, including a procedures trainer for Project Mercury and a proof-of-concept experiment that would help justify the construction of the famous Langley Lunar Landing Research Facility.

Hypersonic Boost/Glide Vehicles

The trajectories used for reentry from space missions for lifting vehicles (such as highly swept delta-wing configurations) involved flight at high angles of attack, which presented many issues regarding the adequacy of stability and control during the final phases of reentry and landing. In support of NACA and NASA research activities, free-flight model tests were conducted in the Full-Scale Tunnel to assess longitudinal and lateral directional characteristics and to develop criteria for satisfactory behavior. Such studies were very detailed and included the effects of artificial stabilization and recommendations for control system design. Hypersonic boost/glide vehicles were characterized by having most of their mass distributed along the fuselage. As a result of the mass distribution and highly swept wings, the configurations usually displayed uncontrollable rolling oscillations at high angles of attack. The Full-Scale Tunnel tests demonstrated that the oscillations could be mitigated with appropriate artificial stabilization systems.[9]

One of the most important contributions to stability and control technology derived in part from this work in the Full-Scale Tunnel was a stability parameter known as "dynamic directional stability," which provided guidance for the relative levels of conventional static lateral directional stability required for satisfactory behavior of slender configurations at high angles of attack.[10] The parameter, which was conceived by Jack Paulson in collaboration with another organization at Langley, proved to be extremely useful by giving a quick prediction of complex dynamic behavior based on conventional wind tunnel data. The parameter was subsequently applied to the emerging fighter aircraft designs for high-angle-of-attack conditions during the 1960s and 1970s.

In addition to the pioneering NACA/NASA studies of hypersonic gliders, the Full-Scale Tunnel was used to support the Air Force's Dyna-Soar program for the development of a

The Full-Scale Tunnel was used for extensive free-flight and force testing of the U.S. Air Force Dyna-Soar reentry glider in 1960. (NASA L-60-8494)

military reentry glider.[11] During the program, a ⅕-scale model of the Dyna-Soar vehicle was used in free-flight tests and supporting force tests to determine the stability and control characteristics of the vehicle for the subsonic phase of reentry flight.[12]

"Spam in a Can"

The decision by the STG to use a crewed capsule for Project Mercury because of reentry heating issues and other concerns brought forth a multitude of technical issues. Many pilots in the astronaut corps bemoaned the fact that they would be piloting a vehicle with no significant cross-range maneuvering capability to a water landing. Some even referred to piloting the capsule concepts as "spam in a can." After the capsule decision was made, virtually all of Langley's facilities and laboratories focused on providing solutions to potential aerodynamic, heating, structural, and control problems for such designs. At subsonic speeds, one of the major issues that had to be addressed was the fact that blunt capsule shapes exhibit static and/or dynamic instabilities during reentry.

A full-scale model of the Mercury capsule is inspected by Clarence D. "Don" Cone during tests in January 1959. (NASA EL-1996-00094)

In early 1959, a full-scale model of a Project Mercury space capsule was tested in the Full-Scale Tunnel to determine the variation of lift, drag, and pitching moment with angle of attack. The tests were conducted for two different parachute canister lengths.[13] The capsule was also tested with a smooth outer surface and with smooth and corrugated canister surfaces. Results were obtained for an angle-of-attack range of –5° to 88.7°. Data obtained in the test program were quickly disseminated to appropriate organizations for analysis and guidance in Project Mercury.

Capsule? Bomb? What Is It?

One of the many unusual tests conducted in the Full-Scale Tunnel has resulted in considerable confusion over the years regarding the test subject and the objectives of the test. At the beginning of 1958, the advent of high-altitude and high–Mach number missiles stimulated concern over potential aerodynamic missile heating problems that are dependent on the character of the boundary layer over the blunted noses of such configurations during flight.[14] The effect of roughness on the aerodynamics of the blunted-nose section was of particular interest; however, at the time there were no wind tunnels that could duplicate the Mach numbers and Reynolds numbers experienced by missile noses. Instead, the scientific community was using very expensive rockets with highly polished noses to obtain such data. Langley researchers in another organization requested an entry into the Full-Scale Tunnel to test a body of revolution of large diameter in low-speed flow in order to determine effects of roughness sizes while retaining the high Reynolds numbers of interest in high-speed conditions.

This photograph of tests of a capsule-like article has caused considerable confusion through the years and has been misinterpreted as a possible space-capsule configuration. In reality, it was a low-speed test in support of supersonic missile technology. (NACA LAL 57-5549)

Previous work had indicated that the critical roughness Reynolds number was independent of Mach number up to a Mach value of 2.0. In addition, the flow behind the detached shock of blunt-nosed bodies in supersonic flight is subsonic near the stagnation point, and the Mach number was not expected to have a large influence on transition. Accordingly, a 10-foot-diameter body with a semi-spherical nose was constructed for testing in the tunnel, and boundary-layer transition was measured for various sizes and shapes of roughness placed from 10° to 30° from the stagnation point at the nose.

Photographs of the test article are often misinterpreted in the media as a test in the Full-Scale Tunnel of an alternate Project Mercury capsule.

Communicating with Astronauts

One particular research activity in the Full-Scale Tunnel building produced especially critical contributions to the embryonic U.S. space program in the late 1950s and early 1960s. As the United States raced to meet the Russian challenge of Sputnik, leading NASA managers and researchers realized how ignorant they were of space and rocket technologies. The massive education process included developing basic understandings of celestial flight; meeting the challenges of designing a rocket/capsule configuration that would be reliable and safe; and developing the technical and political means to communicate, track, and recover astronauts during and after space missions.

Langley's Christopher C. Kraft was assigned the incredibly difficult job of organizing and directing the flight operations for the Nation's first crewed space flights during Project Mercury. As a distinguished veteran of flight research operations, he quickly recognized the special challenges facing his STG team in the areas of procedures and communication. Kraft conceived the idea for flight-procedures trainers that could be tied into a central-command center and provide realistic training for routine and emergency operations for astronauts, controllers, and tracking-station personnel. Under NASA contracts to McDonnell, two Mercury procedures trainers were built by the Link Trainer Company, which was famous for building airplane trainers for pilots in WWII. One of the procedures trainers was installed at Cape Canaveral and the second was located in the Full-Scale Tunnel building at Langley.[15]

A special enclosed room was built beneath the exit cone flooring of the Full-Scale Tunnel (under the tunnel air circuit) to house Mercury Procedures Trainer No. 1, which consisted of a complete mockup of the Mercury capsule with operating instruments and controls interconnected to an analog computer. The trainer provided practice in sequence monitoring and familiarization with the cockpit systems. External reference through the capsule's periscope was simulated by means of a cathode-ray tube display, and provision was included for pressurizing the astronaut's suit and for simulating heat and noise. Also included in the same room, for the equally essential training of the personnel who would crew the 17 Mercury tracking stations across the world, were three ground-control consoles (for a doctor, spacecraft communicator, and systems monitor), which formed the minimum equipment at any one of the tracking stations. The simulators were installed and in use by April 1960. These devices were first called "procedures trainers" and later "Mercury simulators." Here, the astronaut, supine in a mockup capsule, rehearsed the flight plan for a specific mission.

Photograph of the Project Mercury procedures trainer in the Full-Scale Tunnel building shows the setup for simulated missions. Astronaut John Glenn is seated in the simulator and engineer Charles Olasky is stationed at the computer console. (NASA GPN-2002-000044)

By simulating the entire mission, the fledgling team could interact during training sessions, which were held dozens of times a day in the buildup for crewed missions. Virtually all the original seven astronauts used the Langley procedures trainer. With the use of the trainers, the STG team built up the procedural elements during Mercury-Redstone and Mercury-Atlas missions that would become the backbone of future NASA crewed space activities. The Mercury astronauts who flew the suborbital missions claimed that the most useful preflight training for normal and abnormal conditions was obtained in the trainer. The simulators were in use 55 to 60 hours a week during the 3 months preceding the flight of Freedom 7. During his preflight training period, Alan Shepard flew 120 simulated Mercury-Redstone flights.

After the Manned Spacecraft Center was moved from Langley to Houston, TX, Mercury Procedures Trainer No. 1, redesignated the Mercury Simulator, was moved from the Full-Scale Tunnel building to a Manned Spacecraft Center building at Ellington Air Force Base (AFB) in Houston on July 23, 1962.[16] The room that had been occupied by the Mercury simulator at the tunnel building was then remodeled and became a model-preparation room for tests in the Full-Scale Tunnel.

Landing on the Moon

Landing on the surface of the Moon was known to be one of the most critical phases of the Apollo program. The control problem for a human pilot was especially difficult for several reasons: the lunar gravity is only one-sixth that of Earth's; all control of vehicle lift and attitude is provided by rockets, which provide an on-off control rather than a linear variation of control similar to aircraft; and the nature of the lunar surface was not known in detail. One of the foremost issues was how to train the astronauts to perform this difficult landing task.

Langley's distinguished long-time researcher W. Hewitt Phillips was the first to conceive a special facility for practicing lunar landings.[17] His concept in 1961 was to simulate the reduced gravity of the Moon by providing a suspension system for the pilot's vehicle that would exert a constant force in the vertical direction equal to five-sixths of the weight of the vehicle. The force would be measured by a strain-gage balance and used to control the output of a servomechanism to reel the cable in and out to apply the desired constant force to the vehicle. Phillips visualized the use of hydrogen-peroxide rockets to support the weight of the vehicle and provide control moments to maneuver the vehicle.

Phillips's first technical obstacle was to ensure that the servomechanism would work and provide crisp and reliable operations. To test the feasibility of this concept, Phillips worked with the staff of the Full-Scale Tunnel in 1962 to provide a piloted evaluation of the system components. A simplified system was constructed in which a pilot's "chair" was suspended by a vertical cable, with the servomechanism that reeled the cable in or out mounted to the girders of the roof of the west return passage in the Full-Scale Tunnel, 60 feet above the vehicle. The return passage at the tunnel was chosen for the experiment because the tunnel

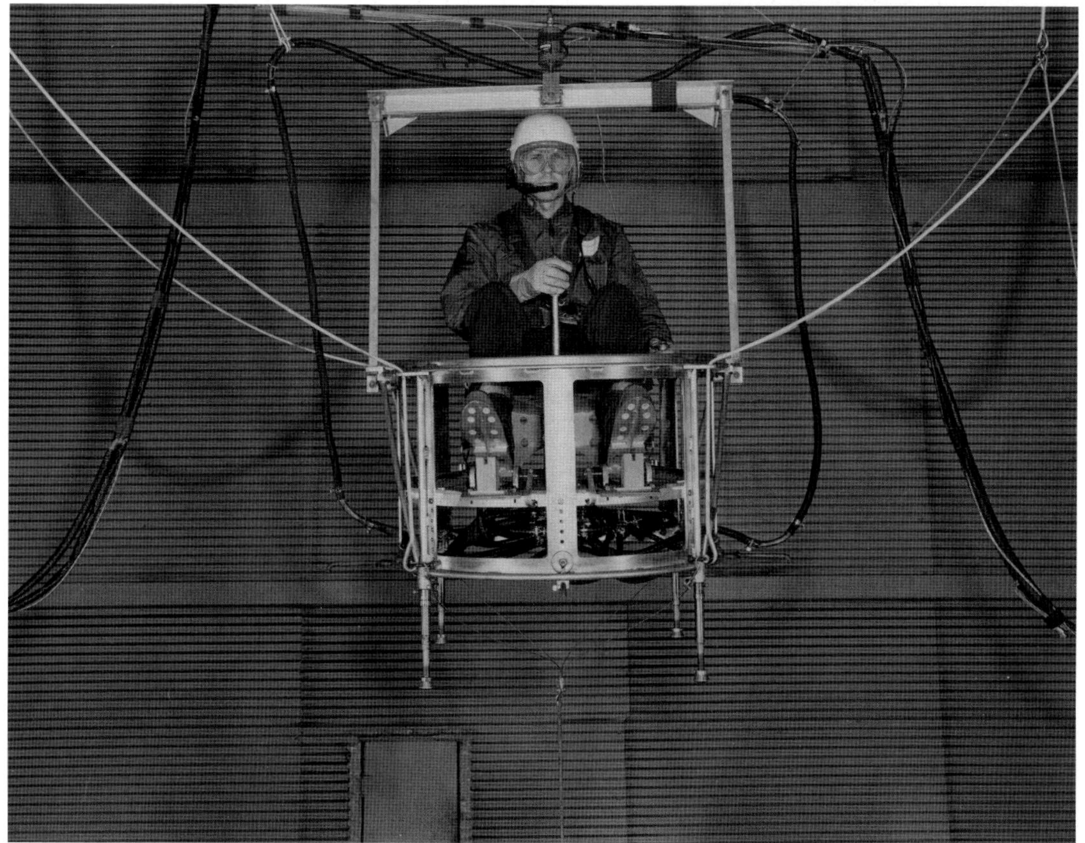

Langley test pilot Bob Champine flies the "flying chair" apparatus during proof-of-concept testing to develop the servomechanism concept required for the development of the famous Langley Landing Research Facility. The evaluation was conducted in a return passage of the Full-Scale Tunnel in 1962. (NASA L-62-1458)

had a powerful air compressor (normally used for free-flight model tests) that could be used to provide vertical thrust with which to simulate the effects of the lunar gravity. An analog computer was used to calculate the signals driving the servomechanism, and a safety system was in place so that the flying chair could be locked in place or lowered to the ground in the event of a malfunction.

The assessments of the proof-of-concept model by Langley test pilots Jack Reeder and Robert A. "Bob" Champine were very favorable—despite the fact that many naysayers claimed the concept would not be feasible. Boosted in confidence by the successful evaluation at the Full-Scale Tunnel, Phillips proceeded to advocate at Langley and NASA Headquarters for the construction of a Lunar Landing Research Facility (LLRF). The project was quickly approved at Headquarters and the LLRF was constructed in 1965. Following its historic contributions to training for the Apollo mission, the LLRF was applied to a wide scope of studies, including use as a crash test facility for full-scale aircraft and helicopters. The LLRF was declared a National Historic Landmark in 1985 and is still used today to simulate water landings for future spacecraft. The research vehicle used during the training at the LLRF is now on display at the Virginia Air and Space Center in Hampton.

Hewitt Phillips never forgot the vital role that the proof-of-concept testing in the Full-Scale Tunnel building played in achieving his vision, and he frequently referred to the test in many presentations and discussions of the LLRF.

Handling Qualities on the Moon

When the simplified mockup of the suspension system for the LLRF was evaluated at the Full-Scale Tunnel, some preliminary information on the handling-quality requirements for a crewed lunar landing vehicle operating in a simulated lunar gravitational field was obtained.[18] The level of control power required for satisfactory characteristics had been identified as a critical design factor, and several organizations had begun to explore the situation. For example, researchers at the NASA Ames Research Center had conducted inflight evaluations of flight behavior using the Bell X-14 jet VTOL airplane flown along several proposed lunar-landing trajectories.

For the Langley tests at the Full-Scale Tunnel, Langley test pilot Bob Champine performed maneuvers and assessments of the control levels and responses desired in the simulated lunar gravitational environment. A critical part of the investigation was to correlate control requirements for moon landers with requirements previously experienced for satisfactory handling qualities for helicopters and VTOL aircraft. Results of the Langley evaluation indicated that larger pitch and bank angles were required for linear acceleration of the vehicle in a reduced-gravity environment than were required for helicopters and VTOL airplanes in Earth's gravitational field. In addition, the minimum control requirements were found to be somewhat higher than those required for helicopters and VTOL aircraft. The pioneering data generated in the brief study were very helpful in the preliminary-design requirements for the LLRF as well as the Apollo configuration.

Houston Has a Problem: The Lunar Landing Training Vehicle

While Hewitt Phillips was developing the groundwork for the LLRF at Langley, other concepts for training the astronauts to land on the Moon surfaced within other organizations.[19] Bell Aerosystems was a major participant in the development of vertical takeoff and landing aircraft technology in the 1950s, and its personnel were well acquainted with controllability and VTOL performance issues. Therefore, Bell was in a position to react quickly to a request from NASA in late 1961 for concepts for a free-flying piloted simulator for astronaut training.[20]

Bell subsequently designed and delivered two Lunar Landing Research Vehicles (LLRVs) to the NASA Flight Research Center (now the NASA Armstrong Flight Research Center) in 1964. The flying bedstead resembled early VTOL prototypes, consisting of a tubular truss construction and an open cockpit with an ejection seat. A turbofan engine mounted vertically in a gimbal provided vertical lift in excess of vehicle weight to propel the craft to a predetermined test altitude. The thrust from the engine was then reduced to support five-sixths of the vehicle's weight, representing the Moon's gravitational field. Hydrogen-peroxide lift rockets were used to modulate rate of descent and translational movement during flight. Thrusters powered by hydrogen peroxide were used to provide pitch, roll, and yaw control.

After 2 years of very productive flight research and pilot training, NASA contracted with Bell to deliver three updated Lunar Landing Training Vehicles (LLTVs). These vehicles had improved capabilities to simulate the Apollo Lunar Excursion Module's (LEM) features, including a cockpit display and control system representative of the LEM. One very significant modification to the original LLRV design consisted of modifying the cockpit arrangement to simulate the enclosed cockpit of a lunar module. The modified configuration had three walls and a roof atop the cockpit, with only the front area exposed.

On December 8, 1968, NASA pilot Joseph S. "Joe" Algranti was conducting a flight of LLTV 1 at Ellington AFB as an acceptance flight following vehicle modifications and before releasing it for astronaut training.[21] During the flight, with wind shear present, the vehicle abruptly yawed off to the side and Algranti's control inputs could not stop it. The vehicle rapidly rolled from 90° to the right to 90° to the left. Algranti safely ejected from the vehicle only 0.3 seconds before impact. A quick review of the flight data revealed that the control

A Super Guppy transport was used to transfer a Lunar Landing Training Vehicle to Langley for tests in the Full-Scale Tunnel in December 1968. Photographs document the arrival and unloading of the vehicle. (NASA L-69-1671 and NASA L-69-1740)

A Bell Lunar Landing Training Vehicle was tested in the Full-Scale Tunnel to provide information on the aerodynamic stability and control of the vehicle following an accident in December 1968. Results of the investigation helped revise the vehicle's geometry and training mission ground rules in time for Neil Armstrong to make several training flights prior to the Moon landing in July 1969. (NASA EL-2000-00449)

required had exceeded the ability of the yaw thrusters to control the vehicle. The ensuing accident investigation board recommended that another LLTV be tested in the NASA Langley Full-Scale Tunnel to investigate the aerodynamic behavior and controllability of the configuration. Accordingly, LLTV 3 was loaded aboard a modified Boeing Stratocruiser "Super Guppy" and flown to Langley for installation in the tunnel.

The results of the test program in the Full-Scale Tunnel in January 1969 indicated that the modified cockpit configuration of the LLTV caused large unstable aerodynamic yawing moments when the vehicle was sideslipped as little as 2°. Basically, the cockpit arrangement acted as a large "sugar scoop" to destabilize the vehicle beyond controllability limits during flight in windy conditions. The cockpit was subsequently modified by removing the roof, thereby venting the destabilizing area. Based on the tunnel data, the flight-training program adopted a preliminary restricted flight envelope for angle of attack, speed, and angle of sideslip for the modified vehicle. After validating the flight envelope with additional flight tests, the training continued and the yaw control problem was resolved. Soon thereafter, Neil Armstrong conducted several training sessions and proficiency flying in the LLTV on June 14 (two flights), June 15 (three flights), and June 16 (two flights), before the Apollo 11 mission on July 16, 1969. The LLTV used for the tunnel tests at Langley (NASA 952) is now on exhibit at the Johnson Space Center in Houston.

Thanks to the availability of an old wind tunnel designed in the 1930s to test biplanes, the United States had resolved a major problem in the Apollo program in a timely fashion.

Flying Bathtubs

Early in the NASA space program, technical candidates for reentry and landings of spacecraft were addressed to extend operational range and landing options.[22] Capsules and other symmetric spacecraft shapes designed for the U.S. Moon mission would follow a ballistic trajectory during reentry, which minimized both potential downrange flexibility and options to seek landing sites other than in water. In contrast, vehicles capable of providing lift for maneuvers during reentry might provide crews the ability to deviate from a ballistic path and glide to a land-based runway landing like an aircraft. The prominent NASA leaders in

Free-flight tests of the Eggers M1 reentry body concept were conducted in 1959 to determine the stability and controllability of the concept. Note the flexible flight cable that provided a compressed-air source for thrust and the control inputs from the remotely located pilots. (NASA L-59-1547)

the conception and development of "lifting bodies" in the 1950s were Harvey Allen and Alfred Eggers of the NASA Ames Research Center. Led by Allen and Eggers, engineers at Ames began conceptual studies of blunted half-cone shapes that could be used as potential lifting bodies. Although the lift-to-drag ratios produced by the configurations (about 1.5 at hypersonic speeds) were much lower than those normally associated with aircraft, the lift would provide a revolutionary capability for control of reentry flight parameters.

The early Ames conceptual studies led to the development of a family of potential reentry configurations. One of the first designs, known as the Eggers M1, resembled a sawed-off rocket nose cone. Researchers at the Full-Scale Tunnel were following the development of the Ames configurations with an interest in determining the dynamic stability and control characteristics of such radical configurations. In 1959, free-flight model tests of the M1 design were conducted in the tunnel with very positive results, further increasing the growing interest in lifting bodies.[23] The Ames staff continued to refine and further develop their configurations, resulting in a half-cone shape with a blunted 13° nose semi-apex angle known as the M2. As the concept continued toward maturity with additional wind tunnel testing over the operational speed range, major deficiencies in stability and control were noted at subsonic speeds, requiring modifications to the aft end of the configuration. The M2 lifting body configuration included stubby vertical fins and horizontal elevons, with trailing-edge flaps on the body to provide longitudinal trim.

Free-flight tests were made in the Full-Scale Tunnel to determine the dynamic stability characteristics of further development of the M2 known as the M2-F1.[24] The first flight tests of the full-scale M2-F1 research vehicle occurred at the Flight Research Center on April 5, 1963. The successful flight testing of the M2-F1—essentially a lightweight, low-cost concept demonstrator—gave NASA the confidence to proceed with a pair of more sophisticated heavyweight rocket-powered lifting bodies for evaluations of the handling qualities of more representative lifting-body configurations across the speed range from supersonic flight to landing. The craft would be launched from a NASA B-52 mother ship and be powered by XLR-11 rocket engines. The two advanced configurations were known as the M2-F2 and the HL-10. Following a NASA review of competitive industry designs, both vehicles were built by Northrop.

Langley Births the HL-10

In 1962, researchers at NASA Langley, led by Eugene S. Love, conceived a lifting body known as the Horizontal Lander (HL)-10.[25] In contrast to the half-cone designs of the M2 series, the HL-10 was a flat-bottomed, inverted-airfoil shape with a split trailing-edge elevon

Free-flight testing of a model of the HL-10 was conducted in 1963 to evaluate the dynamic stability and control characteristics of the configuration. In this photograph, an early version of the configuration has a single vertical tail whereas the final configuration used three vertical fins. (NASA EL-2000-00430)

for pitch and roll control with tip fins and a center fin for directional stability. The extent of research conducted at Langley on the HL-10 was massive. Virtually every wind tunnel at Langley tested the configuration. Models of the HL-10 underwent aerodynamic, heating, launch-vehicle compatibility, dynamic stability, and ground- and water-landing tests; and piloted simulator evaluations were also conducted based on aerodynamic inputs from the tunnel testing.

A larger-than-full-scale model of the HL-10 lifting body undergoes tests to determine the effects of tail configuration on its low-speed performance, stability and control in 1965. (NASA L-65-2436)

In 1964, the low-speed flight characteristics of a three-fin version of the HL-10 were investigated in the Full-Scale Tunnel with a 60-inch-long free-flight model powered with compressed air.[26] Langley researchers found that the design possessed excellent stability and control characteristics, particularly at the angles of attack needed for approach, flare, and landing. In fact, at the low speeds of the tests, the model was controllable to angles of attack as high as 45°, well in excess of the maximum value of 25° predicted for approach, flare, and landing. Rolling oscillations that had been noted in free-flight tests of other highly swept shapes were well damped for the three-finned HL-10 configuration.

One major aerodynamic area of concern during the HL-10 tunnel testing was the discovery that a severe degradation in directional stability occurred at low supersonic speeds. Piloted simulator studies showed that the vehicle would have unsatisfactory handling qualities and that configuration modifications were required to increase directional stability. An extensive study of fin arrangements followed, including tests of a large-scale model in the Full-Scale Tunnel in 1964 and 1965.[27] The model tested was even larger (28 feet long) than the actual vehicle (22.2 feet) in an attempt to test the configuration at the highest feasible value of Reynolds number in the tunnel.[28]

Following several subsequent wind tunnel tests that investigated the effects of variations in the geometry of the tip fins, NASA researchers arrived at a configuration that increased the fin area, toe-in angle, and rollout angle. Throughout the study, researchers were sensitive to the requirement to increase stability without reducing aerodynamic performance. After all this research was accomplished, the HL-10 vehicle finally evolved into a triple-fin configuration, based on the massive amount of data that had been generated with no less

The scope of free-flight model testing in the Full-Scale Tunnel in support of the space program is indicated by this collection of models, including a capsule/parawing configuration, the Dyna-Soar, a generic lifting body, the HL-10, and the M2. (NASA L-64-5096)

than 10 different HL-10 models in various wind tunnels—over 8,000 hours of tunnel testing were involved.[29]

Flight tests of the full-scale HL-10 began on December 22, 1966. The final HL-10 configuration was regarded by all test pilots as having excellent flying qualities, and it completed its subsequent flight testing with outstanding success. HL-10 flights continued throughout the late 1960s and into 1970, with the vehicle setting speed (Mach 1.86) and altitude (90,303 feet) marks in 1970, becoming the fastest and highest-flying lifting body. Between 1966 and 1970, the HL-10 completed 37 research flights. Although none of the lifting-body designs were ever adopted as the eventual Space Shuttle design, they provided a wealth of knowledge and data that contributed to the development of the orbiter. The lessons learned included the experience and knowledge of demonstrating how to make precise unpowered steep approaches and dead-stick landings in a reusable reentry type vehicle.

When NASA began planning the wind tunnel support activities for the Space Shuttle program, it was decided that the workhorse for subsonic testing would be Langley's Low Turbulence Pressure Tunnel. Although the Full-Scale Tunnel had made major contributions to the NASA lifting-body program, funding and support for testing in the tunnel that had contributed to the Nation's most critical aerospace projects was not forthcoming. Thus, the Full-Scale Tunnel played no role in the development or operational problem solving for the Space Shuttle orbiter.

V/STOL Takes Off

The most dominant research area at the Full-Scale Tunnel in this era was the pursuit of concepts for V/STOL aircraft.[30] The Cold War threat of the Soviet Union and the possibility of preemptive attacks on U.S. and allied military airfields in Europe stimulated studies of various aircraft concepts that might be capable of providing offsite operations for avoiding losses of aircraft and crew assets and for deployment of aircraft for retaliatory strikes. Starting in the early 1950s, the NACA had explored concepts with free-flight models, and later NASA conducted valuable research on V/STOL concepts in wind tunnels, piloted simulators, propulsion test stands, and flight. Langley and the Ames Research Center led V/STOL research studies, which reached a peak in the 1960s. John Campbell, Mac McKinney, and Bob Kirby were extremely effective in their advocacy and management of V/STOL research efforts at the Full-Scale Tunnel.[31]

Hundreds of V/STOL projects were conducted during this period in the Full-Scale Tunnel and the 12-Foot Low-Speed Tunnel. Even a brief discussion of those activities far exceeds the envisioned scope of this book. Despite decades of research and development on a variety of V/STOL concepts, the only designs that have reached production and been deployed into U.S. military fleet operations at the present time are the AV-8 Harrier flown by the U.S. Marine Corps and the V-22 Osprey flown by the Marine Corps and the Air Force. The Full-Scale Tunnel was not directly involved in the development of the V-22, and that program will not be discussed here.

The following description briefly summarizes some of the more important contributions of research in the Full-Scale Tunnel to various V/STOL concepts—most of which were limited to prototype aircraft, but some of which still hold promise for future configurations.

The Hawker P.1127

In the late 1950s, the Langley Research Center was recognized worldwide as a leader in fundamental and applied research on vertical takeoff and landing aircraft.[32] Leaders in the Langley research efforts had addressed the critical challenge of providing efficient vertical flight with minimal penalties and adequate payload, resulting in a myriad of candidate concepts that included aircraft-tilting (tail sitters), thrust-tilting (tilt rotors), thrust-deflection (deflected slipstream), and dual-propulsion (lift-cruise engines) concepts. The Langley researchers had accumulated in-depth experience with each concept and had identified the limitations and complexities that constrained the satisfactory growth of V/STOL aircraft.

Meanwhile, in England, Hawker Aircraft Ltd. was privately funding the development of an innovative new V/STOL tactical strike aircraft known as the P.1127. The novel propulsion concept used by the P.1127 consisted of rotatable engine nozzles that vectored engine thrust to provide vertical lift for V/STOL operations. Jet reaction controls at the wingtips and tail of the airplane were powered by engine bleed air for control during low-speed and hovering flight. Initial interest from the British government had been lukewarm, and Hawker aggressively pursued potential funding from NATO's Mutual Weapons Development Program (MWDP) for development of the revolutionary P.1127 engine. This engine utilized four swiveling nozzles to redirect the engine thrust for vertical or forward flight. The American members of the MWDP were particularly impressed with the P.1127 concept, and with their outspoken leadership, critical development funds were provided in June 1958 to the engine manufacturer Bristol Siddeley.

Support for the P.1127 project from the U.S. military (particularly the Marine Corps) and NASA was a key element in the success of the program. When Hawker proceeded in the engineering development of the P.1127 from 1959 to 1960, numerous critical issues arose. These critical issues included the design of the flight control system; whether artificial stabilization was required; the lifting capability of the aircraft in ground effect; and the stability, control, and performance of the P.1127 in conventional flight. However, the most daunting question was whether the aircraft could satisfactorily perform the transition from hovering flight (supported by the vertically directed engine thrust) to conventional wing-borne flight. Many skeptics—particularly in the British government—believed that the transition maneuver would be far too complex for the pilot and that the P.1127 would not maintain adequate lift to permit a safe conversion.

John Stack, who served as an assistant director of Langley and was an active member of the MWDP at the time, regarded the P.1127 as the most significant advancement since the achievement of operational supersonic speeds in fighters. Stack directed John Campbell and his team at the Full-Scale Tunnel to provide full support to the project by conducting free-flight tests of a ⅙-scale, dynamically scaled, powered model to determine the characteristics of the P.1127 in the transition maneuver.[33] The same ⅙-scale model was also used for tests on the Langley Control Line Facility (i.e., the large rotating crane equipped with control lines for testing powered models discussed in the previous chapter) to determine characteristics during rapid transitions to and from hovering flight. Stack put extra pressure on Campbell's staff by decreeing that the tests would be completed prior to the initial flights of the prototype aircraft in 1961.

Ironically, Langley's legendary John Stack, who had spearheaded NACA efforts on transonic and supersonic high-speed aircraft, provided the stimulus for Langley's support of the low-speed versatility of the British P.1127. (NASA EL-2000-00372)

Free-flight testing of the P.1127 configuration included assessments of stability and control characteristics during hovering flight, including vertical takeoffs and landings. The photograph shows free-flight model pilots Bob Schade and Charlie Smith seated behind a protective barrier during hovering flights in the Full-Scale Tunnel's west return passage in 1960. (NASA L-60-2543)

Under the direction of Mac McKinney, the free-flight model tests showed that the P.1127 model behaved extremely well when compared with other V/STOL designs tested by Langley. Transitions to and from forward flight were easily performed and thrust management was relatively simple. Several problems were identified, however, including the fact that the model lacked sufficient lateral control power for satisfactory behavior during the transition. (The control power of the aircraft was increased as a result of these tests.) Also, a tendency to pitch up due to longitudinal instability at high angles of attack was readily apparent in the model flight tests. (This problem was subsequently cured by adding anhedral, or droop, to the horizontal-tail surfaces of the P.1127 and subsequent variants.) Despite these shortcomings, the P.1127 was judged to be an outstanding performer by the Langley researchers at the Full-Scale Tunnel. During the test program, Hawker had stationed an engineering representative at the tunnel, and movies of the transition in the tunnel were immediately shown to the British government, which resulted in resounding approval of Hawker's concept.

On September 12, 1961, transition flights both to and from wing-borne to jet-borne flight of the P.1127 full-scale airplane were accomplished. The overall results of the flight tests agreed remarkably well with the Langley model tests. John Stack witnessed transition flights in gusty conditions a week later and referred to them as the smoothest transition of any of the existing crop of V/STOL machines. However, the most important compliment of the Full-Scale Tunnel contributions prior to the first flights came from Sir Sydney Camm (designer of the famous Hawker Hurricane fighter of World War II), the chief designer of Hawker, who said that the Langley wind tunnel tests were the most important tests for the P.1127 project prior to flight.

The Full-Scale Tunnel did not play a direct role in the development of the follow-on descendants of the P.1127 known as the Kestrel and the Harrier, but its program-saving contribution to the P.1127 project was a key factor in the success of the only family of high-performance VTOL fighters.[35]

Engineer Lysle Parlett poses with the 0.13-scale free-flying model of the German Dornier DO-31 V/STOL transport that was used in Full-Scale Tunnel assessments of the dynamic stability and control of the configuration in 1962. (NASA L-62-9198)

The Dornier DO-31

Other vectored-thrust configurations were studied in the Full-Scale Tunnel, including the German Dornier DO-31 V/STOL transport in the 1960s. The DO-31 was a relatively large (50,000 pounds) two-place transport design that used a mixed-propulsion concept that included eight direct-lift engines buried in pods at the wingtips for VTOL operations and two vectored-thrust turbofan engines mounted in wing nacelles for V/STOL and conventional flight.

Free-flight model tests of a 0.13-scale model were conducted in the Full-Scale Tunnel in 1962.[34] Although results of all the free-flight tests were very positive, the program was terminated in 1970 after three full-scale DO-31 aircraft had been built and flight tested. The DO-31 fell victim to a common deficiency of V/STOL aircraft: the drag, weight, complexity, maintenance, noise, and cost penalties of providing V/STOL capabilities to a civil transport negated its advantages compared to conventional transports.

Tail Sitters: Simple, But Pilots Say No

The simplest of all V/STOL concepts, the tail-sitter VTOL aircraft sat on its tail prior to and after flight with its nose pointing upward in a vertical attitude. To be successful, tail sitters had to possess both a relatively high thrust-to-weight ratio for vertical takeoffs and landings and the ability to perform a controlled transition from hovering and vertical flight to conventional aircraft flight and vice versa. However, as the tail-sitter projects evolved, the unsolvable problem had to do with human factors and the difficulty of the piloting task.

As early as 1949, John Campbell's free-flight group had conducted exploratory free-flight model tests of simple tail-sitter configurations—featuring cylindrical fuselage shapes, rudimentary wings and tail surfaces, and counter-rotating large-diameter propellers to minimize adverse propeller-slipstream effects—to evaluate vertical takeoff, hovering, and

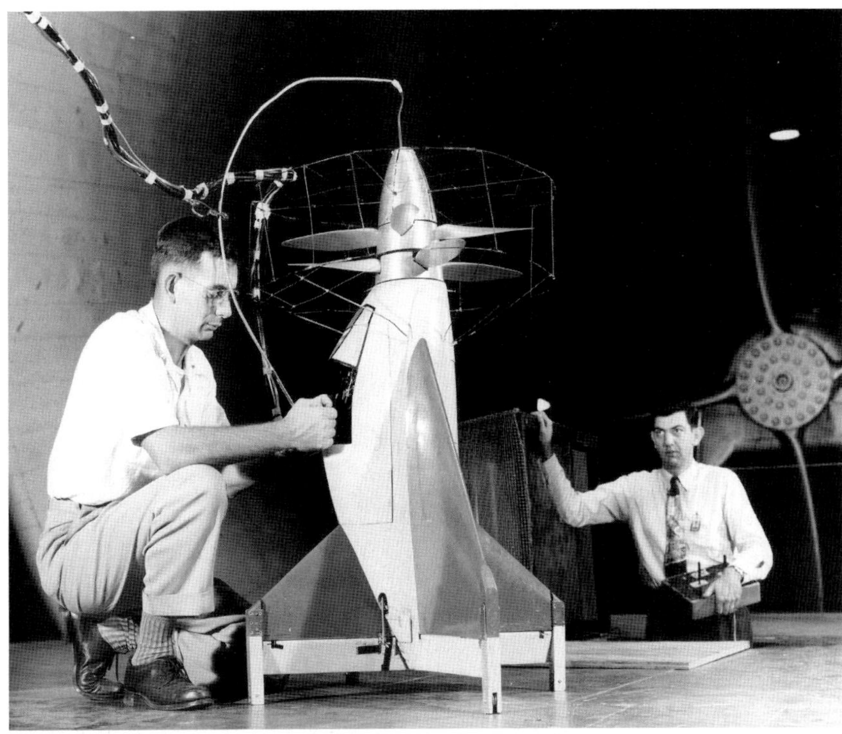

Left: This simple tail-sitter research model was the first VTOL free-flight model used in investigations of dynamic stability and controllability in 1949. (NASA EL-2001-00442) Right: The first free-flight model flown in the test section of the Full-Scale Tunnel was the Convair XFY-1 Pogo in 1952. Langley researchers Charles C. Smith (left) and Robert O. Schade (right) prepare the model for testing. Note the protective propeller shield installed to avoid inadvertent encounters with the flight cable. (NACA LAL-82885)

vertical landing characteristics. These early cooperative studies with the Navy were directed at determining the adequacy of conventional aerodynamic control surfaces on the wings and tail to provide satisfactory control of aircraft motions during hovering flight and low-speed maneuvers. With the control surfaces located in the high-energy slipstream of the propellers, it was anticipated that control levels would be high. These experiments confirmed the adequacy of control power and encouraged the Navy to proceed with flight demonstrations of the tail-sitter concept.[36]

A series of highly successful free-flight experiments in the Full-Scale Tunnel followed in support of specific military tail-sitter VTOL programs in the early 1950s, including the Convair XFY-1 Pogo, the Lockheed XFV-1 Salmon, and the Ryan X-13 Vertijet. The most notable of these designs was the X-13, which was produced under an Air Force contract and made over 100 successful flight tests in which its VTOL capability was dramatically demonstrated. During one spectacular demonstration, the aircraft took off from a special vertically oriented landing platform atop a trailer in front of the Pentagon in Washington, DC, performed a transition to forward flight in front of observers, and then performed a transition back to hovering flight and a vertical landing on its trailer.

Despite demonstrating the ability to convert between hovering and forward flight, the family of tail sitters were maintenance nightmares, and test pilots complained about the lack of adequate vision due to the vertical attitude of the aircraft (especially during attempts at precision vertical landings looking back over their shoulders) and the complexity of vertical landings—even during windless conditions in good weather. After experiences with the XFY-1, XFV-1, and X-13, the services concluded that the tail-sitter concept was not feasible for routine operations on ships or land sites and that acceptable VTOL designs would require a horizontal-fuselage attitude for vertical takeoff and landings. Thus, when interest in V/STOL aircraft peaked in the 1960s, the vertical-attitude concept had been discarded.

An early 1954 Langley concept for a tilt-wing VTOL transport is shown during the conversion from hovering flight to conventional forward flight. Note the tilt angle of the propellers and wing as the wing is rotating down to the conventional position. (NASA L-67-7419)

The Tilt-Wing: McKinney's Concept

As discussed briefly in the previous chapter, Mac McKinney had stimulated the development of the tilt-wing VTOL concept in the early 1950s, resulting in a series of free-flight tests of generic models in the Full-Scale Tunnel. As military interest in the tilt-wing concept increased, several specific configurations emerged with requests for testing in the Full-Scale Tunnel. For example, the Hiller X-18 was supported with free-flight testing in 1956 and continued into full-scale flight-test evaluations by the Air Force.[37]

Inspired by Langley's research in the 1950s, the Army Transportation Corps and the Office of Naval Research collaborated to develop and build the tilt-wing VZ-2 research aircraft to investigate the practicality of the tilt-wing concept. The project was designed around a low-cost research aircraft manufactured by Vertol. A single-turbine engine transmitted power by mechanical shafting to two three-blade propellers and to two tail-control fans that provided yaw and pitch control in hovering flight. Roll control was provided by differential thrust between the two vertically oriented propellers in hovering flight.

The full-scale VZ-2 tilt-wing airplane mounted for testing in the Full-Scale Tunnel in 1961. The original configuration was prone to wing stall and exhibited uncontrollable lateral directional motions during partial-power descending flight. Tests in the Full-Scale Tunnel demonstrated the effectiveness of wing modifications to minimize the problem. (NASA L-61-1767)

The VZ-2 flew for the first time in 1957. It successfully completed the world's first transition from vertical to horizontal flight by a tilt-wing airplane on July 15, 1958, and was flown at Edwards Air Force Base. In 1959, Langley received the VZ-2 from Edwards for extensive in-house research experiments that lasted through 1964. The extent of VZ-2 supporting research at Langley included free-flight model tests and full-scale airplane tests in the Full-Scale Tunnel as well as flight-test evaluations by Langley pilots.

In Langley's early research on the tilt-wing concept, it was learned that the most critical problem of the concept is the tendency of the wing to experience large areas of flow separation and stalling at the high angles of wing incidence required during the transition to and from hovering to forward flight. One of the primary factors that prevents the wing from stalling is the immersion of the wing in the high-energy slipstream of the propellers. Therefore, the probability of wing stall is significantly increased during steeper descents, when power is reduced and the slipstream energy is diminished. When wing stall occurs, tilt-wing aircraft usually exhibit random and uncontrollable rolling, wing-dropping motions, and/or yawing motions, along with generally unsatisfactory flight characteristics. To minimize and prevent this problem requires careful selection of the geometry of wing flaps and leading-edge stall control devices.

During initial VZ-2 flight experiments at Langley, research pilots cited problems with poor low-speed stability and control, and deficient handling qualities during transition. They encountered the anticipated wing-stall problem in the speed range from about 40 to 70 knots, which corresponded to wing incidence angles from about 45° to 25°. In its original configuration, the airplane had no wing flaps or leading-edge high-lift devices and exhibited severe wing stall, heavy structural buffeting, and random rolling and yawing motions, especially in descending flight. The situation in descending flight was so unsatisfactory that pilots considered it an area of hazardous operations.

Following the initial Langley flight tests, the full-scale VZ-2 was mounted in the Full-Scale Tunnel in 1961 for wind tunnel/flight correlation studies, with emphasis on the wing-stall phenomenon.[38] Additional analysis indicated that the use of a large trailing-edge flap would significantly augment the lift required during transition, thereby permitting the wing to operate at lower angles of attack where stalling was less of a problem. In addition to the full-scale airplane test, free-flight model tests were made in the Full-Scale Tunnel to assess the effects of wing modifications on the dynamic stability and control characteristics of a ¼-scale model of the aircraft during simulated descent.[39]

Based on these results, the VZ-2 was subsequently modified with wing changes, including a full-span trailing-edge flap. This modification alleviated much of the undesirable wing stall and unsatisfactory lateral directional motions, and it permitted a significant increase in the useable rate of descent.

The flight-test programs for the basic and modified VZ-2 proved to be successes, with 34 conversions between vertical and horizontal flight. The flight-test program lasted until 1964, and the aircraft, which was designed as a rudimentary research vehicle with no intentions of production, was later donated to the Smithsonian Institution. The data and experiences gathered during the test in the Full-Scale Tunnel provided invaluable design information for the next generation tilt-wing program known as the XC-142A tilt-wing transport.

The Ling-Temco-Vought XC-142A

As interest in V/STOL configurations strengthened, the Army, Air Force, and Navy joined together for a tri-service V/STOL transport project in 1961. Competing proposals were narrowed to a four-engine tilt-wing concept designed by a team of Ling-Temco-Vought, Ryan, and Hiller, and a contract for five aircraft was awarded. The XC-142A was a large (maximum gross weight of 41,500 pounds), fast (top speed of 400 mph), tilt-wing design with extensive cross-shafting of the main engines for safety and a tail-mounted, three-blade, variable-pitch propeller for pitch control at low speeds. In hovering flight, roll control for

This relatively sophisticated free-flight model of the tilt-wing XC-142A V/STOL transport was used in tests to evaluate the dynamic stability and control of the configuration. The model included a programmed horizontal-tail surface that deflected as a function of wing angle to minimize trim changes. (NASA L-62-6326)

the XC-142A was provided by differential deflections of the propeller blade pitch of the four main propellers; yaw control was provided by differentially deflecting the ailerons; and pitch control was provided by varying the pitch of the tail-mounted rotor. These control functions were phased out as the transition to forward flight progressed, resulting in the use of conventional control surfaces.)

In response to military requests, free-flight model tests were conducted in the Full-Scale Tunnel in 1964 using a large 0.11-scale model.[40] The free-flight results showed the existence of unstable control-fixed oscillations in pitch and yaw in hovering flight, similar to results obtained for previous model tests of the VZ-2 and X-18 tilt-wing configurations. However, the unstable oscillations were so slow (a period of about 10 seconds for the full-scale airplane) that they could be easily controlled and the model could be smoothly maneuvered, even without artificial stabilization. No problems were noted during transitions in level flight or in simulated descents, and the minimum control power found to be satisfactory for the model was less than the control power planned for the full-scale aircraft.[41]

After the five XC-142A aircraft underwent military evaluations from 1964 to 1967, the services could not define an operational requirement for a V/STOL transport, and the last XC-142A aircraft was loaned to NASA Langley for general V/STOL flight research studies from 1968 to 1970. The flight activities were conducted at Langley and from the NASA Wallops Island, VA, flight-test facility. The XC-142A flight-test program was successful, showing the V/STOL potential of a large, tilt-wing, propeller-driven transport. At the conclusion of the project, Langley pilots delivered the XC-142A to the Air Force Museum at Dayton, OH, in May 1970.

By the early 1970s and the Vietnam War era, the U.S. military and its leaders showed no further interest in developing V/STOL aircraft other than the helicopter. With no potential mission or interest, tilt-wing V/STOL aircraft research and development was terminated. The tilt-wing V/STOL concept, however, is still regarded as having high potential for military and civil applications and resurfaces from time to time in proposals today.

The Fairchild VZ-5

One of the first VTOL designs conceived by Charles Zimmerman in the 1950s was the idea of using large wing flaps to deflect the slipstream of a propeller-driven aircraft downward to produce higher lift for takeoff and landing. However, the deflected slipstream concept suffered from a basic limitation in that the thrust produced by the engine was substantially reduced by the flow-turning process, whereas many other V/STOL concepts do not experience any loss in thrust. In addition, some type of reaction control would have to be incorporated for low-speed control, and the turning of the slipstream by a wing trailing-edge flap induces a large nose-down diving moment on the aircraft. Finally, the effect of ground proximity can be very detrimental on induced lift. Langley had therefore given up on the feasibility of deflected slipstream VTOL vehicles. Nonetheless, the advent of high power-to-weight turboprop engines stimulated interest from the Army in the concept.

The Fairchild VZ-5 Fledgling was designed under an Army contract as a deflected-slipstream research aircraft in the 1950s. A single turboshaft engine drove four three-blade propellers with cross-shafting. Differential propeller pitch on the outboard propellers provided roll control in hovering flight, while tail fans provided yaw and pitch control. The aircraft was flown in tethered flight in November 1959, but subscale model testing of the configuration had indicated major aerodynamic issues and the Army considered it imperative to test the full-scale airplane in the Langley Full-Scale Tunnel prior to flight tests of the vehicle.

Technician Bob Lindeman views the Fairchild VZ-5 deflected-slipstream V/STOL airplane during tests in the Full-Scale Tunnel in 1961. Note the full-span deflected flaps and the tail-mounted fans. Results of the test were totally unacceptable, and the configuration did not progress to flight. (NASA L-61-2822)

The VZ-5 was tested in the Full-Scale Tunnel in April 1961, and the results of the test program were totally unacceptable.[42] With the aircraft center of gravity in the design position, the aircraft was aerodynamically unstable over the entire speed range, and the vehicle was incapable of being trimmed longitudinally at low speeds or in hovering-flight conditions. In order to obtain stability, it would have been necessary to add an unacceptable ballast of about 700 pounds to the cockpit area. In addition, the aircraft exhibited very large values of effective dihedral, which would have likely resulted in unacceptable lateral directional characteristics. As a result of the tests in the Full-Scale Tunnel, the VZ-5 never flew and the project was terminated.

The Bell X-22A

Another V/STOL concept explored by researchers at NASA Langley was the tilt-duct concept, which used propellers or fans enclosed in tilting ducts that could be rotated between vertical and horizontal attitudes relative to the airframe. The tilt-duct design was pursued because the shrouded propeller offered the promise of benefits such as enhanced thrust for a given propeller diameter (resulting in a more compact design), improved operational safety for air and ground crews because of the protective shroud, and aircraft noise alleviation.

Bell Aerospace had entered the tri-service V/STOL transport competition with a tilt-duct design based on in-house research and development of several tilt-duct layouts. After losing to the XC-142A competition, Bell continued its interest in its tilt-duct design and

Researcher Bill Newsom poses in 1963 with a free-flight model of the Bell X-22A tilt-duct V/STOL configuration prior to free-flight testing in the Full-Scale Tunnel. (NASA L-63-10043)

subsequently was awarded a Navy contract to build and flight test two tilt-duct research aircraft in 1962.

The X-22A was powered by four three-blade ducted propellers, powered through a power-transmission system driven by four turboshaft engines. Control surfaces consisted of four vanes located in the slipstream of each of the ducted propellers. In forward flight, pitch control was provided by differential deflection of the front and rear vanes, while roll control was produced by differential deflections of the right- and left-side vanes. Yaw control in cruise was by differential propeller thrust. In hovering flight, roll control was produced by differential propeller thrust, and pitch and yaw control were provided by vane deflections. During the transition to and from forward flight, the control inputs were mixed as a function of the duct angle.

In 1964, a 0.18-scale free-flight model of the X-22A was tested and flown in the Full-Scale Tunnel.[43] The free-flight model incorporated control system surfaces similar to the full-scale aircraft; however, it was not feasible to replicate the differential propeller pitch mechanisms, and reaction jets powered by compressed air were substituted. The model was also designed with a larger, unscaled inlet lip radius on the propeller ducts because earlier Langley research in other wind tunnels had shown that a scaled inlet lip radius on subscale models could not simulate the flow of air into the inlet without premature stalling of the inlet lip at the low Reynolds number of the free-flight tests. With this modification, the model's aerodynamic behavior more closely represented that of the aircraft.

The free-flight results for the X-22A showed the usual unstable oscillations in pitch and roll that had occurred with controls fixed for other propeller-driven V/STOL designs in hovering flight, and once again the period of oscillatory motion (about 8 seconds full scale) was also slow and easy to control. The minimum control-power level required for satisfactory behavior in all flight modes was found to be equal to or less than half that proposed for the full-scale aircraft. The powerful automatic stabilization system and variable-stability features of the full-scale X-22A completely masked the few issues observed in the model flight tests.

The X-22A flight-test programs for both Bell and the Navy later focused on providing a capability for in-flight simulation of V/STOL aircraft characteristics for research on handling qualities and pilot training. This part of the X-22 program met with great success, and research flights of the X-22A by the Calspan Corporation of Buffalo, NY, continued until 1984. However, the Navy's interest in tilt-duct V/STOL aircraft waned and no X-22As ever entered service.

The Ryan XV-5A

One of the most significant developments in V/STOL technology in the 1960s was the development of advanced lift-fan propulsion concepts by the General Electric Company. By combining tip-driven lift fans with a conventional turbojet, General Electric conceived a dual-propulsion concept for relatively high-speed V/STOL vehicles. The company developed large-scale lift fans through extensive wind tunnel testing at Ames and other ground tests to the point that the Army Transportation Research Command was stimulated to fund the development and flight testing of two research aircraft in 1961.

With General Electric as the prime contractor and Ryan as the airframe partner, work on two research aircraft designated XV-5A (originally VZ-11) began, with support from NASA Langley. The XV-5A was a small fan-in-wing aircraft powered by two J-85 turbojet engines located above and to the rear of the two-place cockpit. Two 5-foot-diameter lift fans were buried in the wing panels, and a smaller 3-foot-diameter fan in the nose was used for pitch trim and control. For cruise flight, the exhaust of the jet engines was routed through conventional tailpipes, and the fans were covered with auxiliary doors to form a smooth outer wing contour. For conversion to hovering flight, the wing's cover doors were opened and a valve was actuated by the pilot to divert the engine exhaust through a ducting system to drive the tip-driven wing fans, as well as the nose fan.

The three fans were capable of producing about three times the total thrust of the two J-85 engines that drove them. Located under each wing fan was a set of louvered vanes that could be deflected rearward to vector the fan thrust and thereby impart forward thrust for transition. When the vanes were deflected differentially between the right and left fans, yaw control could be produced for hover. Finally, if the vanes were used to choke the flow from a lift fan, roll control was produced. The nose fan included a "scoop door" that was deflected to control pitch. In high-speed flight, control was provided by a rudder, ailerons, and an all-moveable horizontal tail in a T-tail arrangement.

Studies of the Ryan XV-5A fan-in-wing V/STOL configuration included free-flight tests of this model. Note the venetian-blind-type louvers under each wing fan, which were used to provide forward thrust when actuated symmetrically and roll and yaw control when actuated differentially. The "scoop doors" in the forward fuselage were used to modulate the net thrust of a nose fan for pitch control. (NASA L-62-9062)

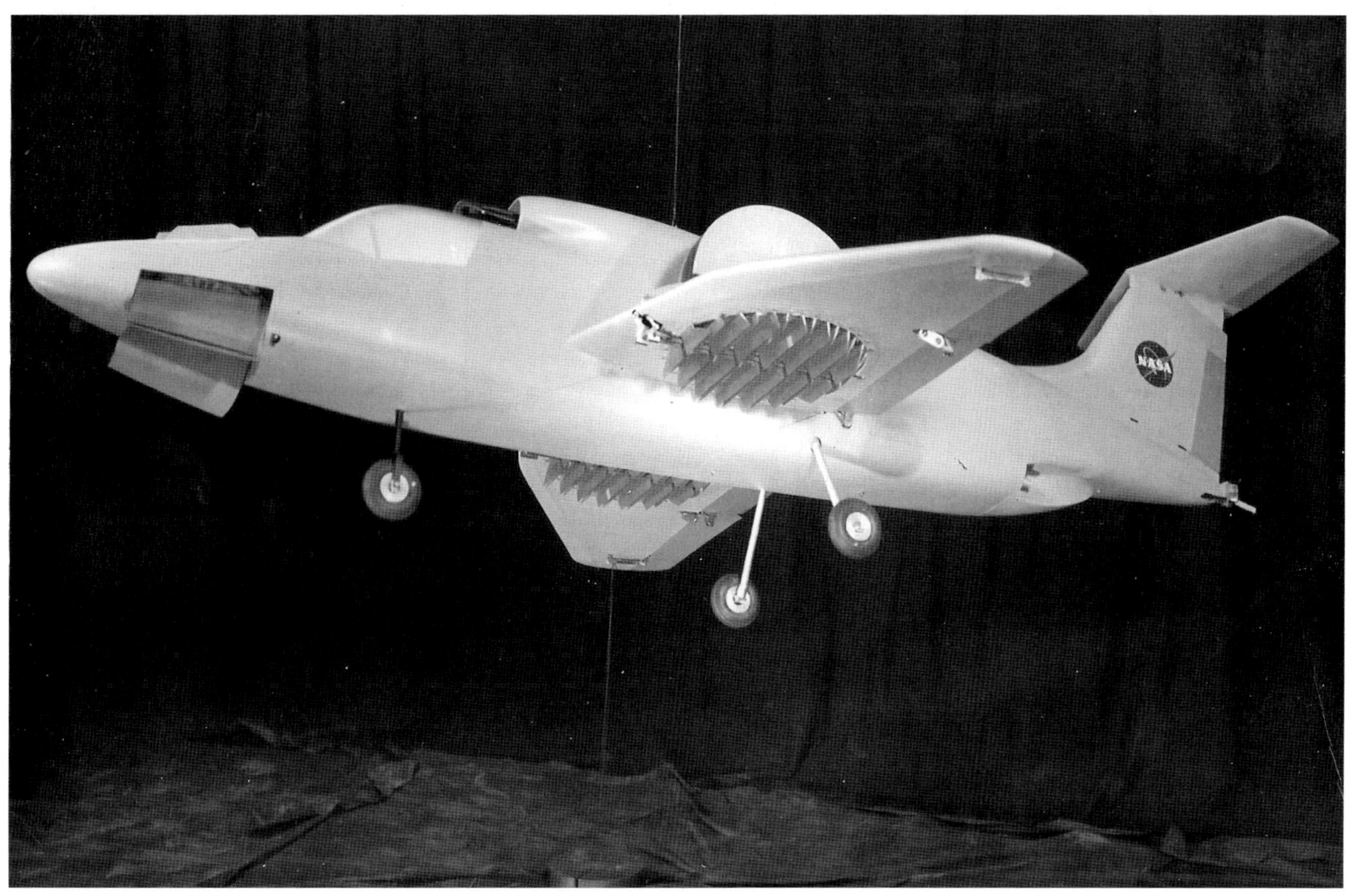

In 1962, Langley fabricated a 0.18-scale free-flight model of the XV-5A at the request of the Army and tested it in the Full-Scale Tunnel.[44] The model was built of composite materials and equipped with sophisticated tip-driven fans and an internal ducting system. The fans were tip-driven with compressed air, and the fan exhaust louvers could be deflected collectively for thrust spoiling (altitude control) or differentially for roll and yaw control. However, for the free-flight investigation, jet-reaction controls at the wingtips and tail were used to simplify the piloting task. The nose-scoop doors were, however, used for pitch trim. The tests were conducted for hovering flight and for transition speeds up to about 97 knots (full scale), at which speed the conversion to wing-borne flight was scheduled. No conversion maneuvers were attempted in the model flight tests.

As was the case for the propeller-driven V/STOL models previously tested, the XV-5A model exhibited unstable pitch and roll oscillations in hovering flight with the controls fixed, but the motions were slow and easy to control. As transition speed increased, the model required more nose-down control from the nose fan, resulting in a lift loss of about 12 percent.

The first of the two XV-5A aircraft flew on May 15, 1964, at Edwards Air Force Base and experienced a fatal crash a year later, apparently caused by an inadvertent pilot control input in a transition maneuver during an official demonstration. Both XV-5A aircraft had been demonstrating the low- and high-speed performance capabilities when the aircraft suddenly

Project engineer Bill Newsom poses with a generic model of a lift-fan transport configuration that was tested with a variety of lift-fan arrangements. (NASA L-67-7754, L-67-7849, and L-67-7850)

nosed over and crashed. The second XV-5A later experienced a fatal crash in October 1966 when a pilot-operated rescue hoist was ingested into one of the wing fans, causing the aircraft to roll and begin descending. The pilot attempted to eject from the aircraft but ejected at an unsurvivable roll angle. The aircraft, which was not destroyed, was rebuilt with landing gear and cockpit modifications as the XV-5B. An extensive XV-5B flight research program on aerodynamics, acoustics, and flying quality investigations was subsequently conducted by NASA at Ames Research Center.

Mac McKinney's group at the Full-Scale Tunnel subsequently conducted many force tests and free-flight tests of generic fan-in-wing models to evaluate the performance, stability, and control characteristics of typical configurations using the concept.[45]

Loss of Lift: Hot-Gas Ingestion

One of the most significant problems encountered by jet VTOL aircraft is hot-gas ingestion, in which the hot engine-exhaust gases and surrounding air heated by the hot exhaust are deflected by the ground and ingested into the engine inlet, resulting in a significant loss of engine thrust. This marked thrust loss is directly caused by the elevated temperature

A large-scale generic model was used to evaluate the effects of configuration variables, wind direction, and wind magnitude on hot-gas ingestion phenomena. The test program included tests in the Full-Scale Tunnel (top) as well as an outdoor test site (bottom). (NASA L-67-2394 and L-66-1971)

of the engine-inlet air and/or an uneven inlet temperature distribution across the engine face. Investigations of the phenomenon had been conducted with relatively small-scale models and did not provide generalized information for applications of the data to different configurations.

In 1966 and 1967, H. Clyde McLemore and Charles C. Smith led a test program in the Full-Scale Tunnel to provide large-scale data on the problem of hot-gas ingestion for fighter-type V/STOL configurations having in-line, rectangular, and single-engine arrangements.[46]

A collection of V/STOL free-flight models on display at the tunnel in 1964 represented a variety of concepts. Clockwise from upper right, the configurations are the Bell X-22A, the LTV XC-142A, the Ryan XV-5A, and the Republic "Alliance" variable-sweep vectored-thrust design. (NASA L-64-5097)

The large generic test model used a turbojet engine operating at a nozzle temperature of 1,200 °F as an exhaust-gas source. In addition to testing in the test section of the Full-Scale Tunnel, the researchers conducted outdoor testing to obtain data for correlation with the tunnel results. The scope of the investigation included test variables of model height above the ground, wing height, engine-inlet position, and wind speed and direction.

The results of the study were extremely revealing and highlighted the variability and sensitivity of the hot-gas ingestion problem. For example, for some exhaust-nozzle configurations, hot-gas ingestion caused an inlet air temperature rise of 200°F over ambient temperature. The ingestion problem was most severe for wind speeds from 0 to 20 knots, and there was virtually no hot-gas ingestion for wind speeds greater than about 30 knots. Deflecting the jet exhaust 25° rearward using vectoring nozzles virtually eliminated the hot-gas ingestion phenomenon.

VTOL Cools Off

After two decades of intense international research and development on VTOL aircraft, worldwide interests and activities rapidly disappeared as the 1960s came to an end. Several factors caused the situation, including the Vietnam War, the lack of a formal military VTOL mission, and a technical recognition that the penalties paid for achieving the VTOL capability were unacceptably high. As will be discussed in the next chapter, in the 1970s, the military began to focus on transport aircraft with short-field capability, resulting in increased interest in short takeoff and landing capability rather than VTOL aircraft concepts.

The relatively abrupt ending of VTOL interest by the military resulted in a dramatic change in the technical thrusts and schedule at the Full-Scale Tunnel. Once again, the old facility provided unique testing capability for unanticipated national topics such as high-angle-of-attack characteristics of high-performance fighters, STOL concepts, second-generation SST configurations, and general aviation.

NASA's Breakthrough: The Variable-Sweep Wing

The very significant aerodynamic advantages of wing sweep were pursued by American and foreign aircraft designers near the end of World War II. The first wind tunnel tests of variable-wing-sweep concepts had been conducted at Langley in the mid-1940s.[47] These tests included the now-familiar symmetric variable-sweep wing, as well as the variable-sweep oblique-wing concept. Results of these early studies revealed that when a single centerline pivot was used for the movable wing panels, the configuration would exhibit excessive

Researcher James L. Hassell, Jr., inspects a free-flight variable-sweep model tested in the Full-Scale Tunnel in 1960. Free-flight demonstrations and force tests of the configuration were very effective in furthering support for the variable-sweep concept. (NASA L-61-525)

longitudinal stability, resulting in marginal maneuverability when the wing was swept back. At the time, it appeared that some kind of variable longitudinal translation of the pivot point was required for a better balance between locations of the aerodynamic center and the aircraft center of gravity—resulting in a weight penalty and concern over complexity.

After flight tests of the variable-sweep Bell X-5 and Grumman XF-10F research airplanes were completed, the military concluded that the weight of the wing-pivot translation feature caused unacceptable penalties compared to a moderately swept fixed wing. During a lull in interest in variable sweep in 1958, Langley's John Stack visited England and saw the "Swallow" configuration—a radical variable-sweep supersonic transport configuration that had been developed by Barnes Wallis, who had apparently solved the structural problems of locating a pivot mechanism in a thin wing, and brought the idea back to Langley. William J. Alford, Jr., and Edward Polhamus then conducted an extensive set of parametric tests that determined that there is an optimum pivot location for swept wings that has essentially the

A series of photographs show the variable-sweep model during a flight in which the wing sweep was varied forward and rearward. (NASA L-60-8573)

same aerodynamic center location at low- and high-sweep angles, thus eliminating the need to translate the wing to maintain balance.[48] The researchers used individual pivots for each wing panel located at positions outboard of the fuselage on a fixed inner-wing surface.[49] The outboard-wing pivot concept proved to be the breakthrough required to implement the variable-sweep concepts without unacceptable weight penalties.

One of the key experiments in demonstrating the feasibility of variable sweep was a free-flight model study in the Full-Scale Tunnel in 1960.[50] During the test program, a generic fighter representative of Navy combat air patrol designs was flown in the tunnel while the wing sweep angle was swept back from 25° to 113°. The results of the investigation revealed stability and control issues for certain wing-sweep angles and also illustrated how artificial stabilization could be used to obtain satisfactory flying characteristics. The study not only provided detailed technical data on dynamic stability and control, but it also provided motion pictures demonstrating the feasibility of the outboard pivot concept and variable sweep. Langley then continued to mature the variable-sweep concept in its suite of wind tunnels in preparation for opportunities to apply the technology to future military aircraft.[51]

This photograph, composed of a sequence of images, shows the free-flight model of the F-111 in 1964. A directional instability at high angles of attack was noted in the test program and was subsequently verified in full-scale airplane flight tests. (NASA L-65-2123)

The General Dynamics F-111

In 1961, Secretary of Defense Robert McNamara ordered the Air Force and the Navy to combine their requirements for a new fighter aircraft into a single design called the Tactical

Fighter Experimental (TFX). After the industry design competition, McNamara personally overruled the source selection board and declared a team of General Dynamics and Grumman the winner to build the TFX. He also stated that the key technical concept that would make the aircraft a success was Langley's variable-sweep wing concept. No other aircraft development program in Langley's history resulted in more intense technical activities across the center (over 15,000 wind tunnel hours in 15 Langley tunnels) or such controversy and ill feelings between Langley researchers and an industry team. Numerous technical problems arose in the program, including excessive transonic drag, inlet and nozzle problems, and stability and control issues. Further, NASA's "watchdog" role of technical experts for the Government during program reviews by Congress infuriated the industry team.

As part of Langley's support for the development program, free-flight model tests of the F-111 were conducted in the Full-Scale Tunnel in October 1964.[52] During the flight tests, the model's wing-sweep angle was varied rearward from 16° to 72.5° and the effects of stability augmentation in roll and pitch were determined. The flight tests were extended to high angles of attack where, with the wings at the 50° and 72.5° sweep conditions, the model exhibited a sudden, uncontrollable yaw divergence prior to attaining maximum lift. Personnel from General Dynamics—including the test pilot scheduled to make the first high-angle-of-attack flights with the aircraft—witnessed the tests and proclaimed the results to be unrealistic and caused by the low Reynolds numbers of the tests. When the full-scale F-111 airplane entered its high-angle-of-attack test program, the model predictions were verified and the test pilot commented that seeing the departure in the tunnel testing had prepared him for the event.

The free-flight model testing of generic variable-sweep models and the F-111 configuration in the Full-Scale Tunnel helped demonstrate the viability of the variable-sweep concept, which was subsequently applied to the United States' F-14 and B-1 aircraft as well as the European Tornado and the Soviet Su-17, Su-24, MiG-23, Tu-22, and Tu-160.

The Need for Speed: Supersonic Civil Transports

NASA's support of the Air Force XB-70 bomber program included testing in most Langley wind tunnels. This free-flight model of an early version of the airplane was flown in the Full-Scale Tunnel in 1957 to evaluate low-speed characteristics. After the bomber program was canceled, NASA began a broad research program on civil supersonic transports. (NACA L-05221)

Langley's first major research efforts in supersonic civil aircraft began in 1958 and lasted until 1971.[53] Two projects dominated these research activities: the supersonic cruise XB-70 bomber and the U.S. Supersonic Transport Program. The XB-70 was the most important early activity to stimulate supersonic transport research at Langley. The bomber program had begun in 1957 and was canceled a few years later, but two

XB-70 aircraft were completed for research flights. In view of the similarity of its relative size and cruise speed to those of a representative supersonic civil transport, the XB-70 evoked considerable interest within NASA for research relevant to civil applications.

At the Full-Scale Tunnel, free-flight model tests were conducted in response to an Air Force request to determine the low-speed dynamic stability control characteristics of an early version of the XB-70 in 1957.[54]

NASA leaders held numerous discussions with the Federal Aviation Administration (FAA) and the DOD to formulate a cooperative SST Program, and President John F. Kennedy subsequently assigned the leadership of the program to the FAA, with NASA providing basic research and technical support. With its cadre of leading experts in supersonic aerodynamics, Langley was poised to propose promising configurations for supersonic transports to the national team. Configuration studies within NASA led to wind tunnel tests beginning in 1959 of 19 different NASA-conceived SST designs, referred to as supersonic commercial air transport (SCAT) configurations. Testing continued for over 7 years on 40 variants of these designs at Langley in its subsonic, transonic, and supersonic wind tunnels.

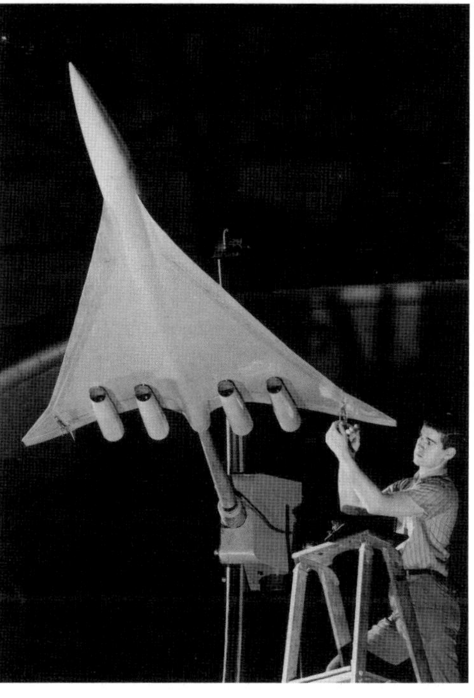

Delma C. Freeman conducted free-flight model tests of the leading U.S. supersonic transport candidates including (top, left to right) the Lockheed L2000 design, the Boeing 733 configuration, and (bottom) the NASA SCAT-15F. Freeman later became director of the Langley Research Center in 2002. (NASA L-66-2764, L-66-1877, and L-65-1360)

A request for proposals was issued by the FAA to industry for a supersonic transport having a cruise Mach number of 2.7, a titanium structure, and a payload of 250 passengers.

Boeing chose a variable-sweep wing configuration as its entry in the competition, and Lockheed chose a fixed-wing double-delta design on the basis that it would be a simpler, lighter airplane. As weight problems began to appear for both designs during the design cycle, the FAA advised both Boeing and Lockheed to explore the Langley-conceived SCAT-15 design—an innovative variable-sweep arrow-wing design that used auxiliary variable-sweep wing panels—as a potential alternate.

During the competitive period, Delma C. Freeman led free-flight model investigations in the Full-Scale Tunnel of the stability and control characteristics of all the designs, including the Boeing variable-sweep configuration, the Lockheed double-delta design, and the SCAT-15 NASA configuration.[55] As might be expected, the free-flight tests were attended by prominent technical leaders for each organization, and the individual results were quickly transmitted to design teams for analysis and guidance in their efforts.

Boeing was declared the competition winner on December 31, 1966, and then changed its design to a fixed-wing double-delta SST design. Boeing did, however, continue its studies of the SCAT-15 design. Meanwhile, in its role as primary advocate for the SCAT-15 design, Langley pursued improved versions of its own original SCAT concept as a potential alternative to the Boeing double-delta design. In 1964, researchers used new Langley-developed computational tools to design an improved derivative, called the SCAT-15F. This computer-generated fixed-wing version of the earlier variable-sweep SCAT-15 configuration was demonstrated by wind tunnel tests to exhibit a lift-to-drag ratio of 9.3 at Mach 2.6, an impressive 25–30 percent improvement over the state of the art at that time.

The Langley Full-Scale Tunnel was a major workhorse for the development of the SCAT-15F configuration. A free-flight model was used to evaluate dynamic stability and control, and aerodynamic data measured for the free-flight model were used as inputs for a piloted ground-based simulator study of the handling qualities of the SCAT-15F during approach and landing.[56]

A major problem identified for the SCAT-15F design involved deficient low-speed stability and control characteristics. The highly swept arrow-wing configuration exhibited a longitudinal instability (pitch-up) typically shown by arrow wings at moderate angles of attack, and the instability was accompanied by the possibility of dangerous, unrecoverable "deep stall" behavior. The deep stall for the SCAT-15F occurred for angles of attack slightly higher than those for the pitch-up tendency, and it was characterized by an abrupt increase in airplane angle of attack to extremely high values (on the order of 60°), where longitudinal controls were ineffective for recovery to conventional flight. Tests in several different Langley tunnels were directed at the unacceptable pitch-up problem and the development of modifications to alleviate it. After extensive testing in the Full-Scale Tunnel, a combination of modifications—including a 60° deflection of the wing leading-edge flap segments forward of the center of gravity, a "notched" wing apex, Fowler flaps, and a small aft horizontal tail—eliminated the deep-stall trim problem.[57] Today, instabilities like that exhibited by the SCAT-15F are routinely mitigated by powerful automatic control systems that prevent the pilot from accidentally entering the condition, but the technology was not ready for civil applications in the late 1960s. Boeing never adopted the SCAT-15F as a viable design for the U.S. supersonic transport.[58]

Faced with extensive domestic controversy, unresolved technical issues, international politics, and a growing public outcry over sonic booms, airport-noise levels, and other environmental concerns, the U.S. Congress canceled the SST program in March 1971. After 8 years of research and development and an expenditure of approximately $1 billion, the United States withdrew from the international supersonic-transport competition. Today, many participants that were close to the program consider the cancellation to have been a wise decision because of the immaturity of technology, bleak profitability outlook, and high risk of failure.

The foregoing activities involving the Full-Scale Tunnel in the U.S. Supersonic Transport Program were precursors to much more intense investigations of the low-speed characteristics of advanced supersonic-transport configurations that would take place over the next 20 years. Although the maturity of computer-based design codes for efficient supersonic configurations rapidly evolved, the low-speed aerodynamic issues involving performance, stability, and control were not amenable to computer predictions, and experimental low-speed wind tunnel work in the Full-Scale Tunnel and other facilities would entail much of the efforts in follow-on activities—as will be discussed in subsequent chapters.

Flexible Wing Concepts: Rogallo's Dream

The parawing concept was conceived by NACA/NASA Langley engineer Francis M. Rogallo in 1947 as an all-flexible, diamond-shaped fabric wing attached to rigid members that formed the leading edge and keel of the vehicle. Some parawing concepts were foldable and could be deployed to a semirigid shape for flight. Rogallo obtained a U.S. patent for his idea in March 1951, but he could find no takers for the concept with the exception of some toy stores, which marketed a small version that became known as the "Rogallo Flexikite."

The use of a deployable parawing to reduce landing speeds and distances of high-performance aircraft was studied in the Full-Scale Tunnel using this model of an early version of the XB-70 in 1960. Results of the investigation were positive and encouraging. (NASA L-60-2517)

Management within the NACA was not initially interested in the concept, but Rogallo pursued his dream of potential aerospace applications of the idea.

One of the earliest research areas for parawing applications involved using deployable parawings to increase the lift and drag of supersonic high-speed aircraft configurations during approach and landing. Exploratory tests were conducted in the Langley 300-MPH 7- by 10-Foot Tunnel in 1960 to determine the benefits of applying a parawing landing device to an early version of the XB-70 bomber (then designated Weapon System 110). The parawing used for the XB-70 investigation had about twice the wing area of the basic model. Researchers anticipated that a large increase in lift could be provided by the auxiliary parawing, resulting in significant reductions in approach speed and landing requirements, and the results of the exploratory study were impressive. The parawing/XB-70 configuration exhibited three times as much lift as the isolated aircraft model, and the static-stability characteristics of the model were improved by adding the parawing.[59]

Encouraged by these conventional wind tunnel tests, researchers turned to an examination of the dynamic-stability characteristics of such configurations in 1960 using a free-flight model in the Full-Scale Tunnel.[60] In the free-flight test project, a model of an early version of the XB-70 was modified with a delta-shaped parawing that had an area about 60 percent greater than the model wing area. The parawing was attached to the aircraft model by several flexible riser lines. The results of the flight tests revealed that the parawing-XB-70 configuration flew much steadier and was more controllable than the isolated XB-70 model. This result was attributed to the very large reduction in airspeed for the combination, less erratic responses to control inputs, and an increase in moments of inertia. These exploratory

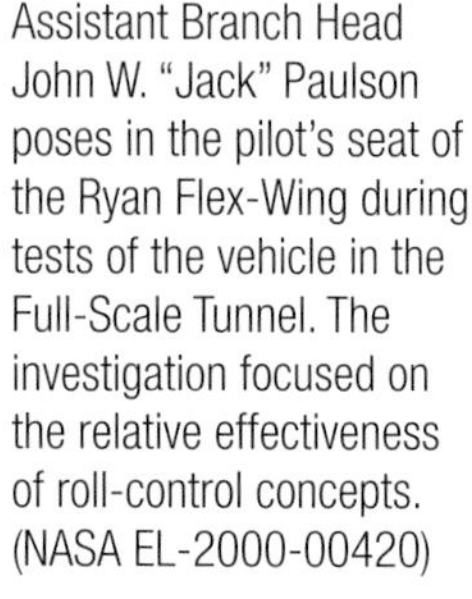

Assistant Branch Head John W. "Jack" Paulson poses in the pilot's seat of the Ryan Flex-Wing during tests of the vehicle in the Full-Scale Tunnel. The investigation focused on the relative effectiveness of roll-control concepts. (NASA EL-2000-00420)

studies in the Full-Scale Tunnel indicated that the use of the parawing as a landing and takeoff aid appeared to be feasible from the standpoint of stability and control, and it offered very large increases in lift that could be used for substantial reductions in takeoff and landing distances.[61]

The Ryan Wings

In 1961, the Ryan Company received a contract from the Army to develop an exploratory parawing utility vehicle known as the Ryan Flex-Wing. Initially powered by a single pusher propeller and a 100 horsepower engine (later upgraded to a 180 horsepower engine), the Flex-Wing consisted of a simple platform/cockpit/parawing arrangement. The parawing was attached to the cargo platform by a truss structure. Control was obtained by banking or pitching the parawing with respect to the platform, and a conventional control wheel and column were used. Directional control was provided by a rudder immersed in the propeller slipstream.

Following some handling-quality issues regarding lateral control that surfaced in the full-scale flight program in 1961, the Army requested that NASA assess the performance, stability, and control characteristics of the Ryan Flex-Wing aircraft in the Langley Full-Scale Tunnel. Accordingly, power-off and power-on tests of the vehicle for speeds from 25 mph to 47 mph were conducted in January 1962.[62]

The results of the Flex-Wing tunnel tests showed that for parawing keel angles of less than 20°, the rear of the flexible parawing experienced flutter; and for keel angles greater than 35°, the vehicle exhibited longitudinal instability (pitch-up). The configuration had adequate longitudinal and directional stability with the rudder on, but the lateral dihedral effect was excessive, creating an undesirable effect on the vehicle's response to lateral control inputs. With its original wing-banking lateral-control concept, the Flex-Wing created only small rolling moments for roll control at high angles of attack, accompanied by large adverse yaw and large lateral stick forces. This undesirable aerodynamic combination resulted in a reduced rolling motion when control inputs were applied and, in some cases, resulted in the vehicle rolling in the direction opposite to that intended by the pilot. The researchers also found that the rudder was more effective as a roll-control device than banking the wing, although a time lag in the vehicle response was involved. Joe Johnson's group conceived a modified roll-control scheme in which the outer wingtip was hinged to permit deflections for roll control instead of banking the wing. The aerodynamic action of the deflected wingtips was similar to aerodynamic tabs used on conventional aircraft surfaces. When the wingtips were deflected differentially, they provided a rolling moment that, in turn, banked the wing. This control concept produced large rolling moments for small pilot forces. This concept, along with several recommendations from NASA based on the Full-Scale Tunnel tests, was under consideration by Ryan for future modifications to the vehicle. After the Full-Scale Tunnel test on July 7, 1962, Ryan test pilot Lou Everett was slightly injured in a crash of the Ryan Flex-Wing while undergoing flight tests at Langley.

The Flex-Wing experience at Ryan and Langley served as valuable guidance when the Army subsequently contracted with Ryan to develop a second-generation parawing vehicle known as the Ryan XV-8A Flexible Wing Aerial Utility Vehicle, or "Fleep" (for Flex-Wing

A model of the Ryan XV-8A Fleep was studied during force and flight tests of the parawing utility vehicle. (NASA L-63-3249)

Jeep). Although the Fleep vaguely resembled the Flex-Wing, it included several changes to correct deficiencies exhibited by the previous design. The wing-banking and pitching concept that had resulted in unsatisfactory roll control of the Flex-Wing was replaced by a system similar to the one devised by the staff at the Full-Scale Tunnel.

In 1963, a ⅓-scale free-flight model of the Fleep underwent flight evaluations in the Full-Scale Tunnel.[63] The longitudinal stability and control characteristics of the model were judged to be satisfactory, and roll control at low angles of attack was also satisfactory. As angle of attack was increased, however, a progressive deterioration in roll-control effectiveness resulted in an unsatisfactory control condition. On several occasions, the model went out of control and diverged out of the tunnel test section. Excessive levels of adverse yawing moments and dihedral effect resulted in a dramatic reduction in roll control at high angles of attack. When an original V-tail was replaced with a conventional vertical tail and rudder in the propeller wake, as had been recommended by the Langley researchers prior to the project, satisfactory lateral control characteristics existed over the entire angle-of-attack range tested.

Flight tests of the full-scale XV-8A concluded that the handling characteristics of the aircraft were good and that control harmony between the longitudinal- and lateral-control systems was excellent, enabling the aircraft to be flown with one hand. Stability in all cases was positive with only light forces required. Takeoff and landing performance demonstrated the STOL capability of the airplane. At maximum gross weight, the takeoff distance over a 50-foot obstacle was 1,000 feet, and landing distance to clear a 50-foot obstacle was 400 feet. Some test operations were conducted from unprepared desert surfaces, establishing the capability for operation from areas other than regular airfields.

Although most of Langley's research on parawing/capsule configurations was conducted out of doors using drop models, a limited number of free-flight tests were conducted in the Full-Scale Tunnel, such as this test in 1961. (NASA L-61-4369)

Ryan also proposed the use of towed parawing gliders for transporting troops and material for the Army. In practice, however, it had been very difficult to achieve an inherently stable tow configuration, and a free-flight model investigation was conducted in the Full-Scale Tunnel in 1962 to determine a satisfactory tow configuration.[64] Robert E. Shanks had been a key researcher at the

12-Foot Free-Flight Tunnel in studies of the dynamic stability of vehicles under tow and was the leader of the tow investigation for the Ryan vehicle. The results of the investigation showed that the basic configuration was unsatisfactory because of a constant-amplitude lateral oscillation that appeared as a sidewise motion. Shanks added a vertical side area beneath the wing, which resulted in satisfactory characteristics under tow.[65]

Parawings and the Space Program

In the early days of the space program, NASA explored many concepts for the landing and recovery of capsules returning astronauts at the end of space missions. The favored approach following reentry into the atmosphere was to use parachutes for deceleration to an uncontrolled water landing and recovery. However, many leaders within the Agency were interested in other, lift-producing concepts that might extend the operational mission footprint to permit landing options at land sites or runways and thus eliminate the complexity and cost of water landings. The Rogallo flexible-wing concept appeared as a candidate recovery system that might be deployed by the capsule crew following reentry at lower altitudes and air speeds, providing a capability for gliding flight to a controlled landing.

During the Gemini program in 1961, interest began to heighten regarding using parawings for land recovery of returning space capsules. As an indication of NASA's interest, industry was requested to propose candidate configurations for a parawing spacecraft-recovery system. North American Aviation was subsequently selected to develop the Gemini Paraglider concept, which used an all-flexible inflatable parawing and called for demonstrations of uncrewed and crewed gliding flights of Gemini/parawing configurations following release from tow behind a helicopter.

Extensive research was conducted on capsule/parawing configurations by Mac McKinney's groups based at the Full-Scale Tunnel during these years; however, most of the work was conducted outdoors using drop models launched from helicopters. A few exploratory free-flight tests were conducted in the Full-Scale Tunnel on generic configurations.

In 1964, NASA made the decision to terminate the paraglider space-vehicle recovery system program in favor of conventional parachute systems. This decision immediately decreased the level of research and interest in parawing applications in the Agency. Despite the lack of NASA applications, Francis Rogallo's parawing concept lives on today as a worldwide favorite of sport aviation in the form of powered and unpowered hang gliders. Spurred on at the end of the 1960s by initial applications within the water-skiing community, a new form of sport had been born that still remains strong to this day, with both land- and water-based applications.

The Limp Parawing and Gliding Parachutes

Even after the cancellation of the Gemini paraglider program, a continuing interest in gliding parachutes as a means of space-vehicle recovery and cargo delivery existed. During the Vietnam War era, several gliding-parachute concepts emerged for such systems. At the Full-Scale Tunnel, several evaluations of different parachute-like devices with gliding capability were conducted in the mid-1960s.

One concept receiving considerable attention was a parawing completely void of any rigid structural members and utilizing only the tension forces produced by the aerodynamic loading to maintain the shape of the canopy. Conceived and developed at Langley, this unique parawing configuration was known as the limp parawing. In 1966, aerodynamic tests of an 18-foot all-flexible parawing were made in the Full-Scale Tunnel to assess the geometric and aerodynamic stability of the configuration over a large range of angles of attack.[66] Results showed that the parawing was stable for angles of attack from about 30° to 40°, but for higher angles of attack it became unstable and for lower angles of attack the nose portion of the wing collapsed.

The Inflatable Micrometeoroid Paraglider (IMP) was tested in the Full-Scale Tunnel in 1962 prior to flight tests. (NASA L-62-5996)

The Inflatable Micrometeoroid Wing

In the early 1960s, Langley's distinguished scientist Dr. William H. Kinard was deeply involved in addressing the question of the density of meteorites and micrometeorites in space and the level of protection required for possible impacts with future space vehicles.[67] A flexible, inflatable parawing equipped with extensive instrumentation, which was deployed after launching from atop a carrier rocket, appeared to Kinard to offer a relatively lightweight approach to obtaining the information. The concept was called IMP for Inflatable Micrometeoroid Paraglider and tested in various Langley wind tunnels up to a Mach number of 8.0.

Tests were conducted in the Full-Scale Tunnel in August 1962 to determine the low-speed aerodynamic characteristics of the IMP configuration.[68] The tests were part of a broader program that included a test on a smaller ⅕-scale model in another tunnel, and outdoor free-flight tests using the same full-scale model used in the Full-Scale Tunnel. Results of the wind tunnel tests indicated that the vehicle had a maximum lift-drag ratio of about 3.0 and positive stability. The outdoor flight test demonstrated that the configuration could be trimmed for steady gliding flight and that it was capable of recovering from launches at zero speed at extreme pitch attitudes and roll attitudes.

The IMP was launched on June 10, 1964, on a sounding rocket to an altitude of 96 miles over White Sands, New Mexico.[69] It then doubled back to Earth at a speed of more than 5,000 mph and was inflated. Unfortunately, after inflation the paraglider turned upside down, the rocket nose-cone-jettison system failed to separate, and the latter began its descent with the nose cone anchor still attached. After righting itself, the IMP flew briefly until aerodynamic pressures and heat caused one wing boom to collapse, and the vehicle dropped to the desert below. Researchers worked for weeks to find the meteoroid collection panels that were torn off the wing.

Tunnel tests of several gliding parachute concepts were conducted in the Full-Scale Tunnel during the 1960s including a sail wing (left) and a ram-air inflated parafoil (right). (NASA L-67-6406 and L-68-3411)

Gliding Parachutes

In the late 1960s, considerable interest from the military and industry resulted in several test entries in the Full-Scale Tunnel for advanced gliding parachutes. Known as parafoils, parasails, and sailwings, these configurations were assessed during relatively short one-week test programs in the tunnel. The scope of most of the projects included a determination of the lift-drag ratios of the configuration and its longitudinal stability characteristics. In many cases, lateral directional stability and the configuration control effectiveness were also determined. Visual observations were made of the structural stability of the canopy, which in some cases collapsed for certain simulated flight conditions.

One of the configurations tested in 1965 was a ram-air-inflated fabric wing known as the Jalbert parafoil, which had a rectangular planform and an airfoil cross-section with an opening at the leading edge to allow ram air to inflate the wing to the desired shape in flight.[70] In 1966, tests were conducted for a multiple-lobe canopy, roughly rectangular in inflated planform and having a span greater than its chord length and an airfoil-type leading edge.[71] In 1967, a twin-keel parawing was tested.[72]

The Princeton University sail-wing concept was evaluated in a cooperative study to determine aerodynamic performance, stability, and control. (NASA L-66-4789)

Princeton Sailwing

Princeton University had been developing a unique semi-flexible foldable-wing concept since 1948. First conceived as an advanced sail for boats, the concept had been converted to aircraft applications as a minimum-structure wing. Known as the sailwing, it was envisioned for applications such as auxiliary wings for air-cushion vehicles, towed cargo gliders, foldable light aircraft, rocket-booster recovery aids, lifting-body reentry vehicles, and compound helicopters.

In 1966, cooperative Princeton/NASA studies of a full-scale airplane model equipped with a sailwing were conducted in the Full-Scale Tunnel.[73] The airplane's wing had a rigid leading-edge spar, rigid root and wing-tip ribs with a trailing-edge cable stretched between these ribs, and a fabric covering stretched between the leading and trailing edges. The airframe had been used for limited flight testing at Princeton before it was modified for testing in the tunnel.[74] Results of the test program showed that the wing airfoil experienced a rapid increase in camber as the angle of attack was increased, with maximum values of lift-drag ratio comparable to those of conventional solid wings. The magnitude of lateral control provided by a wing-warp technique was effective for angles of attack up to maximum lift, where control effectiveness became low and nonlinear. Follow-on testing of an isolated and updated wing was conducted in the Full-Scale Tunnel in 1967.[75]

One of several configurations tested as part of an investigation of the aerodynamic characteristics of a flying ejection seat based on the application of the Princeton University sail-wing concept. (NASA L-70-4554)

Flying Ejection Seats: The AERCAB Program

The Princeton Sailwing concept attracted the interests of the Navy and Air Force during the Vietnam War in 1969 as concern arose over the consequences of captured U.S. pilots in enemy territory. These concerns had resulted in the Goodyear Inflatoplane, as previously discussed, but other concepts were also of interest. In 1969, Langley engineer Sanger M. "Tod" Burk had become interested in applications of the sailwing to a "fly away" ejection seat, which might use a deployable wing and small propulsion device for pilot-initiated rescue following aircraft damage during combat. Burk's study consisted of exploratory tests in the Full-Scale Tunnel to define the stability characteristics of several ejection seat configurations through conventional static tests of a subscale model.[76] Burk was well known as an innovative, curious researcher who became excited over "out-of-box" projects. He saw an ideal answer for the operational requirement in the sailwing.

Burk was always the subject of jokes whenever he began such projects, and his model test of the flying ejection seat in the Full-Scale Tunnel was entered in the test log as "Burk's Folly." However, he had the last laugh as the military was sparked by the idea. The Navy began a program known as the Integrated Aircrew Escape/Rescue System Capability (AERCAB), with contracts awarded for the wind tunnel and limited flight testing of concepts that might be retrofit to the service's F-4 Phantom and A-7 Corsair II aircraft. Contracts were awarded to Kaman and Fairchild-Hiller's Stratos-Western Division, with Kaman studying an autogiro concept and Stratos-Western proposing folding, telescoping flight surfaces on an ejection seat using the Princeton concept. A small turbofan engine would power both designs.[77]

Deployment tests of the Stratos-Western version of the AERCAB concept were conducted in the Full-Scale Tunnel in 1970.[78]

The Last Gasp: Helicopters and Other Rotorcraft

In his summary paper of the history of NACA/NASA rotating-wing aircraft research from 1915 to 1970, Langley's Frederic B. Gustafson noted adverse developments in rotorcraft research within the new Space Agency during the 1960s.[79] He indicated a large negative impact on traditional helicopter research as aeronautics struggled for its identity in the glamorous new world of space research. The emergence of a multitude of unconventional VTOL vehicles in the 1960s became the focal point of vertical flight because of the newness and innovation on display. Meanwhile, in many quarters the perspective that all meaningful helicopter research had been completed began to diminish NASA's interest in resource allotments for the subject.

As evidence of the shift away from helicopters, Gustafson states that at the NACA's 1954 helicopter conference, 32 papers were presented on rotorcraft. By the time NASA held its 1960 conference on V/STOL aircraft, only 5 of the 26 papers were on rotorcraft; and in its 1966 V/STOL and STOL conference, only 3 of the 23 papers were concerned with rotorcraft. Finally, he noted that since the mid-1950s, the NASA Ames Research Center had vigorously pursued rotorcraft research in its 40- by 80-Foot Wind Tunnel and had succeeded in capturing many of the wind tunnel tests of full-scale rotors or vehicles. All of the foregoing factors resulted in a downturn of rotorcraft testing in the Full-Scale Tunnel, which the reader will remember was stripped of many of its helicopter experts by the reorganization of late 1958.

Meanwhile, at Langley advances had been made in miniature pressure pickups that could be used on rotor blades to determine pressure distributions. Preliminary measurements in the Full-Scale Tunnel for both hovering and forward flight on generic rotor test articles demonstrated the accuracy and robustness of the pickups. Measurements of pressure distributions on a tandem-rotor arrangement by Robert J. Huston had revealed marked interference effects from blade-tip vortices and influenced industry designs for vertical spacing on the front and rear rotors of production helicopters.[80]

Other rotorcraft tests in the tunnel included rotor aerodynamic measurements for high-thrust conditions, which showed significantly more lift than simple calculations suggested, and an investigation of rotor performance at high tip-speed ratios representative of future compound helicopters using an auxiliary propulsion system.[81] In addition, Langley's rotorcraft program pioneered the analysis and understanding of the hingeless-rotor concept through an integrated program of wind tunnel and flight tests that included tests of a hingeless rotor in the Full-Scale Tunnel in 1962.[82]

Drag cleanup of helicopters was a topic of interest in the early 1960s, and two notable test entries were directed at providing design information similar to that produced by the legendary drag cleanup tests for fixed-wing aircraft that had been conducted in the tunnel during World War II. In November 1960, tests were conducted in the Full-Scale Tunnel to determine the parasite drag of two full-scale helicopter fuselage models that included appendages, with emphasis on correlation with existing prediction methods.[83] The second major drag cleanup activity involved a 1961 test that was conducted at the request of the

Navy to determine the effect of body shape, engine operation, appendages, and leakage on the drag of the fuselage of a HU2K helicopter fuselage.[84] By far the most significant result of the HU2K test was that the rotor-hub installation increased the parasite drag contributed by the helicopter fuselage by about 80 percent over that of the faired and sealed production body.

One of the last significant helicopter tests in the Full-Scale Tunnel in the 1960s was an entry in 1969 to investigate aerodynamic phenomena that had produced directional-control problems for helicopters with tail rotors in low-speed rearward flight amid ground effect.[85]

In 1960, tests were conducted to identify fuselage drag for a helicopter model. The project was reminiscent of World War II drag cleanup tests conducted in the Full-Scale Tunnel for military aircraft. (NASA L-60-7776)

At the request of the Navy, a Kaman HU2K-1 helicopter underwent a drag cleanup evaluation in 1961. (NASA L-61-4693)

In this landmark study, a subscale helicopter model was mounted close to the tunnel ground board and tested in tailwinds from 0 to 25 knots. The results of the study identified significant adverse effects of the main rotor wake, which included an increase in adverse force on the fan, a decrease in the net tail-rotor thrust, and an increase in torque required for the tail rotor. The adverse effects were found to be the result of the immersion of the tail rotor and fin in a ground vortex generated by the interactions of the main rotor wake and the wind in the presence of ground effect. When rearward airspeed was sufficiently increased, the effects of the ground vortex were diminished and induced away from the tail rotor and fin, with the result that an abrupt change in tail-rotor collective pitch was required. The data analysis derived from the test proved invaluable in guidance on the design of tail-rotor configurations to minimize low-speed control problems.

The Rotor/Wing Concept

In the mid-1960s, the helicopter community was conducting major efforts to combine the high hovering efficiency of the helicopter with the high-cruise speed efficiency of a conventional fixed-wing airplane. Candidate designs included compound helicopters as well as radical new configurations in which the main rotor would be stopped or stowed for conventional airplane flight. A novel concept known as the rotor/wing was conceived by the Hughes Tool Company Aircraft Division in response to an Army competition known as the Composite Aircraft Program. The concept combined the hovering and cruising lift systems into a single lifting surface, which had a large center hub and three hingeless blades in an attempt to reduce the weight penalty associated with independent systems. For vertical flight the single lifting surface and its blades rotated as a helicopter main rotor, whereas for airplane flight the surface was stopped with one of the three blades pointing forward in the direction of flight.

The rotor/wing aircraft used conventional helicopter and airplane flight controls. In the helicopter mode, pitch, roll, and height control were obtained by collective and cyclic pitch control of the blades. Yaw control was provided by a small fan in the vertical tail surface. In the airplane mode, control was provided by a conventional elevon and rudder arrangement.

Conversion from helicopter to fixed-wing flight was conducted by accelerating the aircraft in the helicopter mode to airspeeds above the stall speed for the fixed-wing configuration and then decelerating the rotor until it stopped with one blade forward. Considerable concern existed over the feasibility of the conversion process, and NASA conducted two wind tunnel tests at Langley to gather aerodynamic data and understanding of a representative configuration as well as an extensive analysis of the maneuver.[86]

In September 1966, a complex powered model of a rotor/wing configuration was tested in the Full-Scale Tunnel at the request of the Navy. Small-scale model tests had been conducted in a Navy wind tunnel at the David Taylor Model Basin facility, but the Navy was concerned about the validity of those results. In addition to the Hughes wing design, Langley included two additional wing-hub configurations.

The investigation and subsequent analysis were not intended to solve issues for the conversion process, but rather to provide indications of the magnitude of the potential attitude disturbances that would be encountered during the conversion from wing-borne to rotor-borne

Analysis of the conversion from wing-borne flight to rotor-borne flight was accomplished for the Hughes Rotor/Wing configuration with tests of this powered model in the Full-Scale Tunnel. (NASA L-66-7293)

flight. Based on the data gathered in the study, project engineer Bob Huston concluded that the principal problem associated with the conversion was the possibility of very large oscillating loads and attitude disturbances during the first revolution of the rotor during its startup process. The disturbances were caused by the three-blade configuration because of the rotation of the center of pressure of lift in an elliptical path. Analysis of the results indicated that conventional helicopter cyclic blade inputs could trim the aircraft during the conversion process, but a four-blade configuration was recommended for significantly smaller disturbing moments during the starting and/or stopping of the rotor system.

A second Langley tunnel investigation was conducted in the 12-Foot Low-Speed Tunnel with an unpowered model and fixed wing to help resolve the controllability issues during the conversion process.[87]

Almost 40 years later, the Boeing X-50 Dragonfly was developed as a canard rotor/wing demonstrator under a joint funding agreement with the Defense Advanced Research Projects Agency (DARPA) to demonstrate that a helicopter's rotor could be stopped in flight and act as a fixed wing. Funding was provided for two unmanned prototypes; however, neither of the aircraft ever successfully transitioned to full forward flight. DARPA terminated the program in late 2006.[88]

Flying Jeep

In the late 1950s, the Army became intensely interested in using a type of VTOL aircraft termed a "flying jeep" to perform in the air essentially the same services performed by a Jeep on the ground.[89] Industry responded with 26 proposals, of which 20 involved the use of two or four unshrouded propellers with some type of platform between them. In early 1957, Army representatives from Fort Eustis, Virginia, visited Langley to request support for research on multiple shrouded-propeller configurations. As a result of the Army's request, a series of free-flight and force-test models underwent studies in the Full-Scale Tunnel in late 1959 and the early 1960s. The envisioned full-scale vehicles would be capable of speeds up to about 60 knots and would carry a payload of about 1,000 pounds.

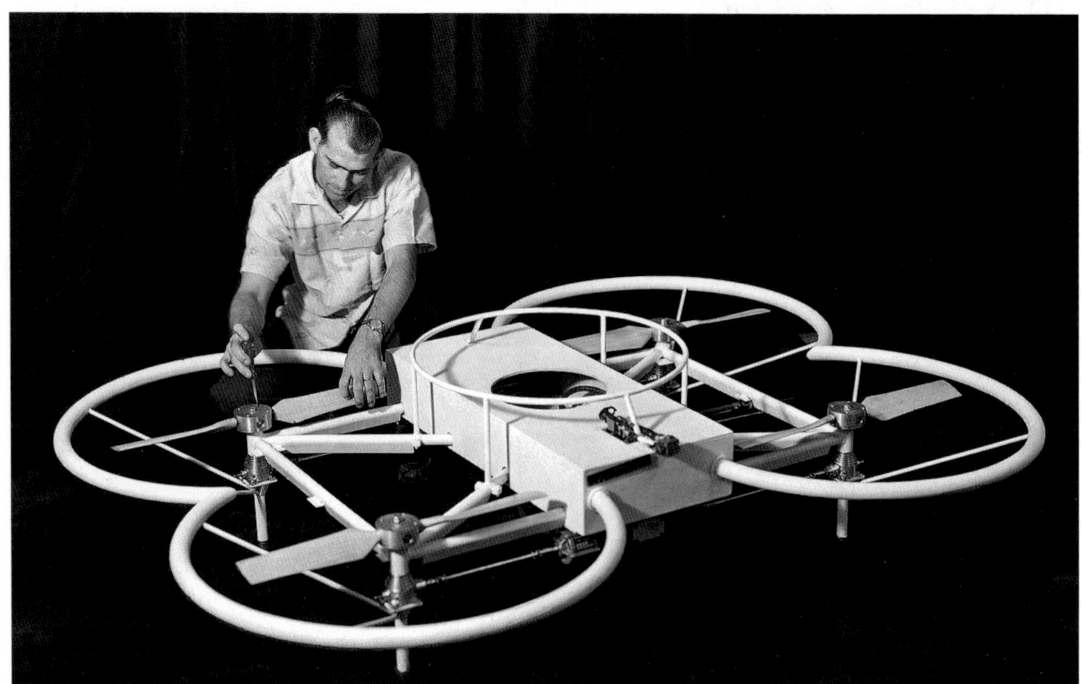

Free-flight and force tests were conducted for a broad variety of flying jeep configurations as a result of an Army request for NASA to institute a research program on such vehicles. This typical research model was flown with four unshrouded propellers. Virtually all the configurations suffered from severe trim problems at forward speeds. (NASA L-59-3433)

The configurations that were studied consisted of a platform for the engine, pilot, and cargo supported by two or more propellers that were either shrouded or unshrouded. Results of several tests in the Full-Scale Tunnel arrived at the same conclusions: that unacceptably large forward tilt angles of the platform were required for high-speed flight, and that excessive nose-up pitching moments overpowered the available control with increasing forward speed.[90]

General Aviation Research

During the early 1960s, the NASA Flight Research Center (now the NASA Armstrong Flight Research Center) conducted a flight research program to evaluate the flying qualities of seven general aviation aircraft.[91] A twin-engine Piper PA-30 aircraft was used as one of the test beds, and after flight evaluations of the general handling characteristics of the aircraft

After an absence of 30 years general aviation aircraft re-entered the Full-Scale Tunnel in 1967 when a Piper PA-30 twin-engine aircraft was tested in support of a flight program at the NASA Flight Research Center. (NASA L-67-7624)

at the Flight Research Center, the PA-30 was tested in the Full-Scale Tunnel to obtain data for correlation with flight results and for further analysis of behavior shown in flight.[92] The tests in the Full-Scale Tunnel in 1968 were the first conducted for a personal-owner-type general aviation aircraft since the testing of the Weick W1 aircraft in the 1930s.

One of the characteristics of the aircraft that was considered unacceptable if encountered by the inexperienced or unsuspecting pilot was power-on stall behavior that culminated in rapid roll offs and/or spins. The results of the test program indicated that rolling and yawing moments greater than those produced by full-opposite control occurred at zero sideslip and high power because of asymmetrical wing stall. As part of its test program, Langley also tested a subscale ⅙-scale model of the PA-30 in the Full-Scale Tunnel and found that the results were in poor correlation with results of the full-scale airplane tests. High–Reynolds number tests were then made in the Langley Low-Turbulence Pressure Tunnel as reported in NASA TN D-7109 (1972), where significant effects of Reynolds number were noted.[93]

After the test program, the manufacturer revised the configuration with counter-rotating engines and modified wing leading edges to mitigate the roll-off tendency. The NASA PA-30 was used after the test program in pioneering efforts at the Flight Research Center to develop testing techniques to be used in flying remotely piloted research vehicles (RPRVs) and several other diverse research programs.

The PA-30 test program was followed by several other tests of general aviation aircraft in the Full-Scale Tunnel. A single-engine PA-24 aircraft that was investigated in the Flight Research Center program was also tested in the tunnel in April 1969.[94] Later that year, a cooperative test of a single-engine Ryan Navion airplane was conducted with representatives from Princeton University, who operated the Navion for research purposes.[95] Many more general aviation tests would be forthcoming in the Full-Scale Tunnel during the next decade, as will be discussed in the next chapter.

Summarizing the Sixties

The Full-Scale Tunnel had clearly demonstrated its value following the critical change-over from the NACA to NASA. Unique new testing techniques and national demands for research on low-speed characteristics of civil and military aircraft had supplied a rich cadre of new challenges. The emergence of V/STOL configurations, and a bewildering number of concepts to achieve an efficient blend of hovering and forward flight capabilities, provided the opportunity to contribute priceless research to the aviation community. Once again, the facility and its staff adapted to new missions and established that NASA's oldest wind tunnel could still play a vital role in aeronautics.

Endnotes

1. "Dr. Dryden Gives Views on the Role of the NACA in Space-Age," *Langley Air Scoop* (January 31, 1958).
2. Robert Hotz, "NACA, The Logical Space Agency," *Aviation Week* (February 3, 1958).
3. Jerry Brewer would later work for his former boss, Hack Wilson, in Wilson's Reentry Program's Office, starting in 1962. After the successful conclusion of Project FIRE in support of the Apollo Program, Brewer worked in the Lunar Orbiter Project in 1964.
4. Campbell had been named head of the Free-Flight Tunnel Section in 1944. On December 23, 1958, he became Head of the Dynamic Stability Branch, which had been managed by Charles Zimmerman until he was assigned to the new NASA Space Task Group.
5. Paulson's section consisted of 6 individuals while McKinney's section included 18. McKinney's section was assigned the responsibility of operations of the Full-Scale Tunnel.
6. George M. Ware, "Full-Scale Tunnel Memories," (Langley Research Center: NASA NP-2010-03-252-LaRC, October 14, 2009), pp. 10–11.
7. Ibid., pp. 10–11.
8. At the time of his retirement, Reid had been serving as an assistant to his successor, Floyd L. Thompson, since May 20, 1960. Reid and Ed Sharp retired within a few months of each other after having managed their respective laboratories for many years.
9. Video scenes of free-flight tests of hypersonic boost-glide models in the Full-Scale Tunnel are available at *http://www.youtube.com/watch?v=1j0q5aKb45g*, accessed January 8, 2012.
10. Martin T. Moul and John W. Paulson, "Dynamic Lateral Behavior of High-Performance Aircraft," NACA RM L58E16 (1958).
11. Robert E. Shanks and George M. Ware, "Investigation of the Flight Characteristics of a 1/5-Scale Model of a Dyna-Soar Glider Configuration at Low Subsonic Speeds," NASA TM X-683 (1962).
12. Video scenes of tests of the Dyna-Soar Model in the Full-Scale Tunnel are available at *http://www.youtube.com/watch?v=o2QJFiafYfY*, accessed January 8, 2012.
13. William I. Scallion, "Full-Scale Wind-Tunnel Investigation of the Low-Speed Static Aerodynamic Characteristics of a Model of a Reentry Capsule," NASA TM X-220 (1959).
14. John B. Peterson, Jr., and Elmer A. Horton, "An Investigation of the Effect of a Highly Favorable Pressure Gradient on Boundary-Layer Transition as Caused by Various Types of Roughnesses on a 10-Foot-Diameter Hemisphere at Subsonic Speeds," NASA TM 2-8-59L (1959).
15. Loyd S. Swenson, Jr., James M. Grimwood, and Charles C. Alexander, *This Island Earth: A History of Project Mercury* (Washington, DC: NASA SP-4201, 1989), chapter 8-4.
16. The Space Task Group was officially renamed the Manned Spacecraft Center while still at Langley in November 1961. At that time the Houston facility was still under design.
17. W. Hewitt Phillips, *Journey into Space Research: Continuation of a Career at NASA Langley Research Center* (Washington, DC: NASA SP-2005-4540, 2005), pp. 38–39. Phillips

later recalled the pilot model testing in the Full-Scale Tunnel in "Next Step: Practice for a Perfect Landing," *Langley Researcher News* (May 19, 1989).

18. Peter C. Boisseau, Robert O. Schade, Robert A. Champine, and Henry C. Elksnin, "Preliminary Investigation of the Handling Qualities of a Vehicle in a Simulated Lunar Gravitational Field," NASA TN D-2636 (1964).
19. Joseph R. Chambers and Mark A. Chambers, *Radical Wings and Wind Tunnels.*
20. Gene J. Matranga, C. Wayne Ottinger, Calvin R. Jarvis, and C. Christian Gelzer, *Unconventional, Contrary, and Ugly: The Lunar Landing Research Vehicle* (Washington, DC: NASA SP-2004-4535, 2005).
21. Ibid., pp. 188–189.
22. Chambers and Chambers, *Radical Wings and Wind Tunnels.*
23. James L. Hassell, Jr., "Investigation of the Low-Subsonic Stability and Control Characteristics of a ⅓-Scale Free-Flying Model of a Lifting-Body Reentry Configuration," NASA TM X-297 (1960).
24. James L. Hassell, Jr., and George M. Ware, "Investigation of the Low-Subsonic Stability and Control Characteristics of a 0.34-Scale Free-Flying Model of a Modified Half-Cone Reentry Vehicle," NASA TM X-665 (1962). Digitized videos of tests of the M2 flights in the Full-Scale Tunnel may be viewed at *http://www.youtube.com/watch?v=xU9PWH9Ynuk*, accessed January 4, 2012.
25. Chambers and Chambers, *Radical Wings and Wind Tunnels.*
26. George M. Ware, "Investigation of the Flight Characteristics of a Model of the HL-10 Manned Lifting Entry Vehicle," NASA TM X-1307 (1967).
27. George M. Ware, "Full-Scale Wind-Tunnel Investigation of the Aerodynamic Characteristics of the HL-10 Manned Lifting Entry Vehicle," NASA TM X-1160 (1965).
28. George M. Ware, interview by author, July 12, 2010.
29. Digitized videos of tests of the HL-10 tests in the Full-Scale Tunnel and the Langley Impact Structures Facility (previously Tow Tank 2) are available at *http://www.youtube.com/watch?v=uyPjNkmACMY*, accessed January 4, 2012.
30. Chambers and Chambers, *Radical Wings and Wind Tunnels.*
31. An excellent video of a 1957 film by the researchers summarizing Langley's work on V/STOL concepts is available at *http://www.youtube.com/watch?v=NKwc40t6YDM*, accessed December 27, 2011.
32. Joseph R. Chambers, *Partners in Freedom: Contributions of the Langley Research Center to U.S. Military Aircraft of the 1990s* (Washington, DC: NASA SP-2000-4519, 2000), pp. 13–18.
33. Charles C. Smith, Jr., "Flight Tests of a ⅙-Scale Model of the Hawker P 1127 Jet VTOL Airplane," NASA TM SX-531 (1961).
34. Charles C. Smith, Jr., and Lysle P. Parlett, "Flight Test of a 0.13-Scale Model of a Vectored-Thrust Jet VTOL Transport Airplane," NASA TN D-2285 (1964).
35. Video scenes of tests of the P.1127 model in the Full-Scale Tunnel are available at *http://www.youtube.com/watch?v=yd0cxIkwtO4*, accessed January 3, 2012.

36. Video scenes of tests of the XFY-1 model in the Full-Scale Tunnel are available at *http://www.youtube.com/watch?v=jXiq2-VUbY4*, accessed January 8, 2012.
37. Louis P. Tosti, "Flight Investigation of Stability and Control Characteristics of a ⅛-Scale Model of a Tilt-Wing Vertical-Take-Off-And-Landing Airplane," NASA TN D-45 (1960). Video of the X-18 tests is available at *http://www.youtube.com/watch?v=KLRyJsJIvNE*, accessed January 14, 2012.
38. Robert G. Mitchell, "Full-Scale Wind-Tunnel Test of the VZ-2 VTOL Airplane with Particular Reference to the Wing Stall Phenomena," NASA TN D- 2013 (1963).
39. Robert O. Schade and Robert H. Kirby, "Effect of Wing Stalling in Transition on a ¼-Scale Model of the VZ-2 Aircraft," NASA TN D- 2381 (1964).
40. William A. Newsom, Jr., and Robert H. Kirby, "Flight Investigation of Stability and Control Characteristics of a 1⁄9-Scale Model of a Four-Propeller Tilt-Wing V/STOL Transport," NASA TN D-2443 (1964).
41. Digitized video segments of free-flight tests of the XC-142A model is available at *http://www.youtube.com/watch?v=G1n7sXCjvM8*, accessed January 13, 2012. A longer version is available at *http://www.youtube.com/watch?v=3vhTpA-gWeY*, accessed January 3, 2012.
42. Marvin P. Fink, "Full-Scale Wind-Tunnel Investigation of the VZ-5 Four-Propeller Deflected-Slipstream VTOL Airplane," NASA TM SX-805 (1963).
43. William A. Newsom, Jr., and Delma C. Freeman, Jr., "Flight Investigation of Stability and Control Characteristics of a 0.18-Scale Model of a Four-Duct Tandem V/STOL Transport," NASA TN D-3055 (1966).
44. Robert H. Kirby and Joseph R. Chambers, "Flight Investigation of Dynamic Stability and Control Characteristics of a 0.18-Scale Model of a Fan-In-Wing VTOL Airplane," NASA TN D-3412 (1966).
45. For example, see William A. Newsom, Jr., "Wind-Tunnel Investigation of a V/STOL Transport Model with Four Pod-Mounted Lift Fans," NASA Technical Note TN D-5942 (1970); and William A. Newsom, Jr., and Sue B. Grafton, "Flight Investigation of a V/STOL Transport Model with Four Pod-Mounted Lift Fans," NASA TN D-6129 (1971).
46. H. Clyde McLemore and Charles C. Smith, Jr., "Hot-Gas Ingestion Investigation of Large-Scale Jet VTOL Fighter-Type Models," NASA TN D- 4609 (1968). See also H. Clyde McLemore, "Jet-Induced Lift Loss of Jet VTOL Configurations in Hovering Condition," NASA TN D-3435 (1966).
47. Chambers, *Partners in Freedom*, pp. 63–64. Specific tests are discussed in M. Leroy Spearman and Paul Comisarow, "An Investigation of the Low-Speed Static Stability Characteristics of Complete Models Having Sweptback and Sweptforward Wings," NACA Research Memorandum L8H31 (1948); and Charles J. Donlan and William C. Sleeman, "Low-Speed Wind-Tunnel Investigation of the Longitudinal Stability Characteristics of a Model Equipped with a Variable-Sweep Wing," NACA RM L9B18 (1949).
48. Personal correspondence from Roy V. Harris, March 10, 2012.
49. Alford and Polhamus were awarded a patent for their concept, which has been successfully applied to several U.S. and foreign military aircraft. For a detailed discussion of the

variable-sweep program, see Edward Polhamus, "Application of Slender Wing Benefits to Military Aircraft," AIAA Paper 83-2566 (1983).

50. James L. Hassell, Jr., "Low-Speed Flight Characteristics of a Variable-Wing-Sweep Fighter Model," NASA TM X-1036 (1965).
51. Video scenes of tests of the variable-sweep model in the Full-Scale Tunnel are available at *http://www.youtube.com/watch?v=hc9Sxg6Sgws*, accessed January 8, 2012. Scenes of a generic variable-sweep transport model are available at *http://www.youtube.com/watch?v=65gsjHhwV_8*, accessed January 6, 2012.
52. Peter C. Boisseau, "Flight Investigation of Dynamic Stability and Control Characteristics of a 1⁄10-Scale Model of a Variable-Wing-Sweep Fighter Airplane Configuration," NASA TM X- 1367 (1967).
53. Joseph R. Chambers, *Innovation in Flight: Research of the NASA Langley Research Center on Revolutionary Advanced Concepts for Aeronautics* (Washington, DC: NASA SP-2005-4539, 2005), p. 7.
54. Joseph L. Johnson, Jr., "Wind-Tunnel Investigation of Low-Subsonic Flight Characteristics of a Model of a Canard Airplane Design for Supersonic Cruise Flight," NASA TM X- 229 (1960).
55. See Delma C. Freeman, "Low Subsonic Flight and Force Investigation of a Supersonic Transport Model with a Variable-Sweep Wing," NASA TN D-4726 (1968); Delma C. Freeman, "Low Subsonic Flight and Force Investigation of a Supersonic Transport Model with a Double-Delta Wing," NASA TN D-4179 (1968); and Delma C. Freeman, "Low Subsonic Flight and Force Investigation of a Supersonic Transport Model with a Highly Swept Arrow Wing," NASA TN D-3887 (1967).
56. William D. Grantham and Perry L. Deal, "A Piloted Fixed-Based Simulator Study of Low-Speed Flight Characteristics of an Arrow-Wing Supersonic Transport Design," NASA TN D-4277 (1968). Grantham was a member of the Simulation and Analysis Section at the Full-Scale Tunnel, which was instituted by Mac McKinney to augment experimental studies.
57. Delma C. Freeman, Jr., " Low-Subsonic Longitudinal Aerodynamic Characteristics in the Deep-Stall Angle-of-Attack Range of a Supersonic Transport Configuration with a Highly Swept Arrow Wing," NASA TM X-2316 (1971).
58. Video scenes of tests of the SCAT 15F model in the Full-Scale Tunnel are available at *http://www.youtube.com/watch?v=Xt-OguMSpAU&lr=1*, accessed January 8, 2012.
59. Video scenes of early parawing research are available at *http://www.youtube.com/watch?v=LlwZ1R_wHZg*, accessed January 8, 2012.
60. Joseph L. Johnson, Jr., "Low-Subsonic Flight Characteristics of a Model of a Supersonic-Airplane Configuration with a Parawing as a Landing Aid," NASA TN D- 2031 (1963).
61. Video scenes of tests of the parawing-XB-70 in the Full-Scale Tunnel are available at *http://www.youtube.com/watch?v=LlwZ1R_wHZg*, accessed January 8, 2012.
62. Joseph L. Johnson, Jr., and James L. Hassell, Jr., "Summary of Results Obtained in Full-Scale Tunnel Investigation of the Ryan Flex-Wing Airplane," NASA TM SX-727 (1962).

63. Joseph L. Johnson, Jr., "Low-Speed Force and Flight Investigation of a Model of a Modified Parawing Utility Vehicle," NASA TN D- 2492 (1965). See also Joseph L. Johnson, Jr., "Low-Speed Force and Flight Investigation of Various Methods for Controlling Parawings," NASA TN D-2998 (1965).
64. Robert E. Shanks, "Experimental Investigation of the Dynamic Stability of a Towed Parawing Glider Air Cargo Delivery System," NASA TN D- 2292 (1964).
65. Video scenes of the tow tests in the Full-Scale Tunnel are available at *http://www.youtube.com/watch?v=kmc_jASzqCM*, accessed January 8, 2012.
66. Charles E. Libbey, George M. Ware, and Roger L. Naeseth, "Wind-Tunnel Investigation of the Static Aerodynamic Characteristics of an 18-Foot (5.49-Meter) All-Flexible Parawing," NASA TN D-3856 (1967).
67. James Oberg, "The Pod People (Inflatable Reentry Vehicles)," *Air and Space* (November 2003), pp. 58–63.
68. Delwin R. Croom and Paul G. Fournier, "Low-Subsonic Wind-Tunnel and Free-Flight Drop-Test Investigation of a Paraglider Configuration Having Large Tapered Leading Edges and Keel," NASA TN D- 3442 (1966).
69. James H. Siviter, Jr., "Flight Investigation of a Capacitance-Type Meteoroid Detector Using an Inflatable Paraglider," NASA TN D-4530 (1968).
70. Sanger M. Burk, Jr., and George M. Ware, "Static Aerodynamic Characteristics of Three Ram-Air-Inflated Low-Aspect-Ratio Fabric Wings," NASA TN D-4182 (1967).
71. George M. Ware and Charles E. Libbey, "Wind-Tunnel Investigation of the Static Aerodynamic Characteristics of a Multilobe Gliding Parachute," NASA TN D-4672 (1968).
72. George M. Ware, "Wind-Tunnel Investigation of the Aerodynamic Characteristics of a Twin-Keel Parawing," NASA TN D-5199 (1969).
73. Marvin P. Fink, "Full-Scale Investigation of the Aerodynamic Characteristics of a Model Employing a Sailwing Concept," NASA TN D-4062 (1967).
74. One noted participant in the sailwing program was a Princeton student named Phil Condit. Condit completed his master's degree in 1965 while working the sailwing and left Princeton before the wind tunnel tests in 1966. In 1965, Condit joined Boeing as a new engineer and was assigned to work with NASA Langley as an aerodynamics, stability, and control expert on an in-flight simulation of supersonic transport handling qualities using the Boeing 707 prototype as an in-flight simulator. He became President of the Boeing Company in 1992, and in 1996 became the company's Chief Executive Officer.
75. Marvin P. Fink, "Full-Scale Investigation of the Aerodynamic Characteristics of a Sail Wing of Aspect Ratio 5.9," NASA TN D-5047 (1969).
76. Sanger M. Burk, Jr., "Wind-Tunnel Investigation of Aerodynamic Characteristics of a ½-Scale Model of an Ejection Seat with a Rigid-Wing Recovery System," NASA TN D-5922 (1970).
77. *Aviation Week and Space Technology* 90, no. 15 (1969), p. 29.
78. Video scenes of tests of the deployment tests of the AERCAB in the Full-Scale Tunnel are available at *http://www.youtube.com/watch?v=xXp9z61Xcj0*, accessed January 8, 2012.

79. Frederic B. Gustafson, "History of NACA/NASA Rotating-Wing Aircraft Research, 1915–1970," *Vertiflite* (1970). The information appeared in a series of articles beginning in September 1970.
80. Robert J. Huston, "Wind-Tunnel Measurements of Performance, Blade Motions, and Blade Air Loads for Tandem-Rotor Configurations with and without Overlap," NASA TN D-1971 (1963).
81. George E. Sweet, Julian L. Jenkins, and Matthew M. Winston, "Results of Wind-Tunnel Measurements on a Lifting Rotor at High Thrust Coefficients and High Tip-Speed Ratios," NASA TN D-2462 (1964). See also Julian L. Jenkins, Jr., "Wind-Tunnel Investigation of a Lifting Rotor Operating at Tip-Speed Ratios from 0.65 to 1.45," NASA TN D-2628 (1965).
82. John F. Ward, "A Summary of Hingeless-Rotor Structural Loads and Dynamics Research," NASA TM X-56726 (1965).
83. Julian L. Jenkins, Jr., Matthew M. Winston, and George E. Sweet, "A Wind-Tunnel Investigation of the Longitudinal Aerodynamic Characteristics of Two Full-Scale Helicopter Fuselage Models with Appendages," NASA TN D-1364 (1962).
84. William I. Scallion, "Full-Scale Wind-Tunnel Investigation of the Drag Characteristics of an HU2K Helicopter Fuselage," NASA TM SX-848 (1963).
85. Robert J. Huston and Charles E.K. Morris, Jr., "A Wind-Tunnel Investigation of Helicopter Directional Control in Rearward Flight in Ground Effect," NASA TN D-6118 (1971).
86. Robert J. Huston and James P. Shivers, "A Wind-Tunnel and Analytical Study of the Conversion from Wing Lift to Rotor Lift on a Composite-Lift VTOL Aircraft," NASA TN D-5256 (1969).
87. Robert J. Huston and James P. Shivers, "The Conversion of the Rotor/Wing Aircraft," AGARD Specialist Meeting on Fluid Dynamics of Rotor and Fan Supported Aircraft at Subsonic Speeds, Gottingen, Germany, September 11–13, 1967.
88. James T. McKenna, "One Step Beyond," *Rotor & Wing* (February 2007), p. 54.
89. The term "flying jeep" caused considerable headaches for NASA, as the Government legal offices declared that American Motors properly retained trademark use of the term "jeep" because of its famous WWII vehicle. Mac McKinney was infuriated when NASA Headquarters promoted the idea of calling the configurations "flying platforms," which, in his opinion, did not portray the proper use of the vehicles. The Society of Automotive Engineers felt so strongly about the legal infringement situation that they refused to publish a 1959 paper by McKinney entitled "Stability and Control of the Aerial Jeep." McKinney finally skirted the issue by referring to the platforms as "aerial vehicles."
90. The relative seriousness of these problems depended on whether the propellers were shrouded or unshrouded and whether the propellers were side-by-side or in tandem. See Lysle P. Parlett, "Stability and Control Characteristics of a Model of an Aerial Vehicle Supported by Four Ducted Fans," NASA TN D- 937 (1961); and Robert H. Kirby, "Flight-Test Investigation of a Model of an Aerial Vehicle Supported by Four Unshrouded Propellers," NASA TN D-1235 (1962).

91. Marvin R. Barber, Charles K. Jones, Thomas R. Sisk, and Fred W. Haise, "An Evaluation of the Handling Qualities of Seven General-Aviation Aircraft," NASA TN D-3726 (1966).
92. Marvin P. Fink and Delma C. Freeman, Jr., "Full-Scale Wind-Tunnel Investigation of Static Longitudinal and Lateral Characteristics of a Light Twin-Engine Airplane," NASA TN D-4983 (1969).
93. Vernon E. Lockwood, "Effect of Reynolds Number and Engine Nacelles on the Stalling Characteristics of a Model of a Twine-Engine Light Airplane," NASA TN D-7109 (1972).
94. Marvin P. Fink, Delma C. Freeman, Jr., and H. Douglas Greer, "Full-Scale Wind-Tunnel Investigation of the Static Longitudinal and Lateral Characteristics of a Light Single-Engine Airplane," NASA TN D-5700 (1970).
95. James P. Shivers, Marvin P. Fink, and George M. Ware, "Full-Scale Wind-Tunnel Investigation of the Static Longitudinal and Lateral Characteristics of a Light Single-Engine Low-Wing Airplane," NASA TN D-5857 (1970).

Intense international interest in civil supersonic transports and the introduction of the European Concorde transport resulted in exhaustive testing of candidate designs throughout NASA. At Langley, the Full-Scale Tunnel focused on the low-speed performance, stability, and control characteristics of such configurations. This arrow-wing model was tested in 1975. (NASA EL-1996-00117)

CHAPTER 7

Faster, Slower, More Maneuverable

1969–1984

Challenging Times

The passing of the 1960s marked the end of the great challenge and excitement that had consumed the Nation and the entire NASA community. The United States had beaten the Soviet Union to the Moon in an extraordinary display of national determination and focus not seen since the end of World War II. The pride and resolve that had characterized the period were now part of the legacy of science and technology, and NASA was now faced with major questions and issues regarding its future goals, funding, and mission.

At Langley, the post-Apollo years began with concern and insecurity. In early 1971, the staff was informed by Center Director Edgar M. Cortright that a reduction in force (RIF) action would be initiated to trim the Langley workforce and that some technical disciplines might experience a major redirection. The resulting personnel reviews were conducted within complicated "bumping" procedures and rules involved in an RIF. The process totally consumed managers of research organizations in efforts to justify and protect key individuals from separation from the Agency. Employees were naturally very concerned about their future as well as about the outlook and stability of the Langley Research Center. Initial planning had called for as many as 150 involuntary RIF separations at Langley, but the final number was reduced to 47 because of resignations and retirements. Within these 47 separations, Langley lost 19 engineers and 19 technicians. The separations were completed by October 2, 1971.

The staff at the Full-Scale Tunnel had come under scrutiny during the RIF procedure, and the tunnel had lost two young engineers.[1] Ironically, as Mac McKinney worried over the possibility of even further reductions in his staff, Langley began to hire new employees and augment its workforce immediately after the RIF of 1971. In addition to new civil service employees, two new sources of manpower were used to grow the level of human resources at the tunnel—the Army and contracting services.

The Army Arrives

In February 1970, Langley entered a bilateral agreement with the Army Aviation Systems Command regarding mutual interests in research and development, including sharing of test facilities. In July 1971, the Army announced the establishment of a new Army Aviation Research and Development Office based at Langley under the direction of Thomas L. Coleman, a noted NASA Langley employee.[2] Later known as the Langley Directorate of the Army Air Mobility R&D Laboratory, the new organization placed its employees within several major facilities in support of NASA and Army programs.

The Army assigned additional staff to the Full-Scale Tunnel, including engineers, technicians, and data-reduction personnel. Although most of their efforts were in non-Army technical projects, these individuals became experts in the operations and procedures used at the tunnel. They quickly became valued members of the staff and made significant contributions to the success of the tunnel's programs. This unique Army/NASA personnel arrangement continued to provide critical capabilities for research programs conducted in the tunnel for the next 25 years.

Onsite Contractors

The significant increase in workload at the Langley Research Center during the space race, without an attendant increase in civil service workforce, had resulted in a major cultural change at the Center with the use of onsite contractors.[3] The use of industry contractors had initially begun with nontechnical maintenance and administrative jobs such as delivering the mail, operating cafeterias, and maintaining warehouses, but these "support-service" contracts rapidly expanded into technical capabilities, including the assignment of contractors within research organizations.[4]

During the 1970s, the staff of the Full-Scale Tunnel was augmented with several onsite contractors who worked alongside the civil service staff during wind tunnel investigations, data reduction, and related piloted simulator studies. The regulations regarding the use of support-service contractors clearly stipulated that they could not be integrated into the civil service workforce and must remain separated; however, the contractors and civil servants worked side-by-side, became close friends, and shared personal interests and dedication throughout Langley facilities.

One-Stop Services for the Nation

Under Mac McKinney and his successors, the Dynamic Stability Branch evolved into a versatile research organization with unique research tools and facilities. Management of the Full-Scale Tunnel, the Langley 12-Foot Low-Speed Tunnel, and the Langley Spin Tunnel provided specialized testing capability and an enthusiastic staff. Special test hardware was developed and implemented to permit measurements of aerodynamic phenomena associated with dynamic motions of aerospace vehicles. The free-flight technique was matured in sophistication and validated by correlation with many full-scale flight experiences. Full-scale wind tunnel testing of aircraft was conducted, and piloted simulator evaluations became a vital component of studies conducted by the branch.

Although limited to the subsonic-speed capabilities of the Full-Scale Tunnel, the foregoing capabilities provided a complete repertoire for research and development in flight dynamics. The reputation of the branch and its contributions to critical programs became well known to the industrial and military communities as a "one-stop-shopping" source for their aerospace interests in these technical areas. Applications of the Full-Scale Tunnel had dramatically changed from conventional static wind tunnel tests of full-scale aircraft into a broad spectrum of static, dynamic, and free-flight tests to evaluate the flight dynamics of aerospace vehicles.

Threat from the West Coast

The new emphasis on flight dynamics at the Full-Scale Tunnel also served to protect the facility from threats of closure in the 1970s. As NASA faced critical reviews of its role in supporting U.S. aerospace industries, serious proposals began to surface within the Agency for closing the oldest NASA wind tunnel. These arguments were based on the fact that the 40- by 80-Foot Tunnel at the NASA Ames Research Center had higher test speeds, better flow quality, and a larger test section compared to capabilities of the Full-Scale Tunnel. At a higher level, the viability of the workload and mission of the Ames Center itself had come under attack. In fact, the question of Ames's continued existence as a NASA installation was a heated topic of conversation in 1969 and 1970.[5]

When Dr. Hans Mark became the aggressive and controversial new director of Ames in 1969, he waged a major campaign to bring focus areas and facilities into the Ames Center, protect it from possible closure, and establish it as a "Center of Excellence" in several technical programs. He led an energetic raid on Langley and other NASA Centers to acquire special programs well suited to the Ames facilities and thereby disarm the competition. In this scenario, the issue of closing the Langley Full-Scale Tunnel surfaced for debate within NASA. After considerable discussions within NASA, its advisory committees, and the industry and military users of the facility, it was concluded that the Full-Scale Tunnel should remain in the wind tunnel inventory because the tunnel had a 2-year backlog of projects and because the research mission of the Full-Scale Tunnel was significantly different from that of the 40- by 80-Foot Tunnel. As a result of this decision, the Full-Scale Tunnel became known within industry and the military as the NASA facility of choice for investigations of flight dynamics, whereas the 40- by 80-Foot Tunnel was the facility of choice for large-scale aerodynamic testing.

The future workload of the 40- by 80-Foot Tunnel was further augmented when, after a furious and still-debated campaign waged by Dr. Mark, NASA Headquarters decided to transfer Langley's helicopter research program to Ames in 1976. Suddenly, Ames found itself leading research in an area that had been pioneered by Langley, but without the expertise of the Langley staff. In any event, the issue of closing the Full-Scale Tunnel during this era was resolved.

New Organizations, New Leaders

On October 4, 1970, Langley Director Edgar Cortright announced a new organization of the Center to accommodate its new responsibilities in projects and systems engineering.[6] The Flight Mechanics and Technology Division was abolished and a new Low-Speed Aircraft Division (LSAD) was established under John Campbell. The Dynamic Stability Branch remained under the continued leadership of Mac McKinney and Jack Paulson within the LSAD. Sections were formed within the branch with Joseph Johnson heading the Full-Scale Tunnel Section, James S. Bowman heading the Spin Tunnel Section, and Joseph R. "Joe" Chambers leading the Simulation and Analysis Section.

Leadership of the branch remained stable within the LSAD until June 1974, when Mac McKinney was elevated to assistant chief of the LSAD and Joe Chambers became the new branch head, with Jack Paulson remaining as assistant branch head. Another major reorganization was immediately encountered when Campbell retired as chief of the LSAD in late June.

Upon Campbell's departure, Director for Aeronautics Robert E. Bower reorganized Langley's aeronautics organizations in order to highlight the Center's expertise in the field of aerodynamics. As part of the reorganization, a new Subsonic-Transonic Aerodynamics Division (STAD) was formed under Richard E. "Dick" Kuhn, with responsibilities for the operation of the Full-Scale Tunnel by the Dynamic Stability Branch under Chambers and Paulson. Jack Paulson retired as assistant branch head at the end of 1974 and was replaced by Joe Johnson in March 1975. In April, researcher and former Co-op student William P. Gilbert became head of the Simulation and Analysis Section.

As branch head and assistant branch head, Chambers and Johnson merged their personalities and talents into an effective working relationship. Chambers maintained a strong personal interest in establishing close communications with the civil and military communities regarding research opportunities and the capabilities of the Full-Scale Tunnel, while Johnson led the day-to-day operations of the facility, including supervision of personnel and co-ops. Arguably, Johnson became one of the most respected and admired leaders to ever manage research at the tunnel. In addition to his leadership and management skills, he was a brilliant researcher with a "can do" attitude and interest in innovation. With Chambers serving as "Mr. Outside" to bring projects to the tunnel, Johnson made his "Mr. Inside" role extremely productive.

After a few years of organizational stability, Dick Kuhn was relieved as Chief of STAD in May 1977 for a special assignment to the Navy, and he was replaced by Percy J. "Bud" Bobbitt. By early 1979, Bob Bower's plan to emphasize Langley's capabilities in aerodynamics was gaining even more momentum. Bower placed a particular emphasis on creating advanced computational aerodynamics expertise within the STAD, and it became obvious that the flight-dynamics orientation of the Dynamic Stability Branch was not a particularly good fit within that organization's focus on computational methods.

Following discussions with Bower, Chambers and Johnson were given approval to move the Dynamic Stability Branch to the Flight Mechanics Division (FMD) headed by Robert Schade. The FMD included Langley's flight-research organization and was a much better fit for the Dynamic Stability Branch. Schade retired in July 1980 and was succeeded by his former assistant, Joseph W. Stickle. As a private pilot and a former member of the flight-research organization at Langley, Stickle brought a focused interest on general aviation and civil aircraft to his management position. Joe Chambers was promoted to assistant chief of the Flight Mechanics Division in early 1981. Later that year, Bower made yet another major reorganization, changing the names of his organizational units to reflect Langley's expertise in aerodynamics. The Flight Mechanics Division became the Low-Speed Aerodynamics Division. Joe Johnson became head of the Dynamic Stability Branch in 1981 with Bill Gilbert serving as the assistant branch head.

Under the steady and capable leadership of Joe Johnson, the organizational structure and research activities at the Full-Scale Tunnel remained stable for the remainder of this era. Several notable changes had taken place at the facility, including a complete emphasis on flight dynamics and a strong cooperative relationship with the military, supersonic civil transport technologists, and the general aviation community. In recognition of the organization's capabilities and mission, the name of the Dynamic Stability Branch was changed in 1983 to the more appropriate Flight Dynamics Branch.

The Return of Abe

On June 28, 1979, Dr. Abe Silverstein returned to visit the old wind tunnel he had helped design and establish as an aeronautical legend. The occasion was a formal ceremony commemorating almost 50 years of research in the Full-Scale Tunnel.[7] Langley Director Donald P. Hearth introduced Silverstein, who had retired as director of the NASA Lewis Research Center. About 80 Langley employees and retirees attended the event held at the tunnel in a manner similar to the many historic conferences of the past. The attendees included other noted personnel with ties to the tunnel, including John Becker, Hack Wilson, Sam Katzoff, Charles Zimmerman, John Campbell, and many others. One emotional highlight was the meeting between Silverstein and Frances Reeder, who had served as Silverstein's secretary at Langley. In his comments as guest speaker, Silverstein expressed his delight in the continual production of major aeronautical contributions from the Full-Scale Tunnel and wished it a prolonged future.

Noted employees and friends of the Full-Scale Tunnel attended a homecoming celebration for Dr. Abe Silverstein at the tunnel on June 28, 1979. Front row, from left: Axel T. Mattson, Langley Director Donald P. Hearth, Dr. Abe Silverstein, Mrs. Silverstein, Mark R. Nichols, Joseph A. Shortal, and Mrs. Shortal. Row two: J. Cabell Messick, Mrs. Messick, Herbert "Hack" Wilson, Mrs. Wilson, Thomas A. Harris, Mrs. Harris, Hartley A. Soule, and Mrs. Soule. Row three: Mrs. Cushman, Ralph Cushman, Robert O. Schade, Mrs. Schade, Frank Lofurno, and Mrs. Lofurno. Row four: I. Edward Garrick, Mrs. Davis, Don D. Davis, Jr., Samuel Katzoff, Jean G. Thompson, Ralph W. May, and Mrs. May. Row five: John W. Paulson, Sr., Mrs. Paulson, H. Clyde McLemore, Mrs. Nelson, William J. Nelson, Mrs. Butler, T. Melvin Butler, Blake W. Corson, and Mrs. Corson. Row six: Helen Stack, Eugene R. Guryansky, Mrs. Guryansky, Langley Deputy Director Oran W. Nicks, Mrs. Nicks, Herbert Roehm, W. Hewitt Phillips, and Mrs. Phillips. Row seven: Henry A. "Hank" Fedziuk, Frances W. Reeder, Donald D. Baals, Mrs. Baals, Laurence K. Loftin, Jr., John V. Becker, Mrs. Becker, and Ray W. Hooker. Row eight: Arthur W. Carter, Charles H. Zimmerman, Joseph R. Chambers, Mrs. Chambers, William P. Gilbert, Mrs. Gilbert, Roland E. Olson, and Mrs. Olson. Row nine: Charles A. Hulcher, John C. Houbolt, Howard B. Edwards, John P. Campbell, Mrs. Campbell, Mrs. Johnson, and Joseph L. Johnson, Jr. (*Langley Researcher News*, July 13, 1979)

In the early 1970s, Langley and Ames were in competition for a new large subsonic wind tunnel facility. Langley's proposal was for a "Super Tunnel" to be built in the NASA West Area adjacent to the Langley 14- by 22-Foot Subsonic Tunnel. The model shown in the photograph was tested in the Full-Scale Tunnel to determine the effects of external winds. The large end would be an open inlet for a large test section that would use the same drive system as a smaller closed-circuit tunnel. Engineer Clyde McLemore inspects the open inlet segment of the model tunnel in 1972. Langley's management cooled to the concept, and Ames was given the go-ahead to modify its 40- by 80-Foot Tunnel with a new 80- by 120-foot test section for a similar layout that became known as the National Full-Scale Aerodynamics Complex (NFAC). (NASA L-72-237)

Modifications, Rehabs, and Upgrades

Acoustic Treatment

In the early 1970s, regulatory issues regarding the operational noise profiles for civil aircraft during takeoff and landing stimulated an effort to use Langley wind tunnels for conducting aeroacoustic tests. The ability of the Full-Scale Tunnel to test full-scale general aviation propeller aircraft offered the promise of evaluating the impact of new concepts on noise signatures as well as aerodynamic performance. Since the tunnel was designed by Smith DeFrance for measurements of aerodynamic properties and not aeroacoustic fields, the ability to adapt the tunnel to this new application required the addition of sound-absorbing material within the test section. Several studies were conducted to measure the basic acoustic properties of the test section and the ability of acoustic materials to lower the ambient noise levels during tunnel operations.[8]

Following a study of the most cost-effective placement of sound-absorbing material, the roof and upper east and west walls of the test section (above the ground board) were treated for sound absorption. A special treatment for the upper surfaces of the ground board was also implemented for acoustic tests. The acoustical treatment consisted of fiberglass insulation encased in panels and covered with perforated sheet metal. The work was completed in early 1974 in time for noise measurements for several investigations, as will be described. The most severe constraint on the use of the Full-Scale Tunnel in noise studies was the location of the noise-producing drive motors and propellers directly behind the test subject. As a result of such limitations, more suitable wind tunnels such as the Langley 14- by 22-Foot

The background of this photograph of an advanced supersonic-transport model shows the acoustic panels that were added to the sidewalls and ceiling of the Full-Scale Tunnel's test section. (NASA L-75-735)

Tunnel were used for noise studies, and acoustic testing was not pursued in the Full-Scale Tunnel after the 1970s. The acoustic treatment remained in place within the tunnel's test section for the remaining life of the facility.

Pause for a Rehab

After 44 years of continual service, the Full-Scale Tunnel ceased operations on January 27, 1975, for its first major rehabilitation program. Although the tunnel schedule traditionally allowed for a brief 2-week annual shutdown for routine maintenance, it was now time for more intense repair and modification activities. The scope of the rehabilitation program centered on inspection, cleaning, maintenance, and rewinding of the drive motors; inspecting and refinishing the tunnel's wooden propeller blades; designing, fabricating, and installing new propeller hub spinners and nacelle fairings; replacing a major part of the roof; and updating the engineering and administrative offices.

During the rehab, a detailed inspection of the structural integrity of the motor supports revealed that some of the rivets securing the east drive motor bed to its mount had apparently failed and fallen through the mount. The west motor support structure had also experienced rivet failures. The motors were fastened to massive, 10-inch-thick rectangular beds with huge bolts, and the beds were riveted to flanges at the top of each of four truss structures. The support structures were braced by tie-rod cross members in an "X" configuration. When the fairings were removed, the repair crew observed that the rivets holding one corner of the east motor bed to the supporting flange were completely missing, allowing the bed to move about a vertical axis in a yawing motion. In addition, several rivets were missing or fractured on other support flanges.[9]

Workmen removed the propellers and motor fairings as part of the first major rehabilitation work on the Full-Scale Tunnel in 1975. They were shocked to find that several rivets were missing in the motor-mount support structure. (NASA L-75-2327)

The failed rivets caused grave concern regarding the safety of high-speed operations of the tunnel and immediately resulted in a reduction in the maximum permissible speed of the tunnel from 100 mph to 62 mph. Several investigations of possible forcing mechanisms, vibratory loads, and recommendations for repair of the motor mount were undertaken by the tunnel staff and several engineering organizations. Analysis suggested that the propeller-drive system was being subjected to massive unsteady loadings experienced by the blades

Left photo is a view looking downstream toward the west drive motor after the motor-mount fairings and nacelle were removed. Note the four support-truss structures braced by tie-rod cross members. The motor bed is the thick component at the top of the structure. The photo on the right shows a view from underneath the motor mount. The two large bolt-and-nut assemblies securely fastened the motor bed to the mounting structure, but three of the rivet heads are missing in the mounting-flange pad in the center of the photo. All of the rivets in another flange on the east motor support pad were missing. The cause of the failures was vibratory loads resulting from loading changes on the propellers during their 360° rotation. (NASA L-75-1612 and NASA L-75-2327)

as they rotated through 360° during normal operations. The failed rivets had permitted the motor bed to relieve stresses through the yawing motions.

The motor beds were then welded to the support pads, the cross-rods were welded to the support structures, and the nacelle-fairing structure was stiffened.[10] The planned rehab activities were completed by the end of 1976, and after a brief tunnel flow survey to determine the quantitative characteristics of airflow in the test section, the first research test (a large supersonic-transport model) began in January 1977. Unfortunately, concern over the robustness of the drive-motor structural system continued through the remainder of the tunnel's lifetime.

The sheet-metal fairings for the motor supports and nacelles had originally been mounted to wooden ribs and support structures, but new fairings were made in which the wooden members were replaced with welded angle iron. Stiffening the fairing assemblies with the angle iron aggravated the transmission of stresses, and cracks and skin-panel failures were noted during inspections of the motor fairings, nacelles, and spinners in 1978, resulting in a month of down time for repair work.

Modernizing the Free-Flight Technique

The wind tunnel free-flight technique had been continually updated and modified through the years since its initial implementation in the old Langley Free-Flight Tunnel in the late 1930s. Control actuators became more powerful, propulsion systems were updated with fans and ejectors, and instrument packages were used for measurements of model motions and flow angles. The location of the remote human pilots in the Full-Scale Tunnel had been finalized and the balcony located on the east wall had been enclosed for test crews. However, with the advent of aerodynamically unstable, control-configured vehicles such as the General Dynamics YF-16 prototype fighter, it became obvious that a rigorous simulation of the critical elements of full-scale aircraft flight control systems would have to be implemented in the free-flight testing technique in the near future.

In the early 1980s, the free-flight technique was updated with a digital computer that processed onboard sensor data and pilot-control inputs to generate command signals to drive the pneumatic control-surface actuators in the model.[11] With this upgrade, it became possible to simulate critical stability-augmentation systems and enhanced control-feedback features similar to those used by full-scale aircraft. Instrumentation carried on board included potentiometers to measure control position, three-axis rate gyroscopes to measure angular rates, linear accelerometers to measure accelerations, and boom-mounted vanes to measure angle of attack and angle of sideslip. The development and implementation of the advanced free-flight capability was a key accomplishment in maintaining NASA's ability to contribute to future advanced aircraft programs. The contribution was especially valuable in the case of the X-29 forward-swept-wing research aircraft program, in which the full-scale vehicle was extremely aerodynamically unstable.[12] When the X-29 model was flown in the tunnel, it had an aerodynamic longitudinal instability of 30-percent negative static margin—the largest instability ever flown in Langley's history of free-flight model testing. With the computer engaged, the model was easy to fly and the effectiveness of the test hardware and software had been dramatically demonstrated.

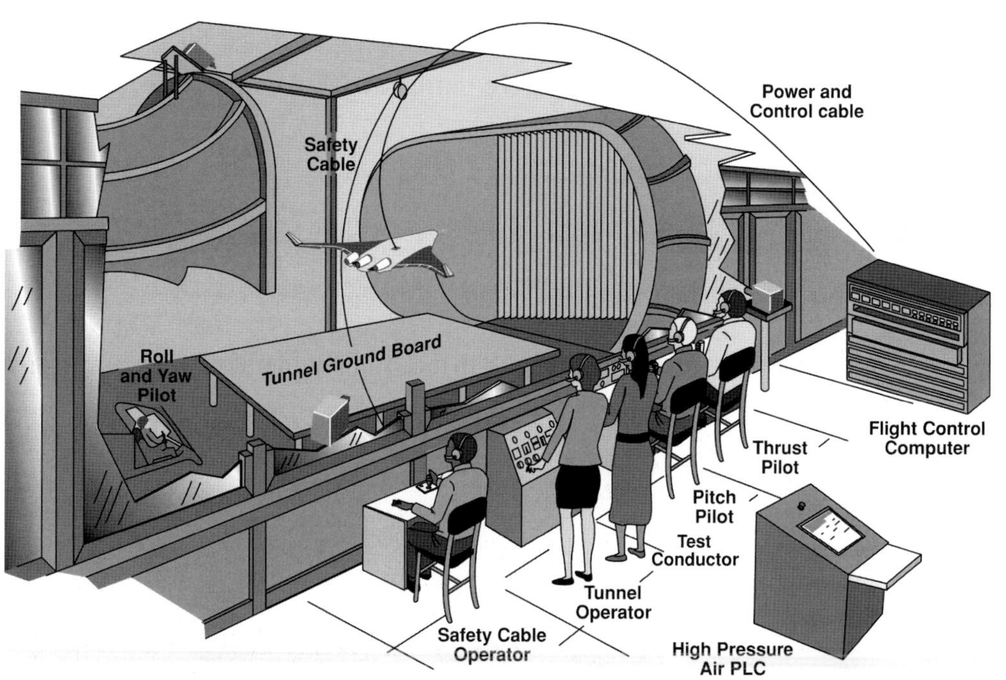

Sketch of test setup for the modernized free-flight test technique in the Full-Scale Tunnel. (NASA/Langley)

Free-flight test pilot Dan Murri poses in the seat of the roll-yaw pilot in a special room in the exit cone of the Full-Scale Tunnel. (NASA, identification unknown)

Retiring the Toledo Scales

After 48 years of service, the original Toledo scales used to measure aerodynamic loads on test aircraft via the strut and balance frame were retired in 1979 and replaced with new electronic scales that greatly enhanced data productivity.

One of the original Toledo scales at the Full-Scale Tunnel is shown in 1945 on the left, and some of the electronic scales that replaced them are shown in 1979 on the right. The pivot bar on the front of the older scales was used by technicians to reset internal counterweights during a test, thereby maintaining the ability to measure extremely small loads. (NACA LMAL 42050 and NASA L-79-8495)

Tunnel Aerostructural Problems Revisited

The concern that had arisen over the structural motor-mount issues discovered during the 1975–1977 rehabilitation program of the Full-Scale Tunnel had resulted in an imposed limit on maximum tunnel speed (of 62 mph) and a continuing investigation by engineering and fabrication specialists at Langley regarding the forcing function and characteristics of the phenomena causing the problems.

In July 1983, engineer Frank L. Jordan, Jr., made numerous static and dynamic flow measurements across two stations forward of the propellers in the exit cone. Supporting engineering organizations conducted analytical studies of the structural dynamics of the system, and a dynamic model of the propeller-motor-mount system was built and used for analysis. Jordan's measurements were made with an aircraft nose boom–type device consisting of a moveable angle-of-attack probe that included an airspeed-measuring propeller. The probe device was mounted on a two-cable support system, and measurements were made in both vertical and horizontal directions across the entire exit cone.[13]

The results of the survey revealed that the flow into the propellers was relatively uniform across the center section of the exit cone; however, a marked reduction in inflow velocity was measured around the exit cone as the outer walls were approached. The flow deficit began almost 10 feet from the wall, and the flow velocity was reduced by almost 25 percent at the walls. As a result of the unsymmetrical flow-velocity profile, massive unsteady air loads could be produced by the propellers and transmitted to the supporting structure. As each of the four propeller blades rotated from the high-velocity area near the exit-cone splitter in the center of the tunnel to the low-velocity area near the outer wall, its blade loading suddenly decreased; the blade loading then increased as the rotation continued back to the

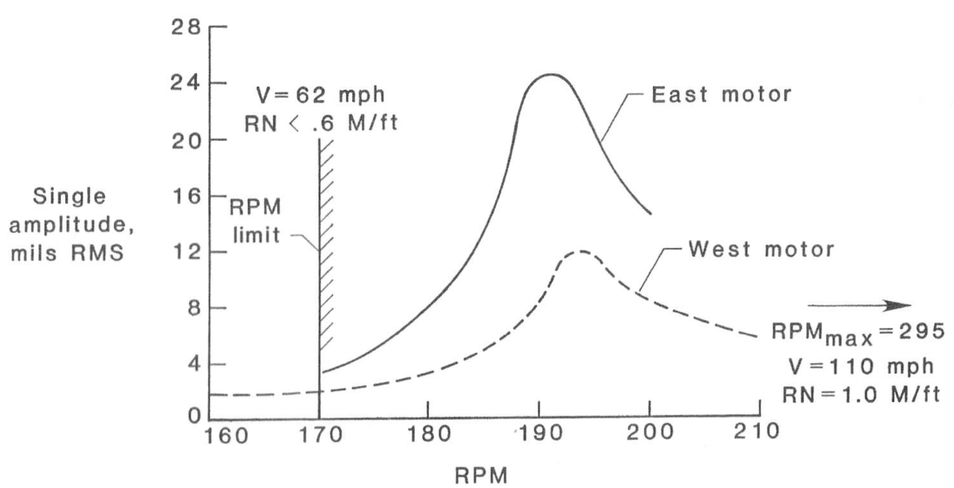

The characteristics of the severe motor-support platform vibration problem at the Full-Scale Tunnel are illustrated by measurements made by Frank Jordan of the variation of the vibratory amplitude magnitude of the support bed with drive-fan speed in revolutions per minute. As indicated by the data, the vibration of the east motor platform was much greater than the west motor platform. Both motor platforms reached a resonant condition with maximum vibratory amplitudes near 192 rpm. As a result of extreme concern over the vibratory loads, normal operations were limited to 170 rpm (62 mph). On special occasions requiring higher speeds, runs were made at 210 rpm where the vibratory amplitudes decreased. (Frank L. Jordan, Jr.)

centerline, resulting in a once-per-revolution torque (per blade) that was transmitted to the motor mount and created vibratory loads about a vertical axis. The peak amplitude of the torsional vibration of the east motor was twice that of the west motor.[14] Although strain-gage instrumentation was not applied to the blades to measure loads, Jordan had the outer portion of a single propeller blade on each motor painted black, affixed wool tufts to the blade, and used a strobe light to view the state of flow as the propeller blade progressed around a revolution. The visualization disclosed that the flow on the blade was attached during passage through the high-velocity area near the exit-cone splitter, but the tufts indicated disturbed flow when the blade approached the flow deficit near the wall.

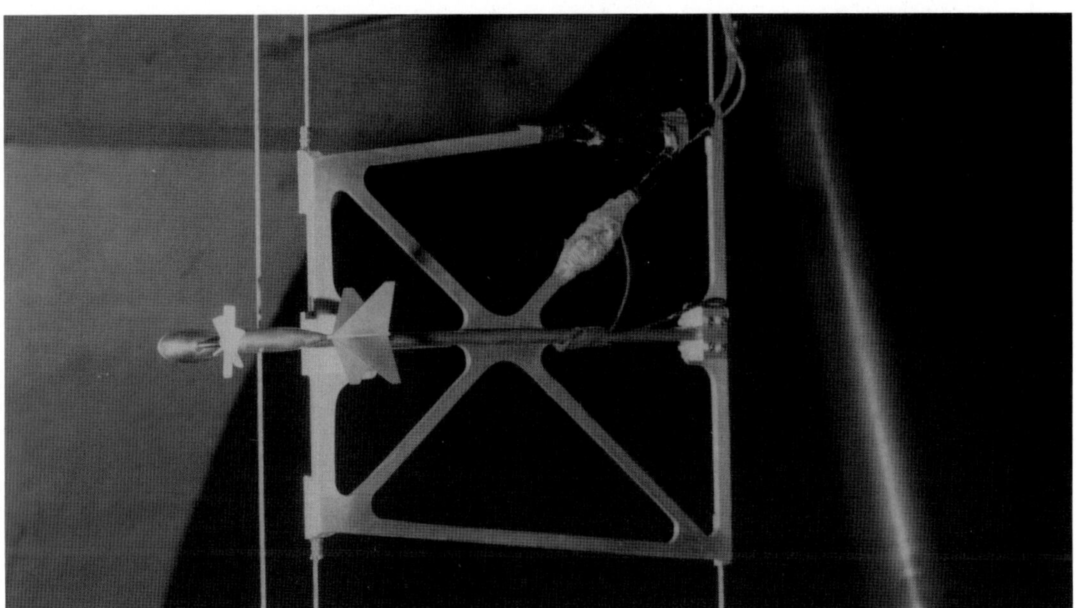

This airspeed/angularity probe used in the 1983 flow surveys revealed a marked lack of uniformity in the velocity distribution into the drive propellers near the exit-cone walls. The probe was mounted to a two-cable support system and could be traversed across the area in a vertical or horizontal direction while data were recorded from its sensors. (NASA L-83-8158)

Jordan considered several concepts to fill in the flow deficit, including the use of auxiliary flow-deflector vanes that might redirect some of the high-energy flow into the velocity-deficit area near the walls. Exploratory tests were made in which an existing high-aspect-ratio rectangular wing was attached to the tunnel's overhead survey carriage and positioned at various locations within the exit cone while the tunnel was running and vibratory motions were

Following exploratory evaluations of the ability of flow deflectors to mitigate the vibratory motions of the drive motors, two flow deflector vanes were installed in the exit cone of the Full-Scale Tunnel. Note the black paint on a single blade of each propeller and wool tufts attached to the exit cone for flow-visualization studies by Frank Jordan. (NASA L-85-6726)

Exploratory tests of a trailing-edge flap on the rear of the ground board were conducted during the attempts to reduce motor-mount vibratory loads. The flap location is shown on the left and the partially deflected flap is shown on the right. (NASA L-83-8148 and NASA L-83-8147)

measured for the drive motors. Promising locations were noted for which the maximum vibratory motions of the drive motors were significantly reduced. Following the test, two flow deflectors were installed vertically on the sidewalls of the exit cone. Additional vanes were later added, as will be discussed in the next chapter.[15]

Other concepts were also explored, such as raising the tunnel's ground board and deflecting a flap on the trailing edge of the ground board to deflect air downward into the area of flow deficit. Measurements of the vibratory amplitude indicated that the two flow deflectors and the ground board flap reduced the peak magnitude of vibration on the east motor by 50 percent. Although the deflected flap produced improvements in the exit-cone flow, it was not pursued because of impacts on the flow in the test-section area.

It would have been very helpful to conduct flow testing in a model of the tunnel to further investigate the problem and identify possible solutions. However, as previously noted, the original 1⁄15-scale model of the Full-Scale Tunnel had been given to Portugal and was unavailable for such tests in the 1970s. Langley requested that NASA Headquarters attempt to have the model tunnel returned to Langley for additional studies of altering the flow properties in the full-scale version, but the return of the tunnel was resisted (the problem actually was discussed at levels as high as Secretary of State Henry Kissinger) and the tunnel remained in Portugal. A second 1⁄15-scale model of the Full-Scale Tunnel test section would ultimately be built in the 1990s.

An Unwanted Name Change

In 1981, NASA management became infatuated with the concept of emphasizing the use of International System of Units (SI) in all NASA technical documents. Although this was standard practice in the space program, the 64-year-old NACA/NASA aeronautics program had traditionally used English units such as feet, miles per hour, and pounds to describe its work. Following this decree, NASA's aeronautical documents used SI units to denote primary measurements with the traditional English units in parentheses. Within the aeronautical community and for its customers, the changeover was needless, confusing, and highly controversial.

Management further stipulated that the names of facilities—especially wind tunnels—be changed to reflect the new metric policy. For a brief time in 1981, the name of the Full-Scale Tunnel became the Langley 9- by 18-Meter (30- by 60-Foot) Wind Tunnel. As might be expected, the new name was regarded as an example of bureaucratic nonsense, and the loss of the classic Full-Scale Tunnel name was considered a major disaster by many. After a few months, this questionable action was thankfully aborted, and the facility was renamed the Langley 30- by 60-Foot (Full-Scale) Tunnel, which it remained for 15 years until its operations came under the management of Old Dominion University. ODU reverted to the facility's traditional name, the Langley Full-Scale Tunnel, for the remaining 13 years of its existence.

Research Activities

The NASA Supersonic Cruise Research Program

Following the termination of the U.S. SST Program by Congress in 1971, NASA conducted a new program between 1971 and 1981 known as the NASA Supersonic Cruise Research (SCR) Program to continue research on the problems and technology of supersonic flight.[16] The new program's goals were to build on the knowledge gained during the SST program and to provide the supersonic technology base that would permit the United States to keep its options open for proceeding with the development of an advanced supersonic transport, if and when it was determined to be in the national interest. At that time, supersonic aerodynamic design methods were well in hand, as evidenced by Langley's use of computational fluid dynamics in the design of the remarkably efficient SCAT-15F.

Langley researcher Jim Shivers inspects an over-the-wing-blowing supersonic transport configuration during tests in the Full-Scale Tunnel in 1974. (NASA L-74-2728)

Unfortunately, the highly swept supersonic configurations exhibited deficiencies in stability, control, and performance at subsonic speeds. Therefore, a major effort was directed to improve the low-speed behavior of supersonic transports during the SCR Program, and only a relatively limited effort was expended on improving supersonic cruise efficiency. The Full-Scale Tunnel became a workhorse during the program as researchers explored wing-planform effects, various wing leading-edge devices, and other innovative technical concepts in efforts to augment low-speed lift while providing satisfactory stability and control. Collaborative efforts with industry teams led to solutions for many of these problems.

During the SCR program, research activities in the tunnel consisted of conventional static tests to measure the aerodynamic characteristics of large-scale models of advanced configurations, with an emphasis on improving low-speed performance. For example, in one study in 1975, the use of upper-surface blowing over the wing and thrust vectoring of the engine exhausts was evaluated as concepts to increase lift for a configuration that had engine nacelles mounted above the highly swept wing.[17] The results of this particular investigation demonstrated that inducing additional lift on the highly swept wing was unsuccessful and that large untrimmable nose-down pitching moments were produced by thrust vectoring.

Technicians prepare the "platypus-nose" SCR model for testing in 1973. Note the two-dimensional thrust-vectoring engine nacelles. (NASA L-73-7547)

Another test involved an evaluation of an advanced in-house supersonic-cruise configuration featuring a "platypus-nose" all-wing planform and a propulsion system that used thrust vectoring.[18] Results of the study revealed the difficulty of providing sufficient lift at low speeds. In addition, data measured during tests with the model in a sideslipped condition indicated that the levels of roll control provided by conventional trailing-edge surfaces would be insufficient to trim the configuration in a cross wind during takeoff and landing.

Tests were also conducted on a large model of a supersonic transport configuration with a variable-sweep wing and a T-tail in 1973.[19] The variable-sweep concept had been abandoned by Boeing in the earlier SST Program in the 1960s because of unsatisfactory longitudinal stability at high angles of attack that resulted from the T-tail, and because of unacceptable tail heating and buffeting with a low-tail position. Results of the study showed the severe pitch-up characteristic that had been expected at moderate angles of attack and insufficient control to recover from a potential deep-stall condition. By 1973, the emergence of new control-configured vehicles (CCVs) offered the potential to solve the T-tail problem for a variable-sweep SST.

As the 1970s drew to a close, the timing was wrong for any advocacy for supersonic transports. Anti-SST feelings still ran high within environmental groups; fuel prices soared, making supersonic flight extremely expensive; the Concorde had proven to be an economic disaster; and low-fare availability on wide-body subsonic transports was the public rage, rather than exorbitantly expensive tickets for supersonic junkets. In view of these factors, as well as the mounting funding problems for its Space Shuttle program, NASA terminated the SCR Program in 1982.

Test director Clyde McLemore poses with the variable-sweep SCR model during tests in 1973. (NASA L-73-6059)

Civil Concepts Turned Military

In the mid-1970s, the potential lethality of air combat over Europe had increased so dramatically that the Air Force became interested in fighter aircraft with the ability to cruise at supersonic speeds without the use of an afterburner. In response to this interest, American industry design teams promoted new fighter concepts with highly swept wing platforms for efficient supersonic performance. In view of NASA's extensive efforts in designing efficient supersonic transport configurations, the design teams consulted with Langley researchers for guidance and design methodology. Thus, the products of the earlier SCR efforts were transferred from a potential civil application to vehicles with military missions.

The SCIF Program

In 1975, Langley conceived and initiated a new research program on supersonic-cruise fighters known as the Supersonic Cruise Integrated Fighter (SCIF) Program. The objective of the program was to apply powerful computational tools to advanced fighter configurations designed by an in-house team of Langley experts. Several different configurations were pursued to the stage of conducting low- and high-speed wind tunnel testing and computational analyses. The Full-Scale Tunnel continued its role of evaluating and optimizing the low-speed, high-angle-of-attack characteristics of the unconventional configurations. A highly swept cranked-arrow concept known as SCIF-IV was tested in 1977, and a canard configuration known as SCIF-II was studied in 1978, providing the staff of the Full-Scale Tunnel and industry partners with a view of both the problems that might be encountered and potential solutions.

Langley-conceived designs for supersonic-cruise fighters included the SCIF-II, viewed by researcher Bill Newsom (left), and the SCIF IV (right). Models of both configurations were tested in the Full-Scale Tunnel in 1977 and 1978. (NASA L-78-5129 and NASA L-77-6904)

The F-16XL

As NASA-industry interactions intensified during the SCIF activities, Lockheed Martin became interested in developing a supersonic-cruise version of its F-16 Fighting Falcon featuring a cranked-arrow wing similar to a configuration studied by Langley in its fighter studies. In 1977, Langley and Lockheed Martin agreed to a cooperative study to design a new supersonic wing for the F-16. The vision of Lockheed Martin was to develop a modified F-16 supersonic-demonstrator aircraft in response to the interest displayed by the Air Force. The new swept wing would be inserted on the existing fuselage and structure of the original F-16 configuration, modified with fuselage inserts. The development of this advanced version of the F-16 offered an opportunity for NASA researchers to apply their tools and technology in a real-time technology transfer arrangement. A cooperative program was initiated with Langley to participate in the development of what was initially called the Supersonic Cruise and Maneuver Prototype (SCAMP).

At this point, the scope of Langley involvement broadened considerably. Testing of the SCAMP configuration took place in several facilities, including the Full-Scale Tunnel to study dynamic stability and control at high angles of attack.[20] Although the cranked-arrow wing planform derived in the cooperative study was highly effective in increasing the supersonic performance of the basic F-16, the projected low-speed aerodynamic characteristics measured in the Full-Scale Tunnel and Lockheed Martin tunnels revealed problem areas that required modifications to the configuration during the development process. Originally, the SCAMP design used an all-moveable vertical tail and all-moving outer wing panels for roll

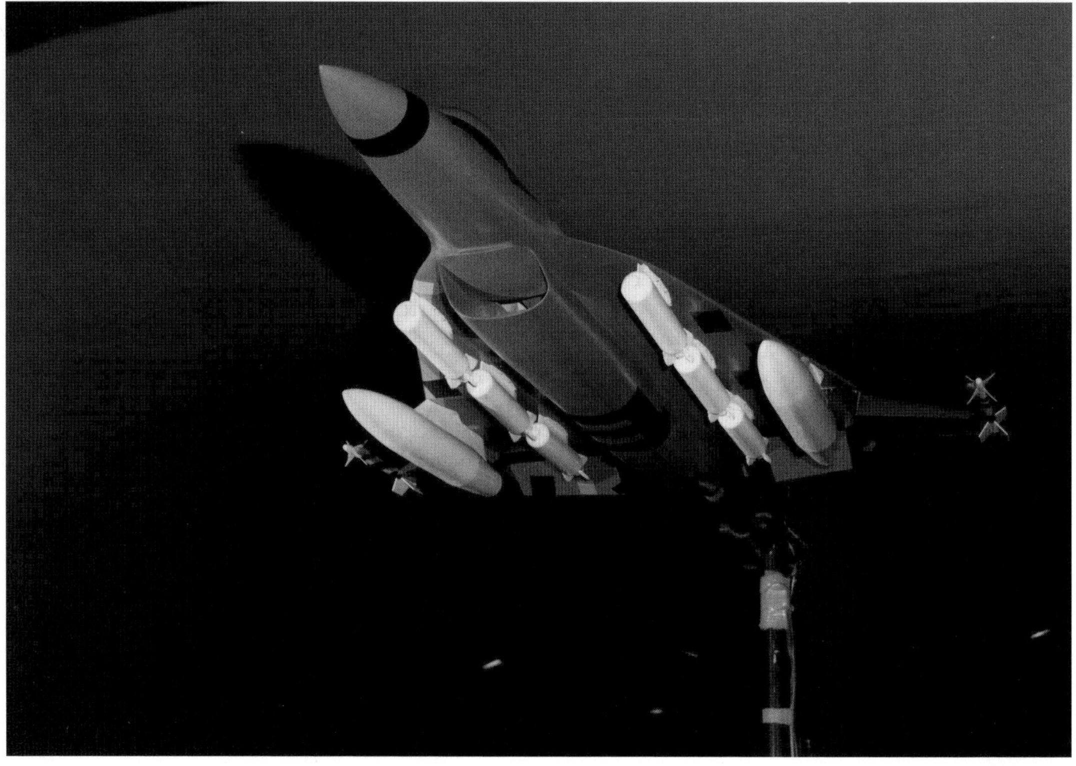

Project engineer Sue Grafton poses with the free-flight model of the F-16XL in 1981 (top). At this time, the original all-flying wingtips and all-moveable vertical tail had been eliminated and replaced by more conventional surfaces. The photograph on the bottom shows the "S-shaped" wing-apex region recommended by Joe Johnson. (NASA L-81-11836 and NASA L-82-4740)

control, but these concepts proved ineffective at high angles of attack. The configuration was also susceptible to loss of longitudinal stability and an uncontrollable pitch up at moderate angles of attack because of strong vortical flow emanating from a highly swept wing apex. Wingspan and angle of sweep were varied in attempts to resolve the low-speed issues, and the NASA–Lockheed Martin team arrived at fixes that included Joe Johnson's suggestion of modifying the shape of the wing apex with a rounded planform.

The final configuration became known as the F-16XL (later designated the F-16E), which displayed an excellent combination of reduced supersonic wave drag, vortex lift for transonic and low-speed maneuvers, low structural weight, and good transonic performance. Subsequently, two F-16XL demonstrator aircraft (a one-seat and a two-seat) were built from modified F-16 aircraft and entered flight tests in mid-1982. Highly successful flight tests of the aircraft at Edwards Air Force Base confirmed the accuracy of the wing-design procedures, wind tunnel predictions, and control-system designs that had resulted from the cooperative program. Unfortunately, initial Air Force interest in a supersonic-cruise version of the F-16 diminished when the service's priority became the development of a dual-role fighter with ground strike capability. Although the F-16XL could efficiently carry a significant amount of weapons for ground-strike missions under its large wing, the Air Force ultimately selected the F-15E in 1983 for developmental funding and terminated its interest in the F-16XL.

Increase in Military Support

The very productive relationship between the military services and the Full-Scale Tunnel that had begun with the very first tunnel tests in 1931 greatly intensified during the 1970s and 1980s through many outstanding and extremely productive joint programs. During this time, the services initiated several critical developmental programs for high-performance fighters, bombers, and advanced subsonic transports.

Even though many uninformed managers within the NASA system believed that military support consisted of simply allowing the military and their contractors to conduct tests in NASA facilities, the true scope of these activities went far beyond that perception. The industry greatly valued the high-risk, long-term research conducted by NASA, and it incorporated many of the fundamental concepts conceived by the research into its applications in critical aircraft programs. When a new aircraft development program requested support from the staff at the Full-Scale Tunnel, it received much more than a wind tunnel entry. The scope of activities included conventional tests to define aerodynamic characteristics over a wide range of configuration and flight variables; special dynamic force tests to measure the effects of angular rates on aerodynamics;[21] free-flight model tests to establish dynamic stability and control for an extended range of flight conditions; and, in many cases, piloted simulator evaluations of full-scale aircraft handling qualities based on aerodynamic inputs derived from tests in the tunnel. When results from all these test techniques were collated and correlated, potential problem areas were identified and potential solutions were recommended in early designs.

Providing support to the military in their development programs provided the staff of the Full-Scale Tunnel with very significant benefits.[22] By responding to military requests, the tunnel maintained its relevance in national programs and received test assets, invaluable data, and validation of research concepts. The military typically provided funding and information for fabrication of wind tunnel models, and after the immediate objectives of specific aircraft development programs were met, the models were frequently retained by the tunnel's staff for generic research and assessments of advanced concepts that could not be funded within the NASA aeronautics budget. By participating in aircraft programs, the researchers were exposed to a myriad of real-world requirements and constraints, thereby gaining valuable awareness of the limitations in designing and conducting research in flight dynamics. Finally, by becoming a team member in military aircraft development programs, Langley researchers obtained highly valued flight-test data and feedback that was otherwise unaffordable within NASA budget constraints. In many cases, researchers from the Full-Scale Tunnel who had participated in early developmental efforts were invited to military flight-test facilities to witness and participate in flight testing of the full-scale article.

Mutual respect and work experience between military personnel and the staff at the Full-Scale Tunnel continued to grow during this period. A special relationship of trust, dedication, and cooperation permeated the partnership. For example, a noted working-level manager within the Navy was so pleased with the high-quality, reliable efforts of the NASA staff at the Full-Scale Tunnel that he routinely provided annual unspecified funding to support the tunnel's efforts on behalf of his organization, noting informally that he was willing to authorize funding without specifying any tasks in advance because of his confidence in the tunnel's personnel.

Specific Military Aircraft Programs

Free-flight tests of fighter-type aircraft at the Full-Scale Tunnel had been preceded by a long history of similar tests in the old Free-Flight Tunnel. When the technique was transferred to the Full-Scale Tunnel in the 1950s, it was used to support several fighter development programs in assessing dynamic stability and control for high-angle-of-attack conditions. As previously noted, these tests had included support for the F-111 program in the mid-1960s. However, during the period from 1969 to 1984, investigations of fighter aircraft became a major component of the research program at the tunnel in support of the large number of fighter aircraft developed by the military in that era.[23]

Special Contributions for High Angles of Attack

During the Vietnam War, U.S. pilots flying F-4 and F-105 aircraft faced highly maneuverable MiG-17 and MiG-19 aircraft, and the unanticipated return of the close-in dogfight demanded maneuverability that had not been envisioned or required during the design and initial entry of these U.S. aircraft into operational service. Unfortunately, the F-4 exhibited a marked deterioration in lateral directional aerodynamic stability and control characteristics when flown at the high angles of attack required for maneuvering against agile enemies. As a result of these aerodynamic deficiencies, inadvertent loss of control and spins became major issues for the F-4 fleet, with over 100 Navy and Air Force versions of the aircraft lost in combat and training accidents. A request for analysis of the F-4 problem by representatives of the Air Force Aeronautical Systems Division was submitted in 1967 to the Full-Scale Tunnel for action. The NASA response resulted in an extensive analysis of the high-angle-of-attack deficiencies of the aircraft using wind tunnel, free-flight model, and piloted-simulator studies.[24] The efforts also included free-flight model demonstrations of the effectiveness of wing leading-edge slats in mitigating the stability problem.[25]

Research on stall/departure problems of the F-4 Phantom in 1969 led to a multiyear program in the Full-Scale Tunnel on subsequent military aircraft. In the photo on the left, Sue Grafton inspects fixed-wing leading-edge slats that mitigated the problem in 1970. Shown on the right is a free-flight test in 1983 to evaluate the effectiveness of pneumatic span-wise blowing on the F-4 at high angles of attack. (NASA L-70-3315 and NASA L-83-7326)

The F-4 experience is especially noteworthy for the Full-Scale Tunnel's contributions to high-angle-of-attack technology. Based on the successful demonstrations of analysis and design tools by NASA for the F-4 and additional configurations, management within the Air Force, Navy, and NASA strongly supported the active participation of the Agency in military high-angle-of-attack technology. As a key facility of this effort, the Full-Scale Tunnel was frequently requested for similar NASA involvement in virtually all subsequent DOD high-performance aircraft development programs over the next 25 years.

The Full-Scale Tunnel's role in developing high-angle-of-attack technology rapidly accelerated beginning in 1971. Active participation in the B-1, F-14, and F-15 programs was quickly followed by similar research for the YF-16 and YF-17 Lightweight Fighter prototypes, as well as later efforts for the F-16, F-16XL, F/A-18, XFV-12, and X-29 programs. In each case, a timely assessment of potential high-angle-of-attack problems was conducted early in the design and development program, and solutions involving airframe or flight control system modifications were recommended.[26]

Puzzling Answers for High-Priority Programs

Several interesting events experienced within these high-priority test programs highlighted the importance of the contributions of the Full-Scale Tunnel test activities.[27] For example, during the test program supporting the development of the Air Force's F-16, researchers at the Full-Scale Tunnel conducted early static wind tunnel tests of the configuration over

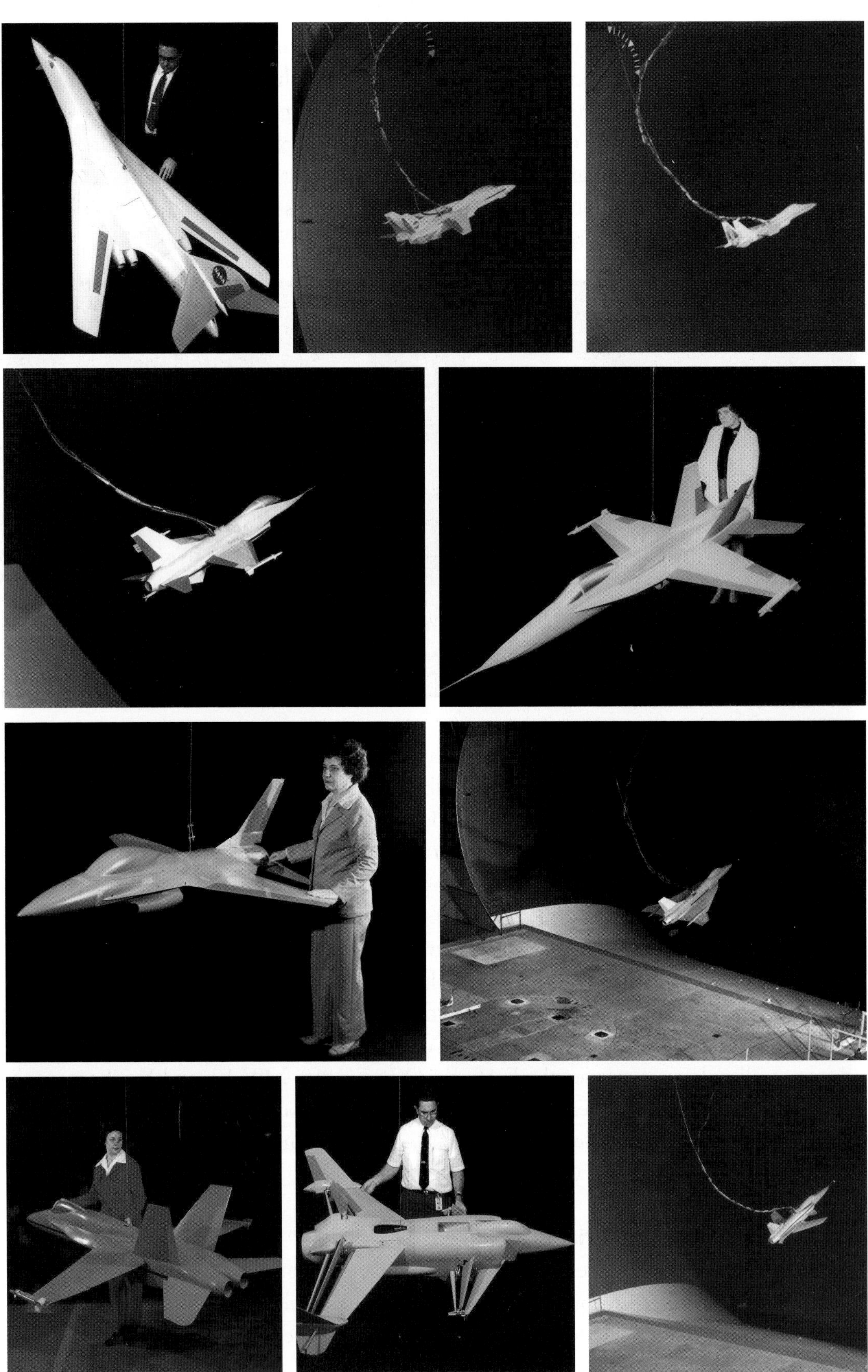

The Full-Scale Tunnel provided aerodynamic data and predictions of the high-angle-of-attack behavior of every U.S. high-performance military aircraft developed during the 1970s and 1980s. Shown are free-flight models or tests under way for (top, left to right and top to bottom) Rockwell B-1A, Grumman F-14, McDonnell Douglas F-15, General Dynamics YF-16, Northrop YF-17, General Dynamics F-16, General Dynamics F-16XL, McDonnell Douglas F/A-18, Rockwell XFV-12, and Grumman X-29. (NASA L-72-953, NASA L-71-7573, NASA L-71-3003, NASA L-73-2395, NASA L-73-1593, NASA L-80-4155, NASA L-82-915, NASA L-78-792, NASA L-74-5303, and NASA L-82-2759)

large values of angles of attack and sideslip. Analysis of this data indicated that a potentially unrecoverable "deep stall" condition could be encountered during rapid rolling maneuvers. Such maneuvers could saturate the nose-down aerodynamic control capability of the flight control system and result in the inherently unstable airplane pitching up to an extreme angle of attack with insufficient nose-down aerodynamic control to recover to normal flight. The wind tunnel data derived from the Full-Scale Tunnel were programmed as inputs into the NASA Langley air-combat simulator known as the Differential Maneuvering Simulator (DMS) and flown by Langley pilots to verify the potential of the original F-16 to enter this dangerous unrecoverable condition.

Pilots from General Dynamics and the Air Force flew the Langley simulator and were impressed by the severity of the problem, but aerodynamic data obtained in other NASA and industry wind tunnel tests did not indicate the existence of such a problem. As a result of the disagreement between data, the Full-Scale Tunnel data were dismissed as contaminated with "scale effects," and concerns over the potential existence of a deep stall were minimal when the aircraft entered flight testing at Edwards Air Force Base. However, during zoom climbs with combined rolling motions, the specially equipped F-16 high-angle-of-attack test airplane entered a stabilized deep-stall condition as predicted by the Langley results, and after finding no effective control for recovery, the pilot was forced to use a special emergency spin-recovery parachute to recover the aircraft to normal flight. The motions and flight variables were virtually identical to the Langley predictions.

Since the Full-Scale Tunnel aerodynamic data for the F-16 provided extremely realistic inputs for the incident encountered in flight, a joint NASA, General Dynamics, and Air Force team aggressively used the Langley simulator to develop a piloting strategy for recovery from the deep stall. Under Langley's leadership, the team conceived a "pitch-rocker" technique in which the pilot pumped the control stick fore and aft to set up oscillatory pitching motions that broke the stabilized deep-stall condition and allowed the aircraft to return to normal flight. The concept was demonstrated during F-16 flight evaluations and was incorporated in the early flight control systems as a pilot-selectable emergency mode. Ultimately, the deep stall was eliminated by an increase in the size of the horizontal tail (for other reasons) on later production models of the F-16.

The highly successful contribution of the Full-Scale Tunnel to the prediction and resolution of the F-16 deep-stall problem naturally resulted in extreme concerns over the validity of high-angle-of-attack wind tunnel testing techniques, as well as questions regarding why a relatively large-scale model tested at low speeds—and correspondingly low values of Reynolds number—should provide more accurate predictions than data from facilities at much higher values of Reynolds number. This issue was never totally resolved following the F-16 program, and the results from the Full-Scale Tunnel remain controversial. Nonetheless, the Air Force was appreciative of the fact that the old Full-Scale Tunnel provided critical data for their high-priority fighter.

A similar case of issues regarding data correlation from wind tunnels occurred during Full-Scale Tunnel tests in support of the Navy's F/A-18 fighter/attack aircraft. During free-flight tests in the tunnel, the model was observed to have a marked deterioration in lateral directional stability at high angles of attack with the wing leading-edge flap deflected at the same angles used in test programs at other wind tunnels. When the free-flight model was subjected to force tests, the deficient aerodynamic behavior was noted, in agreement with the characteristics observed during the free-flight tests. Additional tests were conducted to evaluate the effect of variations in the leading-edge flap deflection angle for various angles of attack, and it was found that the problem could be eliminated with an additional 5° deflection of the flap. The free-flight results were regarded as misleading due to the low Reynolds number of the study.

After the free-flight tests had been completed, a full-scale F/A-18 test aircraft unexpectedly encountered a departure from controlled flight during a windup turn during the test program at the Navy Patuxent River Flight Test Center. Several concepts were evaluated to eliminate the problem; however, when the aircraft's leading-edge flap deflection (which was automatically programmed with angle of attack) was increased by about 5°, the problem was eliminated. With the pressure of the aircraft development schedules, the recurring issue of why the Full-Scale Tunnel produced such accurate results remained unresolved. In any event, the second occurrence of exceptionally good correlation with full-scale flight results heightened the Navy's level of interest and respect for results obtained in the Langley Full-Scale Tunnel.

Cutting Edge: The NASA High-Angle-of-Attack Technology Program

As the 1970s came to an end, the U.S. military fleet of high-performance fighter aircraft had been transformed from departure-prone designs such as the F-4 to new configurations that had outstanding stability and departure resistance at high angles of attack. Thanks to the national research and development efforts of industry and Government, the F-14, F-15, F-16, and F/A-18 demonstrated that the danger of high-angle-of-attack departure exhibited by the previous generation of fighters was no longer a critical concern. Rather, flight at high angles of attack under certain tactical conditions could be exploited by the pilot without fear of loss of control and spins. At air shows and public demonstrations, the new "supermaneuverable" fighters wowed the crowds with high-angle-of-attack flybys, but more importantly, the capability provided pilots with new options for air combat. High-angle-of-attack technology had progressed from concerns over stall characteristics to demonstrated spin resistance and was moving into a focus on post-stall agility and precision maneuverability.

Reflecting on the advances in high-angle-of-attack technology of the 1970s and related emerging research thrusts, technical managers at Langley, Dryden, and Ames began to advocate for a cohesive, integrated high-angle-of-attack research program focused on advancing predictive technologies and demonstrating innovative new control concepts. The Agency was in an excellent position to initiate such a program thanks to the unique ground- and flight-testing capabilities that had been developed at the three Centers and the expertise that had been gathered by interactions of the NASA researchers with the "real-world" challenges of specific aircraft programs. At Langley, researchers had been intimately involved in high-angle-of-attack/departure/spin activities in the development of all the new fighters and had accumulated in-depth knowledge of the characteristics of the configurations, including aerodynamics, flight-control architecture, and handling characteristics at high angles of attack. Technical expertise and facilities at Langley included subscale static and dynamic free-flight model wind tunnel testing, piloted simulators, advanced control-law synthesis, and computational aerodynamics.

At NASA Dryden, that Center's world-class flight-test facilities and technical expertise for high-performance fighter aircraft had been continually demonstrated in highly successful flight-test programs in which potentially hazardous testing had been handled in an extremely professional manner. Meanwhile, at NASA Ames, the aeronautical research staff had aggressively led developments in high-performance computing facilities and computational aerodynamics. Computational fluid dynamics (CFD) codes developed at Ames and Langley had shown powerful analysis capability during applications to traditional aerodynamic predictions such as cruise performance and the analysis of flow-field phenomena, and the time had arrived to apply these methods to the challenging environment of high angles of attack.

During the early 1980s, research managers at Langley, Dryden, and Ames conceived and obtained Headquarters approval for a research project known as the NASA High-Angle-of-Attack Technology Program (HATP). The content of the program was initially focused on an assessment and development of computational and experimental methods to improve the designer's toolbox of methods for earlier prediction and, if necessary, modifications for high-angle-of-attack conditions. The HATP planners designed a series of closely coordinated efforts involving wind tunnel tests in several facilities, applications of CFD, and aircraft flight tests. The issue of whether a full-scale aircraft was required for verification of the ground-based predictions was resolved by agreement with NASA Headquarters to acquire a special research vehicle. After evaluating the known advantages and disadvantages of available aircraft that could serve as the configuration of interest for the research efforts, the Navy F/A-18 was chosen as the baseline for the program.

Langley's relations with the Navy were very positive due to the Center's many contributions during the aircraft's development program, and Joe Chambers carried the advocacy for the aircraft acquisition to the Naval Air Systems Command. One of the prototype F/A-18A aircraft that has been used for spin testing by McDonnell Douglas and the Navy in the F/A-18 development program was acquired on loan from the Navy, refurbished, and modified for flight testing at Dryden. The NASA F/A-18 research aircraft was known as the High Alpha Research Vehicle (HARV). The aircraft was delivered to NASA Dryden for refurbishment and instrumentation work in 1984.

At Langley, the selection of the F/A-18 as the baseline configuration resulted in extensive wind tunnel and computational studies by several different organizations. Pressure distributions over the configuration were measured in several wind tunnels using highly instrumented models, and data were analyzed and prepared for correlation with results obtained in subsequent flight tests with the full-scale aircraft at Dryden. Complementing the wind tunnel tests were a series of CFD studies for similar flight conditions. One of the most unique contributions of the HATP activities would be aerodynamic data measured for similar conditions and instrumentation between computational, experimental, and flight sources.

Most of the experimental aerodynamic studies were conducted in facilities other than the Full-Scale Tunnel, because that facility had already produced extensive low-speed high-angle-of attack data on the configuration during its participation in the Navy's development program for the F/A-18 during the 1970s. In addition, the staff of the Full-Scale Tunnel was concentrating its efforts on the development of new concepts that might provide impressive levels of control for high-angle-of-attack conditions.

Thrust-Vectoring Concepts

Two events occurred in early 1980 that resulted in Full-Scale Tunnel demonstrations of a concept—multiaxis thrust vectoring—that provided major improvements in high-angle-of-attack maneuverability and control. The first event was a cooperative effort between the members of the Simulation and Analysis section at the Full-Scale Tunnel and researchers from the Navy's David Taylor Model Basin. The Navy group was interested in conducting a cooperative simulator investigation at Langley of the potential impact of providing increased levels of yaw control for fighter aircraft. At the time, directional controllability for engine-out conditions for the F-14 during low-speed carrier operations had become an issue within the Navy, and the potential benefit of vectoring the exhaust flow from the fighter's twin engines was under investigation as a possible solution. The results of the joint NASA-Navy simulation were very impressive, indicating that a revolutionary improvement in low-speed, high-angle-of-attack control was at hand. The Navy pursued concepts to provide vectoring and outfitted an F-14 with a paddle-type vane between each engine to provide yaw vectoring. Although never flown for high-angle-attack conditions, the experimental vanes performed well and demonstrated the structural integrity required for the concept.

The second event that led to research at the Full-Scale Tunnel on thrust vectoring followed the joint Navy study. Spurred on by the potential applications of the Navy concept for enhanced yaw control at high angles of attack, Joe Johnson and his staff began in-depth discussions with the propulsion integration experts at Langley. Researchers at the Langley 16-Foot Transonic Tunnel had led the United States' efforts on the research and development of thrust-vectoring concepts for decades, and their interactions with the staff of the Full-Scale Tunnel stimulated discussions to demonstrate the effectiveness of control augmentation using powered free-flight model tests.

The existing F/A-18 free-flight model was the first to be modified with thrust-vectoring vanes. The first vane concept consisted of two single-axis yaw-vectoring vanes placed between the engine exhausts and deflected symmetrically as a single unit. By deflecting the vanes in compressed airstreams used for simulated engine thrust, in early 1983, researchers demonstrated the ability to fly the F/A-18 model to extreme angles of attack (beyond 55°) in level flight without loss of control. The results of the Full-Scale Tunnel tests indicated that relatively simple thrust-turning concepts such as paddles placed in the exhaust stream could be used very effectively for yaw control at high angles of attack where the rudders were almost totally ineffective. The next step in the vane investigations involved using multiaxis

F/A-18 model without vertical tails flies at extreme angle of attack using thrust vectoring for yaw control. (NASA L-83-5176)

vectoring (pitch and yaw) vanes. For this configuration, an additional single pair of vanes was located above the engine nozzles to provide pitch control. The model could be flown with crisp control at angles of attack as high as 70°.

In 1984, as the intensity of thrust-vectoring research increased within the Flight Dynamics Branch, its unit responsible for outdoor helicopter drop-model operation took on the task of demonstrating the benefits of thrust vectoring on post-stall maneuvering capability. A large F/A-18 model used in the Navy F/A-18 development program was modified with a vectoring rocket attached to its rear end and dropped from a helicopter for extreme maneuvers. The results were impressive and augmented the case for thrust vectoring at high angles of attack in the Full-Scale Tunnel.

While the F/A-18 HARV was being readied for aerodynamic studies at Dryden, Joe Johnson spurred on his staff for more demonstrations of the generic effectiveness of thrust vectoring for other aircraft configurations. Next up was an X-29 forward-swept-wing free-flight model, outfitted with a two-axis vectoring vane system similar to the General Electric variable-cycle augmented deflector exhaust nozzle (ADEN) concept developed for Short Takeoff and Vertical Landing (STOVL) aircraft. The model was powered with a compressed-air ejector, and the exhaust was vectored by sidewall vane deflectors and upper- and lower-exhaust vanes. With vectoring off, the model could be flown to angles of attack of about 40°, where lateral control was markedly reduced; but with vectoring on, flights were routinely made to 55° with crisp response to lateral inputs. Some vectoring flights were made to angles of attack as high as 75°.

The next configuration selected for the additional thrust-vectoring demonstrations was the original F-16XL SCAMP that had exhibited unacceptable longitudinal characteristics at moderate angles of attack. The modifications made to the original design during the development of the F-16XL had solved the pitch-up problem but with a penalty to supersonic performance. Free-flight tests of the model, which had been reconfigured back to its original swept-back wing apex, were conducted with a two-axis thrust-vectoring scheme consisting of a single exhaust vane providing nose-down pitch control and yaw vanes for directional control. Back-to-back flight tests vividly demonstrated that the configuration could not be flown at angles of attack above about 18° because of a severe pitch-up against full corrective control using aerodynamic surfaces. On every attempt, the model would rapidly pitch up out of control, requiring immediate extraction from the tunnel airstream via the flight safety cable attached to the top of the model. However, when the thrust-vectoring system was engaged (including feedback from an angle-of-attack probe mounted on a nose boom), the model was easily controlled to angles of attack beyond 70°. It was obvious that thrust vectoring was a powerful new tool for combatting loss of control and improving maneuverability at high angles of attack.

As was the case for most of the studies conducted at the Full-Scale Tunnel during this era, the experimental work on the impact of thrust vectoring on high-angle-of-attack flight dynamics for the F-16XL was accompanied by a piloted simulated study using the Langley Differential Maneuvering Simulator.[28] In this study, the aerodynamic data gathered from static and dynamic testing in the Full-Scale Tunnel served as the input data for a simulated air-to-air engagement of a hypothetical thrust-vectoring-equipped F-16XL with a conventional opponent. In addition to obtaining evaluation comments from Langley and military research pilots, the staff began to establish expertise in control-law requirements and handling-quality requirements for vectoring aircraft at high angles of attack.

Still not satisfied with these impressive demonstrations with the F/A-18, X-29, and F-16XL, Johnson pressed on with additional demonstrations of the effectiveness of thrust vectoring with free-flight tests of the existing free-flight model used in the F-16 fighter development program. The model was equipped with a similar two-axis vane-vectoring arrangement, and flights could routinely be made to angles of attack approaching 70°.[29] During many of these remarkable tests, local commanders of the Langley Air Force Tactical Air Command (now Air Combat Command) were invited to witness the demonstrations, and the visitors were duly impressed by the ability of thrust vectoring to provide carefree maneuverability at high angles of attack. Whether the demonstrations had an influence on the subsequent decision of the Air Force to adapt thrust vectoring for the F-22 Raptor is debatable, but the word quickly spread among high-level Air Force managers to visit the Full-Scale Tunnel and observe the free-flight tests.

In any event, the results of the Full-Scale Tunnel's demonstrations of thrust vectoring were briefed to members of the Intercenter HATP team and were received with intense interest. Proposals were submitted to NASA Headquarters to modify the F/A-18 HARV airplane with a research vectoring-vane control system for more in-depth studies of controls system requirements and the handling qualities of vectored aircraft. Funding for the activity, however, would not be forthcoming until fiscal year 1990, after the first aerodynamic phase of the program was completed.

Vortex Flap: The Legend of the "F-53"

The primary objective of the joint Langley–General Dynamics SCAMP wing design study was to develop a supersonic-efficient wing while retaining transonic-maneuver capability comparable to the basic F-16.[30] Providing this supersonic-cruise capability quickly led the project participants to propose a relatively thin cranked-arrow wing similar to those studied by NASA and industry for civil supersonic transport configurations. Unfortunately, highly swept arrow wings exhibit flow separation around the swept leading edge of the wing at moderate angles of attack, resulting in the formation of vortices that cause excessive drag and greatly diminished aerodynamic efficiency for maneuvering flight. Langley's experts in vortical-flow technologies accepted the challenge of identifying concepts that could be used to improve the transonic maneuver capability of the new wing.

Sophisticated wing designs were identified in wind tunnel and computational studies for suppressing and controlling the path of leading-edge vortices at high angles of attack for the desired level of transonic maneuverability. The more efficient wing design, however, involved an extraordinary distribution of wing camber and shaping that produced serious concerns regarding the manufacturing and structural design of such wing shapes. The Langley staff then focused on a concept known as the "vortex flap," which uses a potentially simpler, specially designed, deflectable leading-edge flap to capture and control the wing leading-edge vortices shed by highly swept wings.

In 1983, NASA undertook an aircraft flight project to demonstrate the benefits of the vortex-flap concept and to study other factors, such as the potential impact of maneuvers on aircraft handling qualities. For example, one concern was whether dynamic aircraft rolling motions would cause disruption of the vortex-control capability of the vortex flap. With a delta-wing F-106B aircraft in its inventory, Langley was poised to begin a flight project to provide answers to these and other issues. A series of subscale wind tunnel tests from subsonic through transonic speeds was undertaken at Langley and Ames to provide aerodynamic data on a specific vortex-flap configuration that had been designed at Langley using Langley-developed computational codes.

At the Full-Scale Tunnel, tests of the Langley-designed flap configuration included a full-scale test of an actual F-106 airframe. Testing a complete F-106 in the Full-Scale Tunnel was not feasible because of the relative size of the aircraft, so the Langley team literally sawed a second F-106 in half and tested one half of the aircraft, which was jokingly referred to as the "F-53."

Free-flight tests of a model of the F-106B were also conducted in the Full-Scale Tunnel to evaluate effects of lateral motion on dynamic stability and control, and piloted simulator evaluations using the Langley Differential Maneuvering Simulator were used to establish the general handling qualities of aircraft before flight.[31] Design and fabrication of the vortex flap was led by an in-house Langley team in Langley shops. The flap was designed to be a "bolt on" configuration, fixed at an angle prior to flight and evaluated for several flap settings.

After a series of flow-visualization test flights for the basic airplane in 1985, researchers conducted detailed analyses of the results in preparation for the vortex-flap investigation.

Researcher Long P. Yip views the huge "F-53" semispan version of a NASA F-106 aircraft during tests to evaluate the effectiveness of a Langley-designed wing leading-edge vortex flap in 1984. (NASA L-84-11339)

Hal Baber poses with the free-flight model of the F-106B vortex-flap configuration prior to flight tests in the Full-Scale Tunnel in 1984. (NASA L-84-6726)

In 1988, the first flights of the F-106B vortex-flap experiments began in a program that would last for about 2 years. Flights were conducted for vortex-flap angles of 30° and 40° for Mach numbers from 0.3 to 0.9 and for altitudes up to 40,000 feet. The results of the flight study were very impressive. For example, the aircraft's sustained-g capability was increased by about 28 percent at a Mach number of 0.7. In addition to the dramatic demonstration of improved performance, the flight project provided a valuable validation of the vortex-flap design process and experimental prediction capabilities.

Rise of General Aviation Research

As discussed in the previous chapter, studies of general aviation aircraft had been initiated in the Full-Scale Tunnel with a test of the twin-engine Piper PA-30 in the early 1960s. By the 1970s, two factors had increased the level of general aviation research and development in the facility. The first stimulus was the creation of a General Aviation Technology Office within NASA Headquarters. Led by enthusiastic and aggressive management, the office quickly became popular and supported by the general aviation industries and private-pilot associations. A wide variety of potential NASA research areas was identified, including stall/spin technology, propulsion systems, performance-enhancing concepts, and aerial applications.

The second factor that sparked interest in general aviation research was the managerial impact of Joe Stickle, whose personal interest in general aviation and numerous contacts within industry resulted in several cooperative programs conducted in the Full-Scale Tunnel.

Shrouded Propellers

One of the research interests within the general aviation community in the early 1970s was the development of propulsion systems that might reduce the noise and engine-emissions pollution of civil airplanes without severely penalizing aerodynamic performance. One proposed method of accomplishing this objective for light, propeller-driven, general aviation aircraft was the use of a small-diameter shrouded propeller with a direct-drive rotary engine. This propulsive concept was expected to result in a more compact propulsion package, shielding of propeller noise, minimization of pollutants, and lower weight.

Following the installation of sound-absorbing material on the walls and ceiling of the test section of the Full-Scale Tunnel, as previously discussed, aerodynamic and aero-acoustic tests of a modified Cessna 327 twin-boom pusher airplane were conducted in 1974 and 1978.[32] The test program included measurements for four different propeller arrangements, including a two-blade free propeller, two three-blade shrouded propellers, and a five-blade shrouded propeller. In addition to the acoustically treated test section, the airplane support struts were wrapped with special sound-absorbent matting and a special sound-absorbing ground board was used. The ground board consisted of a 4-inch-thick layer of fiberglass covered with a porous, perforated plate.

Somewhat surprisingly, the results of the test programs showed that the free propeller provided the best overall aerodynamic propulsive performance and lower noise levels in simulated forward-flight conditions. In static conditions, the free-propeller noise levels were as low as those for the shrouded propellers except for the propeller in-plane noise, where the shrouded-propeller noise levels were lower.

Clyde McLemore views a modified Cessna 327 during studies of the performance and acoustic effects of a shrouded-propeller concept. The photo on the right shows details of the installation. (NASA L-78-7475 and NASA L-78-3973)

Lessons Learned: Drag Cleanup

In 1976, the general aviation community invited NASA to participate in several national conferences devoted to technologies of interest to the industry. In preparation for a conference on aerodynamics, Mac McKinney initiated an activity to summarize the results of the classic drag cleanup tests that were conducted in the Langley Full-Scale Tunnel during the period from 1935 to 1945, with a view toward potential application to state-of-the-art general aviation airplanes. Data from applicable Full-Scale Tunnel tests on 23 aircraft were collated and summarized for presentation at the meeting, as well as for a subsequent technical report.[33]

In addition to giving examples of drag-producing aircraft components, the report stressed that although the drag contributions from features such as air leakage, cockpit canopies, control surface gaps, antenna installations, and power-plant installations may not be large, the sum of the incremental contributions to the total drag level is significant. It also appeared that considerable reduction in drag could be obtained by proper attention to details in aerodynamic design and by adherence to the guidelines and experiences learned in the Full-Scale Tunnel during the war years.

A Leap in Technology: The ATLIT

Following the major rehabilitation of the Full-Scale Tunnel in the mid-1970s, a 2-month test was conducted to determine the aerodynamic characteristics of a research aircraft known as the Advanced Technology Light Twin-Engine (ATLIT) airplane.[34] The ATLIT was developed by the University of Kansas Flight Research Laboratory as a project sponsored by Langley. The objective of the project was to apply jet-transport wing technology and advanced airfoil technology to general aviation airplanes to improve safety, efficiency, and utility. The aircraft was a significant modification to the Piper PA-34 Seneca airplane that involved wing planform modifications (taper, increased aspect ratio, and reduced area) for improved cruise efficiency, an advanced general aviation airfoil (17-percent-thick GA(W)-1 airfoil) to improve high-lift and induced-drag characteristics, full-span Fowler flaps, a wing-spoiler lateral-control system, winglets, and advanced technology propellers incorporating a supercritical airfoil. The scope of the investigation at Langley included flight tests of the aircraft as well as wind tunnel tests in the Full-Scale Tunnel.

Results of the study showed that performance was seriously degraded by excess drag during climbing flight conditions. Premature flow separation at the wing-fuselage juncture and leakage through the wing were the primary causes of the drag. Installation of a wing-fuselage fillet provided a significant drag reduction. Stalling of the horizontal tail produced longitudinal instability with the wing flap deflected and a full-power wave-off condition.

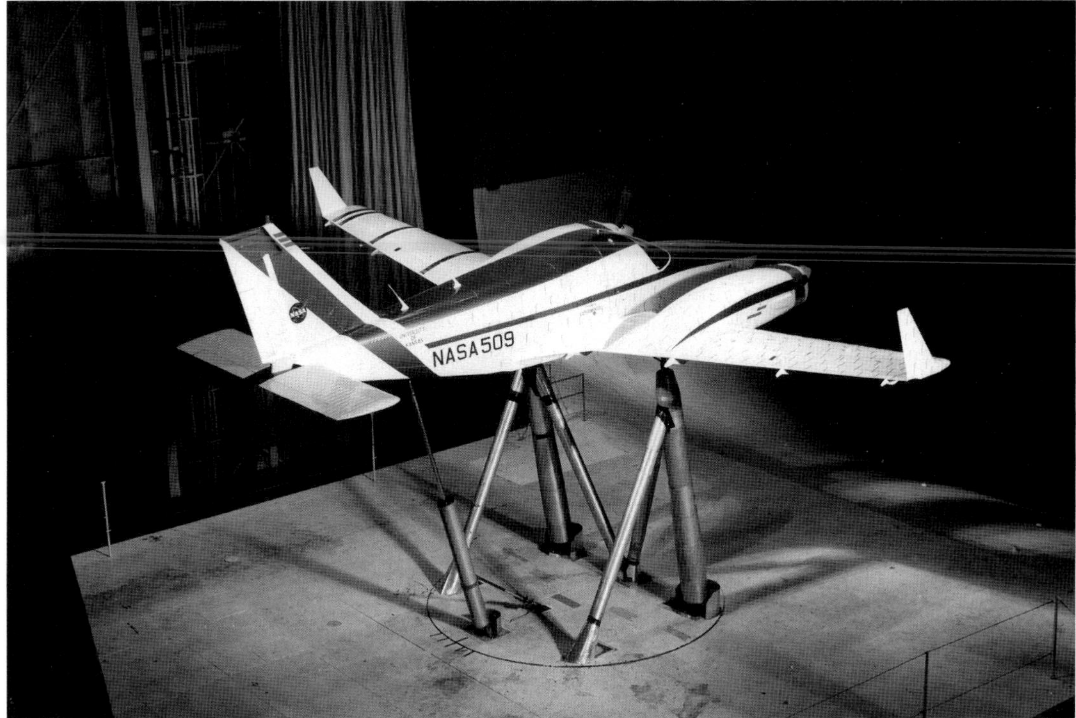

The Advanced Technology Light Twin-Engine research aircraft mounted for tests in the Full-Scale tunnel in 1977. (NASA L-77-2768)

Aerial Applications

As part of its expanded general aviation research program, NASA Headquarters met frequently with leaders in agriculture for a possible role in advancing technology for the aerial application of sprays and materials. Several national workshops subsequently resulted from these interactions. The use of aircraft in agriculture had begun in the 1920s, when the concept demonstrated advantages in the speed of application and the ability to apply material when ground applications were impossible or impractical. The world fleet of agricultural aircraft in the 1970s consisted of over 24,000 fixed- and rotary-winged vehicles.

Until about 1950, agricultural aircraft had been designed for other purposes, such as military training. With surplus WWII aircraft plentiful and available for minimal cost, the aircraft were adapted to aerial applications with spreaders and spraying equipment. Although the safety features of agricultural airplanes have improved, the aerodynamic performance and configuration features of modern-day aircraft retained technology typical of aircraft designs of the 1930s. Such aircraft featured uncowled radial engines, no aerodynamic fairings or fillets at the wing-fuselage juncture, and no design methodology for integrating dry material spreaders and spraying systems for maximum dispersal efficiency and aircraft performance.

Two operational problems faced the industry: the drift of toxic chemicals from treated areas and nonuniform coverage on crops, which result in a large loss of productivity, waste of chemicals, and environmental or health hazards. A major cause of drift and nonuniform coverage is the interaction of the aircraft wake with the dispersed material. With these issues in mind, Langley formulated an experimental and analytical investigation of aerial applications.

In the 1950s, tests of a subscale crop-duster model were conducted by the Full-Scale Tunnel staff in an area out of the tunnel circuit to provide qualitative information on the aerodynamic wake characteristics of a typical crop-duster airplane.[35]

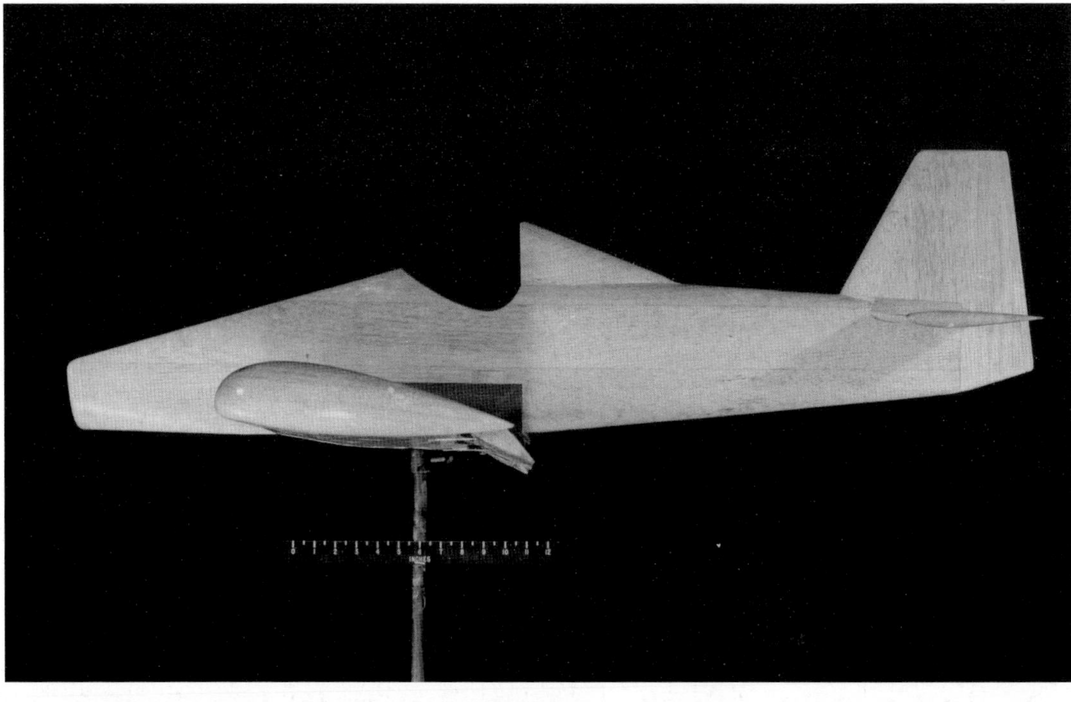

Photos of flow-visualization tests of a generic crop-duster model during tests by the staff of the Full-Scale Tunnel in November 1950. The unpowered model was apparently launched by a bungee-cord system and flew through balsa dust for visualization of the vortical flow in the trailing wake. (NACA LAL 68159)

Technical reports describing the test and its results are apparently unavailable, but surviving photographs show that the model was launched by a bungee-cord system and allowed to fly through balsa-dust particles to visualize the flow character of the model's wake. As frequently occurs in research, the concept of flying a model through a flow-visualization setup would once again be used a quarter-century later in the NASA program.

A ground-based program was conducted in the Langley Vortex Facility and the Full-Scale Tunnel, and flight tests were conducted at the NASA Wallops Flight Facility.[36] The Vortex Facility was a modern adaptation of a towing tank and impact basin that was originally used by the NACA in the 1940s and 1950s to study the stability, control, and performance of seaplanes and the ditching characteristics of land planes. NASA converted the facility in the 1960s to study the upset hazard associated with the strong vortices generated by large jet-transport aircraft. In 1976, the facility was further modified to permit testing of models of agricultural aircraft, including the dispersal of simulated agricultural materials.

Ayres Thrush Commander agricultural aircraft under test in the Full-Scale Tunnel in 1978. The geometric features of the airplane were very similar to the McDonnell Doodlebug tested in the tunnel in the 1930s. (NASA L-78-3637)

Researchers investigated the characteristics of an Ayres Thrush Commander 800 agricultural airplane in the Full-Scale Tunnel in February 1978. The test program was extremely broad in scope, with studies of the aircraft's performance, stability and control, and evaluations of a number of modifications designed to improve aerodynamic performance such as leading-edge slats, a ring cowling, wing-fuselage and canopy fairings, and wake-modification devices. Tests were made for the isolated airplane as well as the aircraft with the dispersal systems installed.

Near-field spray characteristics were determined in the Full-Scale Tunnel by spraying water from the airplane's dispersal system. The spray-bar arrangement is shown on the left, and dispersal testing showing the water spray is shown on the right. (NASA L-78-4148 and NASA L-78-4061)

During the dispersal systems tests, the various systems were operated—or such operation was simulated—to evaluate the performance and efficiency of the process. With dry material spreaders, pressure surveys were made to evaluate the internal flow characteristics with and without perforated blockage plates to simulate the transport of material. Liquid dispersal systems were tested to document near-field spray characteristics and spray-wake interactions. Spray dispersal systems installed on the aircraft were operated using water as the dispersal medium, and droplet-size distributions and concentrations were measured in the aircraft's wake using laser droplet-spectrometer probes. The liquid-dispersal systems included nozzles designed for improved drift control and rotary atomizers. The measurements were made quite successfully, providing confidence in the testing techniques.

The dispersal of water in the Full-Scale Tunnel was approached with considerable concern over the possible harmful effects on the drive motors and structure of the tunnel, but careful inspection of the facility avoided major issues. However, Frank Jordan recalled one humorous event that occurred during prolonged testing of the Thrush Commander's liquid-dispersal system: "After an extended test run, water apparently pooled downstream of the tunnel-drive propellers in an area directly over the offices of the staff at the south end of the building. Engineer James L. Hassell had been to a meeting in another building for several hours, only to return and find that the water had seeped through the floor of the exit cone onto his desk and ruined his laborious hand-plotted data for the final report of a high-priority job!"[37]

Unfortunately, opportunities to apply new technology within the niche role of agricultural aviation did not crystallize. Major companies were not interested in marketing such specialized aircraft, and the operators were in no position to procure more expensive assets. In response, NASA's interest in pursuing opportunities in agricultural aircraft quickly diminished and funding to continue the program was canceled.

Spin Resistance

In 1973, NASA Langley initiated a General Aviation Stall/Spin Program in response to an alarming increase in fatal accidents that was occurring in the general aviation community.

At one time, over 26 percent of the fatalities in this sector of aviation were caused by inadvertent stalls and spins. The NASA program was closely coordinated and involved tests of several representative configurations, with free-spinning and rotary-balance testing in the Langley Spin Tunnel, outdoor flight testing of radio-controlled models, force tests in the 12-Foot Low-Speed Tunnel, free-flight model tests and full-scale aircraft tests in the Full-Scale Tunnel, and full-scale aircraft flight tests at the NASA Wallops Flight Facility.

Testing in the Full-Scale Tunnel concentrated on assessments of concepts that would enhance the stall characteristics and spin resistance of general aviation aircraft, with an emphasis on three approaches: active systems to limit angle of attack; wing modifications that would enhance stall behavior; and configuration features, such as canards, that might inherently "stall-proof" the configuration.

The concept of preventing aircraft stall by limiting pitch-control power was demonstrated by NACA researcher Fred Weick at Langley in the 1930s, and he carried it into his highly successful Erco Ercoupe design. This early approach, which utilized elevator-travel limits, proved to be technically unfeasible for modern aircraft designs, which typically exhibit large effects of flap deflection and power-on aerodynamics and a wide range of center-of-gravity travel. In 1981, the staff of the Full-Scale Tunnel participated in a cooperative test of an active stall-prevention concept conceived by Howard L. Chevalier at Texas A&M University.[38] Chevalier's concept used angle-of-attack information to automatically deflect a tail spoiler to limit the nose-up trim capability of the elevator to an angle of attack below the angle of wing stall.

A highly modified full-scale model of a representative low-wing general aviation airplane was used in the test program, which explored the effectiveness of the tail-spoiler concept

A full-scale model of a low-wing airplane used in the Langley stall/spin program was used for 1979 evaluations of a tail-spoiler concept for limiting angle of attack. (NASA L-79-07120)

to limit angle of attack. The results of this exploratory program indicated that active stall-prevention systems for modern general aviation aircraft must be capable of accommodating large effects of flap and power for a wide range of center-of-gravity positions. The tail spoiler was effective for the entire angle-of-attack range investigated, regardless of whether the horizontal tail was providing an upload or a download.

The second major thrust of the Full-Scale Tunnel activities on spin resistance involved documentation of the effectiveness of wing leading-edge modifications on the high-angle-of-attack aerodynamic characteristics of the same low-wing configuration.[39] During the progress of the overall NASA general aviation program, a unique discontinuous drooped leading-edge configuration had been determined to greatly increase the spin resistance of typical general aviation airplanes and had been evaluated both in flight and with radio-controlled models.[40] Additional testing was conducted in the Full-Scale Tunnel with the full-scale aircraft used for the tail-spoiler research in order to obtain more detailed aerodynamic information on the concept at relatively high Reynolds numbers. The results indicated that the droop maintained attached flow on the outer wing to very high angles of attack, with minimal increase in drag for cruise angles of attack. Free-flight model tests were also conducted in the tunnel.

Researcher Bill Newsom poses with the free-flight model of the low-wing configuration. Note the discontinuous drooped leading-edge segments on the outer wing panels. (NASA L-78-3372)

Canard configurations were of special interest in the research program because properly designed canard aircraft can inherently limit the angle of attack to values below that for wing stall. Joe Chambers and Joe Johnson had maintained communications with famous airplane designer Elbert "Burt" Rutan, who had conceived several canard-type general aviation aircraft noted for their outstanding stall/spin behavior and stall resistance. NASA was extremely interested in analyzing the aerodynamic characteristics of Rutan's Varieze homebuilt aircraft design, and a cooperative agreement was arranged for tests in the Full-Scale Tunnel. The first test in 1980 consisted of free-flight tests of a 0.36-scale model of the aircraft, during which its stall-resistant features were documented and analyzed.[41]

Free-flight model of the Rutan Varieze during stall-resistance testing in the Full-Scale Tunnel in 1980. Note the propeller-guard assembly used to protect the flight cable during the investigation. (NASA L-80-5673)

Following the model flight investigation, a full-scale Varieze was fabricated in a Langley shop and prepared for testing in the Full-Scale Tunnel in 1981.[42] A 3-month test program was conducted, with a primary emphasis on evaluating the aerodynamic performance, stability, and control characteristics of the basic configuration. However, the scope of the program was extensive, and included documentation of the effects of Reynolds number; canard; outboard wing leading-edge droop; center-of-gravity location; elevator trim; landing

Researcher Long Yip inspects the full-scale model of the Varieze during tests in 1981. The left side of the model has been painted to enhance flow visualization, and wool tufts were affixed to the wing upper surface for flow-visualization tests. Note the discontinuous leading-edge droop segments added to the wing for the study. (NASA L-81-7333)

gear; power; fixed transition of the boundary layer; water spray on the canard surface; canard incidence, position, and airfoil section; and the effects of winglets.

Results of the full-scale aircraft investigation verified that the canard was effective in providing increased stall-departure resistance because the canard surface stalled before the wing stalled. Interactions of the canard flow field on the wing decreased the inboard loading of the wing, and the addition of leading-edge droop increased longitudinal stability and stall angle of attack. A chemical-sublimation technique revealed that extensive laminar flow occurred on the canard, with boundary-layer transition occurring at the 55-percent chord location. Variations in the canard airfoil demonstrated that the selection of airfoil could strongly affect the aircraft stall and post-stall characteristics. The test data also included measurements of loads on the canard, pressure distributions, propeller thrust-torque loads, and flow visualization using tufts and sublimating chemicals.

A Better Concept: Short Takeoff and Landing

While the international aeronautical community of the 1950s and 1960s attempted to arrive at feasible VTOL configurations amongst hundreds of concepts, NACA and NASA researchers also directed their attention to the less demanding task of short-takeoff-and-landing missions. It became clear that the penalties of VTOL (e.g., weight, complexity, etc.) were excessive for most applications, but providing STOL capability would be less demanding. History has indeed shown that the many years of V/STOL research in the United States and abroad has paid off in operational STOL configurations, while VTOL efforts have resulted in minimal applications.

Beginning in the 1950s, Langley researchers pursued the concept of using combinations of wing leading- and trailing-edge high-lift devices together with the redirection of propeller wakes or jet-engine exhausts onto wing trailing-edge flap systems to achieve unprecedented performance in takeoff and landing. This approach, known as "powered lift," provided dramatic increases in lift available for STOL applications. The magnitude of maximum lift achieved was three to four times larger than those exhibited by conventional configurations, permitting impressive reductions in field-length requirements and approach speeds.

Two of the powered-lift concepts—the externally blown flap (EBF) and the upper-surface blown (USB) flap—were pioneered, developed, and demonstrated through testing in the Langley Full-Scale Tunnel. In addition to obtaining detailed design information on the aerodynamic, stability, and control characteristics of powered-lift transports, the efforts included extensive free-flight model tests and obtained aerodynamic input data for sophisticated piloted-simulator evaluations of the handling qualities of transport aircraft equipped with powered-lift systems.

The Externally Blown Flap

Researchers in several organizations at Langley had originally pursued wing boundary-layer control concepts to increase lift and enhance takeoff and landing capabilities. Such systems bled high-pressure air from the vehicle's engines and redirected the flow over a trailing-edge flap system. After extended research, it became obvious that internally blown flap aircraft would probably have unacceptable propulsion-system penalties, and researchers began work in the 1950s on externally blown systems that used innovative redirection of external engine flows over high-lift devices to augment lift.

Technician Joe Block prepares the first Langley EBF free-flight model for tests in the Full-Scale Tunnel in 1956. At the time, John Campbell's group was still located at the 12-Foot Free-Flight Tunnel. (NACA LAL 96823)

While head of the 12-Foot Free-Flight Tunnel in the 1950s, John Campbell conceived and patented the externally blown flap concept, which used the approach of tilting wing-mounted engine pod nacelles in a nose-down direction so that the engine exhausts impinged directly on large slotted wing trailing-edge flaps for increased aerodynamic circulation and lift at low speeds. When Campbell and his group moved from the 12-Foot Free-Flight Tunnel to the Full-Scale Tunnel, he continued to refine the concept, directing attention to the issues of controllability during engine-out conditions and the stability and controllability of EBF configurations.

Unfortunately, when Campbell conceived the EBF concept, the only jet engines available were turbojets, which produced unacceptable high exhaust temperatures and relatively inefficient small mass flows for the concept to be considered feasible. With the advent of turbofan engines that had relatively cool exhaust flow and large quantities of air available for increased airflow through the flaps in the 1970s, the concept quickly garnered interest in the military community. Led by Joe Johnson, the staff at the Full-Scale Tunnel studied

Full-Scale Tunnel investigations of the effectiveness of the EBF concept included this Air Force–sponsored study of an application to the C-5 transport in 1967. (NASA L-67-9567)

John P. Campbell (left) of Langley and Gerald Kayten of NASA Headquarters discuss the features of a generic four-engine transport free-flight model used to evaluate dynamic stability and control in 1971. Note the nose-down attitude of the engine nacelles, the deflected wing leading- and trailing-edge flaps, and the high T-tail required for satisfactory stability. (NASA L-71-2955)

the aerodynamic performance and potential control problems for two- and four-engine EBF configurations, resulting in general design guidelines.[43]

The Simulation and Analysis Section at the tunnel was provided with aerodynamic data from the tunnel tests to be used as inputs to piloted-simulator studies, during which detailed studies of handling qualities and control system requirements were conducted.[44]

In 1972, the Air Force initiated a program known as the Advanced Medium STOL Transport (AMST) to seek replacements for aging C-130 transports. McDonnell Douglas received a contract for flight evaluations of a four-engine, 150,000-pound EBF transport known as the YC-15, which subsequently demonstrated exceptional STOL performance with an approach speed of only 98 mph and a field length of only 2,000 feet. Although the YC-15 was not placed into production, the experience gained by McDonnell Douglas with the EBF concept gave the company confidence in a future application of the concept to the C-17 transport.

The Air Force C-17 transport uses several concepts that are based on fundamental research conducted at Langley—including a supercritical wing, winglets, composite materials, and advanced glass cockpit displays—but the most important contribution from Langley is arguably John Campbell's EBF concept, which enables the transport to make slow, steep approaches with heavy cargo loads as large 160,000 pounds, into runways as short as 3,000 feet.

The Upper-Surface Blown Flap

Langley researchers first studied the upper-surface blown flap concept in 1957 by investigating the effects of exhausting jet-engine efflux over the wing's upper surface such that the flow attached to the wing and turned downward over the trailing-edge flap for lift augmentation. Compared to the EBF concept, the USB offers an advantage in thrust recovery, since the engine exhaust does not directly impinge on the lower surface of the flap. In addition, since the engine exhaust flow is over the upper surface of the wing, the concept offers substantial advantages for noise shielding during powered-lift operations.[45]

In the 1970s, Langley's new deputy director, Oran W. Nicks, became intrigued by the potential capabilities of the USB concept and strongly suggested that the aeronautics program increase its efforts to develop the technology required to mature the idea. The primary task to develop the USB was assigned by John Campbell to Joe Johnson and his research team at the Full-Scale Tunnel. Initial exploratory testing in the 12-Foot Low-Speed Tunnel resulted in extremely impressive preliminary data in 1971, and more sophisticated evaluations were scheduled for the Full-Scale Tunnel, including model force tests, free-flight evaluations, and large-scale studies with turbofan engines to obtain design information.[46]

The powerful flow-turning capability of the upper-surface blowing concept is illustrated by wool tufts attached to the trailing-edge flap segments behind the upper-surface engines of this free-flight model in 1973. (NASA L-73-6289)

In 1974, a multidisciplinary team participated in a Full-Scale Tunnel test program to obtain quantitative large-scale data for USB configurations. The team used a modified Rockwell Aero Commander airframe with JT15D-1 turbofan engines arranged in a USB configuration. Following component engine/flap testing at the outdoor test stand between the Full-Scale Tunnel and the Back River—the same location used for engine-cooling tests during WWII—the Aero Commander entered the tunnel for testing to determine its aerodynamic performance, steady and unsteady aerodynamic loads, surface temperatures, and acoustic characteristics.

The large-scale results greatly matured the technology base required to reduce risk for USB applications. The aerodynamic turning performance of the USB system on the Aero Commander resulted in a humorous incident that occurred during powered testing. Joe Johnson recalled that, "At one point, the test team was in the process of evaluating the effectiveness of a horizontal T-tail for providing trim at high-power, low-speed conditions. When no detectable change occurred in aerodynamic pitching moments with the tail installed [even with large tail deflections], the test crew surveyed the airflow in the region of the tail and found that for high-power, low-speed conditions, the turning of the USB concept was so powerful that it turned the entire airflow from the wing down through the open-throat test section to the floor of the Full-Scale Tunnel building!"[47]

In addition to the detailed force tests, the staff of the Full-Scale Tunnel conducted numerous free-flight tests of USB transport configurations as well as piloted-simulator studies.

A highly modified Rockwell Aero Commander airframe was used to obtain design data such as aerodynamic performance, static and dynamic loads, acoustics, and temperatures for USB configurations. Researchers found that at high-power conditions, the USB system turned the entire tunnel flow stream downward, rendering the high T-tail ineffective. (NASA L-74-8728)

Industry had, of course, tracked the progress of USB technology at Langley, and when the Air Force AMST competition was initiated, the Boeing Company selected a USB-configured design known as the YC-14 as its candidate. Boeing technical representatives had maintained close communications with Joe Johnson and others at Langley during the design phase for the YC-14, and extensive exchanges of data were obtained from Boeing during the design process. Langley and Boeing engaged in cooperative studies of the YC-14 USB design, including outdoor test-stand evaluations at Boeing and Langley. The full-scale YC-14 aircraft met all of its design requirements and impressed international crowds at the Paris Air Show in 1977.

Unfortunately, the anticipated mission requirements for the AMST did not agree with Air Force funding priorities at the end of the flight evaluations, and the program was canceled. In 1981, the Air Force again became interested in another transport—one having less STOL capability but more strategic airlift capability than the YC-14 and YC-15 prototypes. That airplane ultimately became today's C-17, which uses the EBF concept. The USB approach to STOL missions continues to gain attention for potential civil-transport applications and is regarded as one of the most promising concepts for future aircraft.

Engineer Jim Hassell poses with a large-scale component model of the YC-14 USB installation at the outdoor test site of the Full-Scale Tunnel. Note the vortex generators placed behind the engine to enhance turning of the flow. (NASA L-73-7729)

In summary, the research conducted at the Full-Scale Tunnel for powered-lift transport aircraft contributed a solid foundation for industry to advance STOL transports with unprecedented short-field capabilities. The combination of detailed aerodynamic and multidisciplinary experiments in the wind tunnel, assessments of dynamic stability and control using free-flight models, and extensive assessments of handling qualities in piloted simulators were impressive examples of the value of fundamental research conducted by NASA.

Power Generators: Wind Turbines

Vertical-Axis Windmill

After the Oil Embargo of the 1970s, the Nation became extremely interested in alternate power sources, including windmills. The Langley engineering organization requested test time in the Full-Scale Tunnel in 1974 to evaluate the effectiveness of a vertical-axis windmill. Whereas conventional windmills are arranged with a horizontal rotational axis and use sails or propeller blades to capture wind forces for the generation of electricity, the vertical-axis windmill has its axis of revolution perpendicular to the wind. The concept was patented in 1931 and had several potential cost-saving advantages: the vertical axis of the windmill allows for direct shafting to electrical generating equipment located on the ground, allowing for a lighter support structure; although an airfoil shape is required for the blades to obtain good performance, the blades do not require twisting, and mass production of the blades may be economically more feasible; the windmill is omnidirectional with respect to wind direction and does not require a separate mechanism to maintain a particular orientation with respect to the wind direction; and the major blade loads are caused by centrifugal forces that are steady in nature, thereby avoiding major material fatigue problems. The vertical-axis concept has one major disadvantage in that it is not self-starting and requires some device to restart it when prevailing winds drop below the threshold level for operation.

Jim Shivers (left) and Ralph Muraca (right) inspect the test set up for a 14-foot-diameter, two-blade vertical-axis windmill in the Full-Scale Tunnel in 1974. The windmill and support structure are mounted to beams atop the balance and scale system of the tunnel. (NASA L-07588)

The test in the Full-Scale Tunnel was made with an existing 14-foot-diameter, two-blade vertical-axis windmill, which was mounted on beams to the tunnel balance system.[48] Aerodynamic loads were measured with the tunnel scale system, and output data included wind velocity, shaft torque, shaft rotation rate, and the drag and yawing moment of the windmill and supporting structure. A velocity survey of the flow field downstream of the windmill was also made for one operating condition.

Although good correlation was obtained between the experimental results and results from analytical studies, the data indicated that the efficiency of the vertical-axis windmill was about 25 percent lower than that of a high-performance horizontal-axis windmill. The degradation in performance in this particular test was attributed to fabrication errors, including variations in the blade profile from the NACA 0012 design and misalignment of the blades with respect to the shaft. It was estimated that with better manufacturing procedures, the difference in efficiency could be reduced to less than 10 percent.

Wind Turbine Section with Ailerons

In the 1970s, the NASA Lewis Research Center (now the NASA Glenn Research Center) became NASA's focal point for wind-energy projects. The Lewis organization included the NASA Wind Energy Project Office, which was sponsored by the U.S. Department of Energy to assess promising concepts for wind-generated power sources. One of the projects managed by the office was an evaluation of the use of ailerons as a method for controlling the output power of large propeller-type wind turbines (over 100 feet in diameter), protecting the rotors from over-speed conditions, and reducing the rotational speed to a low value (preferably to a complete stop) in the event of an emergency.

In mid-1983, the Project Office sponsored the fabrication of three 24-foot-long blades with ailerons over 20 feet of the span. Two of the three blades were mounted on the ends of a two-blade rotor to form a large rotor measuring 128 feet from tip to tip. The resulting rotor was field tested to determine the control and shutdown characteristics for various aileron configurations. The third 24-foot blade was built for testing in the Full-Scale Tunnel under steady uniform wind conditions to measure the lift, drag, and moment characteristics of the blade; the aileron hinge-moment loads; and chordwise surface-pressure distributions at one spanwise station.[49]

Technician Dave Brooks (left) and researcher Frank Jordan (right) discuss the test plan for a 24-foot-long outer blade of a windmill in 1984. The project was sponsored by the Wind Energy Project Office at the NASA Lewis Research Center with the objective of investigating the effects of ailerons on the controllability of large windmill blades. (NASA L-02783)

Results of the test indicated that a plain aileron produced better rotor control and aerodynamic braking characteristics than a balanced aileron. In addition, vortex generators produced a significant increase (up to 19 percent) in the chordwise force, which would produce an increase in rotor performance. Gaps between the aileron and the main blade produced degradation in the aerodynamic lift and an increase in drag. When the leading edge of the blade was roughened, the aerodynamic effects are negligible, and the effect of Reynolds number was negligible over the speed range tested.

Versatility and Value

During the 1970s and early 1980s, the Full-Scale Tunnel contributed critical data and demonstrations of the viability of a wide variety of aerospace vehicles. Supersonic transports and their inherent stability, control, and performance issues for takeoff and landing conditions continued long after the demise of the national SST program. Cooperative NASA work with industry brought forth the F-16XL, and the Full-Scale Tunnel was the site of many key problem-solving exercises for the configuration. Free-flight models played key roles in supplying industry and DOD with predictions of the flight behavior of advanced fighters for high-angle-of-attack conditions. An Intercenter NASA program on high-angle-of-attack technology was initiated, and the dramatic benefit of vectoring on high-angle-of-attack maneuverability was demonstrated. General aviation activities were expanded, including assessments of the application of advanced design features to a twin-engine research airplane, evaluation of current technology for aerial applications, and advanced concepts for personal-owner and commuter configurations. Finally, the research required to mature concepts for STOL operations was conceived and conducted for externally blown flap and upper-surface blown flap concepts.

After pausing for its first major rehabilitation work in 1975, the Full-Scale Tunnel emerged ready to take on more opportunities to contribute to the Nation's needs.

Endnotes

1. One of the engineers was separated directly in the RIF; he became concerned to the point that he left Langley for a non-aerospace job.
2. Thomas L. Coleman was a well-known expert in aircraft loads technology who had served as technical assistant to the director for aeronautics, Laurence K. Loftin, Jr.
3. Steven T. and Sarah W. Corneliussen, "NASA Langley Research Center Support Services Contracting, Historical Summary" (unpublished document, May 29, 1991), LHA.
4. An excellent review of the growth of contractor operations at Langley is discussed in James R. Hansen, *Spaceflight Revolution: NASA Langley Research Center from Sputnik to Apollo* (Washington, DC: NASA SP-4308, 1995), pp. 109–111.
5. Elizabeth A. Muenger, *Searching the Horizon: A History of Ames Research Center, 1940–1976* (Washington, DC: NASA SP-4304, 1985), chapter 7.
6. "Major Organizational Changes Announced," *Langley Researcher* (October 2, 1970).
7. "Full-Scale Tunnel Ceremony Held at Langley," *Langley Researcher* (July 13, 1979).
8. A.L. Abrahamson, P.K. Kasper, and R.S. Pappa, "Acoustical Modeling of the Test Section of the NASA Langley Research Center's Full-Scale Wind Tunnel," Bolt, Beranek and Newman, Inc., Report Number 2280 (November 1971).
9. H. Clyde McLemore and Joseph W. Block, interviews by author, September 4, 2011.
10. The fairing support structure was originally a wooden construction. The decision was made to replace the structure with steel components.
11. D. Bruce Owens, Jay M. Brandon, Mark A. Croom, C. Michael Fremaux, Eugene H. Heim, and Dan D. Vicroy, "Overview of Dynamic Test Techniques for Flight Dynamics Research at NASA LaRC," AIAA Paper 2006-3146 (2006).
12. Daniel G. Murri, Luat T. Nguyen, and Sue B. Grafton, "Wind-Tunnel Free-Flight Investigation of a Model of a Forward-Swept-Wing Fighter Configuration," NASA Technical Paper 2230 (1984).
13. According to Jordan, the probe had been used successfully during surveys of the Langley 14- by 22-Foot Tunnel.
14. Frank L. Jordan, Jr., interview by author, September 18, 2011. See also data presented by Frank Jordan at the Integrated Systems Review for Modifications to the 30- by 60-Foot Tunnel (August 17, 1988). The single-amplitude torsional vibration of the east motor was about 24 thousands of an inch (mils) at a resonant condition near a propeller speed of 192 rotations per minute (rpm).
15. Clyde McLemore recalled that Eugene Guryansky had experimented with screens immediately behind the propellers in the early 1940s in an attempt to fill out the flow deficits and obtain better inflow characteristics. H. Clyde McLemore, communication with author, September 21, 2011.
16. Joseph R. Chambers, *Innovation in Flight: Research of the NASA Langley Research Center on Revolutionary Advanced Concepts for Aeronautics* (Washington, DC: NASA SP-2005-4539, 2005).

17. Paul L. Coe, Jr., H. Clyde McLemore, and James P. Shivers, "Effects of Upper-Surface Bowling and Thrust Vectoring on Low-Speed Aerodynamic Characteristics of a Large-Scale Supersonic Transport Model," NASA TN D-8296 (1976).
18. H. Clyde McLemore and Lysle P. Parlett, "Low-Speed Wind-Tunnel Test of a 1⁄10-Scale Model of a Blended-Arrow Supersonic Cruise Aircraft," NASA TN D-8410 (1977).
19. H. Clyde McLemore, Lysle P. Parlett, and William G. Sewall, "Low-Speed Wind-Tunnel Tests of a 1⁄9-Scale Model of a Variable-Sweep Supersonic Cruise Aircraft," NASA TN D-8380 (1977).
20. Sue B. Grafton, "Low-Speed Wind-Tunnel Study of the High-Angle-of-Attack Stability and Control Characteristics of a Cranked-Arrow-Wing Fighter Configuration," NASA TM 85776 (1984).
21. Dynamic force tests were a specialty of the Full-Scale Tunnel's stable of test techniques. Data from such tests formed a realistic basis for dynamic aerodynamic properties at high-angle-of-attack conditions, where analytical methods and computer-based techniques were not reliable.
22. Joseph R. Chambers, *Partners in Freedom: Contributions of the Langley Research Center to U.S. Military Aircraft of the 1990s* (Washington, DC: NASA SP-2000-4519, 2000).
23. For detailed discussions of the contributions of NASA Langley and its partners to high-angle-of-attack technology for current military aircraft, see Chambers, *Partners in Freedom.*
24. Joseph R. Chambers and Ernie L. Anglin, "Analysis of Lateral-Directional Stability Characteristics of a Twin-Jet Fighter Airplane at High Angles of Attack," NASA TN D-5361 (1969); Frederick L. Moore, Ernie L. Anglin, Mary S. Adams, Perry L. Deal, and Lee H. Person, Jr., "Utilization of a Fixed-Base Simulator to Study the Stall and Spin Characteristics of Fighter Airplanes," NASA TN D-6117 (1971); William A. Newsom, Jr., and Sue B. Grafton, "Free-Flight Investigation of the Effects of Slats on Lateral-Directional Stability of a 0.13-Scale Model of the F-4E Airplane," NASA TM SX-2337 (1971).
25. Video scenes of the free-flight tests of slats are available at *http://www.youtube.com/watch?v=BfEH9gconW0*, accessed on January 5, 2012.
26. See Joseph Chambers, "Care-Free Maneuverability at High Angles of Attack," *NASA's Contributions to Aeronautics*, vol. 2 (Washington, DC: NASA SP-2010-570, 2010), pp. 763–816. Selected reports on specific investigations in the Full-Scale Tunnel include William P. Gilbert, Luat T. Nguyen, and Roger W. Van Gunst, "Simulator Study of Automatic Departure and Spin-Prevention Concepts to a Variable-Sweep Fighter Airplane," NASA TM-X-2928 (1973); William P. Gilbert, "Free-Flight Investigation of Lateral-Directional Characteristics of a 0.10-Scale Model of the F-15 Airplane at High Angles of Attack," NASA TM-SX-2807 (1973); William A. Newsom, Ernie L. Anglin, and Sue B. Grafton, "Free-Flight Investigation of a 0.15-Scale Model of the YF-16 Airplane at High Angle of Attack," NASA TM-SX-3279 (1975); Luat T. Nguyen, Marilyn E. Ogburn, William P. Gilbert, Kemper S. Kibler, Philip W. Brown, and Perry L. Deal, "Simulator Study of Stall/Post Stall Characteristics of a Fighter Airplane With Relaxed Longitudinal Stability," NASA TP 1538 (1979); Daniel G. Murri, Luat T. Nguyen, and Sue B. Grafton, "Wind-Tunnel Free-Flight Investigation of a Model of a

Forward-Swept Wing Fighter Configuration," NASA TP 2230 (1984); Sue B. Grafton, "Low-Speed Wind-Tunnel Study of the High-Angle-of-Attack Stability and Control Characteristics of a Cranked-Arrow-Wing Fighter Configuration," NASA TM-85776 (1984); and Luat T. Nguyen, Marilyn E. Ogburn, William P. Gilbert, Kemper S. Kibler, Philip W. Brown, and Perry L. Deal, "Simulator Study of Stall/Post Stall Characteristics of a Fighter Airplane With Relaxed Longitudinal Stability," NASA TP 1538 (1979).

27. Chambers, *Partners in Freedom.*
28. Marilyn E. Ogburn, Luat T. Nguyen, Alfred J. Wunschel, Philip W. Brown, and Susan W. Barzoo, "Simulator Study of Flight Dynamics of a Fighter Configuration with Thrust-Vectoring Controls at Low Speeds and High Angles of Attack," NASA TP 2750 (1988).
29. The full-scale operational F-16 fighter is limited by its flight control system to angles of attack less than about 25°.
30. Chambers, *Innovation in Flight.*
31. Long P. Yip, "Wind-Tunnel Free-Flight Investigation of a 0.15-Scale Model of the F-106B Airplane with Vortex Flaps," NASA TP 2700 (1987).
32. The Cessna 327 Baby Skymaster was a scaled-down four-seat version of the Cessna 337. Only one prototype was built before the project was cancelled, and it was used for the subject tests. Results of the NASA tests are summarized in H.H. Clyde McLemore and Robert J. Pegg, "Aeroacoustic Wind-Tunnel Tests of a Light Twin-Boom General-Aviation Airplane with Free or Shrouded-Pusher Propellers," NASA TM 80203 (1980).
33. Paul L. Coe, Jr., "Review of Drag Cleanup Tests in Langley Full-Scale Tunnel (From 1935 to 1945) Applicable to Current General Aviation Airplanes," NASA TN D-8206 (1976).
34. Bruce J. Holmes, "Flight Evaluation of an Advanced Technology Light Twin-Engine Airplane (ATLIT)," NASA CR-2832 (1977); Long P. Yip, "Comparison of Wing Pressure Distributions and Boundary-Layer Characteristics for the Advanced Technology Light Twin-Engine Airplane (ATLIT) With Two-Dimensional Section Data for the GA(W)-1 Airfoil," *NASA Advanced Technology Airfoil Research*, vol. 2, NASA CP-2046 (1979), pp. 33–44; James L. Hassell, Jr., William A. Newsom, Jr., and Long P. Yip, "Full-Scale Wind-Tunnel Investigation of the Advanced Technology Light Twin-Engine Airplane (ATLIT)," NASA TP 1591 (1980).
35. Attempts to locate reports dealing with this test were not successful, but retiree Mike Fink believes that the Langley expert on balsa-flow visualization techniques, Marion Taylor, was probably involved. Retiree Wilmer H. "Bill" Reed believes the activity could have been in response to a 1950 visit to Langley by Fred Weick, who at that time worked at Texas A&M. Weick came to Langley and met with Langley manager Charles Donlan for support of his AG-1 agricultural plane activities.
36. Frank L. Jordan, Jr., and H. Clyde McLemore, "Status of Aerial Applications Research in the Langley Vortex Research Facility and the Langley Full-Scale Tunnel," NASA Technical Memorandum 78760 (1978).
37. Frank L. Jordan, interview by author, February 15, 2011.
38. Dale R. Satran, "Wind-Tunnel Investigation of the Tail-Spoiler Concept for Stall Prevention on General Aviation Airplanes," NASA TM 83208 (1981).

39. William A. Newsom, Jr., Dale R. Satran, and Joseph L. Johnson, Jr., "Effects of Wing-Leading-Edge Modifications on a Full-Scale, Low-Wing General Aviation Airplane," NASA TP 2011 (1982); see also Joseph L. Johnson, Jr., William A. Newsom, Jr., and Dale R. Satran, "Full-Scale Wind-Tunnel Investigation of the Effects of Wing Leading-Edge Modifications on the High Angle-of-Attack Aerodynamic Characteristics of a Low-Wing General Aviation Airplane," AIAA Paper 80-1844 (1980).
40. Staff of Langley Research Center, "Exploratory Study of the Effects of Wing-Leading-Edge Modifications on the Stall/Spin Behavior of a Light General Aviation Airplane," NASA TP 1589 (1979).
41. Dale R. Satran, "Wind-Tunnel Investigation of the Flight Characteristics of a Canard General-Aviation Airplane Configuration," NASA TP 2623 (1986).
42. Long P. Yip, "Wind-Tunnel Investigation of a Full-Scale Canard-Configured General Aviation Airplane," NASA TP-2382 (1985).
43. John P. Campbell and Joseph L. Johnson, Jr., "Wind-Tunnel Investigation of an External-Flow Jet-Augmented Slotted Flap Suitable for Application to Airplanes With Pod-Mounted Engines," NACA TN 3898 (1956); Joseph L. Johnson, Jr., "Wind-Tunnel Investigation at Low Speeds of Flight Characteristics of a Sweptback-Wing Jet-Transport Airplane Model Equipped With an External-Flow Jet-Augmented Slotted Flap," NACA TN 4255 (1958); Lysle P. Parlett, Delma C. Freeman, Jr., and Charles C. Smith, Jr., "Wind-Tunnel Investigation of a Jet Transport Airplane Configuration with High Thrust-Weight Ratio and an External-Flow Jet Flap," NASA TN D-6058 (1970); Lysle P. Parlett, Sandy J. Emerling, and Arthur E. Phelps, III, "Free-Flight Investigation of the Stability and Control Characteristics of a STOL Model With an Externally Blown Jet Flap," NASA TN D-7411 (1974).
44. William D. Grantham, Luat T. Nguyen, James M. Patton, Jr., Perry L. Deal, Robert A. Champine, and Robert C. Carter, "Fixed-Base Simulator Study of an Externally Blown Flap STOL Transport Airplane During Approach and Landing," NASA TN D-6898 (1972); William D. Grantham, Luat T. Nguyen, and Perry L. Deal, "Simulation of Decelerating Landing Approaches on an Externally Blown Flap STOL Transport Airplane," NASA TN D-7463 (1974).
45. Chambers, *Innovation in Flight.*
46. Arthur E. Phelps, William Letko, and Robert L. Henderson, "Low-Speed Wind-Tunnel Investigation of a Semispan STOL Jet Transport Wing-Body with an Upper-Surface Blown Jet Flap," NASA TN D-7183 (1973); Charles C. Smith, Jr., Arthur E. Phelps III, and W. Latham Copeland, "Wind-Tunnel Investigation of a Large-Scale Semi-Span Model with an Un-Sweptwing and an Upper-Surface Blown Jet Flap," NASA TN D-7526 (1974); Staff of the Langley Research Center, "Wind-Tunnel Investigation of the Aerodynamic Performance, Steady and Vibratory Loads, Surface Temperatures and Acoustic Characteristics of a Large-Scale Twin-Engine Upper-Surface Blown Jet-Flap Configuration," NASA TN D-8235 (1976); Lysle P. Parlett, "Free-Flight Wind-Tunnel Investigation of a Four-Engine Sweptwing Upper-Surface Blown Transport Configuration," NASA TM X-71932 (1974).

47. Joseph L. Johnson, interview by author, February 26, 2009. This flow-entrainment phenomenon had been encountered once before in the Full-Scale Tunnel during full-power tests of a large-scale six-propeller VTOL tilt-wing model in 1960. Test engineer Bob Huston recalled, "The tunnel flow had been turned downward completely out of the exit cone collector."
48. R.J. Muraca and R.J. Guiliotte, "Wind-Tunnel Investigation of a 14 Foot Vertical Axis Windmill," NASA TM X-72663 (1976).
49. J.M. Savino, T.W. Nyland, A.G. Birchenough, F.L. Jordan, and N.K. Campbell, "Reflection Plane Test of a Wind Turbine Blade Tip Section with Ailerons," DOE/NASA/20320-65 NASA TM-87018 (1985).

On October 27, 1995, NASA operations at the Full-Scale Tunnel were terminated and the tunnel was deactivated. Active and retired friends and staff of the tunnel gathered for picture at the closing ceremony. (NASA L-95-6377)

CHAPTER 8
The Final NASA Programs
1985–1995

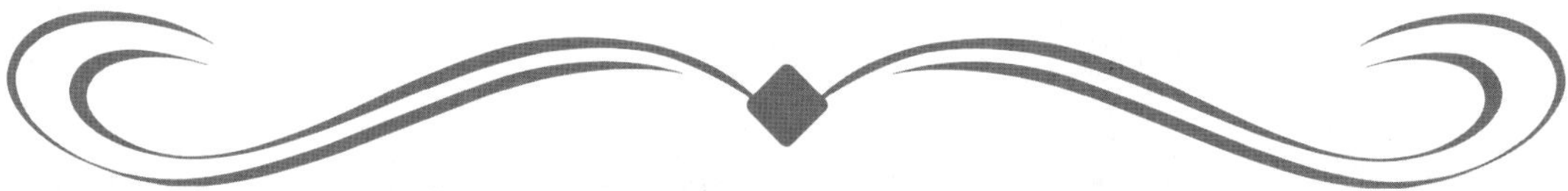

The Best of Times, the Worst of Times

By the mid-1980s, the reputation of the Full-Scale Tunnel had been solidly established as the Nation's "go-to" test facility for technologies associated with flight at high-angle-of-attack conditions. Extensive correlations between preflight predictions and full-scale flight tests had been obtained for numerous aircraft configurations and the results indicated extremely good agreement. Throughout the decade, the tunnel schedule was dominated by military requests for assessments of specific high-priority fighter designs, and industry had joined the staff in several cooperative investigations of a generic nature. The NASA High-Angle-of-Attack Technology Program produced unprecedented data to advance control concepts and design methodology, paced by test results coming from the Full-Scale Tunnel. The tunnel played a major role in the development of the international X-31 research aircraft program, which stressed air-combat maneuvers at post-stall conditions. In the 1990s, over half of the test activity in the Full-Scale Tunnel was classified at very high levels, and detailed results of those studies remain secret today. The final activities of the decade included testing in support of the Air Force's Advanced Tactical Fighter Program, which resulted in the F-22 Raptor. The Navy also continued its traditional close working relationship with the tunnel's staff in test activities for the EA-6B Prowler and the F/A-18E Super Hornet.

Civil aircraft testing included continuing efforts to improve the low-speed performance, stability, and control of supersonic-transport configurations; assessments of the aerodynamic characteristics of general aviation designs with advanced airfoils; and advanced propeller-driven configurations. The staff also conducted static and dynamic force tests for the ill-fated National Aero-Space Plane (NASP) Program and for generic hypersonic vehicles. A few large-scale-model and aircraft tests were conducted in the Full-Scale Tunnel, but free-flight model testing and its associated static force tests, dynamic force tests, and simulation composed over 80 percent of the test program.

This plaque was posted at the entrance to the Full-Scale Tunnel building in 1985 in honor of the facility being named a National Historic Landmark. (Mary Gainer)

Historical Recognition

In 1985, the U.S. Department of the Interior designated the Langley Full-Scale Tunnel and four other Langley facilities as National Historic Landmarks in recognition of their roles in developing the technological base from which the early space program was initiated. A historic plaque was mounted near the entry door of the building, reminding visitors of the historic legacy of the facility.

Agency Issues

Even as the tunnel's test activities and technical output intensified in the early 1990s, NASA was facing major funding issues for its Space Shuttle and Space Station programs, and the entire NASA program was examined by senior management for potential savings and closures. In 1992, Daniel S. Goldin became the administrator of NASA with a philosophy of "faster, better, cheaper" and a mandate for change from the previous NASA culture. The cultural changes at Langley were widespread and controversial, including a massive organizational change that completely modified the research center's mode of operation. The call for change also included close scrutiny of existing Langley facilities, and as will be discussed, the impact of Goldin's directions weighed heavily in NASA's plans for continued use of its Full-Scale Tunnel. Despite a full workload and protests from traditional customers, the tunnel was deactivated by NASA in 1995.

In addition to the unexpected closure of the Full-Scale Tunnel and the massive cultural changes stimulated by Goldin, the year 1995 was chaotic for the entire NASA Langley Research Center. As the NASA FY 1996 budget woes continued under President Bill Clinton, rumors of closure of entire NASA research centers became rampant, and political efforts to save targeted facilities began. On July 10, 1995, the Congressional Veterans Affairs, Housing and Urban Development, and Independent Agencies Appropriations Subcommittees proposed that the NASA Langley Research Center, the NASA Marshall Space Flight Center, and the NASA Goddard Space Flight Center all be shut down by October 1998. The proposal was regarded as a political ploy by Representatives from California, Ohio, and Texas, and it brought immediate reaction from Virginia Representatives Robert Scott and Herbert Bateman.[1] On July 18, the full House Appropriations Committee removed the proposal; however, the blunted political threat severely shook Langley staff members—most of whom had first discovered the proposal in their morning newspapers—with a genuine concern for the future of the Center. It was generally acknowledged (with great alarm) that Langley was more vulnerable than any other NASA Center to congressional actions, despite its acknowledged past and current contributions to the Nation.

Organizational Changes

The management of the Full-Scale Tunnel had remained stable during the early 1980s, with Joe Johnson as head of the Flight Dynamics Branch and Bill Gilbert as his assistant. In 1985, however, Gilbert left the branch to join the emerging NASP Program and was replaced by Luat T. Nguyen, who had led the Simulation and Analysis Section at the tunnel. The facility continued to operate as part of the Low-Speed Aerodynamics Division under Joe Stickle.

In July 1985, Roy V. Harris replaced Robert Bower as Director for Aeronautics and Joe Stickle became Langley's chief engineer. Harris reorganized the aeronautics divisions and created a new Flight Applications Division headed by Joe Chambers, with Bill Gilbert as an assistant for special programs.[2]

Joe Johnson spent his entire 46-year career with the NACA and NASA in only two organizations. He had joined John Campbell's group at the old 12-Foot Free-Flight Tunnel in 1944, and he moved next door to the Full-Scale Tunnel when the group moved there in late 1958. His technical and managerial skills in leading the staff of the Full-Scale Tunnel were outstanding, but in 1990 he felt it was time to provide opportunities to the next generation of leaders, and with great emotion, "J.J." retired.

With the retirement of Johnson, the Full-Scale Tunnel was headed by Luat Nguyen, with Dana J. Dunham as his assistant. Although neither individual had ever conducted a large-scale test in the Full-Scale Tunnel, they brought outstanding credentials for their assignments. Nguyen began his career at the Full-Scale Tunnel in the Simulation and Analysis Section and had led many critical piloted simulation efforts that were tightly coordinated with data extracted from the tunnel, receiving accolades from industry and the military for his superb analysis and contributions. Dunham came to the tunnel with experience in both experimental aerodynamics and computational fluid dynamics, including testing in the Langley 14- by 22-Foot Subsonic Tunnel.

In 1992, Nguyen left the Flight Dynamics Branch for other duties and Dana Dunham became manager of the Full-Scale Tunnel. She would also be the last NASA manager of the facility before its operations were transferred to Old Dominion University. Dunham was assisted by Dr. Paul L. Coe, who had originally been a member of the tunnel's staff in the 1970s before transferring to the 14- by 22-Foot Tunnel. Coe had been a visiting professor from Hofstra University and was hired by NASA in the 1970s. His upbeat response to inquiries as to his status was always, "Never had it so good!"

Dana Dunham was an exceptionally gifted manager who not only grasped technical details and objectives but also had outstanding capabilities in personnel management. Her tenure as branch head was especially challenging, filled with tragedies and organizational upheavals that resulted in extreme stress for the staff. In 1993, the branch's popular young secretary, Krista Bullock, was killed in a murder/suicide at the hands of her husband. The event devastated the staff of the Full-Scale Tunnel and the local community. Dunham was immediately confronted with maintaining the branch's activities while providing solace and inspiration to the staff. Incredibly, a week later her assistant, Dr. Coe, perished in a tragic automobile accident. Coming on the heels of the previous week's tragedy, Coe's death brought a new wave of shock and despair to the organization. The successful recovery of technical projects at the Full-Scale Tunnel and the emotional recovery of its staff were direct results of the leadership and contributions of Dana Dunham.

Together, Dana Dunham and Paul Coe brought an unprecedented interest in computational fluid dynamics technology to the branch. During their tenure, three full-time specialists in CFD were hired for efforts within the NASA High-Speed Research Program and other research programs. The consolidation of these experts at the Full-Scale Tunnel brought a rebirth of analytical methods for aerodynamic analyses not seen since the 1930s and the efforts of the section under Sam Katzoff.

The immediate and subsequent impacts on the Full-Scale Tunnel and its staff of an unprecedented Langley organizational upheaval in 1994 will be discussed later in this chapter.

Facility Improvements

The Full-Scale Tunnel ceased operations in September 1985 for the start of a $4.4 million Construction of Facilities (CofF) project to implement modifications in the test section. The project included installing a new turntable and strut fairings fashioned after the Ames 40- by 80-foot tunnel; designing and installing new aircraft-mount struts; lowering the ground board to a level flush with the inlet cone lip; installing additional flow deflectors;

and installing a new high-angle-of-attack strut support system for subscale models at the left rear of the ground plane.[3] A brief flow survey was conducted in July 1986 and the ground-board change took place in August, followed by checkout of a new data-acquisition system, the new turntable checkout, and a calibration of the balance.

The tunnel was put back into research testing with a full-scale semispan model of the Gulfstream Peregrine wing in December 1986 for 3 months, after which the subscale model support system was installed. The CofF project was officially completed on July 6, 1987, and a year of testing ensued.

Installation of the final annular flow-deflector layout in the exit cone took place in July 1988. The deflector configuration consisted of the two existing 12-foot-span sidewall deflectors and eight additional 10-foot-span deflectors mounted on support struts on the upper and lower lips of the exit cone. The orientation and height of the deflectors could be varied with a remotely controlled electric drive system.

The deflectors and their effectiveness in suppressing motor-mount vibrations remained a debated and controversial issue through the final years of the facility. Recently, data and interviews have clarified the situation regarding this interpretation, and the following discussion hopefully provides some insight into the controversy.

The previous chapter discussed research conducted by Frank Jordan in 1983 regarding the mechanism driving the vibrations and the effect of drive-fan rotational speed on the amplitude of vibration. The fact that use of sidewall deflectors and a ground-board flap decreased the magnitudes was also discussed. The following figure presents data measured during runs made with a combination of sidewall deflectors and a deflected flap at the trailing edge of the ground board.

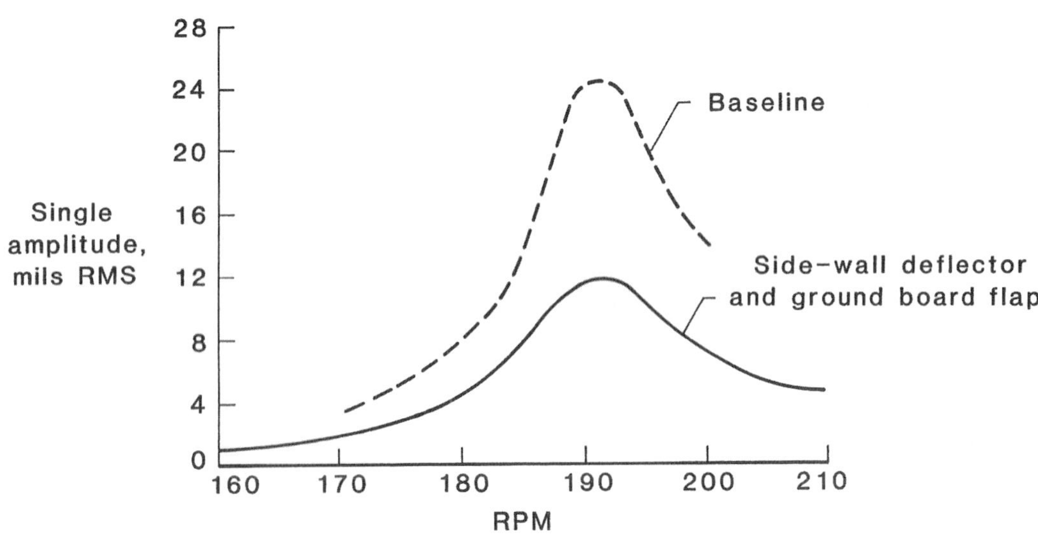

Data measured in 1983 indicated the effect of two sidewall flow deflectors and a ground-board flap on the amplitude of vibrations for the east motor support platform. The ground-board flap was not acceptable because of test-section flow effects. (Frank Jordan, Jr.)

The data show an almost 50-percent reduction in the maximum vibration amplitude at the resonant condition near 195 rpm, but the magnitude was almost an order higher than the amplitude experienced at normal operations near 170 rpm. As previously discussed, the ground-board flap was unacceptable because of its undesirable effects on the quality of flow in the test section. When the sidewall deflectors and the new upper and lower deflectors were tested, unpublished data showed a 70-percent reduction in the magnitude at 195 rpm. Although significantly reduced, the magnitude of vibration was considered unacceptable for routine operations at that speed. Normal operations of the tunnel were therefore maintained at 170 rpm; however, on occasion the staff accelerated the tunnel through the 195-rpm resonant condition to higher values of rpm where the vibratory motions subsided. At speeds up to 170 rpm and higher than about 210 rpm, the data indicated that the effects of the top, bottom, and side deflectors were dramatically reduced. Researchers who operated the tunnel under these guidelines after the deflector modification rightfully referred to the deflectors as "having no effect" at speeds removed from the resonant condition. However, the measurements leave no doubt that the deflectors were indeed effective at the resonant condition. In summary, both Jordan and the tunnel operators were correct in their interpretation of the effectiveness of the deflectors.

The desire for a model of the Full-Scale Tunnel to study the effects of further modifications on flow characteristics in the tunnel's test section continued. In 1990, seed money became available for construction of a new ⅟15-scale model tunnel, and NASA proceeded to have a second model tunnel constructed under contract to DSMA International in Canada. After a difficult contractual effort, the tunnel was located in the hangar annex of the Full-Scale Tunnel.

The final upgrade to equipment during NASA's management of the Full-Scale Tunnel occurred during a 2-month CofF project in 1992, when the overhead flow-survey apparatus and its instrumentation were upgraded with automated systems.

Research Activities

NASA HATP Goes to Flight

The Full-Scale Tunnel support for the development of the Navy's F/A-18 Hornet had been completed in the 1970s, and the staff had gathered extensive data on the aerodynamic, stability, and control characteristics of the airplane at high angles of attack. The strong vortical flows emanating from the wing-fuselage leading-edge extension (LEX) surfaces were of extreme importance in maintaining lift and stability, and the low-speed testing had provided a good understanding of the flow fields of interest. In 1987, the NASA High-Angle-of-Attack Technology Program proceeded to flight for detailed studies of aerodynamic flows on the High Alpha Research Vehicle at Dryden. The objectives of the initial flight investigation focused on correlation of general results with data from several wind tunnels and CFD. This first phase of the HATP flight studies continued through 1989, with 101 research flights of the HARV for angles of attack up to 55°. Special flow-visualization and pressure instrumentation were used to compare aerodynamic phenomena obtained in flight, tunnels, and CFD for specific flight conditions. Data obtained in the earlier Full-Scale Tunnel studies were of great value in planning the experiments.[4]

While other organizations in the HATP concentrated on aerodynamics, the staffs of the Full-Scale Tunnel and the Langley 16-Foot Transonic Tunnel were working together to mature the concept of a simple paddle-type thrust-vectoring system that might be implemented on the HARV and used for assessments of the impact of multi-axis vectoring on maneuverability at high angles of attack.

Thrust Vectoring: Quick and Dirty

Having demonstrated the results of thrust-vectoring tests of free-flight models in the Full-Scale Tunnel to NASA management, the military, and industry, the HATP planners decided to implement a relatively low-cost, rudimentary, three-paddle-per-engine thrust-vectoring system on the HARV at Dryden for flight evaluations and demonstrations. The installation was to use external paddle actuators and Inconel paddle surfaces. McDonnell Douglas was awarded a contract to modify the HARV aircraft with both a paddle-based mechanical

Details of the rear end of the F/A-18 free-flight model in 1989 configured to represent the HARV vectoring system. A three-vane system was used to provide pitch and yaw control by deflecting the exhaust flow of ejector-type engine simulators. Note that the divergent nozzles were removed from the engines and that the vane actuation system was mounted externally. The rectangular box at the top of the fuselage represents the emergency spin-recovery parachute container carried by the HARV. (NASA L-89-1358)

thrust-vectoring system and a special research flight-control system to accommodate the vectoring capability. Major contributions from the staffs of Dryden and Langley were also required for this effort. At Langley, supporting activities in the Full-Scale Tunnel included

Spectacular engine test run for the HARV in 1994 with nose-up pitch control applied. (NASA EC 91-075-38)

free-flight tests of the F/A-18 model configured with a vectoring-vane system similar to that designed for the full-scale aircraft.

In 1991, the HARV thrust-vectoring system became operational, kicking off the second major phase of the HATP in which thrust-vectoring capability was demonstrated and evaluated. The powerful controls provided by vectoring were used for two purposes. In the first application, the aircraft was stabilized at extreme angles of attack for aerodynamic studies. The second application was to demonstrate the improved low-speed, high-angle-of-attack maneuverability predicted by the Langley model testing in the Full-Scale Tunnel and in piloted-simulator studies. The HARV was the first aircraft to demonstrate multi-axis vectoring, and the simplified approach to vectoring achieved all the goals of the NASA program.[5]

Harnessing Forebody Vortices

After disseminating the impressive Full-Scale Tunnel results for thrust vectoring at high angles of attack, Joe Johnson and his researchers pursued yet another concept to increase controllability at high angles of attack. The engineering community had long known that strong vortex flow fields emanating from long pointed bodies at high angles of attack could create extremely large yawing moments. In fact, several operational aircraft had exhibited asymmetric shedding of such vortices, resulting in uncontrollable yawing and rolling motions, including departures and spins. Johnson's team, led by Daniel G. Murri, theorized that forcing the separation of vortices from one side of the nose in a controlled manner might significantly increase yaw control at high-angle-of-attack conditions, where the effectiveness of conventional rudders typically diminishes. At high angles of attack, creating sideslip by yawing an aircraft couples with dihedral effect to cause the aircraft to roll in the desired direction. The scope of the research conducted in the Full-Scale Tunnel included evaluations of the concept on generic models and specific applications to the F/A-18 free-flight model for possible application to the full-scale HARV airplane.[6]

Results of the Full-Scale Tunnel free-flight tests of the F/A-18 model with forebody strakes were impressive, especially when the strakes were used in combination with thrust vectoring. Control power was dramatically increased and maneuverability was significantly improved. After the Full-Scale Tunnel tests, conventional wind tunnel testing at subsonic and transonic speeds in other wind tunnels was accomplished with F/A-18 models to determine the effectiveness of the strake concept over the subsonic and transonic speed ranges. A Dryden team was subsequently formed to design and implement this concept on the HARV aircraft under a project known as Actuated Nose Strakes for Enhanced Rolling (ANSER).[7]

After several years of research on generic models, the staff of the Full-Scale Tunnel applied the fuselage-strake concept to the F/A-18 free-flight model (left). Note the relatively small size of the strake surface on the nose. In 1991, detailed aerodynamic tests were made of a full-scale forebody (right). (NASA L-88-06485 and NASA I-91-12913)

The ANSER effort consisted of the design and fabrication of a special radome at Langley, followed by the installation of the forebody, actuators, and instrumentation on the HARV at Dryden. The development program included force and moment testing of a full-scale F/A-18 forebody in the Full-Scale Tunnel in 1991 to address detailed aerodynamic issues on the location and size of the strakes, aerodynamic loads on the strakes, and linearity of the strake-induced moments with strake deflection angle.

Head-on view of the F/A-18 HARV at Dryden in 1995 shows the configuration of the forebody strakes used for flight evaluations. The photo on the right illustrates the strong vortex produced on the left side of the nose when the left strake was deflected. The photo was taken by a wingtip camera and the vortex was traced with smoke injected into the flow by a port near the strake. (NASA EC 95-03419 and NASA EC 96-01598)

The HARV ANSER flight investigations at Dryden began in 1995 and continued into 1996, with the operational effectiveness and utility of the strakes being evaluated and demonstrated both individually and in combination with the aircraft's thrust-vectoring system capability. The flight results verified the results obtained in the free-flight tests at low subsonic speeds, but they also showed that the strakes were powerful control devices at higher subsonic speeds, where close-in dogfights typically occur.[8]

The output of the HATP activities was extremely broad, resulting in hundreds of technical reports and presentations by the staffs of Langley, Dryden, Ames, and Lewis (now Glenn) Research Centers. The HARV research aircraft completed 385 research flights with angles of attack up to 70°, including demonstrations of the potential benefits of the thrust-vectoring and forebody flow controls that had been conceived and developed through research in the Langley Full-Scale Tunnel. Results were rapidly disseminated to the industry through biannual conferences held at Langley and Dryden until the program ended in 1996. Most of the data and results of the program are currently classified or restricted.

The role played by the Full-Scale Tunnel and its staff in the conception and research activities of the HATP cannot be understated. Many of the investigations conducted in the program were direct results of fundamental studies that had been completed years before being applied on the HARV. In addition, the results of the facility's unique participation in military aircraft development programs provided invaluable guidance in the technical goals of the project. The scope of the studies undertaken by the group included static and dynamic force tests, free-flight model investigations, piloted simulator studies, drop-model tests, and onsite participation during flights of the HARV at Dryden.

Technician Jim Chiaramida poses with the original Rockwell SNAKE configuration, which had been designed analytically with a minimum amount of wind tunnel tests. Tests in the Full-Scale Tunnel revealed that the design was unstable in all axes. Note the down-turned wingtips and the thrust-vectoring vanes. (NASA L-85-3093)

International X-Plane: The X-31

Interest on an international scale in advancing the state of the art for high-angle-of-attack technology and assessing the potential impact of "carefree maneuverability" at extreme angles of attack reached a peak in the mid-1980s. One of the most active American companies was Rockwell International (now Boeing), which had previously won a contract to develop and flight test a twin-tail, canard-configured, remotely piloted research vehicle known as the Highly Maneuverable Aircraft Technology (HiMAT) demonstrator. In 1984, Rockwell proposed a joint program with the staff of the Full-Scale Tunnel to explore a similar configuration known as the Super Normal Attitude Kinetic Enhancement (SNAKE) design. Unlike the HiMAT, the SNAKE project was focused on flight at extreme angles of attack, and the aircraft design included thrust-vectoring vanes for controllability at high angles of attack.)

The Rockwell SNAKE configuration had been designed using computer-based technology with a minimum amount of wind tunnel testing, and preliminary testing of the design in the Full-Scale Tunnel revealed that the model was unstable in pitch, roll, and yaw for all angles of attack. Under the leadership of Joe Johnson and Mark A. Croom, testing in the Full-Scale Tunnel resulted in configuration modifications that enabled the modified SNAKE model to be flown successfully to angles of attack as high as 80°.

Meanwhile, Dr. Wolfgang Herbst of the German Messerschmitt-Bolkow-Blohm (MBB) company had led European interest in the use of post-stall maneuvers for enhanced close-in combat. Collaboration between Rockwell, MBB, and the U.S. Defense Advanced Research Projects Agency led to the first international X-plane project, known as the X-31. In view of the expertise and value demonstrated at the Full-Scale Tunnel during the Rockwell SNAKE

An early version of the X-31 configuration is readied for free-flight tests in 1988. The staff of the Full-Scale Tunnel was a major contributor to the success of the research project and provided rapid-response solutions to problems that were encountered during aircraft flight tests at Dryden. (NASA L-88-13225)

program, in 1986 the partners requested that Langley participate in the development of the X-31. Langley's support through the staff of the Full-Scale Tunnel included static and dynamic force tests, free-flight wind tunnel studies, spin-tunnel tests, and outdoor drop-model studies of post-stall characteristics. In addition to providing wide-ranging aerodynamic data for the X-31 team, the results of the free-flight tests in the Full-Scale Tunnel revealed that the configuration might have marginal nose-down control at high angles of attack and that it might exhibit severe, unstable lateral oscillations (i.e., wing rock). The results also demonstrated that a simple control-law concept would eliminate the potential problem.

Two X-31 aircraft begin flight tests at Dryden in 1992, and early solutions to problems involving lack of nose-down control and yaw departures were provided by quick-response testing by Croom and his team in the Full-Scale Tunnel. The X-31 concluded a highly successful test program that included spectacular displays of unprecedented maneuvers at low speeds and demonstrations of increased maneuverability over conventional fighters in 1995. It also performed a series of post-stall maneuvers at the Paris Air Show in 1995 that astounded the attendees.

Near the end of its fighter maneuverability test program at Dryden, the X-31 team advocated for a possible follow-on program to demonstrate the capability of the thrust-vectoring concept to enable a reduction in tail size. In 1994, software was installed in the X-31 to demonstrate the feasibility of stabilizing a tailless aircraft at supersonic speed using thrust vectoring, and "quasi-tailless" tests began that year. During these flights, the aircraft

The Full-Scale Tunnel was used to obtain unique aerodynamic data as interests in a tailless version of the X-31 increased. Here, the tailless model is mounted on an apparatus to measure the effect of motion (dynamic derivatives) during yawing oscillations in 1994. Such data were used as inputs for analysis and piloted simulators. Rotary motions of a flywheel driven by an electric motor on a shelf at the base of the apparatus were converted to yawing motions of the model by a connecting rod arrangement. The Full-Scale Tunnel provided the only U.S. source for such measurements at high angles of attack. (NASA L-94-08995)

was destabilized with the rudder to stability levels that would be encountered if the aircraft had a reduced-size vertical tail. At Langley, support for the growing interest in a tailless X-31 experiment included static and dynamic force testing in the Full-Scale Tunnel. In addition, studies were made using differential deflections of the canards for yaw control with and without the vertical tail.

In another aspect of the X-31 tailless activity, the Flight Dynamics Branch used the X-31 tests to transition its drop-model operations to the NASA Wallops Flight Facility in 1995. In two drop tests, the model was flown with and without the vertical tail, with differential canard deflections used for yaw control. The impact of the activity was to prove the feasibility of conducting drop operations at Wallops as well as to provide information on the characteristics of the tailless configuration.

View of the tailless X-31 drop model as it falls away from the helicopter during a test at the NASA Wallops Flight Facility. The model was retrieved from a water landing following the use of an onboard recovery parachute. (NASA L-94-08995)

Three-Surface Fighters: The F-15 STOL and Maneuver Demonstrator

As the staff of the Langley 16-Foot Transonic Tunnel captured the attention of industry by maturing thrust-vectoring technology in the mid-1970s, it also stimulated interest in "three-surface" fighter designs. In this concept, a fighter aircraft equipped with vectoring engine nozzles uses three aerodynamic surfaces: a wing, an aft horizontal tail, and canard surfaces on the forebody. Such configurations offered options in lift-sharing between the aerodynamic surfaces that might optimize the thrust-vectoring capability and produce maximum lift performance.

The staff of the Full-Scale Tunnel and their peers in industry became interested in the characteristics of the radical three-surface designs at high angles of attack and embarked on exploratory in-house and cooperative programs using modified free-flight models. The

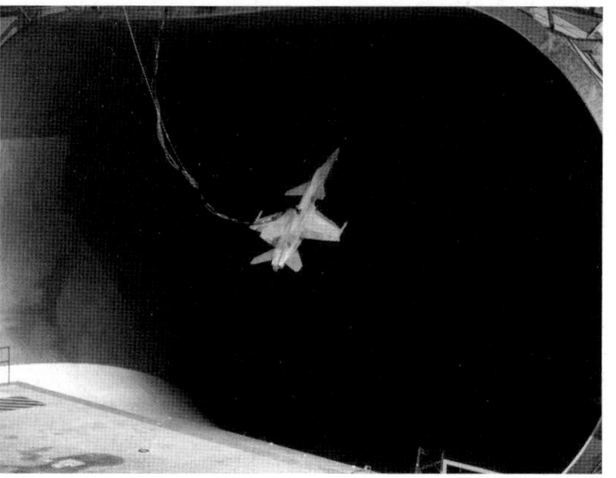

Mark Croom inspects a three-surface version of an F/A-18 free-flight model in 1982. In addition to forebody canards, the model was equipped with vanes to provide thrust vectoring for yaw control at high angles of attack. The three-surface F/A-18 research model flies at high angles of attack in the Full-Scale Tunnel in 1983. (NASA L-82-7534 and NASA L-83-4692)

objective of the research was to create a database for the designs and to determine whether general problems might exist in dynamic stability and control.

One of the first studies of three-surface fighters at the Full-Scale Tunnel was an investigation in 1981 of a three-surface derivative of the F/A-18 for which the wing-body strake was replaced with an all-moveable canard. The scope of the study included conventional static wind tunnel tests, dynamic force tests, and free-flight tests in the Full-Scale Tunnel, as well as outdoor drop-model testing. First-ever high-angle-of-attack data for configurations of this type were generated by the study, and the staff at the tunnel accumulated experience with the key factors that dominated high-angle-of-attack characteristics of three-surface designs.[9]

Also in the early 1980s, a joint program with McDonnell Douglas on three-surface configurations was undertaken that resulted in exploratory experiments with an F-15 free-flight model.[10] The activity provided early data to an emerging McDonnell Douglas project, under contract to the Air Force in 1984, to develop a short takeoff and landing/maneuver technology demonstrator (STOL/MTD) using an F-15 equipped with canards (F/A-18 horizontal-tail surfaces), vectoring two-dimensional nozzles (single-axis pitch vectoring), and thrust reversing. As a result of mutual interests in developing the technologies for three-surface aircraft, a cooperative project was initiated between the staff of the Full-Scale Tunnel, the Air Force Wright Research and Development Center, and the McDonnell Douglas Corporation to investigate the high-angle-of-attack characteristics of the F-15 STOL/MTD configuration. The scope of the studies, which were conducted from 1985 to 1987, included static and dynamic force tests to define aerodynamic data and wind tunnel free-flight tests

Technician Dave Robelen poses with the free-flight model of the F-15 STOL/MTD aircraft prior to free-flight testing in 1985. (NASA L-85-7469)

The F-15 STOL/MTD configuration was modified by the staff at the Full-Scale Tunnel to include vanes for yaw vectoring and demonstrate the general benefit of yaw vectoring for increased control at high angles of attack. The model was flown without vertical tails during some of the tests as shown here. (NASA L-88-01061)

to assess the 1-g departure resistance of the configuration at high angles of attack.[11] The aerodynamic data from the wind tunnel tests were used as inputs for the development of the aircraft flight-control laws by McDonnell Douglas.

Following the tunnel tests to evaluate the high-angle-of-attack characteristics of the F-15 STOL/MTD in its basic configuration, the staff modified the design in 1988 by adding additional vanes to provide yaw vectoring, and proceeded to demonstrate the powerful effect of these vanes on dynamic stability and control. Once again, the marked improvement in controllability with yaw vectoring at high angles of attack was impressively displayed by flights of the model, including several in which the vertical tails were removed without adverse effect at extreme angles.

After its flight program with the Air Force and McDonnell Douglas was completed in 1991, the full-scale F-15 STOL/MD aircraft was acquired by the Dryden Flight Research Center in 1993 and equipped with axisymmetric, three-dimensional pitch/yaw vectoring nozzles for thrust-vectoring research and other controls-related research. Conventional static force tests were conducted in the Full-Scale Tunnel in 1995 with the modified F-15 STOL/MTD model in support of the program.

A Significant Upgrade: The EA-6B

The Grumman EA-6B Prowler was the Navy's primary electronic warfare aircraft from 1971 through the turn of the century. Continual growth in takeoff, combat, and landing weights through the years, coupled with no increase in lifting capability, severely limited the maneuvering capability of the aircraft during military operations. In fact, the constraints imposed by the growth factors resulted in operational angles of attack very near the stall, leaving little lift available for pulling g's during combat operations. A request from the Navy to explore technology to improve the aircraft's performance and maneuvering capabilities was received by the Langley Research Center in 1984. In response to the request, studies were conducted in several Langley facilities, including the Langley Low Turbulence Pressure Tunnel, the Langley 14- by 22-Foot Tunnel, the Langley 16-Foot Transonic Tunnel, the Langley 12-Foot Low-Speed Tunnel, the Langley National Transonic Facility (NTF), and the Langley Full-Scale Tunnel. The scope of Langley's activities in these facilities covered high-lift characteristics, dynamic stability and control at high angles of attack, and the development of new airfoils for improved performance.[12]

At the Full-Scale Tunnel and the 12-Foot Low-Speed Tunnel, the test activities focused on factors contributing to a directional instability exhibited by the EA-6B at high angles of attack and the development of airframe modifications that would eliminate or delay these instabilities to angles of attack further removed from the flight envelope.[13] The test program included active onsite participation by the engineering staff of Grumman. Detailed static force tests and flow-visualization tests revealed that the configuration experienced a marked loss of directional stability at high angles of attack because of an adverse side wash at the aft fuselage and vertical-tail location. The undesirable side wash was produced by a vortex system originating near the wing-fuselage juncture. In addition, the EA-6B model experienced a loss of effective dihedral caused by stalling of the leading wing panel during sideslip at high angles of attack.

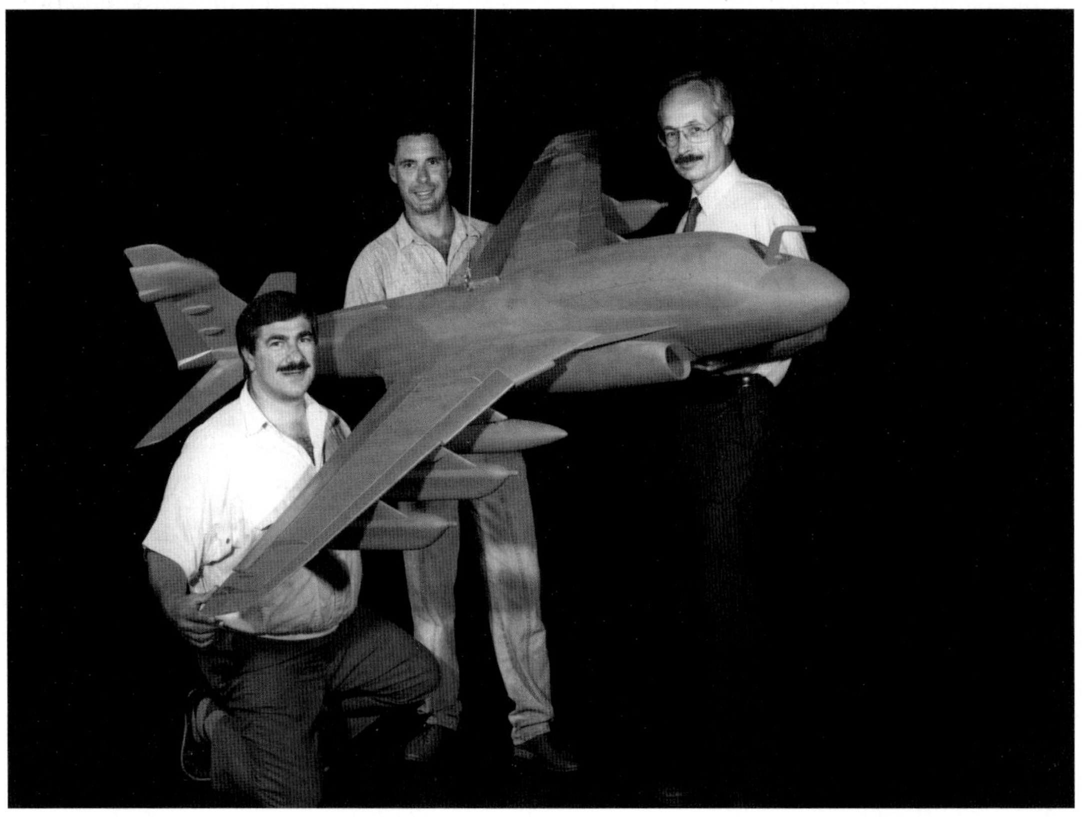

The joint NASA-Grumman-Navy project to improve the performance and maneuver capabilities of the EA-6B Prowler electronics warfare aircraft included participation by the Flight Dynamics Branch and testing in the 12-Foot Low-Speed Tunnel and the Full-Scale Tunnel. In this 1989 photograph, team members (left to right) Matt Masiello and Bill Gato of Grumman pose with Langley's Frank Jordan with the free-flight model of the advanced EA-6B configuration. Modifications, including an extension to the vertical tail, a wing-fuselage strake, a drooped leading-edge inner-wing, and use of wingtip speed brakes as ailerons, significantly improved the high-angle-of-attack characteristics of the aircraft and were implemented on a full-scale test aircraft. (NASA L-89-08564)

Following extensive tunnel testing, flow visualization, and exploratory evaluations of modifications to the airframe, modifications that promised to alleviate the stability problem were identified, including an inboard wing leading-edge droop, a glove strake for the wing-fuselage juncture, and an extension of the top of the vertical tail. Results of free-flight tests of the 0.12-scale EA-6B model in the Full-Scale Tunnel showed that the modified configuration exhibited good dynamic stability characteristics and could be flown at angles of attack significantly higher than those of the unmodified configuration.

In addition to the subscale model investigations, a full-scale semispan wing test was conducted in the Full-Scale Tunnel to evaluate the effectiveness of using the split-surface wingtip surfaces (normally used as speed brakes) as conventional ailerons to improve handling qualities during carrier approaches. Used in this unconventional manner, the speed brakes were effective for roll-control augmentation at moderate and high angles of attack where the wing trailing-edge flaperons lost their effectiveness. The full-scale test supplemented the small-model tests and increased confidence in the data prior to full-scale airplane flight tests. The test matrix included deflecting only one speed-brake panel as well as both.

A full-scale semispan of an EA-6B aircraft was tested in the Full-Scale Tunnel in 1991 to measure detailed aerodynamic data for a modification to use the wingtip speed-brake surfaces as ailerons. The conventional EA-6B aircraft used split upper and lower surfaces to control airspeed during carrier approaches. In the Full-Scale Tunnel studies, the surfaces were coupled to permit deflections as a single conventional aileron for improved controllability at high angles of attack. (NASA L-91-11228)

A Navy program known as the Advanced Capability (ADVCAP) EA-6B was initiated to integrate the NASA results and other advances in avionics into a full-scale EA-6B demonstrator aircraft. The modified aircraft included the leading-edge strakes, vertical-tail extension, and ailerons derived from the Full-Scale Tunnel testing, as well as recontoured leading-edge slats and trailing-edge flaps from other Langley tests. Results of the flight tests of the new EA-6B ADVCAP configuration from 1992 to 1994 showed greatly improved flying qualities, and the Navy intended to modify all EA-6B aircraft into the ADVCAP configuration. However, competition for funding in the DOD budget resulted in the program being eliminated in the FY 1995 budget.

The collective results of all the Langley efforts to improve the EA-6B were extremely impressive and resulted in a special technical session at a major meeting of the American Institute of Aeronautics and Astronautics in August 1987.[14]

V/STOL's Last Gasp at the Full-Scale Tunnel: The E-7A V/STOL

In the 1980s, European and American interests in V/STOL aircraft turned to supersonic fighters with short takeoff and vertical landing (STOVL) capabilities. The lead NASA Center for vertical takeoff and landing aircraft was the Ames Research Center, which sponsored many contractor studies of advanced STOVL configurations for possible Navy or Marine Corps applications. General Dynamics responded to this interest with the E-7A design, which used an ejector-augmenter system in the wing-root section of a large clipped delta wing and a vectorable core nozzle located aft of the center of gravity to provide the lift and balance required for STOVL capability. Attitude control would be provided by reaction-control jets at the wingtips, nose, and tail of the configuration. For hovering flight, all engine-bypass air was ducted forward to the wing ejectors, where a large secondary mass flow

Researcher Don Riley poses with the 0.15-scale free-flight model of the General Dynamics E-7 STOVL fighter configuration used for tests in a cooperative program with the Flight Dynamics Branch. The covers for the downward-directed ejector nozzles are open, as they would appear for hovering and transition flight. The aircraft would have used components from the F-16 for its fuselage. The photo on the bottom shows the E-7A model during hovering flight in the Full-Scale Tunnel in preparation for transition to forward flight. Note the wool tufts attached to the lower fuselage indicating the vertical exhaust of the ejector units and the rotated engine nozzle at the rear of the model. The flight cable attached to the model is unusually large and indicative of the large mass flow required for operation of the ejector system. The model of the F-15 STOL/MTD mounted to the force test apparatus in the background represents a typical condition at the Full-Scale Tunnel, where different models were tested on day and night shifts. (NASA L-85-5629 and NASA L-85-8463)

would be induced to increase the thrust output of the system. The core nozzle was deflected to turn the core flow 90° downward. During transition, the core-nozzle angle was varied and the engine bypass air was modulated between the wing ejector system and the aft nozzle. In conventional flight, all engine-bypass air was directed rearward through the aft nozzle.

The ejector-based propulsion scheme had previously been pursued without success, first by Lockheed in the XV-4A Hummingbird aircraft of the 1960s and then by Rockwell in the failed XFV-12A program; therefore, considerable skepticism faced the General Dynamics E-7A program.

A cooperative program between General Dynamics and the staff at the Full-Scale Tunnel was formulated in 1985 to evaluate the dynamic stability and control characteristics of this unique aircraft. The test program included conventional force testing and free-flight testing in the Full-Scale Tunnel as well as force tests in the 12-Foot Low-Speed Tunnel.[15] A 0.15-scale model of the E-7A was fabricated for force- and flight-test programs in the Full-Scale Tunnel in 1985. For hovering and transition flight, the ejector inlet covers were open and the diffuser units were powered. The testing in the 12-Foot Tunnel consisted of an investigation of the loss of roll control effectiveness by the wing-mounted reaction jets as forward speed was increased from hovering flight. The loss of roll control effectiveness had been widely reported in the literature for VTOL designs and was not unexpected for the E-7A configuration.

The E-7 project would be the last vertical-landing configuration tested by NASA in the Full-Scale Tunnel after over 30 years of research on V/STOL configurations. Data gathered in the numerous studies were useful inputs to NASA collaborative studies with industry, the military, and selected European partners in the development of technologies for the Joint Strike Fighter Program that later evolved into the F-35B, a STOVL variant of the F-35 fighter for the U.S. Marine Corps.

Vectoring to the Extreme

During the 1980s, Langley's organizational structure included a systems-oriented organization known as the Aeronautical Systems Division (ASD) under Cornelius "Neil" Driver. The mission of the organization was to push the envelope of aircraft design with radical new concepts for military and civil applications. Supersonic transports, multifuselage transports, span-loader configurations, and advanced fighters were included in its charter. The group had begun to revive the old "tailsitter" concept for VTOL aircraft in exploratory studies in the 1980s, but with a new objective of integrating thrust vectoring and supersonic persistence into the design.

One of the ASD research efforts was the conceptual design of a vertical-attitude, tailless fighter with outstanding supersonic performance. Although the concept was tempered by interests in the conventional-attitude STOL designs in favor at the time, discussions between ASD and the staff at the Full-Scale Tunnel led to a joint investigation of the dynamic stability and control of a tailless thrust-vector control (TVC) configuration. The design featured an inlet above the cockpit similar to the North American F-107 of the 1950s. Control was provided by yaw and pitch vanes (similar to the ADEN-type

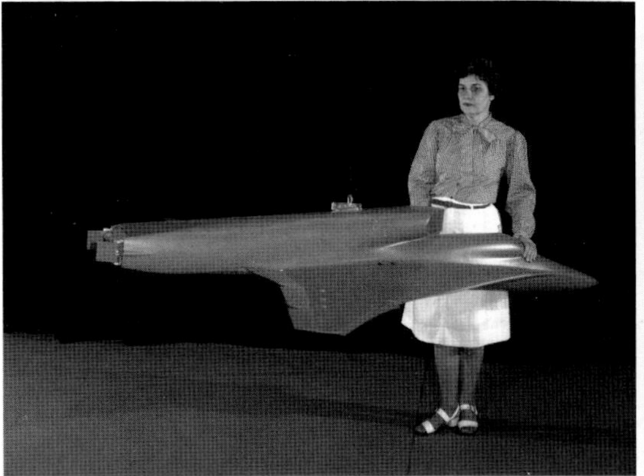

Project engineer Sue Grafton inspects the free-flight model of a hypothetical supersonic tailless V/STOL fighter tested in the Full-Scale Tunnel in 1985. Flight tests of the tailless thrust-vector control model were extremely impressive. In the right photograph, the model is flying in a fully controllable condition at an angle of attack of about 80°. Note the nose boom with angle-of-attack probe used for feedback of flight data for artificial stabilization. (NASA L-85-7388 and NASA L-85-9308)

arrangement used on the X-29 model, discussed in the previous chapter) in the exhaust of a simulated single-engine aircraft, and wing leading- and trailing-edge surfaces were used for roll control. A nose boom was used for feedback of angle-of-attack and sideslip data, which provided inputs to control laws generated within the balcony-based computer in the Full-Scale Tunnel. The flight tests were a tribute to the sophistication used in the 1990s for control-law simulation in the Langley testing technique. Pitch control included inputs to the wing trailing-edge controls and pitch vanes based on trim requirements, pitch rate, and angle of attack. Roll control inputs to the wing ailerons were based on roll trim, roll rate, and angle of sideslip. The yaw control inputs to the vectoring vanes were based on roll rate, yaw rate, and angle of sideslip.

Flight tests of the model in 1985 were nothing less than spectacular. Joe Johnson took great satisfaction in remarking that even a barn door could be flown with thrust vectoring.

Flight tests of a model of a generic supersonic fighter were conducted as part of a cooperative study between Langley and McDonnell in 1990. The thrust-vectoring vanes used by the model cannot be seen in this view, but the unconventional tiperons can be distinguished at the wingtips. (NASA L-90-08317)

Supersonic Persistence

In the mid-1980s, the NASA Langley Research Center and the McDonnell Aircraft Company initiated a cooperative research program to develop a low-speed design database for supersonic-cruise configurations. A wide variety of fuselage shapes and vertical-tail geometries were evaluated, including advanced-control concepts. Early phases of the study involved the evolution of candidate supersonic-wing designs with efforts by Langley organizations involved in high-speed aerodynamic research. The staff at the Full-Scale Tunnel participated in the effort with static and dynamic force tests for a large variety of configurations, followed by free-flight evaluations of a candidate design.[16]

The supersonic persistence fighter model had a 65° arrow wing, twin vertical tails, and a canard. It was also equipped with unconventional controls, including deflectable wingtips (tiperons) and pitch and yaw thrust vectoring. Two multiport ejectors supplied with compressed air were used to generate thrust, and secondary air from the model's engine inlets was entrained and mixed with the high-pressure air from the ejectors. The control laws devised for the flight control system programmed on the digital flight computer efficiently blended the responses of the conventional and unconventional control surfaces with emphasis on high-angle-of-attack flight conditions.

The advanced supersonic fighter exhibited good flying characteristics and was flown up to an angle of attack of about 80°. The highly successful free-flight test indicated that it was possible to effectively blend conventional and unconventional controls to achieve carefree maneuvering well into the post-stall angle-of-attack region.

Forebody Blowing for High Angles of Attack

The need for large yawing moments for coordinated rolling maneuvers at high angles of attack had been pursued at the Full-Scale Tunnel using two major approaches: mechanical systems (deflectable forebody strakes) and pneumatic systems (forebody blowing). Mechanical systems had progressed from initial concept development to flight demonstrations with the HARV research aircraft, as previously discussed. Pneumatic forebody controls had been investigated in cooperative studies between the Air Force and the Dryden using the X-29 forward-swept-wing flight demonstrator. The X-29 investigation used a system where the pilot manually opened and closed valves to control blowing on the forebody. Although the experiment demonstrated the validity of wind tunnel–derived yawing moments obtained with forebody blowing, it did not encompass the issues of integrating advanced-control system features.

In the mid-1990s, the staff of the Full-Scale Tunnel demonstrated forebody blowing for yaw control at high angles of attack when it was integrated into a flight control system for stability augmentation as well as control. Free-flight

This generic free-flight model was used in 1995 to demonstrate the integration of advanced control laws and fuselage forebody blowing for augmentation of yaw control at high angles of attack. Note the wingtip boom used for measurement of flow conditions for feedback in control laws. (NASA L-95-04112)

model tests for the first-ever flight demonstration of the integrated control system used a generic fighter configuration equipped with conventional, thrust-vectoring, and pneumatic-forebody controls. The free-flight model computer used at the Full-Scale Tunnel was programmed with appropriate control laws, and the effectiveness of pneumatic forebody controls was demonstrated without the use of rudders or thrust vectoring for additional yaw control.[17]

The free-flight tests were preceded by conventional static and dynamic force tests in which the relative location and configuration of slotted nozzles on the fuselage forebody were determined. During the free-flight tests, detailed studies were made of the possible impacts of flow lags involved with a pneumatic controller of this type. The model was successfully flown with forebody blowing and without rudder or thrust-vectoring yaw controls to angles of attack of about 45°.

This generic in-house study was typical of fundamental engineering research conducted at the Full-Scale Tunnel using free-flight models and a variety of associated test techniques to establish a foundation for future designers of advanced military aircraft.

Helping Preserve the Eagle: The F-15

Researchers at the Full-Scale Tunnel made significant contributions to the Air Force F-15 Eagle for over 30 years. The research activities began with the initial development of the aircraft and continued after the airplane was deployed to operational organizations and new variants emerged for new missions. Free-flight model tests and associated force and moment tests were conducted to evaluate the high-angle-of-attack characteristics of the configuration during the early phases of development in the 1970s.[18]

As frequently happened in NASA's support activities for high-performance aircraft, operational aircraft configurations returned to NASA wind tunnels for additional support when changes in the aircraft's design mission occurred, when configurations were significantly modified, or when unanticipated problems arose during normal operations. In addition, the high-angle-of-attack characteristics of some configurations had become well known (and verified by flight experience), which made them well suited as test subjects for assessments of advanced technologies in cooperative research programs with industry and DOD.

In the mid-1990s, the team at the Full-Scale Tunnel was requested to participate in an Air Force project known as Keep Eagle, which focused on the high-angle-of-attack behavior of the F-15E Strike Eagle variant of the F-15 design. The program was initiated in January 1993 as the result of concern over several incidents of unintentional loss of control and departures of F-15Es from controlled flight at high angles of attack. Objectives of the program were to implement enhancements to the F-15E that would improve pilot situational awareness, departure resistance, and spin recovery. Full-scale flight tests of an F-15E were scheduled in mid-1994, and subscale model tests were requested for several reasons, including documenting the high-angle-of-attack stability and control characteristics of the basic F-15E configuration, the configuration with external stores, and the configuration with altered fuselage-forebody geometric characteristics. In addition, flow surveys in the wake behind the aircraft at high angles of attack were requested prior to full-scale flight tests to mitigate risk in the design of an emergency spin-recovery parachute system.

Under the supervision of Sue Grafton, the 23-year-old, 1/10-scale F-15A model that had been previously used for free-flight tests in the 1970s was updated with configuration changes to represent the F-15E for testing in the Full-Scale Tunnel. The modification required many removable pieces, external stores, nose-tip shapes, store pylons, antennas, refueling probes, conformal fuel tanks, and surface protuberances for the forebody. Free-flight tests were not required, but a wide range of force and moment measurements in static and dynamic force tests were made. The results were used as inputs to piloted-simulator studies to assess the high-angle-of-attack behavior of the airplane and to modify its flight-control system to enhance departure resistance. Another challenging aspect of the test program was the construction of a grid to be used behind the model to measure total and static pressures and help understand the flow pattern at different distances behind the model.

Langley updated its 23-year-old F-15 free-flight model to respond to an Air Force request for high-angle-of-attack testing in support of the F-15E Keep Eagle program. The photograph shows the 1/10-scale model during tests in the Full-Scale Tunnel in 1994. (NASA EL-1996-00064)

The results of the study and subsequent piloted-simulator studies by McDonnell Douglas showed that the improved blending of roll and yaw control significantly enhanced the maneuverability of the F-15E at high angles of attack. Several members of the Vehicle Dynamics Branch, including Dana Dunham, Sue Grafton, Dan Murri, Mike Fremaux, and Ray Whipple, received individual letters of appreciation from the Air Force as members of the Keep Eagle Integrated Product Team (IPT).[19] Dana Dunham also received a certificate of commendation from the Air Force for management contributions to the project. The Keep Eagle Team was awarded the coveted Air Force Association's National Test Team of the

Year Award in 1995 in recognition of its superior effort in this critical project. The team's outstanding efforts significantly enhanced the F-15E operational utility and removed the requirement for special operational restrictions. In addition, the team completed the project on schedule and below cost.[20]

In other interesting activities of the 1990s, multiyear cooperative investigations with the Air Force Wright Laboratory and McDonnell Douglas were conducted to determine the effectiveness of several forebody controls for the F-15E. The control concepts included mechanical concepts such as the deflectable strakes used on the NASA F/A-18 HARV, rotating-radome devices, and pneumatic concepts such as slotted-nozzle blowing. All the concepts were designed to improve maneuverability at high angles of attack by increasing yaw control where conventional rudders become ineffective.

Air Force Super Fighter: The F-22

YF-22

Emerging advanced fighter threats from the Soviet Union in the early 1980s resulted in the Air Force developing a requirement for a new air-superiority fighter to be known as the Advanced Tactical Fighter. A Lockheed-Boeing–General Dynamics team and Northrop–McDonnell Douglas team were selected in October 1986 to develop prototypes to be known as the YF-22 and YF-23. The Air Force asked that Langley provide support for the YF-22 and YF-23 development programs on an as-requested, equal basis. Within their program funds and interests, the Lockheed team requested Langley's support in the areas of supersonic-cruise performance, high-angle-of-attack dynamic stability and control, and spin-tunnel testing; meanwhile, the Northrop team requested high-angle-of-attack and spin-tunnel testing.

The YF-22 and YF-23 were conducted under tight security that required extraordinary efforts from the Langley Security Office to secure and patrol the massive Full-Scale Tunnel building. Throughout its history, the tunnel operations demanded a secure site on occasions, but in the 1990s, a large influx of classified projects raised the requirements to new levels. The response of the Langley organizations to security considerations was remarkable, and the Nation's cutting-edge technologies were securely protected. All data and most photographs and documents relating to the tests were gathered by the manufacturers, and no documents were retained by NASA. As a result, there are no NASA reports of the test results of the YF-22 or YF-23 in the tunnel.

In mid-1989, static and dynamic tests of a model of the YF-22 began in the Full-Scale Tunnel to determine stability and control at high angles of attack. Data were also obtained to develop aircraft control laws. Free-flight tests to determine low-speed longitudinal and lateral directional response characteristics and departure resistance were also conducted. The model exhibited outstanding behavior at high angles of attack, and the Lockheed team was encouraged about upcoming full-scale flight tests. Testing of the Northrop YF-23 configuration in the Full-Scale Tunnel consisted of static and dynamic force tests.

The first flight of the YF-23 prototype occurred on August 27, 1990, and the first YF-22 flight took place on September 29, 1990. The flight-test phase of the competition ended on December 28, 1990, and the YF-22 was announced as the winner in April 1991.

Langley's overall contributions to the development and demonstration of the YF-22 were cited in a letter of appreciation from the Lockheed vice president for ATF, James A. "Micky" Blackwell, Jr., to Langley's Center director, Richard H. Petersen, in March 1991. Blackwell praised Langley's efforts and support of the YF-22 and cited the accuracies of Langley wind tunnel predictions and the dramatic demonstrations of the performance and agility of the prototype at high angles of attack:

> The highlight of the flight test program was the high-angle-of-attack flying qualities. We relied on: aerodynamic data obtained in the Full-Scale Wind Tunnel to define the low-speed, high-angle-of-attack static and dynamic aerodynamic derivatives; rotary derivatives from your Spin Tunnel; and free-flight demonstrations in the Full-Scale Wind Tunnel. We expanded the flight envelope from 20° to 60° angle-of-attack, demonstrating pitch attitude changes and full-stick rolls around the velocity vector in 7 calendar days, December 11 to December 17. The reason for this rapid envelope expansion was the quality of the aerodynamic data used in the control law design and pre-flight simulations.[21]

Free-flight model testing of the YF-22 in the Full-Scale Tunnel in 1991 indicated that the aircraft would have outstanding high-angle-of-attack characteristics, which were subsequently verified by flight tests of the full-scale aircraft. (NASA L-91-13622)

F-22

The geometry of the final F-22 configuration changed significantly from the YF-22 prototype. Specifically, the wingspan was increased, the engine inlets were shortened (based in part on results from tests in the Full-Scale Tunnel), the wing leading-edge sweep was decreased, the vertical tails were reduced in area and moved aft, and the horizontal-tail surfaces were reconfigured. These changes were considered significant enough to warrant specific tests of the F-22 in the same unique Langley facilities as the YF-22. However, during the F-22 program, the Government changed its procurement procedure. In the past, the DOD had basically dictated to industry the types of testing and information that industry would provide to NASA for independent analysis and testing of new aircraft concepts. At the same time, the DOD would provide funding and NASA would supply expertise and facilities for testing during the development program. Under the new procurement regulations (known as the "total-system" procurement process), the decision about NASA interfaces was left up to industry.

A force-test model of the production version of the F-22 Raptor was installed in the Full-Scale Tunnel for conventional static force tests in 1992. (NASA L-95-04112)

As the F-22 development program was planned, Lockheed Martin made the decision to fund only static and dynamic force tests of a large force-test model in the Full-Scale Tunnel in 1992—a free-flight model was not requested or built. Instead, only high-angle-of-attack static and dynamic force tests were conducted in the Full-Scale Tunnel to obtain data for analysis and use in simulators. The scope of testing included power-on tests with open inlets to determine the effect of power on characteristics at high angles of attack.

The first production aircraft of the F-22 was delivered to Nellis Air Force Base, NV, on January 7, 2003—over a decade after the testing in the Full-Scale Tunnel had been completed.

End of a Tradition: The F/A-18E/F

The Last Navy Test

The Full-Scale Tunnel had been closely associated with the Navy and its programs since the first tunnel test in 1931, and it had made critical contributions to fleet aircraft for almost 65 years. As discussed earlier, the Navy had dominated the tunnel's test schedule until early in WWII, when Army protests over the situation brought more equality in test time between the Navy and the Army Air Forces. In modern times, the relationship between the staff of the Full-Scale Tunnel and their peers within the Naval Air Systems Command was special, formed on mutual respect and a team spirit fostered by numerous examples of "getting the job done."

As will be discussed, 1995 brought about a NASA decision to deactivate the tunnel and terminate NASA's official onsite technical presence at the facility. The last chapter of the bond between the Navy and the Full-Scale Tunnel sadly ended with a project in support of the F/A-18E/F during the aircraft's development program.

The tunnel's staff had participated in the entire life history of the Navy's F/A-18, from early tests during the development of the YF-17 in early 1973, through the "legacy" F/A-18A/B and C/D Hornets in the mid-1970s, to the productive NASA High-Angle-of-Attack Technology Program of the 1980s and 1990s, which had been focused on the F/A-18 configuration. As a result of detailed experiences with the high-angle-of-attack characteristics of the configuration, the staff was in a position to rapidly respond to a Navy request in 1992 to support the development of a new version of the aircraft to be known as the F/A-18E Super Hornet.

The F/A-18E/F Super Hornet is a larger version of the F/A-18C/D Hornet with extended mission capabilities. The E/F version is roughly 25 percent larger than the C/D versions, with a 25 percent increase in operating radius and a 22 percent increase in weapons-load capability. To accommodate the growth in aircraft size, a number of changes were required, including a redesign of the wing-fuselage strake (leading-edge extension). Redesigning the LEX was the job of a 15-member team of industry, DOD, and NASA experts, which included Langley researcher Daniel G. Murri of the Flight Dynamics Branch. Murri had accumulated a vast knowledge of the high-angle-of-attack characteristics of the earlier F/A-18s during tests at the Full-Scale Tunnel and had led numerous studies of the HARV in the NASA HATP.

The review team, which was active for the first 6 months of 1993, initially explored small modifications to the size and shape of the original F/A-18C LEX to help provide the required lift and improve lateral directional stability of the F/A-18E.[22] However, subsequent wind tunnel tests showed that this incremental approach would not be successful and that much larger changes to the LEX configuration would be required. Based on his prior research with other configurations, Murri proposed more radical LEX candidates that would potentially satisfy these requirements. One of the LEX configurations recommended by Langley was accepted for further refinements and met all design goals. This configuration was the basis for the final design adopted as the wing LEX configuration for the production F/A-18E/F.

The 0.15-scale free-flight model of the F/A-18E prepares to undergo conventional static force tests in the Full-Scale Tunnel in September 1995. As part of an extensive investigation that included static and dynamic force tests as well as free-flight model studies, this test was the last military project conducted in the tunnel under NASA management. (NASA L-95-5377)

Stability and control characteristics of the F/A-18E/F at high-angle-of-attack flight conditions were evaluated in numerous wind tunnel tests at Langley. In the Full-Scale Tunnel, Gautam Shah and Sue Grafton led a combination of static, dynamic, and free-flight tests conducted to define and develop a database for the high-angle-of-attack aerodynamic, stability, and control characteristics of the aircraft. In addition to tests of the basic configuration, investigations were conducted to study the impact of fuselage-mounted and wing-mounted stores on aerodynamic and stability characteristics, to assess aerodynamic damping characteristics, and to assess the magnitude of thrust-induced aerodynamic effects on the configuration. Free-flight tests were also conducted to provide confirmation of predicted dynamic stability and control characteristics. The database generated by tests in the Full-Scale Tunnel was used by McDonnell Douglas and the follow-on Boeing organization, NASA, and the Navy to conduct flight-simulation studies and to aid in the development of the F/A-18E/F flight-control system.

The staff of the Flight Dynamics Branch was based at the Full-Scale Tunnel and the Langley Spin Tunnel, and although this document is limited to activities conducted in the Full-Scale tunnel, the wide expertise, facilities, and testing techniques of the organization also played key roles in the development of the F/A-18E/F. For example, Mark Croom led critical drop-model investigations of the departure, spin susceptibility, and post-stall behavior of the configuration; Charles M. "Mike" Fremaux led tests in the Langley Spin Tunnel to define developed spins and spin-recovery behavior and to develop the emergency spin-recovery parachute required for flight tests of the full-scale airplane; John Foster contributed analyses of an unexpected "falling leaf" post-stall condition that had been encountered with the earlier Hornets and assisted the Navy and Boeing in eliminating the problem for the Super Hornet; and Bruce Owens led the application of free-to-roll wind tunnel test techniques to help resolve the shortcomings in methodology for predicting a transonic wing-drop problem that plagued the F/A-18E/F in its early development.

The end of testing of the F/A-18E/F in the Full-Scale Tunnel in 1995 was a landmark point in the fabled history of the tunnel and its 64-year support for the U.S. Navy. Arguably, the staff of the Full-Scale Tunnel had participated in more testing and accumulated more knowledge of the high-angle-of-attack characteristics of the F/A-18 aircraft series than any other NASA organization.

Supersonic Civil Transports Revisited

The NASA High-Speed Research (HSR) Program

In his 1986 State of the Union address, President Ronald Reagan discussed visions of a Mach 25 hypersonic civil transport known as the "Orient Express." The results of industry and NASA studies concluded that a Mach 25 transport was not economically feasible, but a supersonic transport with a cruise Mach number between 2.0 and 3.2 was technically realizable if the well-known environmental issues of noise, sonic boom, and emissions could be solved. With congressional support, NASA initiated the NASA High-Speed Research (HSR) Program in 1990 to identify and develop solutions to the many environmental concerns surrounding a second-generation supersonic transport superior to the Concorde. Initially, the research focus was on the most important barriers to acceptance of supersonic transports, including issues on depletion of Earth's ozone layers, excessive airport and community noise, and the unacceptable sonic boom. The hypothetical aircraft would carry 300 passengers at a cruise speed of Mach 2.4 and could cross the Pacific or Atlantic in less than half the time possible on existing modern subsonic, wide-bodied jets—at an affordable ticket price (estimated at less than 20 percent above comparable subsonic flights)—and be environmentally friendly.

Researcher Dave Hahne poses in 1992 with a large-scale model of a Mach 2.2 supersonic transport concept designed by the Douglas Aircraft Company. The test produced a large database of aerodynamic data for low-speed performance, stability, and control, as well as pressures on the wing leading edge. The primary goal of the test was to compare the aerodynamic performance of a 20-year-old constant-chord leading-edge flap with that obtained with a tapered flap designed by computational methods. (NASA L92-12858)

Disciplinary research on supersonic transports was reenergized during the early 1990s, and the continued concern over low-speed performance, stability, and control of such designs led to another series of test activities at the Full-Scale Tunnel. Initially, the activities used generic configurations or those from the earlier NASA SCR program.

In 1992, a 1/10-scale model of a Mach 2.2 supersonic transport developed and tested as part of the SCR activities in the mid-1970s was tested in the Full-Scale Tunnel in cooperation with the Douglas Aircraft Company.[23] In addition to measurements of aerodynamic data of a general nature, a primary objective of the study was to compare the performance of two wing leading-edge flap configurations—a constant-chord flap from the original design

Activities of the NASA High Speed Research Program at the Full-Scale Tunnel included testing of the Reference H supersonic transport concept designed by Boeing for the HSR Program. Shown on the left is a cranked-wing model used to assess planform effects; on the right, Bruce Owens inspects a static force-test model used in a study of the application of boundary-layer control to suppress the formation of wing leading-edge vortices. (NASA L-92-07792 and NASA L-93-5277)

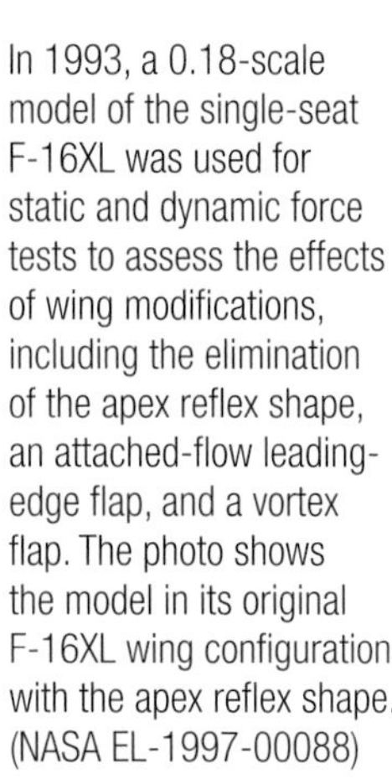

In 1993, a 0.18-scale model of the single-seat F-16XL was used for static and dynamic force tests to assess the effects of wing modifications, including the elimination of the apex reflex shape, an attached-flow leading-edge flap, and a vortex flap. The photo shows the model in its original F-16XL wing configuration with the apex reflex shape. (NASA EL-1997-00088)

and a new tapered flap designed by the Douglas Aircraft Company using computer codes developed at Langley.

Conclusions reached in the study included the observation that although the computational code provided a good indication of potential untrimmed lift capabilities, experimental tunnel data improved the results. Deflection of leading- and trailing-edge flaps provided improvements in wing performance compared to the undeflected case and also improved crosswind-landing performance.

In order to focus the research efforts, the HSR participants agreed to a baseline Mach 2.4 vehicle concept known as Reference H designed by Boeing. Extensive wind tunnel testing of the configuration was conducted in all Langley's tunnels including subsonic, transonic, and supersonic facilities. At the Full-Scale Tunnel, studies of the Reference H concept included wing planform studies in 1992 and a 1993 study of using suction-type boundary-layer control to suppress the formation of wing leading-edge vortices for enhanced performance at moderate angles of attack. In particular, the objective of the investigation was to maximize lift-to-drag ratio while reducing nose-up pitching moments. The model featured a cranked-delta wing and inboard- and outboard-leading-edge flap segments.[24]

The single-seat F-16XL (Ship 1) had been used for supersonic laminar flow experiments at Dryden. The aircraft was brought to Langley for proposed high-lift/takeoff-noise flight experiments in 1993. The activity was cancelled after the aircraft had been painted to enhance results of planned flow-visualization experiments. (NASA EL-2000-00568)

Another notable test program within the interests of the HSR Program focused on determining the effects of wing leading-edge configuration on a 0.18-scale model of the F-16XL Ship 1 aircraft.[25] As mentioned previously, the original development of the F-16XL required a modification in the wing apex area to obtain satisfactory low-speed flight characteristics. In particular, a reflex shape replaced the original continuous apex configuration in order to eliminate longitudinal instability in the form of pitch-up at moderate angles of attack. Although it improved the low-speed behavior of the aircraft, the modification resulted in increased drag at supersonic cruise conditions.

In order to build a database for future supersonic wing applications, a new F-16XL model was fabricated with a continuous 70° leading-edge sweep on the inboard portion of the wing. Wing concepts tested included an integral attached-flow leading-edge flap and a deployable vortex-flap concept. In addition to static force tests, a second entry in the Full-Scale Tunnel was used to measure aerodynamic damping characteristics in roll, yaw, and pitch.

As the HSR Program planning progressed in various technical disciplines, Joe Chambers proposed that the single-seat F-16XL (Ship 1) that had been used at Dryden for supersonic laminar-flow experiments be brought to Langley for flight tests to evaluate the effects of wing leading-edge devices on high-lift characteristics and noise during takeoffs. The aircraft was transferred to Langley in 1993 and prepared for initial flight tests when NASA Headquarters made the decision to eliminate the plan from HSR activities and transfer the aircraft back to Dryden.[26] The foregoing tunnel test program in the Full-Scale Tunnel would have served to support the flight experiments.

By late 1995, results in all technical discipline areas of the HSR Program continued to advance the state of the art, and a new updated baseline known as the Technology-Concept Airplane (TCA) was adopted and studied for a few years. However, in the late 1990s, Boeing's interest in continuing its level of effort in the HSR Program dramatically diminished because of high-risk technical issues, subsonic commercial transport business demands, and marketability issues. The company subsequently announced its withdrawal from the HSR effort, stating its perspective that supersonic transports could not reasonably be ready for the marketplace before 2020. In response to the loss of its major partner in supersonic transport research, NASA terminated the HSR Program in February 1999.

Project engineer Sue Grafton inspects the 1/5-scale force-test model of the HL-20 space-plane concept. As sometimes happens with dynamically scaled free-flight models, mandatory scaling procedures of the mass and inertia characteristics of the dense full-scale vehicle resulted in projected model flight speeds that were far in excess of the operating capabilities of the Full-Scale Tunnel. As a result, only static and dynamic force tests were conducted. The photograph on the right shows the HL-20 model during static force tests in the tunnel in 1990. (NASA L-89-03305 and NASA L-90-4915)

Hypersonic Vehicles

The HL-20

In the mid-1980s, the human space flight community at Langley pursued a space-plane concept known as the HL-20 Personnel Launch System (PLS) whose mission consisted of transporting human beings and relatively small cargo to and from low-Earth orbit. Approximately 29 feet long, the compact, lifting-body design would complement the Space Shuttle with safe, reliable, and relatively low-cost access to space. With its rear fins folded, the HL-20 could fit within the payload bay of the Shuttle. It was designed to play a major role in servicing Space Station Freedom and providing emergency return to Earth missions for its occupants, terminating in a horizontal landing on a runway. In 1990, a full-scale engineering research model of the HL-20 was fabricated by the North Carolina State University and North Carolina A&T University for studies of human factors such as seating arrangements, crew ingress and egress procedures (a payload of two crew members and eight passengers was envisioned), and equipment layout.

Several Langley wind tunnels were used in the aerodynamic development of the HL-20 from subsonic to hypersonic flight conditions.[27] At the Full-Scale Tunnel, compliance with dynamic-model scaling laws for a typical free-flight model of the relatively dense full-scale HL-20 indicated that the flying speed of such a model would be far beyond the rigorous speed constraints in effect at the tunnel. Therefore, the scope of support for the program consisted of static and dynamic force tests of a ⅕-scale model of the HL-20 to obtain data for analysis and as inputs for piloted simulator studies of the vehicle's flying characteristics.[28]

The HL-20 concept was not pursued into hardware development and was cancelled in 1993. However, the legacy of the HL-20 has lived on, and the design was pursued in 2005 as the Sierra Nevada Corporation's Dream Chaser (which was part of NASA's Vision for Space Exploration program) the Commercial Orbital Transportation Services Program, and the Commercial Crew Development Program. Orbital Sciences Corporation also proposed an HL-20 derivative known as Prometheus for the second round of the Commercial Crew Development Program funding, but it was not selected for funding. Both vehicles were envisioned to be launched atop a human-rated Atlas V launch vehicle.

Air-Breathing Transatmospheric Vehicles

Studies of the hypersonic vehicle referred to as the "Orient Express" in President Reagan's 1986 State of the Union address stimulated interest in proposed single-stage-to-orbit vehicles such as the X-30 National Aero-Space Plane. The NASP Program had continued until 1993, when it was terminated, but widespread interest developed (particularly at Langley with hypersonic configuration studies) to gather and collate aerodynamic databases across the operational speed range of such vehicles. Within the Flight Dynamics Branch, issues related to flying qualities at low subsonic speeds, takeoff, and landing were studied with three configurations in the 12-Foot Low-Speed Tunnel and the Full-Scale Tunnel.[29]

The low-speed design challenges for slender hypersonic vehicles include: avoiding significant trimmed-lift penalties that result in raising takeoff and landing speeds to unacceptable

The 9.5-foot-long Test Technique Demonstrator model is shown mounted to a dynamic force test apparatus to measure its aerodynamic damping in yaw in the Full-Scale Tunnel in 1991. An electronic strain gage mounted within the model measured aerodynamic forces and moments while the model was forced to oscillate over ±5° in yaw by a pushrod/flywheel system that was powered by an electric motor at the base of the strut. Historians should note that this photograph was graphically altered to depict a large-scale TTD model in the Full-Scale Tunnel for the cover of the NASA publication *Winds of Change*, by James Shultz. Such a test never occurred in the Full-Scale Tunnel. (NASA L-91-2241)

levels; controlling vortical flows shed by slender forebodies and highly swept wings so as to avoid detrimental effects on static and dynamic stability and control; and providing adequate lateral-control power to trim the vehicle during cross-wind landings.

Early studies within the branch focused on accelerator and wave-rider configurations while later studies investigated the characteristics of a distinct wing-body configuration. The 8-foot-long accelerator model had a 6° conical forebody and a truncated aftbody with a 70° swept wing. The wave-rider model, which was also about 8 feet long, had a sharp-edged lip where the upper fuselage overhung the underside of the body. The model was equipped with canard and wing surfaces having 63.5° of sweep, a vertical tail, and trailing-edge flaps that could be used as elevons for roll control. The wing-body configuration, also referred to as the generic Test Technique Demonstrator (TTD) configuration, was 9.5 feet long and incorporated a 75° clipped delta wing, wingtip vertical tails, and a small trapezoidal canard. The scope of tests included conventional static force and moment measurements, dynamic force tests to measure aerodynamic damping characteristics, flow-visualization studies of the vortical-flow phenomena, and free-flight tests of the TTD model. In the case of the TTD model, extensive studies were also made of how to simulate power effects of the propulsion unit and how to implement the system within the model.

The results obtained for all three configurations showed that low-speed stability and control characteristics for landing and takeoff conditions were strongly influenced by vortical flows, which produced complex aerodynamic phenomena that dominated both static and dynamic stability characteristics. The dynamic lateral-stability characteristics of the TTD during free-flight testing revealed that the configuration exhibited undesirable limited-cycle lateral oscillations (i.e., "wing-rock"), similar to previously discussed motions exhibited by highly swept reentry-vehicle models studied in the Full-Scale Tunnel in the late 1950s and early 1960s. The free-flight tests also demonstrated that a simple stability augmentation system could eliminate the problem.

The LoFLYTE configuration undergoes static force tests in the Full-Scale Tunnel in 1995. (NASA L-95-04052)

Subscale Flight Demonstrator: LoFLYTE

A wave-rider configuration uses its self-generated shock waves to enhance lift-to-drag ratio for hypersonic flight by riding, or "surfing," on the waves. Proposed configurations have included conical shapes with wings and blended-wing shapes with lower-surface engine inlets. In 1995, Langley joined with the Air Force to study the aerodynamic characteristics of a wave-rider configuration in a project known as the Low Observable Flight Test Experiment (LoFLYTE). The project resulted from Small Business Innovation Research funding from NASA, the Air Force, the Navy, and the National Aerospace Plane Joint Program Office. The configuration was designed by the Accurate Automation Corporation of Chattanooga, TN, and was optimized for a Mach number of 5.5 with a blended-wing-body shape and wing and tail surfaces that were swept 75°.

The vortical flows that dominate the aerodynamics of such highly swept configurations would be expected to dictate the dynamic stability and control characteristics of the vehicle. Investigations were therefore conducted in the 12-Foot Low-Speed Tunnel and the Full-Scale Tunnel to document the low-speed aerodynamic characteristics of the LoFLYTE configuration, with emphasis on an analysis of static stability and control and the complex vortical interactions exhibited by the design.[30]

A 0.062-scale model of LoFLYTE was tested in both wind tunnels during detailed studies of the effects of configuration variables such as location of the vertical tail and pitch control effectiveness of "tiperons."

An uncrewed, 8.3-foot-long, 70-pound LoFLYTE model powered by a small turbine engine was subsequently flown in 1996 at Edwards Air Force Base using advanced neural-network computer technology developed by Accurate Automation Corporation in its flight-control system. A LoFLYTE model is currently on display at the National Museum of the U.S. Air Force at Dayton, OH.

The Last NASA General Aviation Tests: Natural Laminar Flow Wings

Cessna 210 NLF

In the 1970s and 1980s, Langley conducted a broad research program to develop airfoils for use by general aviation aircraft ranging from relatively low-speed personal-owner airplanes to relatively high-speed (Mach 0.7) business jets and commuters. Richard Whitcomb had led the way with his development of special General Aviation Whitcomb (GAW) airfoils, and the airfoil group at the Langley Low-Turbulence Pressure Tunnel (LTPT) had generated a family of natural-laminar-flow (NLF) airfoils designed to significantly enhance the cruise performance of general aviation configurations. Of course, issues were raised regarding the impact of the new airfoil shapes on stalling characteristics of three-dimensional wings, control effectiveness of ailerons, and high-lift performance. Perhaps more challenging was the issue of whether laminar flow could be obtained in flight. Opponents of NLF pointed to the "overhyped" laminar-flow airfoil of the North American P-51 Mustang in WWII, since little laminar flow was apparently obtained in service condition at cruise speeds of over 400 mph. Advocates of the NLF airfoils stressed that the relatively low cruise speed of general aviation personal-owner aircraft and the advanced manufacturing techniques (especially smooth composite wings) might enhance the probability of obtaining laminar flow in flight.

In 1985, a cooperative test between Langley and the Cessna Aircraft Company was conducted in the Full-Scale Tunnel and in flight using a full-scale modified Cessna T-210 airplane.[31] The airplane's modified wing used the NASA NLF(1)-0414F airfoil and an increased aspect ratio compared to the production Cessna 210. The primary objectives of the test were to document the characteristics of the airfoil (including the effects of premature boundary-layer transition), determine the effects of power and flap deflections, and evaluate the effects of discontinuous drooped leading-edge modifications

The laminar-flow characteristics of the modified Cessna T-210 natural-laminar-flow airplane were investigated in the Full-Scale Tunnel in 1985. In the photograph, white residue indicating areas of laminar flow is seen on the upper surface of the right wing. (NASA L-85-10429)

The Cessna NLF airplane was flown in July 1985 with sublimating chemical techniques to visualize laminar flow in flight. Note the similarity of the white residue pattern with that obtained in the Full-Scale Tunnel. (NASA L-85-13291)

designed to enhance spin resistance. For this application, the drooped sections used another NLF airfoil.

In addition to the measurement of aerodynamic forces and moments using the tunnel external scale balance system, flow-visualization and boundary-layer studies were also conducted. Wool tufts were used on the wing upper surface for flow-visualization information when boundary-layer transition was artificially fixed near the leading edge of the wing. The tufts were removed from the model when free boundary-layer transition was studied. Boundary-layer transition was measured using both sublimating chemical and hot-film techniques.

Results of the Full-Scale Tunnel investigation showed that large regions of natural laminar flow existed on the wing that would significantly enhance the cruise performance of the configuration. Artificially tripping the flow to a turbulent condition did not significantly affect the lift, stability, or control characteristics. The addition of a leading-edge droop arrangement was found to increase the stall angle of attack at the wingtips and was considered to be effective in improving stall/departure resistance of the configuration without significantly affecting drag.

In flight, the NLF modification cut the drag of the wing by about 30 percent and produced an overall aircraft drag reduction of about 12 percent. Cessna did not produce production versions of the modified T-210, but the research was invaluable in designing a wing for the Cessna Citation Jet.[32] In addition, the results helped provide confidence in NLF and spin resistance for future generations of advanced general aviation aircraft such as the highly popular SR series of aircraft produced by Cirrus.

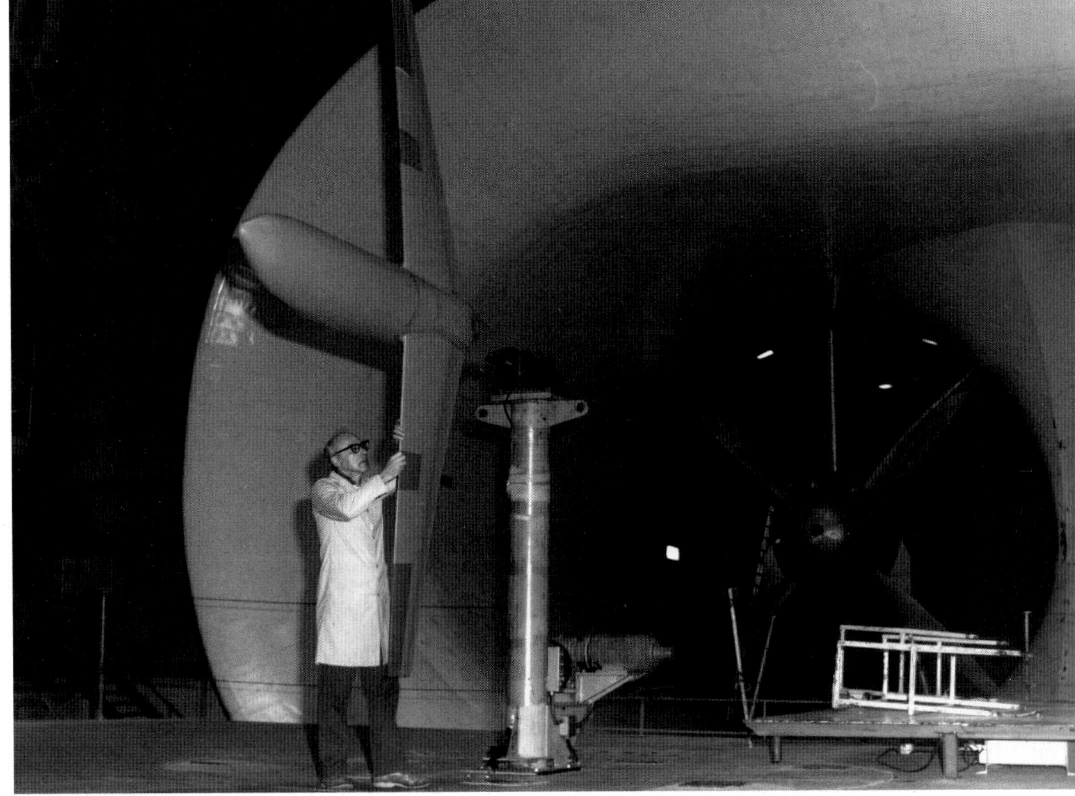

A full-span subscale force-test model of the Gulfstream Peregrine design is shown mounted to the forced-oscillation test apparatus in the Full-Scale Tunnel to determine the effects of leading-edge modifications on aerodynamic damping in roll for angles of attack near stall. The orange segments along the wing leading edges are discontinuous, drooped airfoil segments. The test setup is typical of that used to measure roll damping on free-flight-size models. (NASA L-84-10439)

Gulfstream Peregrine

The scope of testing for modern-day general aviation aircraft in the Full-Scale Tunnel also included low-speed, high-lift studies for advanced business-jet class wings that used NASA NLF airfoils for improved cruise performance. While much research on NLF airfoils had been conducted for enhanced performance, few studies had addressed the implementation and use of high-lift systems for wings incorporating these airfoils for takeoff and landing. In 1984, a cooperative test with Gulfstream was undertaken to investigate the low-speed stability and performance of the company's Peregrine business-jet wing, which incorporated

Frank Jordan (right) confers with technician Jim Staples during tests of a full-scale semispan model of the Peregrine wing in the Full-Scale Tunnel in 1987. Note the wool tufts attached to the wing and trailing-edge flap for flow-visualization studies. (NASA L-87-1236)

an NLF airfoil. The Peregrine was designed as a single-engine business jet and had made its first flight in January 1983.

The first tests of the configuration were studies of a full-span subscale model in the 12-Foot Low-Speed Tunnel and the Full-Scale Tunnel. After exploratory testing in the 12-Foot Tunnel, the model was tested in September 1984 in the Full-Scale Tunnel to evaluate its stall characteristics and, in particular, to evaluate the effectiveness of discontinuous leading-edge droop concepts similar to those developed in the 1970s during the NASA General Aviation Stall/Span Program, which sought to increase spin resistance. The testing included dynamic force tests to evaluate the aerodynamic damping in roll of the baseline wing and the wing with outboard and segmented leading-edge droop. The results of the dynamic tests showed that the subscale model exhibited unstable damping near the stall, but outboard and segmented droops eliminated the instability. Flow-visualization tests with sublimating chemicals indicated that the discontinuous droops did not adversely affect the laminar flow for the NLF wing.

Full-scale tests of the low-speed, high-lift characteristics of the Peregrine were conducted in the Full-Scale Tunnel in 1987 on a semispan wing model, with the main objective of evaluating and documenting the characteristics of the wing when it was equipped with a single-slotted flap system that was designed using a computer code.[33] In addition to the high-lift studies, boundary-layer transition effects were examined, a segmented leading-edge droop for improved stall/spin resistance was studied, and two roll-control devices were evaluated.

The Gulfstream Peregrine business jet made its first flight on January 14, 1983. Gulfstream had anticipated deliveries starting in 1987, but the program was cancelled for lack of orders.

Joe Johnson's Legacy: Advanced General Aviation Configurations

Joe Johnson was unquestionably the most innovative head of the Langley Full-Scale Tunnel. He adhered to the philosophy that aircraft design had a myriad of fundamental variables and

Researcher Bruce Owens inspects a free-flight model of an advanced turboprop aircraft in 1989. As head of the Flight Dynamics Branch, Joe Johnson inspired and challenged young engineers to conceive of innovative aircraft configurations that might offer improved performance and safety. (NASA L-89-04162)

that the shape of the airplane was dictated by the mission characteristics required. Johnson's experiences with extremely unconventional configurations during his days as a researcher at the 12-Foot Free-Flight Tunnel shaped his technical interests and enthusiasm in radical new designs. Young engineers in particular enjoyed his enthusiasm for new concepts. He expressed an unending interest in teaching co-op students about aerodynamics, stability, and control—virtually every student that worked at the Full-Scale Tunnel had a story to tell of something critical they learned from him. He often returned to a number of favorite mottos: get there first with fresh ideas; get 80 percent of the answers, then move on to the next opportunity; and do not get mired down with details.

With Johnson's technical guidance and leadership, the Flight Dynamics Branch embarked on a program of advanced general aviation designs in the mid-1980s, with the objective of identifying advantages and problems of unconventional designs and establishing first-ever databases for the configurations. The program included exploratory research on the dynamic stability and control characteristics of configurations with swept-forward wings, pusher props, NLF airfoils, canards, and other advanced features.

An example of an advanced configuration study conducted in the Full-Scale Tunnel in 1988 was an advanced turboprop business/commuter aircraft design that used twin-engine pusher propellers that were pylon-mounted on the rear of the fuselage.[34] The investigation included what had become traditional project elements, including conventional static force tests, dynamic force tests, and free-flight tests. In addition, tests were conducted using a free-to-roll test apparatus previously employed to investigate lateral oscillations near stall.

The model was stable and easy to fly for angles of attack below stall; however, at the stall the flight was quickly terminated because of an abrupt wing-drop against full corrective roll control. The force tests revealed that the wing-drop was caused by an abrupt asymmetric wing stall. The flight tests included a study of the effects of the NASA-developed outboard wing leading-edge droop concept, which resulted in a significant improvement in roll control and roll damping at extreme angles of attack. Even at post-stall angles of attack (i.e., angles greater than 20°), the overall flying qualities were acceptable with no significant stability or control problems evident. Engine-out trim conditions were conducted and analyzed, including the effects of asymmetric power on wing stall.

Riding a Tornado: Wake-Vortex Encounter

In the mid-1990s, NASA conducted research to enable safe improvements in the capacity of the air transportation system. One part of this program was the Terminal Area Productivity (TAP) program, which had the goal of safely increasing airport capacity levels during instrument flight conditions to those achievable under visual flight conditions. One critical element of the program, known as Reduced Spacing Operations, focused on the potential for reducing aircraft arrival spacing requirements—especially important when an aircraft is following a larger aircraft on approach. In such conditions, the danger of uncontrollable upsets to the following aircraft caused by encounters with the powerful

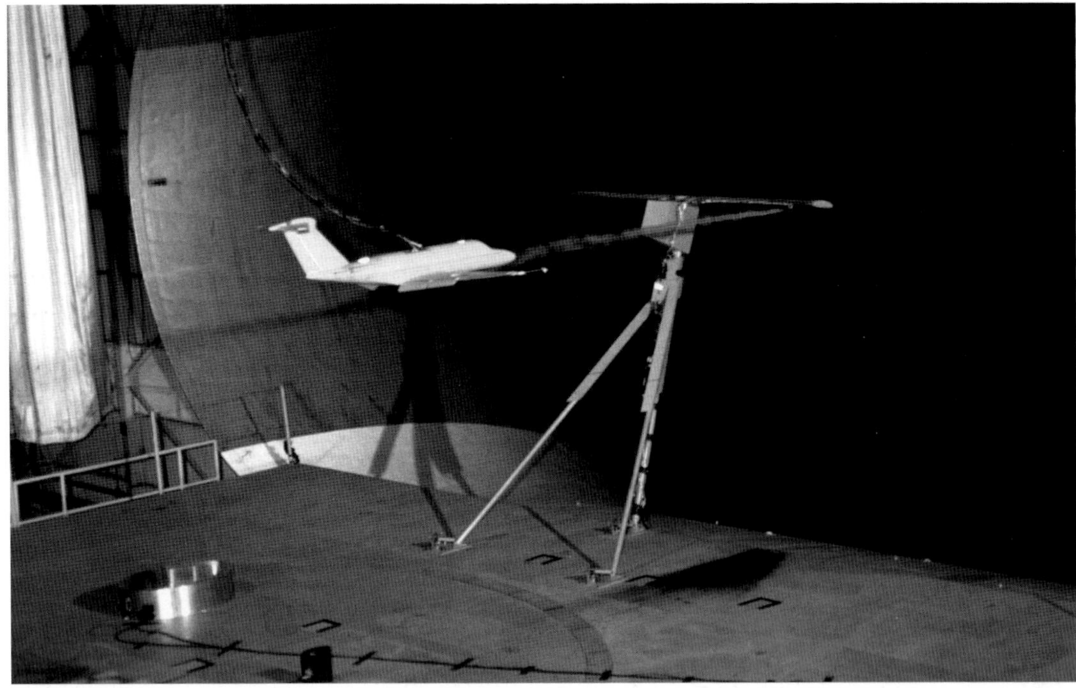

In 1994, the free-flight test technique was used in studies of encounters between a generic business-jet model and the wingtip vortex of a simple wing model. (NASA L-94-01277)

wake vortices of the preceding aircraft is of paramount concern. Experimental NASA research had been conducted in flight and in wind tunnels to provide information on wake-vortex flow fields produced by various aircraft configurations, and to investigate loads imposed by a vortex encounter.

In 1993, the staff of the Full-Scale Tunnel conducted the first attempt to study the dynamic response characteristics of a follower aircraft during wake-vortex encounters using the free-flight technique.[35] The study was viewed as a feasibility effort to determine whether the test technique could be a useful research tool in wake-vortex encounter research. Objectives included determining whether a free-flight model could be flown safely and maneuvered accurately into a wake-vortex flow field of specific strength; developing instrumentation techniques for measuring the position of the model relative to the vortex; developing techniques to estimate rolling moments imposed on the model during the encounter; and exploring qualitative evaluations for various encounter trajectories. In order to permit such research, flight-control system simulation in the free-flight computer had to be developed to enable the encounter scenarios.

A generic business-class jet airplane model was instrumented and flown in the vicinity of a wake vortex generated by a simple wing. The strength of the vortex was varied by adjusting the angle of attack of the generating wing. The study showed that the free-flight test technique was a viable and useful tool in the study of the wake-vortex encounters—combining vortex-flow fields, airplane flight dynamics, sensors, and flight-control requirements. The data obtained during the test included qualitative as well as quantitative results. Steady-state limits of controllability were documented as a function of vortex strength. By flying several vortex-encounter trajectories at high-vortex strengths, the data matrix was conducted of roll angle, roll rate, lateral velocity, and vortex-induced roll-rate acceleration. The results

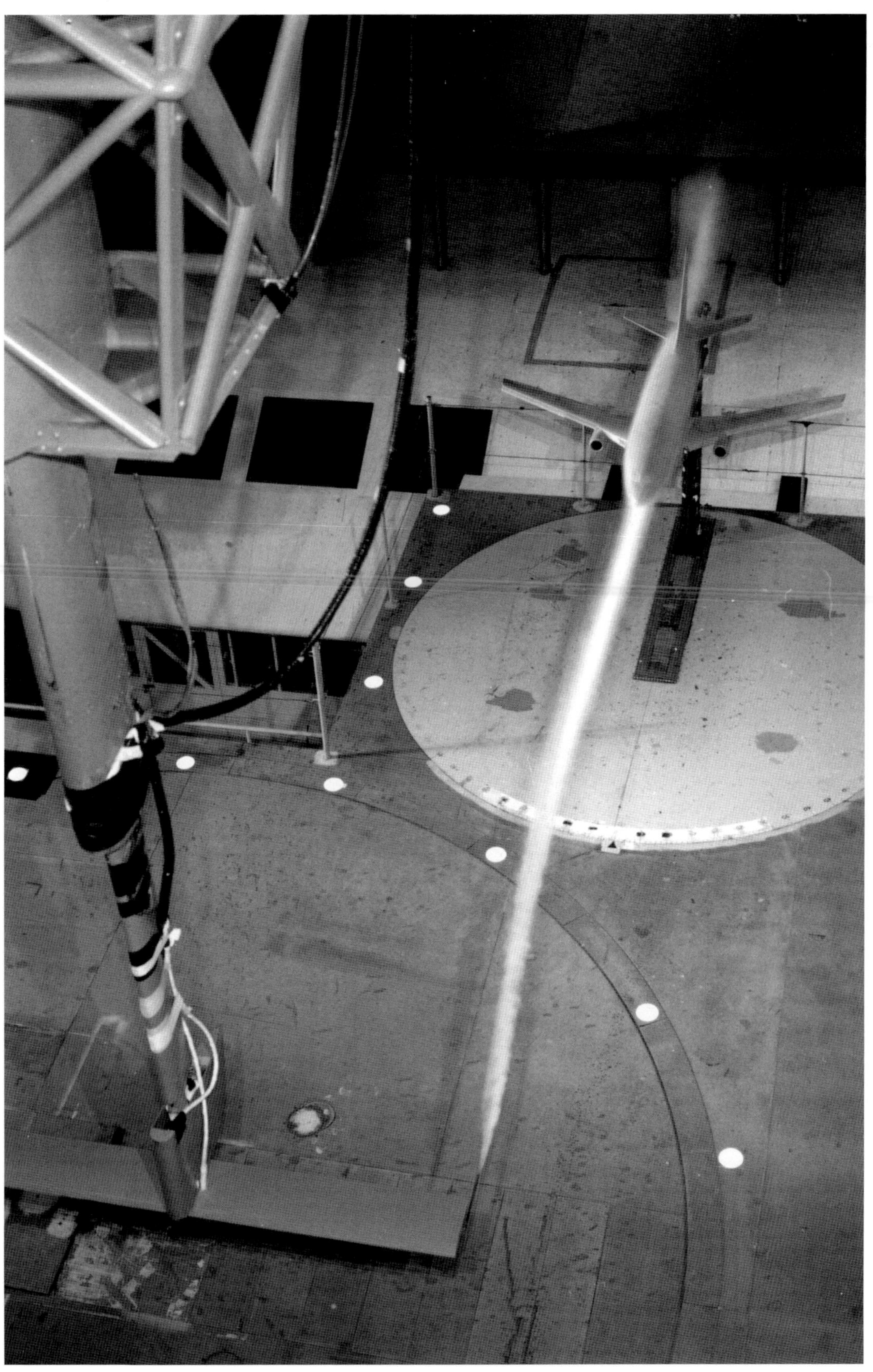

Static force tests were conducted to determine the magnitudes of forces and moments encountered by a B-737-100 transport model during simulated flight into a trailing vortex. Note the vortex-generating wing model mounted to the variable-position overhead survey carriage in the Full-Scale Tunnel. By changing the angle of attack of the wing and its position in the tunnel, researchers were able to map out characteristics of encounter scenarios. (NASA L-95-06044)

were viewed as very positive and encouraging for applications of the technique in additional experiments on wake-vortex encounters.

The wake-vortex upset hazard is such a dominant factor in establishing the minimum safe spacing between aircraft during landing and takeoff operations that NASA conducted a broad spectrum of static and free-flight wind tunnel tests, piloted-simulation studies, and aircraft flight tests to establish a first-order hazard metric and determine the limits of an operationally acceptable wake-induced upset. Conducted within the TAP program, the efforts included tests in the NASA Ames Research Center 80- by 120-Foot Tunnel and the Full-Scale Tunnel at Langley.

The tests conducted in the Full-Scale Tunnel to characterize the hazard included a free-flight test similar to that used in the generic model testing previously discussed, as well as a

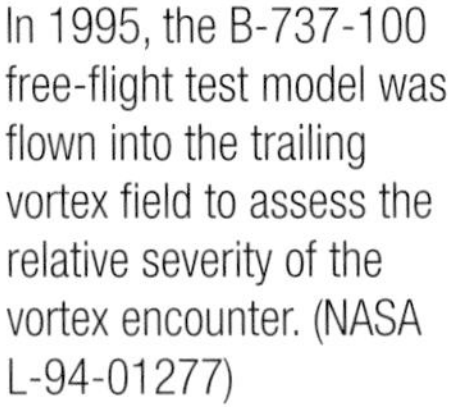

In 1995, the B-737-100 free-flight test model was flown into the trailing vortex field to assess the relative severity of the vortex encounter. (NASA L-94-01277)

static test setup in which the magnitudes of the forces and moments induced by the trailing vortex were measured on a sting-mounted model.[36] A vortex-generating wing was mounted to the movable tunnel survey carriage upstream of the test model and was positioned at various locations to map the effects of the vortex on the downstream model. A five-hole survey probe was also used to quantify the wake-velocity characteristics at the model's position.

After the tests of the generic business-jet model previously discussed were completed, a team of researchers at the Full-Scale Tunnel tested a 1/10-scale free-flight model of the NASA Langley B-737-100 research airplane.[37] The configuration was an ideal selection as a test subject, since high-fidelity simulator models and flight-test data were readily available for comparative evaluation studies. Detailed data were obtained for the configuration relative to response characteristics and vortex-flow fields, aerodynamic characteristics with tails on and off, and effects of entry trajectory. Plans included correlation of data from the Full-Scale Tunnel tests with results from full-scale aircraft flight tests, but the program's funding was canceled before the correlations could be addressed.

The B-737-100 free-flight wake-vortex study in 1995 would be the last test in the Langley Full-Scale Tunnel while the tunnel was under NASA's direct management.

Upheavals: International, Agency, Organizational, and Cultural

When the efforts of President Ronald Reagan resulted in the end of the Cold War in 1991, the sudden elimination of the Soviet Union as a powerful threat to the United States created an impact that filtered down to the missions of research establishments throughout the Nation. The advanced research and development efforts that were under way to develop weapons systems to combat the highly sophisticated Soviet equipment would come under close scrutiny, and there were many cancellations and redirections. Almost immediately, NASA shifted its interest in aeronautics from a balance between military and civil applications to an almost total focus on the challenges of civil applications, with particular emphasis on international economics, productivity of the domestic transportation system, and ensuring the safety of civil-aircraft operations. The long-time association between NASA and the military services dramatically diminished to the point that NASA no longer included military technology in its mission statement. This landmark change had a profound effect on the technical program conducted at the Full-Scale Tunnel and other Langley facilities.

The period from 1993 through 1995 would prove to be one of the most chaotic times in the history of the Full-Scale Tunnel. By early 1992, Richard H. Truly, a former naval aviator and astronaut, had served as NASA's Administrator for 3 years, adhering to a philosophy of a relatively stable organization and culture. Under Truly's leadership, the NASA aeronautics program continued in a business-as-usual mode, especially for projects related to support for military aircraft. However, in April 1992, Truly was succeeded by Daniel S. Goldin, who brought aggressive leadership and a call for change within the Agency. Challenged by major funding issues for the International Space Station and other space-related issues, Goldin brought a mandate to change the NASA culture. At Langley, the changes were comparable to the dramatic organizational and cultural changes brought about when Edgar M. Cortright was appointed Center director in 1968 with an objective of preparing the Center to manage large projects.[38]

At Langley, Goldin's call for change introduced significant cultural and organizational changes. The business world's philosophy of total quality management (TQM) was forcefully introduced into Langley organizations by Langley Director Paul F. Holloway on the premise that the quality of products and processes was the responsibility of everyone in order to meet or exceed customer expectations. Under this philosophy, the individual technical leaders in various disciplines at Langley would step back while their subordinates were empowered to make decisions. Following hard on the heels of this TQM effort, a concentrated focus on the nonaerospace applications of Langley's technology created new organizations and opportunities that diluted human resources from the aeronautics areas.

Following an extended study of organizational options by a special management-appointed team, in November 1993, Holloway unveiled a major reorganization of Langley that eliminated the distinct disciplinary organizational structure it had used for 76 years and replaced it with separate groups managing the scientific programs in aeronautics and space, the traditional research organizations, and technology transfers. Under the new organization, managers in the Aeronautics Programs Group would be given the responsibility of managing funding and technical thrusts of the research programs. The Research and Technology Group would retain the old disciplinary research organizations; however, the heads of those organizations would no longer control funding or the major technical direction of their subordinates and facilities. Instead, the technical specialists would seek sponsorship and funding from the Programs Group in a matrix-type interaction. The fundamental change in Langley's operations and technical management was particularly controversial within the aeronautics program, and it significantly impacted the direction of the program and its relationship with traditional industry and military partners for years to come.

When the new organization was implemented on February 20, 1994, Dana Dunham and her Flight Dynamics Branch were combined with another organization within the Flight Dynamics and Control Division (FDCD) under Dr. Willard W. Anderson and renamed the Vehicle Dynamics Branch. The mission of the FDCD included research in flight dynamics, aircraft and spacecraft guidance and control, flight management, and flight research.[39]

End of the Line: Decommissioning the Full-Scale Tunnel

The Message Arrives

In early 1993, Joe Chambers, then chief of the Flight Applications Division with responsibility for operations of the Full-Scale Tunnel, received a somber telephone call from Roy Harris, Langley's director of Aeronautics. Harris's message was short and to the point: NASA was going to close the Full-Scale Tunnel in order to cut costs by abandoning older facilities and consolidating diminishing workloads in newer wind tunnels.[40] In view of the unique testing techniques associated with the tunnel, it was agreed that the closure would not take place until it could be demonstrated that free-flight model testing was feasible in Langley's other large subsonic wind tunnel, the 14- by 22-Foot Tunnel. In addition, the transfer of specialized equipment such as the test apparatus used for dynamic (i.e., forced-oscillation) force tests would be addressed and appropriate plans defined for the relocation of the testing technique.

A New Home for Test Techniques

A team of researchers from both wind tunnels and from supporting organizations was formed to conduct an analysis of the feasibility, costs, and time required for the move. The study included considerations of all the testing techniques used in the Full-Scale Tunnel and potential alternatives, including piloted simulators and outdoor drop models. On October 7, 1993, the team briefed Roy Harris on the results of its study and recommended proceeding with a proof-of-concept (POC) free-flight test in the 14- by 22-Foot Tunnel.

The movement of the testing techniques would require a concentrated effort from the entire staff at the Full-Scale Tunnel, including its supporting technicians, as well as the staff of the 14- by 22-Foot Tunnel and other supporting organizations. Mark Croom, Dan Murri, and Sue Grafton conducted preliminary analyses of what had to be done and developed an approach for conducting the POC test. Sue Grafton led the transfer as a result of her extensive experience and knowledge of operations at the Full-Scale Tunnel.

An existing free-flight model of the X-29 forward-swept wing aircraft was used to evaluate the feasibility of conducting free-flight tests in the Langley 14- by 22-Foot Tunnel. The photograph shows the model in flight in the open-throat test section. The remote flightcrew was housed in a temporary enclosure to the right of the picture. (NASA L-94-03947)

The challenge of assembling a free-flight testing technique in the 14- by 22-Foot Tunnel was significant. A huge effort was required to move the cabling, air-compressor valves, flight computer, and model instrumentation from the Full-Scale Tunnel. In addition to a marked reduction in test-section size, the transfer team had to address first-order issues regarding the location of the remote pilots and how to secure the model in the event of loss of control.[41] Another significant factor was the compatibility of dynamically scaled models with the space constraints within the 14- by 22-Foot Tunnel; that is, if the free-flight models had to be smaller, could they be fabricated within the requirements of strict dynamic-scaling laws? The pneumatic control actuators used in previous free-flight models might not be appropriate for the smaller flight space, which might then require electronic actuators. Finally, the impact to the test schedule and changes to equipment at the 14- by 22-Foot Tunnel had to be resolved with the joint cooperation of that tunnel's staff.

In April 1994, the POC test was conducted using the existing free-flight model of the X-29 that had been used in previous years in the Full-Scale Tunnel to support the X-29 development program. The model was chosen because it had exhibited large-amplitude wing-rocking motions at high angles of attack, and similar results would validate the technique as far as accuracy was concerned.

The X-29 model was successfully flown in the POC investigation within the open-throat configuration of the 14- by 22-Foot Tunnel using a relatively small "house" enclosure for the flightcrew at the side of the test section. In contrast to the test setup at the Full-Scale Tunnel, the roll/yaw pilot was also stationed at the side of the tunnel but was provided cues of the model's attitude and motions via a television system.

On June 15, the team presented its summary briefing on results of the POC experience to Langley's senior staff. The "bottom lines" of the briefing were that free-flight testing could provide some meaningful technical results if the Full-Scale Tunnel was decommissioned, but significant limitations were inherent in testing in the 14- by 22-Foot Tunnel. Of special concern was the ability to fabricate certain models within required scale-factor relationships—a situation that would later be encountered for free-flight studies of the NASA-Boeing–Air Force Blended-Wing-Body configuration. From the team's perspective, retaining the free-flight capability at the Full-Scale Tunnel was a necessity.

Following the POC evaluation, the researchers had to return the free-flight equipment to the Full-Scale Tunnel to support ongoing commitments such as the B-737-100 free-flight investigation of the wake-vortex hazard mentioned in the previous section. As mentioned, the B-737 free-flight test was the last test conducted in 1995 before NASA decommissioned the tunnel.

The Curtain Falls

As expected, the POC testing and other events of 1994 did not change the decision of NASA Headquarters to decommission the Full-Scale Tunnel. The facility's old claim to fame—testing full-scale aircraft—had long been surpassed by the superior capabilities of the Ames NFAC complex, and the fact that its unique free-flight tests could be accomplished in the 14- by 22-Foot Tunnel (regardless of technical issues) were more than enough ammunition to support the decommissioning decision. By the fall of 1994, Langley's management had established a closure date of September 1995.[42]

Once the decision to deactivate the tunnel was made, a multitude of issues were addressed by management. Since the Full-Scale Tunnel had been declared a National Historic Landmark (NHL) in 1985, the Center was obligated to notify the Virginia Department of Historic Resources regarding the termination of operations and potential future plans for the facility.[43] In mid-1995, Langley's plans included the transfer of certain equipment to the new tunnel, but there were no plans to disrupt the historic major interior features. Due to the large size of the Full-Scale Tunnel and the presence of asbestos in siding and roofing panels, the facility would be maintained in a mothballed status rather than being demolished.

On October 27, 1995, the 64-year-old Full-Scale Tunnel was deactivated at a formal closing ceremony. Note the B-737 model mounted for the last test in the tunnel under direct NASA management. (NASA L-95-6377)

The notification also defined three possible scenarios that would require further consultation with the Department of Historic Resources:

1. Since the Full-Scale Tunnel (and all other NASA facilities) was "permitted" at Langley Air Force Base based on an agreement with the Air Force, the termination of research and testing could result in the Air Force requiring NASA to remove associated structures and utilities when the permitted activity ceased.
2. Private industry and/or universities or a consortium thereof might assume operation of the Full-Scale Tunnel (with the Air Force's agreement as landlord).
3. The Air Force at Langley Air Force Base could acquire the deactivated facility for Air Force purposes (shops, warehousing, etc.)

As part of its responsibilities and actions to preserve the unique history of the Full-Scale Tunnel, Langley contracted the Historic American Engineering Record of the Department of the Interior to record the Full-Scale Tunnel for the Library of Congress and for other documentary or interpretive applications.

In August, an announcement of the plans to deactivate the Full-Scale Tunnel on September 29, 1995, was released to the media by Langley's Office of Public Affairs. The press release stated that the future of the tunnel was uncertain, but there were no plans to demolish it or change its external appearance. Possible uses were under study, including the possibility of making certain components available to the Smithsonian National Air and Space Museum. Meanwhile, planning for a formal closing ceremony in October was under way.

The Closing Ceremony

After 64 years of continuous operation as NASA's oldest operational wind tunnel, the deactivated Full-Scale Tunnel was honored with a closing ceremony on October 27, 1995. About 200 attendees consisting of Langley employees, retirees, and well-wishers gathered in the west return passage of the tunnel and posed on bleachers for a group photograph reminiscent of the early NACA conferences. Center Director Paul Holloway served as master of ceremonies at the event, and Joe Chambers spoke on the history and accomplishments of the tunnel. Dr. Harry Butowsky, historian for the National Park Service, addressed the contributions of the facility to the aerospace community.

Aftermath

Movement of Equipment and Staff

The staff of the deactivated Full-Scale Tunnel began 1996 with unsettling questions regarding the future of the Vehicle Dynamics Branch within Langley's new project-oriented matrix organizational structure and the urgency to transition the testing techniques and tools from the Full-Scale Tunnel to the 14- by 22-Foot Tunnel. Researchers had to scramble to acquire funding and support from managers outside the branch (many of whom knew nothing of the technical expertise and history of the branch) while others were consumed by moving equipment to the new tunnel, and previous commitments to research projects had to be met in the 12-Foot Low-Speed Tunnel and with outdoor drop models and piloted simulators.

The summer of 1996 was especially hectic as interfaces with the staff members of the 14- by 22-Foot Tunnel, engineering organizations, fabrication units, instrumentation shops, and safety groups intensified. As leader of the transfer of the free-flight testing technique and associated forced-oscillation test equipment from the Full-Scale Tunnel, Sue Grafton was especially challenged by the task. The highly successful completion of the equipment transfer (including many upgrades and adaptations to the 14- by 22-Foot Tunnel) was a result of her outstanding expertise and dedication. As will be discussed in the next chapter, 10 years later she would be faced with the task of reinstituting free-flight tests back at the Full-Scale Tunnel.

A special free-flight investigation of the high-angle-of-attack behavior of the F/A-18E was conducted in the 14- by 22-Foot Tunnel in 1998 to establish correlation with results previously obtained from the Full-Scale Tunnel. (NASA L-2003-1661)

The Vehicle Dynamics Branch continued to reside in the Full-Scale Tunnel building during most of 1996 until October, when the entire staff was relocated to the Langley West Area to an office building near the 14- by 22-Foot Tunnel.

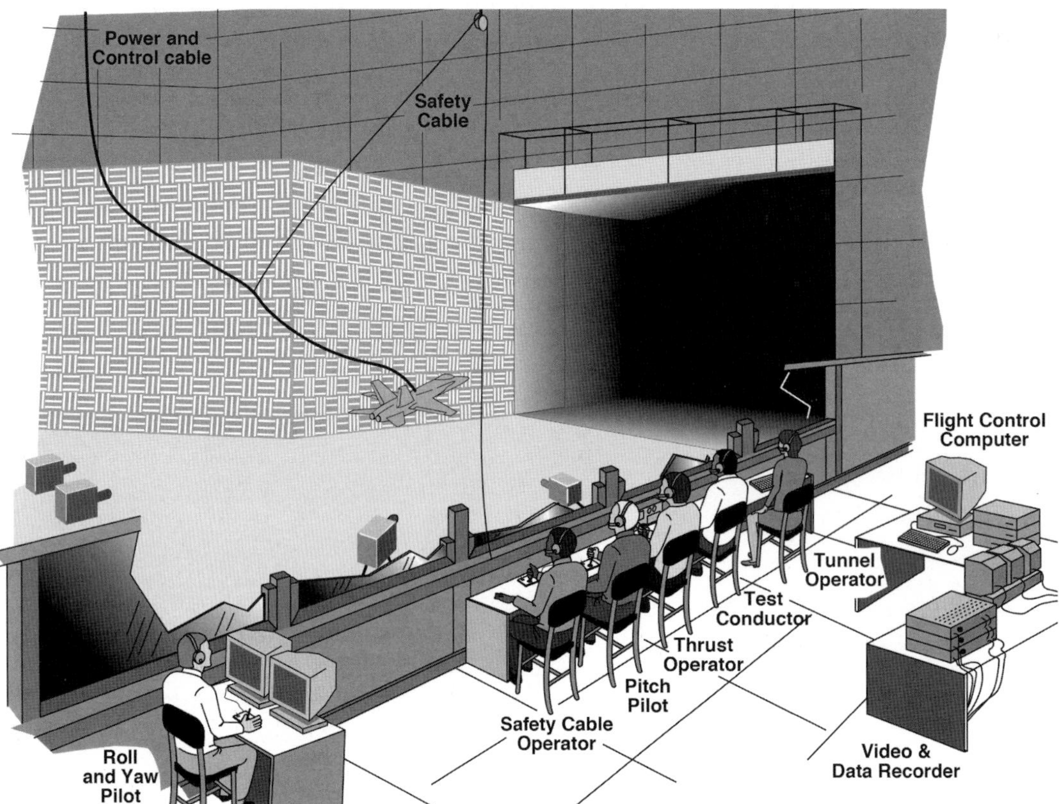

The final test setup used for free-flight tests in the 14- by 22-Foot Tunnel was very similar to the setup used in the Full-Scale Tunnel, with the notable exception of the location of the roll/yaw pilot, who used a television system for cues to fly the model. (NASA/Langley)

An Unhappy Customer

Throughout the life of the Full-Scale Tunnel, its NACA/NASA staff had enjoyed a close cooperative relationship with the majority of its customers—especially from the military community. After the Langley reorganization of 1994, those relationships had become somewhat strained as customers discovered they could no longer directly approach the head of the Full-Scale Tunnel for commitments involving the facility. In the past, the procedure had been as simple as having in-depth technical discussions with the branch head, who had the authority and responsibility to commit funds, workforce, and facility operations to the requested project. Instead, in the new organization the customer had to negotiate with a program manager who, in most cases, had no technical background in the subject of the proposed test but controlled the funds and commitment authority. Friction over this arrangement was apparent and had a decidedly negative impact on outside collaborations.

NASA's decision to deactivate the Full-Scale Tunnel came as a surprise to many uninformed customers who had established long-term relationships and dependency on the facility. Perhaps the most concerned organization was the Navy's Naval Air Systems Command (NAVAIR), which had sponsored numerous programs in the Full-Scale Tunnel and established special confidence in the results obtained in the Langley facility based on correlation with full-scale aircraft flight results.

NAVAIR leaders expressed great concern over moving the dynamic stability and control test techniques from the Full-Scale Tunnel to a new tunnel. Their concern was expressed as a lack of confidence in Langley's transfer of the testing techniques until correlation with previous results from the Full-Scale Tunnel could be established. In order to obtain correlation data, the Navy funded the fabrication of a new F/A-18E free-flight model for tests in the 14- by 22-Foot Tunnel.[44]

Results of the flight tests revealed that the 0.15-scale F/A-18E model was flyable in the 14- by 22-Foot Tunnel and that its dynamic stability characteristics were similar to the larger model flown in the Full-Scale Tunnel. However, the model was very difficult to control in the smaller test section of the 14- by 22-Foot Tunnel.

Gone but Not Forgotten

As the turbulence of the events in 1995 began to clear, the issue of the future plans for the deactivated Full-Scale Tunnel remained unclear. However, a plan was under way to maintain the tunnel in operation and extend the life of the Full-Scale Tunnel for another 14 years. As discussed in the next chapter, operations of the tunnel would be taken over by the Old Dominion University.

Endnotes

1. Paul F. Holloway, "Langley Endures, Thanks to You," *Langley Researcher News 9*, no. 15 (July 28, 1995); Robert Scott, "Scott Addresses Potential Closing," *Langley Researcher News 9*, no. 15 (1995); Herbert Bateman, "Bateman Speaks on Closure Bill," *Langley Researcher News 9*, no. 15 (1995). The subcommittee was chaired by Rep. Jerry Lewis of California and included Rep. Louis Stokes of Ohio and Rep. Tom Delay of Texas. Virginia was not represented. Of the full committee's 56 members, only one was from Virginia. See also Lee Rich, "The Potential Langley-Closing Bill Explained," *Langley Researcher News 9*, no. 15 (1995).
2. At the time, Gilbert was heavily involved in management of day-to-day operations of the NASA High-Angle-of-Attack Technology Program.
3. The strut system was overdesigned and never used in testing.
4. Lane Wallace, *Nose Up: High Angle-of-Attack and Thrust Vectoring Research at NASA Dryden 1979–2001*, (Washington, DC: NASA SP-2009-4534, 2009).
5. Flight tests of the Rockwell X-31 aircraft had demonstrated thrust vectoring in pitch a few weeks before the HARV demonstrated vectoring in pitch and yaw.
6. Jay M. Brandon, "Low-Speed Wind-Tunnel Investigation of the Effect of Strakes and Nose Chines on Lateral-Directional Stability of a Fighter Configuration," NASA TM-87641 (1986); Daniel G. Murri and Dhanvada M. Rao, "Exploratory Studies of Actuated Forebody Strakes for Yaw Control at High Angles of Attack," AIAA Paper 87-2557 (1987); Daniel G. Murri, Gautam H. Shah, Daniel J. DiCarlo, and Todd W. Trilling, "Actuated Forebody Strake Controls for the F-18 High-Alpha Research Vehicle," *Journal of Aircraft* 32, no. 3 (May–June 1995): pp. 555–562.
7. Daniel G. Murri, Gautam Shah, and Daniel J. DiCarlo, "Preparations for Flight Research to Evaluate Actuated Forebody Strakes on the F-18 High-Alpha Research Vehicle," NASA Document N95-14257, NASA High-Angle-of-Attack Projects and Technology Conference, NASA Dryden, Edwards, CA, July 12–14, 1994.
8. Daniel G. Murri, David F. Fisher, and Wendy R. Lanser, "Flight-Test Results of Actuated Forebody Strake Controls on the F-18 High-Alpha Research Vehicle," NASA High-Angle-of-Attack Technology Conference, Hampton, VA, September 17–19, 1996.
9. M.A. Croom, S.B. Grafton, and L.T. Nguyen, "High Angle-of-Attack Characteristics of Three-Surface Fighter Aircraft," AIAA Paper 82-0245, AIAA 20th Aerospace Sciences Meeting, Orlando, FL, January 11–14, 1982.
10. Sue B. Grafton, Mark A. Croom, and Luat T. Nguyen, "High-Angle-of-Attack Stability Characteristics of a Three-Surface Fighter Configuration," NASA TM-84584 (1983); J.W. Agnew, G.W. Lyerla, and S.B. Grafton, "The Linear and Non-Linear Aerodynamics of Three-Surface Aircraft Concepts," AIAA Atmospheric Flight Mechanics Conference, Danvers, MA, 1980.
11. Daniel G. Murri, Sue B. Grafton, and Keith D. Hoffler, "Wind-Tunnel Investigation and Free-Flight Evaluation of a Model of the F-15 STOL and Maneuver Technology Demonstrator," NASA Technical Paper 3003 (1990), ITAR Restricted Distribution.

12. Chambers, *Partners in Freedom*.
13. Frank L. Jordan, Jr., and David E. Hahne, "Wind-Tunnel Static and Free-Flight Investigation of High-Angle-of-Attack Stability and Control Characteristics of a Model of the EA-6B Airplane," NASA TP 3194 (1992).
14. Robert J. Hanley, "Development of an Airframe Modification to Improve the Mission Effectiveness of the EA-6B Airplane"; Frank L. Jordan, Jr., David E. Hahne, Matthew F. Masiello, and William Gato, "High-Angle-of-Attack Stability and Control Improvements for the EA-6B Prowler," AIAA Paper 87-2358 and AIAA Paper 87-2361, *A Collection of Technical Papers: AIAA 5th Applied Aerodynamics Conference* (August 1987).
15. Donald R. Riley, Gautam H. Shah, and Richard E. Kuhn, "Wind-Tunnel Study of Reaction Control-Jet Effectiveness for Hover and Transition of a STOVL Fighter Concept," NASA TM-4147 (1989); Donald R. Riley, Mark A. Croom, and Gautam H. Shah, "Wind-Tunnel Free-Flight Investigation of an E-7A STOVL Fighter Model in Hover, Transition, and Conventional Flight," NASA TP 3076 (1991), ITAR Restricted Distribution.
16. David E. Hahne, "Low-Speed Static and Dynamic Force Tests of a Generic Supersonic Cruise Fighter Configuration," NASA TM- 4138 (1989). See also David E. Hahne, Thomas R. Wendel, and Joseph R. Boland, "Wind-Tunnel Free-Flight Investigation of a Supersonic Persistence Fighter," NASA TP 3258 (1993).
17. Jay M. Brandon, James M. Simon, D. Bruce Owens, and Jason S. Kiddy, "Free-Flight Investigation of Forebody Blowing for Stability and Control," AIAA Paper 96-3444, AIAA Atmospheric Flight Mechanics Conference, San Diego, CA, July 29–31, 1996.
18. William P. Gilbert, "Free-Flight Investigation of Lateral-Directional Characteristics of a 0.10-Scale Model of the F-15 Airplane at High Angles of Attack," NASA TM-SX-2807 (1973).
19. Extensive support for the Keep Eagle project was also provided by the Langley Spin Tunnel as part of the Vehicle Dynamics Branch effort.
20. Commendation letter from General James L. DeStout, F-15 System Program Office director, to Dr. Douglas L. Dwoyer, Research and Technology Group director, NASA Langley Research Center, January 17, 1996.
21. Letter from James A. Blackwell, Lockheed Corp.'s vice president for the Advanced Tactical Fighter Program, to Richard H. Petersen, director of NASA Langley Research Center, March 15, 1991, LHA.
22. An example of the LEX studies is Gautam H. Shah and Samantha B. Clemons, "Parametric Wind-Tunnel Investigation of High-Alpha Aerodynamics and Stability Characteristics of a Prototype F-18E Configuration," AIAA Paper 95-3503, AIAA Atmospheric Flight Mechanics Conference, Baltimore, MD, August 1995.
23. David E. Hahne and Louis J. Glaab, "Experimental and Numerical Optimization of a High-lift System to Improve Low-Speed Performance, Stability, and Control of an Arrow-Wing Supersonic Transport," NASA TP 1999-209539 (1999).
24. D. Bruce Owens and John N. Perkins, "Vortex Suppression on Highly-Swept Wings by Suction Boundary-Layer Control," AIAA Paper 95-0683 (1995).

25. David E. Hahne, "Low-Speed Aerodynamic Data for and 0.18-Scale Model of an F-16XL with Various Leading-Edge Modifications," NASA TM-1999-209703 (1999).
26. After arriving at Langley, the aircraft's wing was painted black as a preliminary step for plans to use wool tufts for flow visualization. Soon after being painted in October 1993, the aircraft was returned to Dryden. The aircraft was subsequently used in the NASA Cranked Arrow Wing Aerodynamic Project, which provided correlations of experimental and computational results with a focus on vortex flows.
27. William I. Scallion, "Aerodynamic Characteristics and Control Effectiveness of the HL-20 Lifting Body Configuration at Mach 10 in Air," NASA TM-1999-209357 (1999); George M. Ware and Christopher I. Cruz, "Aerodynamic Characteristics of the HL-20," AIAA *Journal of Spacecraft and Rockets 30*, no. 5 (September–October, 1993): pp. 529–536.
28. Christopher I. Cruz, George M. Ware, Sue B. Grafton, William C. Woods, and James C. Young, "Aerodynamic Characteristics of a Proposed Personnel Launch System (PLS) Lifting-Body Configuration at Mach Numbers from 0.05 to 20.3," NASA TM-101643 (1989).
29. D.E. Hahne and Paul L. Coe, "The Low-Speed Stability and Control of Three Air Breathing Transatmospheric Vehicles," AIAA Paper 94-0506, 32nd Aerospace Sciences Meeting and Exhibit, Reno, NV, January 10–13, 1994.
30. David E. Hahne, "Evaluation of the Low-Speed Stability and Control Characteristics of a Mach 5.5 Waverider Concept," NASA TM-4756 (1997).
31. Daniel G. Murri and Frank L. Jordan, Jr., "Wind-Tunnel Investigation of a Full-Scale General Aviation Airplane Equipped With an Advanced Natural Laminar Flow Wing," NASA TP 2772 (1987).
32. J. "Mac" McClellan, "Mac Eats Cirrus Propwash in Full-Power Dash," *Flying* 119, No. 9 (September 1992): pp. 85–86.
33. D.E. Hahne and Frank L. Jordan, Jr., "Full-Scale Semispan Tests of a Business-Jet Wing With a Natural Laminar Flow Airfoil," NASA TP 3133 (1991).
34. Paul L. Coe, Jr., Stephen G. Turner, and D. Bruce Owens, "Low-Speed Wind-Tunnel Investigation of the Flight Dynamic Characteristics of an Advanced Turboprop Business/Commuter Aircraft Configuration," NASA TP 2982 (1990).
35. Jay M. Brandon, Frank L. Jordan, Jr., and Robert A. Stuever, "Application of Wind Tunnel Free-Flight Technique for Wake Vortex Encounters," NASA TP 3672 (1997).
36. Dan D. Vicroy, Jay Brandon, George Greene, Robert Rivers, Gautam Shah, Eric Stewart, and Robert Stuever, "Characterizing the Hazard of a Wake Vortex Encounter," AIAA Paper 97-0055, AIAA 35th Aerospace Sciences Meeting and Exhibit, Reno, NV, January 6–9, 1997.
37. Langley's B-737-100 research airplane was retired and is now on exhibit at the Boeing Museum of Flight in Seattle, WA.
38. James R. Hansen, *Spaceflight Revolution: NASA Langley Research Center From Sputnik to Apollo* (Washington, DC: NASA SP-4308, 1995).
39. "Dwoyer Speaks out about New RTG," *Langley Researcher News 8*, no 6 (March 25, 1994): p. 5.

40. NASA Headquarters regarded the closure of the Full-Scale Tunnel an action in response to a 1993 Center Roles and Missions study led by Roy Estess. Closure was considered an appropriate action in view of the budget environment and efforts to reduce facility infrastructure by 25 percent. See Wesley L. Harris, associate administrator for aeronautics, internal NASA memorandum to NASA deputy administrator, December 8, 1994.
41. The useable area in the 14- by 22-Foot Tunnel's test section for free-flight testing is about 12 by 18 feet.
42. H. Lee Beach, Jr., Langley deputy director, internal NASA memorandum to Paul F. Holloway, Langley director, October 25, 1994.
43. H. Lee Beach, Jr., Langley deputy director, letter to David H. Dutton, director of projects review, Department of Historic Resources, July 3, 1995.
44. The model tested in the Full-Scale Tunnel was a 0.18-scale F/A-18E, and the model tested in the 14- by 22-Foot Tunnel was a 0.15-scale model.

The Old Dominion University conducted its last test in the Full-Scale Tunnel in September 2009 involving a remotely piloted X-48C research vehicle. The X-48 configuration is a revolutionary concept for an advanced subsonic transport under study by NASA, Boeing, and the U.S. Air Force. (U.S. Air Force photo)

CHAPTER 9

The ODU Era

1996–2009

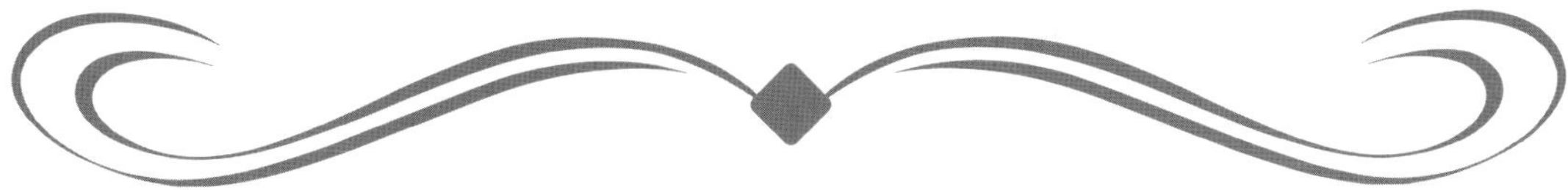

Rebirth of the Full-Scale Tunnel

Dr. Ernest J. "Jim" Cross, Jr., served as dean of the Old Dominion University's (ODU) College of Engineering and Technology from 1984 until 1997. Cross had an intense interest in aerospace engineering. He brought with him a strong background in experimental aerodynamics and wind tunnel operations acquired during previous positions at the Air Force Flight Dynamics Laboratory, Mississippi State University, and Texas A&M University. Cross had maintained a long and close relationship with technical peers and management at the NASA Langley Research Center throughout his career and was intimately knowledgeable of Langley's activities at the Full-Scale Tunnel and of many other wind tunnel facilities. He would become a key figure in the 13-year period of operations of the Langley Full-Scale Tunnel under ODU management.

In the early 1990s, Langley had already begun to deal with the challenge of downsizing its infrastructure and determining the fates of its aging wind tunnels. The Langley 7- by 10-Foot High-Speed Tunnel had been one of the most productive Langley facilities from its first operations in 1945 until an accident caused by catastrophic fatigue failure of its wooden fan blades in 1985. The expense of repairing the damage (new wooden fan blades and repairing the bent drive shaft) was estimated at over $1.7 million, which sensitized management to the potential costs of maintaining old facilities. After repairs, Langley operated the tunnel for a few years before turning it into a popular wind tunnel exhibit for the visiting public (which could be done prior to the increased security following the terrorist attacks of September 11, 2001). The 7- by 10-Foot High-Speed Tunnel then became a target for closure.[1]

While attending a conference at Langley in 1992, Jim Cross was approached by a Langley manager who discussed the gloomy outlook for the future of the 7- by 10-Foot Tunnel and encouraged a proposal from ODU for the university to manage and operate the tunnel. Cross submitted a proposal that was reviewed for almost a year before being rejected in 1993 on the basis of legal regulations that disallowed the cohabitation of contractor and civil-service personnel.[2] Langley Director Paul Holloway then suggested that Cross submit a similar proposal for the privatization of the Full-Scale Tunnel. After ODU submitted a proposal to Langley, the Langley legal office once again objected, but an arrangement was forthcoming based on the authority of the 1958 NASA Space Act. Cross estimated that a sufficient market existed for nontraditional use of the tunnel to justify a business development venture—even without aircraft testing. Under the agreement, ODU would be responsible for the interior and roof maintenance of the Full-Scale Tunnel building, while NASA would be responsible for the maintenance and corrosion control of the exterior of the building. The Air Force, as owner of the property, reviewed the proposal and was initially against the arrangement on the basis of security concerns but later approved and supported the agreement, which stated that ODU would not assign foreign nationals to Air Force property or permit them to enter the facility.[3] The tunnel was envisioned by ODU to become the primary facility in an enterprise center to support industrial development, university research, and education.

Changing of the Guard

Birthing Pains and Rocky Relations

By late 1995, Holloway had arranged for Jim Cross and a few ODU personnel to move into a small office in a remote location within the Full-Scale Tunnel building (removed from the NASA civil service personnel). The first task for the group prior to the signing of a formal Space Act Agreement between NASA and ODU was a nonreimbursable agreement with NASA signed in July 1996 for wind tunnel tests of a full-scale F-15 forebody, which will be discussed later. The funding for the project was provided by the Air Force through NASA. ODU operated the tunnel and Bihrle Applied Research conducted the test while the NASA Vehicle Dynamics Branch was still in the building; as might be expected, the already frayed nerves of the NASA personnel were even more aggravated by the presence of outside organizations that were taking over their facility while they were being told that the facility was closing because it was no longer needed. The ODU group was naturally viewed as a competitor by the longtime NASA occupants.

In October 1996, the Vehicle Dynamics Branch moved from the Full-Scale Tunnel to its new home in NASA Building 1192C in the NASA West Area. The group had been working for a year, moving critical test equipment from the tunnel to the 14- by 22-Foot Tunnel. The orders from NASA management were to take all necessary items needed for the transfer of the free-flight technique, subscale model force tests, and forced-oscillation tests in the Full-Scale Tunnel to the new tunnel, and the branch moved virtually all the hardware, instrumentation, and data systems associated with those tests. Since full-scale airplane tests would not be part of the branch's mission in the 14- by 22-Foot Tunnel, all equipment used in such tests (e.g., tunnel scale system, struts, high-pressure air station, flow-survey rig, data-acquisition system for full-scale tests, etc.) was left behind at the Full-Scale Tunnel.

From the perspective of the new ODU tenants, they had inherited a gigantic empty building with no equipment to conduct subscale tests.[4] Jim Cross immediately obtained a $417,000 line of credit with the university's Research Foundation for salaries and equipment, including the acquisition of a new data system.[5]

Legal Agreements

The first Space Act Agreement (SAA) on the use of the Full-Scale Tunnel was signed between NASA and ODU on August 19, 1997, for a 10-year period with automatic renewals in 3-year increments.[6] The wording of the agreement displayed the unrest and sensitivity of the Langley staff to the potentially competitive arrangement, as indicated by the following excerpt from the agreement:

> The purpose of ODU's operation of the facility shall be for commercial testing of non-aerospace vehicles and structures, student instruction, and commercial testing of aerospace vehicles. Aerospace vehicle testing in the wind tunnel shall be only for the purposes of providing vehicle data reports to customers on the low angle-of-attack aerodynamics of aircraft or to test models too large for more conventional wind tunnels. It is a specific goal, and ODU intends to transition from testing predominately aerospace vehicles to primarily non-aerospace vehicles within a period of 3 years after the start date of this Agreement.... Langley currently has no plans for testing in the tunnel. Should NASA have requirements to test, however, ODU shall consider NASA's needs and the testing needs of the Department of Defense as the highest priority for tunnel occupancy time.... Langley will compensate ODU at ODU-established customer rates for NASA-sponsored testing.[7]

With the signing of the agreement, the Full-Scale Tunnel became the largest university-operated wind tunnel in the world. ODU renamed the facility, returning to the traditional Langley Full-Scale Tunnel (LFST) designation, and commenced operations in February 1998. Since NASA had specifically called for an exit from traditional aerospace testing within 3 years, Jim Cross would have to quickly find nonaerospace customers to fund the tunnel's operations.[8]

The original 1997 SAA was followed by five modifications, one of the most important of which was signed 3 years later in April 2000 by Langley Director Jeremiah F. Creedon, ODU President James V. Koch, and ODU Research Foundation Acting Director Jerald B. Jones. The modification revised the constraints of the original SAA and permitted ODU to test aircraft or aircraft components in the Full-Scale Tunnel.

The fifth and final modification to the SAA was signed on August 4, 2009, by Langley Director Lesa B. Roe, ODU Vice President of Research Mohammed Karim, and ODU Research Foundation Executive Director Ruth B. Smith. The brief 2-page modification stated that operations of the Langley Full-Scale Tunnel would be terminated no later than September 30, 2009, at which time ODU would remove its property and equipment and vacate the Full-Scale Tunnel.

While all parties worked diligently to abide by the terms of the SAA and its modifications, nerves quickly unraveled and working relationships between the departing and arriving staffs at the Full-Scale Tunnel were initially cold, although the situation improved in the mid-2000s. Today, past participants from both NASA and ODU agree that the approach used in transferring operations of a major NASA wind tunnel to another operator under the specific circumstances of the situation of the Full-Scale Tunnel was very stressful.

Staff and Management During the ODU Years

Jim Cross left his position as dean of Old Dominion University's College of Engineering and Technology in 1997 to become head of Special Projects, which included responsibilities for operation of the Full-Scale Tunnel. During the period, he spent equal time at the tunnel and teaching classes at the Norfolk, VA, campus of ODU. In 2000, Cross was named full-time director of the tunnel operation. During the tenure of ODU at the Full-Scale Tunnel, no one had a larger impact on the startup and success of the operation than Cross. His knowledge of experimental aerodynamics and his aggressive marketing of the tunnel's capabilities to the motorsports community, ground transportation companies, nonaerospace organizations, and academic interactions with NASA were the vital factors that brought the Full-Scale Tunnel back to life. After the strenuous task of rebuilding the test capability at the tunnel, he worked diligently to identify and encourage new business in fields that had never been associated with NASA's wind tunnel expertise and facilities. Cross was awarded NASA's highest nonemployee medal, the NASA Distinguished Public Service Medal, for his contributions to the NASA mission. In 2005, he was relieved of his duties as director of the wind tunnel but retained his participation in tunnel activities while

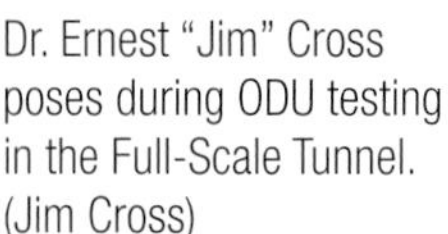

Dr. Ernest "Jim" Cross poses during ODU testing in the Full-Scale Tunnel. (Jim Cross)

teaching classes at ODU. After a distinguished 13-year career at the Full-Scale Tunnel, Jim Cross retired in 2008.

The first staff members hired by Cross included Colin Britcher, Drew Landman, and Earl Conkling. Landman and Britcher were experts in applied aerodynamics and associate professors in the Department of Aerospace Engineering at ODU, and Conkling was a retired, highly skilled Langley technician with extensive experience at the Full-Scale Tunnel. Augmented by graduate students, this core team built the nonprofit organization into a productive undertaking. Amazingly, the early development was accomplished by Cross and his team of only two faculty members and one full-time technician.[9]

Colin Britcher's title was director of research and education for the Langley Full-Scale Tunnel. He had previously completed a 2-year residence at NASA Langley as a National Research Council (NRC) Associate in which he conducted extensive research in experimental methods, aerodynamics, and magnetic-suspension systems. Britcher proved to be one of the most versatile members of the staff, demonstrating extensive expertise in all aspects of experimental wind tunnel testing methods and hardware. He was also an expert on the tunnel's drive system and motor controls. After the closure of the tunnel in 2009, Britcher returned to the ODU campus as a professor of aerospace engineering.

Drew Landman served as the chief engineer of the Full-Scale Tunnel from the first ODU operations until the tunnel was closed. He became Cross's protégé and participated in all aspects of the business venture, with steadily increasing responsibilities and expertise in the day-to-day operation of the facility and relationships with customers. His interactions with students conducting research and experimental testing were particularly valuable. Landman became the spokesman and major point of contact during test activities and visits. In addition to his duties as chief engineer, he became the facility assistant manager from 2002 to 2005 and was promoted to manager from 2005 until the facility closed in 2009. Like Colin Britcher, he also returned to continue his academic career at ODU as a professor.

The formidable task of replacing, maintaining, and upgrading the hardware required for tests in the tunnel fell to Earl Conkling, an outstanding retired technician who had completed a career at Langley that included years of experience in NASA's operation of the Full-Scale Tunnel. Conkling's expertise and contacts within the Langley organization would prove to be invaluable, especially in arranging loan agreements with NASA for equipment needed for subscale model tests. He also was a skilled fabricator of hardware and made the tunnel's shop into a productive in-house component of the operation. Conkling retired from the ODU operation in 2006.

In 1999, Jim Cross acquired exceptional talent from the motorsports community when he hired Eric Koster, who drove racecars, built racecars, and specialized in building racing engines. Koster was initially hired as director of marketing and later also became director of motorsports operations. When he joined the ODU wind tunnel staff, he brought a level of motorsports racing knowledge and enthusiasm that was contagious, especially for students. After NASA-funded tests in the Langley 14- by 22-Foot Tunnel, with its unique moving ground belt, proved too expensive for the racing community, Koster strongly advocated for ODU to build and operate a rolling-road wind tunnel in Southside, VA. He also made significant contributions to the vision for and development of the Virginia Motorsport Technology Park, bringing university research to the commercial sector. His biggest contribution was in support of the Society of Automotive Engineers (SAE) competitions, which teams of ODU engineering students travelled across North America to enter. Koster was also pivotal in the effort to bring a Formula SAE student competition to Virginia International Raceway (VIR). Koster remained on the tunnel staff until its closing. He tragically died of cancer in 2010 at the age of 59.

Other ODU staff members at the Full-Scale Tunnel included Dr. Stan Miley, Whitney Seay, and John Bledsoe. Bledsoe was an electronic engineer with superb skills and expertise in automated systems. He personally updated the turntable and survey apparatus of the Full-Scale Tunnel with automated systems, providing the capability to automate entire test runs. ODU was also able to retain outstanding graduate students as staff members. After earning his Ph.D., Ilhan Bayraktar was retained as a postdoctoral staff member specializing in the application of CFD methods for guidance in experimental tunnel testing and correlation with experimental results.

After Jim Cross stepped down as director of the tunnel enterprise in 2005, ODU selected Professor Robert L. "Bob" Ash to replace him. Ash was a research engineer and faculty member at the facility during the 2000s. Among his most valuable contributions were being instrumental in bringing the Wright Experience team to the tunnel and heading up the Wright propeller test program while working with Stan Miley. Ash was a liaison to NASA Langley during numerous interactions, including the original SAA proceedings.

In addition to the foregoing ODU personnel, the successful operations of the Full-Scale Tunnel involved the technical and business capabilities of other key organizations. One of the most important contributors was the staff of Bihrle Applied Research (BAR), who were specialists in applied aircraft research including static and dynamic wind tunnel testing techniques and piloted-simulation methodology. BAR had provided support for NASA activities in the Langley Spin Tunnel since the mid-1970s, and the company was familiar with operations at the Full-Scale Tunnel. However, BAR brought a more important factor into the ODU business capability—numerous contacts and successful contracts for the aircraft industry and DOD. During the time period covered by this chapter, BAR was the prime ODU business partner for aerospace testing in the Full-Scale Tunnel while ODU focused on motorsports and other nonaerospace projects. On most occasions BAR successfully initiated contracts and test crews for aircraft testing in the tunnel while ODU provided staffing and operation of the tunnel. BAR brought its own data-acquisition systems for the testing, directed the test schedule, and provided analysis of the results to the customer.

Cross and his associates were particularly appreciative of the aerospace business that BAR brought to the tunnel. Whereas the motorsports testing projects typically lasted 1 or 2 days during quick "in-and-out" entries, the aircraft-related testing lasted at least a week, providing stability to the business interests of the facility.

Hugging the Ground: Modifications to the Tunnel

Jim Cross had his strategy for attracting automotive customers in mind, but first he had to examine the auto-testing capabilities of an old wind tunnel that had been built in the 1930s for testing biplanes. After an inspection and calibration of the hardware in the test section and completion of flow surveys near the ground board, it became evident that some major changes to the tunnel and its support components would be required for the motorsports market.

Balance and Turntable

As discussed in previous chapters, the external-balance system installed in the Full-Scale Tunnel in the 1930s had consisted of a movable frame supported by four columns with links that transferred forces to weigh-beam scales. A turntable was not used, and tests of aircraft in sideslip conditions required laborious adjustments of the huge mounting struts. In the 1950s, Bill Scallion and Joe Walker conceived of and installed a turntable to permit more efficiency and range for sideslip tests. During a facility update in the 1980s, the load frame was changed from rectangular to circular and the scales were fitted with strain-gage load cells to permit measurements with modern data-acquisition systems. When ODU began operations, the vertical-force capability of the balance system was 20,000 pounds, more than enough for even the heaviest cars and light trucks. However, the excess load capability caused concern over potential mechanical hysteresis, and the external balance proved to be insensitive to measuring the small changes in down force required by automotive teams.[10]

Drew Landman was tasked with designing a sensitive modular balance that could be inserted into the turntable for automotive testing. This first "prototype" was used for years before replacement by Landman and Britcher's second design, a full 6-degrees-of-freedom balance. The new load-cell-based system was fabricated by Eric Koster and students. The system permitted measurement of lift, drag, and vehicle-specific loads such as sideforce and down-force loads at wheel locations. A state-of-the-art, automated turntable control system was also developed by John Bledsoe and installed for more efficient testing.

Lift System for Cars

After many years of operations, the overhead crane system (nicknamed "Annie") used in the Full-Scale Tunnel to lift and position full-scale aircraft and models a vertical distance of about 21 feet atop the balance mounting struts had become unreliable and was a major cause of delays in test programs. Since tests of motorsport cars and other ground transportation vehicles only lasted 1 or 2 days, a breakdown in the crane unit was of considerable consequence. Therefore, Landman designed and implemented a new lift system with a 15,000-pound capability.

An Active Ground Board

One of the most critical features of any wind tunnel test for ground vehicles is an accurate representation of the flow at and beneath the test subject. The ground board of the original

Full-Scale Tunnel was installed primarily to provide a work surface for personnel to more easily work on aircraft mounted near the vertical centerline of the test section. In the wind tunnel environment, the boundary-layer flow on the ground board thickens as the flow moves along the test-section floor, resulting in unrealistic flow conditions under the test article. When certain test articles, such as semispan wings, were tested by the NACA and NASA, a dummy half-fuselage was usually mounted on the ground board in order to raise the test article to a higher position above the ground board, thereby avoiding the boundary-layer effects. Flow surveys made by Britcher and others indicated that the boundary-layer growth from the leading edge of the ground board in the Full-Scale Tunnel was extremely rapid, resulting in a boundary-layer displacement thickness at the center of the automobile balance location of about 1.1 inches—much larger than is commonly accepted for automotive testing in wind tunnels.[11]

The flow-survey investigations by ODU also disclosed that the leading edge of the ground board was experiencing a flow-separation bubble that aggravated the boundary-layer characteristics. Before a solution to the boundary-layer buildup could be considered, the separation bubble had to be eliminated. Detailed assessments of the leading-edge flow (including the use of CFD methods) indicated that the semicircular leading edge was experiencing laminar-flow separation as a result of both the low local Reynolds number and the inclination of the oncoming flow as it exited the entrance cone adjacent to the ground board. It was decided that the leading edge of the ground board would be recontoured, and a leading-edge cuff with camber was affixed to alleviate the problem.

After consideration of several alternatives to minimize boundary-layer growth, Britcher designed a suction-type boundary-layer control system that used a forward-facing ramped slot installed in front of the turntable. A commercial axial-flow blower was used to vent boundary-layer flow into a ducted plenum chamber with external discharge. The slot was located about 30 inches ahead of the air dam of a typical NASCAR vehicle.

Results of flow surveys at a location about 13.5 inches forward of a typical vehicle air-dam location demonstrated that the boundary-layer displacement thickness with suction was less than one-third the size of the uncontrolled case (reduced to 0.173 inches from 0.621 inches). The higher-speed underbody flow with suction greatly affected the lift and drag exhibited by test subjects, depending on the ground clearance of the specific vehicle. As might be expected, one effect of using the active boundary-layer control system was an increase in drag force because of the increased airspeed of the underbody flow.

Survey-Rig Update

The overhead survey apparatus was an integral part of the instrumentation in the Full-Scale Tunnel test section and had been used since the tunnel began operations in 1931 for detailed flow surveys of aircraft and other test subjects. The flow survey probe and associated equipment had been updated by NASA in the 1990s, but the operation of the survey equipment was still a laborious, time-consuming manual procedure.

As part of the facility modernization undertaken by Jim Cross's staff, John Bledsoe designed and implemented a modern, computer-controlled auto-positioning system for the survey rig that permitted preprogrammed test sequences with precision positioning of the survey probe at specific locations near the test subject.

Model Tunnel

Since it was not required for its new home at the 14- by 22-Foot Tunnel, the NASA Vehicle Dynamics Branch left the 1⁄15-scale model of the Full-Scale Tunnel at the facility. The model, which was located in the hangar annex of the Full-Scale Tunnel, was subsequently used at various times during research activities sponsored by ODU. When the Full-Scale Tunnel was demolished between 2010 and 2011, the National Institute for Aerospace in Hampton, VA, acquired the model, which is now in storage and intended for educational use by the Institute in the future.

Research Activities

Initially faced with the challenging task of identifying nontraditional customers to support operations in the Full-Scale Tunnel, Jim Cross and his staff conducted an extraordinary campaign to attract interest from a broad community that would benefit from aerodynamic testing in the unique facility. The success with which the customer market was captured was remarkable. Clients who had previously been totally unaware of the existence of the Full-Scale Tunnel became some of its most enthusiastic customers and advocates. After the first 4 years of wind tunnel operations, a positive spirit of optimism prevailed in the ODU business venture. The final payment on the $417,000 line of credit Jim Cross had obtained from the ODU Research Foundation was paid in full in October 2001. Revenues from tunnel tests rose from $680,000 in FY 2000 to $940,000 in FY 2001.[12]

The following examples of studies conducted in the tunnel provide an appreciation of the scope of contributions made within the rejuvenated Full-Scale Tunnel. Many other projects were conducted, but the proprietary nature of the work prohibits discussion herein.

Breath of Life: Motorsports

In the 64 years that the NACA and NASA operated the Full-Scale Tunnel, the research projects had been almost exclusively devoted to aerospace interests. The tunnel schedule was always backed up with planned work and scheduled for at least 2 years in advance with these traditional activities, leaving no time for nonaerospace entries. With the threat of closure continually present after World War II because of the more capable 40- by 80-Foot Tunnel at the Ames Research Center, diverting the tunnel to other applications—such as the aerodynamics of ground-transportation vehicles—would have been an admission of a lack of aerospace customers and a reason to terminate operations. Mac McKinney was particularly sensitive to this situation during his days as manager of the Full-Scale Tunnel, and he emphatically turned down numerous requests for nonaerospace testing, ranging from proposed tests on stadium light configurations to advanced sails for marine applications.

Selling the Capability

When Jim Cross mapped his initial strategy of how to comply with the NASA-ODU agreement, which called for a cessation of aerospace testing after 3 years, he turned to his personal area of interest, the NASCAR community. In 1997, the only wind tunnels in North America involved in testing full-scale automobiles were the tunnels operated by General Motors in Detroit, MI; Lockheed Martin in Marietta, GA; and the National Research Council in Ottawa, Canada. None of these tunnels at that time had the special features of particular interest to NASCAR racers, especially a system to measure the forces at each wheel; neither did the Full-Scale Tunnel.

Cross began intensive visits and communications with the NASCAR community, beginning with the Daytona 500 race in 1996. He visited everyone from drivers to pit crews to stir up interest and awareness of his plans. In April 1997, he arranged a demonstration of testing capability for representatives of NASCAR participants and the local press.[13] The demonstration was the first automobile test in the wind tunnel in its 66-year history. The event was attended by a local motorsport promoter and the vice president of competition of NASCAR, Dennis Huth. The car placed in the test section of the Full-Scale Tunnel for the demonstration was a NASCAR Monte Carlo owned by the Falk family of Norfolk, VA, and driven by well-known racer Mike Wallace. The demonstration provided some valuable publicity for Cross's program and the capabilities of the facility. During the demonstration, a driver was seated in the car and the tunnel was run to a maximum speed to about 80 mph.

ODU charged a rate of $1,200 to $1,500 an hour for tunnel time—about half of what other tunnels charged, which was a result of the lower labor rates for graduate students and faculty compared to industry staff. Huth's reaction was not unexpected: "Most racing teams make their own wind-tunnel arrangements, so ODU's marketing will have to extend beyond the NASCAR brass. They have quite a facility here, but a lot people out there are like me. I didn't even know that this wind tunnel existed."[14]

NASCAR Responds

The NASCAR teams received the message and responded with great enthusiasm. At one point, over 55 teams had pursued tunnel entries to take advantage of the new testing capability. As might be expected, the use of the tunnel by NASCAR teams and the details of test results were highly proprietary, and most activities have not been published. In late 1997, one racing team began a series of test entries at the Full-Scale Tunnel ranging from tests of prototypes to basic studies on flow fields on individual racing cars, as well as mutual interference effects experienced during two-car passing or drafting situations. The tunnel's unique data-acquisition capability with the ODU load-cell measurement was of particular interest, especially for measurements of aerodynamic drag and downforce on the wheels. Teams used the tunnel about four times per year beginning in January 1998 for investigations that typically lasted 2 to 3 days. By mid-1998, four NASCAR Winston Cup Series test teams had made multiple entries. Test objectives varied according to specific interests. For example, the impact of configuration modifications on drag was particularly important for stock racing such as the Daytona events ("speedway testing"), whereas the focus of interests for shorter track races such as the Richmond International Raceway Sprint Cup Series centered on wheel downforces ("downforce testing"). Three NASCAR Craftsman Series truck teams also made tunnel entries, and a World Sports Car–class racecar team tested extensive configuration effects.

Motorsports test activity in the Full-Scale Tunnel. Note the slot for the boundary-layer control system immediately ahead of the car at the upper left. (ODU)

Revenues from motorsport testing during the late 1990s brought in over $1 million as 35 of the 50 NASCAR teams participated in testing in the Full-Scale Tunnel, with many teams having two or three entries per year. It was during this upsurge of interest that Jim Cross hired Eric Koster as a marketing manager for the tunnel, resulting in even more test activity from motorsports road racing, dragsters, Indiana Racing League cars, truck designers, and truck-stack-emission testing. Ultimately, an entire variety of ground-vehicle applications were involved in tunnel testing of large trucks, pickup trucks, motorcycles, and solar-powered cars.

Typical ¼-scale truck aerodynamic study in the Full-Scale Tunnel. (ODU)

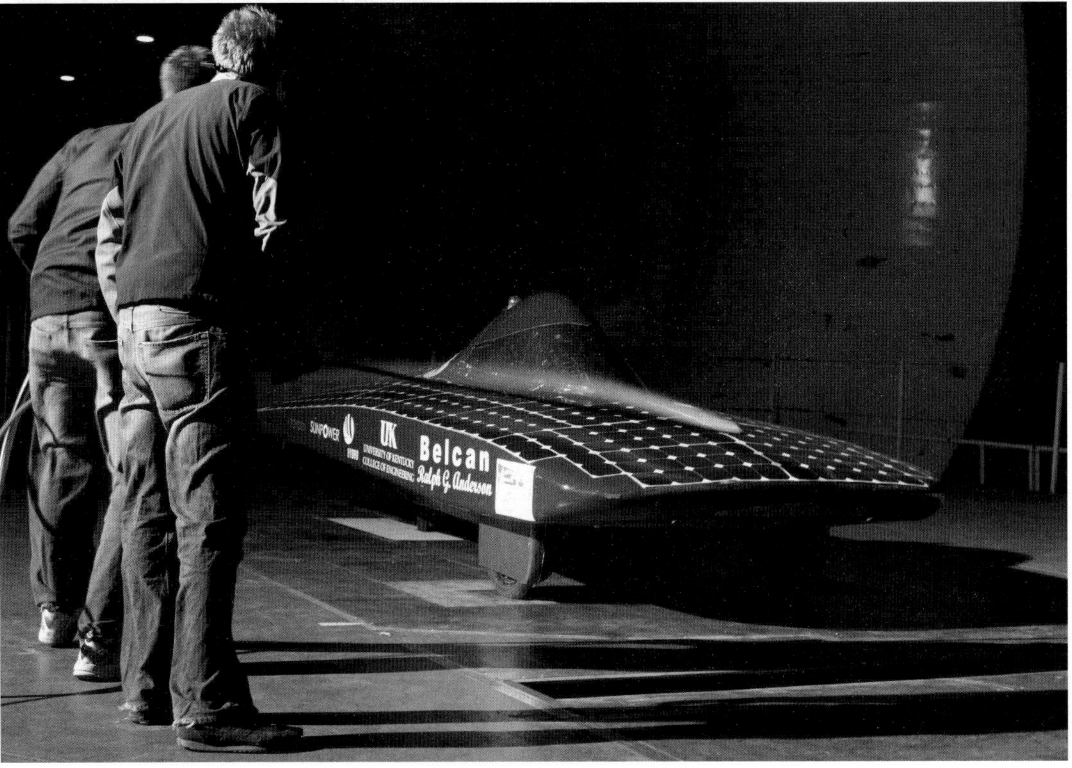

Flow-visualization tests of a solar-powered car designed by students at the University of Kentucky. (ODU)

Priceless Education

The ODU staff at the Full-Scale Tunnel quickly provided enthusiasm, technical material, lectures, and experimental projects appropriate for a major educational curriculum in motorsports technology. The Aerospace Engineering Department offered a motorsports engineering minor for students in mechanical engineering and engineering technology, which introduced

Members of an ODU ground-vehicles aerodynamics class pose with a Chase Daytona prototype racecar. Professor Drew Landman is second from right in the back row. (ODU)

ODU students in a design-of-experiments class pose in the control room of the Full-Scale Tunnel with Professor Drew Landman (seated at right). (ODU)

critical technology such as aerodynamics, chassis dynamics, piston engines, and racecar performance. The focused objective of the educational program was to prepare students for entry-level positions in motorsports or in the automotive industries.

In 2008, the department also initiated a master's of engineering degree in motorsports engineering, which included experimental content involving the Full-Scale Tunnel, a motion-based simulator, and onboard engine performance. The program also included activities at the Virginia Institute for Performance Engineering and Research, located close to the Virginia International Raceway near Danville, VA.

Enduring Versatility: Nonaerospace Projects

During the 1950s, the Full-Scale Tunnel had become widely known for a series of nonaerospace test projects that clearly demonstrated the value of a large, low-speed wind tunnel as a research tool. The scope of test subjects included a submarine, a large wind tunnel concept, and buildings. The availability of the tunnel for nonaerospace testing under ODU management continued this earlier display of versatility.

Return of Submarines

One of the many legendary tests conducted in the Full-Scale Tunnel was the famous 1950 investigation of the performance, stability, and control of a large model of the Albacore submarine, as discussed in chapter 5. Fifty-five years later, submarines returned to the test section when tests were conducted on the Newport News Experimental Model-1 (NNemo-1), a large, radio-controlled submarine model designed by Northrop Grumman's local Newport

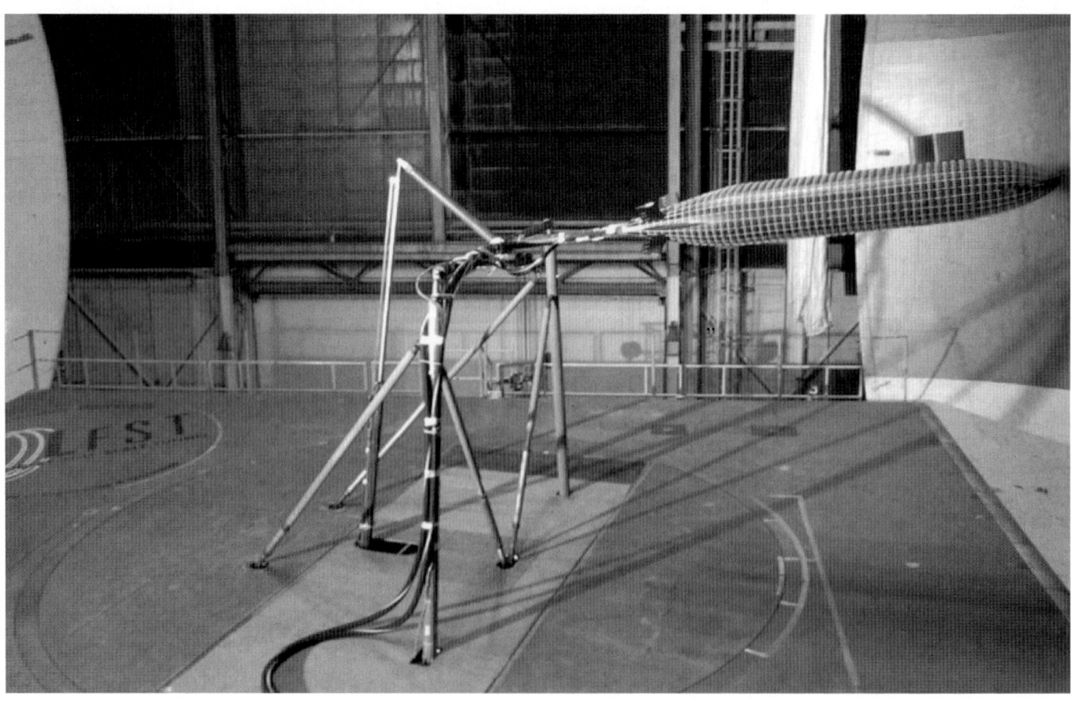

The Northrop Grumman NNemo-1 submarine research model mounted for an investigation of a circulation-control sail concept in the Full-Scale Tunnel in December 2005. Note the high-pressure air lines used to provide flow to the dual-slot sail configuration. (U.S. Navy)

News sector in 2003 to explore advanced technology for future naval applications. The scope of the December 2005 test program was similar to that of the earlier Albacore project, with variations in geometric parameters and hull details.

At the end of the test program, a 1-day evaluation of an advanced sail (i.e., bridge fairwater) concept using circulation control to enhance maneuverability was conducted with the support of the Office of Naval Research (ONR).[15] Circulation control has long been pursued and demonstrated for enhanced aircraft aerodynamic performance by ejecting flow near the trailing edge of a lifting surface and using the Coanda Effect for controlling the lift force that develops. In the case of submarine applications, the concept is of interest for improving maneuvering capabilities in a horizontal plane—especially for reducing turn diameter at low speeds. In addition, the ability to vary control forces without mobile surfaces has the potential to reduce cost and complexity.

In the experiment, the 14.5-foot-long NNemo-1 vehicle was equipped with a dual-slot circulation-control concept using compressed air. By operating the wind tunnel at a dynamic pressure of 6 pounds per square foot, the airspeed resulted in the same Reynolds number obtained by Northrop Grumman in previous underwater tests. In the Langley Full-Scale Tunnel, forces and moments were measured with a sting-mounted, six-component internal strain-gage balance. Results of the investigation indicated that the circulation-control sail concept produced very high lift forces on the sail that were virtually independent of flow incidence angle, and that the forces would produce high levels of maneuverability for horizontal turns. The control effectiveness was maintained over a range of drift angles of ±30°.

Communication Towers

The proliferation of transmission towers and other support structures for cellular and communication systems has resulted in numerous community objections to the environmental

Tests of a cell-phone tower disguised as a tree were conducted in 2007 to determine aerodynamic loads. Note the protective net on the front of the exit cone. The inset shows another test configuration. (ODU)

and scenic effects of these structures. Arguably, one of the most unusual tests ever undertaken in the Full-Scale Tunnel occurred when an organization requested a brief investigation of the air loads experienced by a camouflaged communication tower. In the study, artificial foliage in the form of leafed branches, palm leaves, and other arrangements were attached to a series of representative tower structures in order to obscure the tower configurations.

Nature's Apocalypse: Hurricane Isabel

By 2003, ODU operations at the Full-Scale Tunnel were proceeding very well. The NASCAR motorsports community was now aware of the relatively cost-effective use of the facility, and results obtained from testing entries had established credibility during competition. Race teams were scheduling multiple entries per year, and competition for tunnel entries during the racing off-season was heated. The Bihrle staff had established a solid tie-in for the tunnel to military aircraft projects, and the facility had been upgraded with hardware and data-acquisition systems. Although the threat of severe weather in the form of nor'easters and hurricanes was ever-present, only minor flooding had impacted operations during the first 7 years of operation. That would soon change.

On September 6, 2003, Hurricane Isabel was formed from a tropical wave in the Atlantic Ocean. Moving northwest within a storm-feeding environment of light wind shear and warm waters, it steadily strengthened to reach peak winds of 165 mph on September 11. After fluctuating in intensity for 4 days, Isabel made landfall near the Outer Banks of North Carolina, with winds of 105 mph, on September 18. At Langley Air Force Base, about 6,000 workers were ordered to evacuate elsewhere due to anticipated flooding. In some locations in Hampton Roads, the storm surge exceeded that produced by the 1933 Chesapeake-Potomac Hurricane. Although it was classified as a hurricane, sustained winds only reached about 70 mph; however, hurricane-force wind gusts were recorded, with unofficial reports peaking at 107 mph near the Northern Peninsula. Hurricane Isabel proved to be the costliest disaster in Virginia's history.

Today, Jim Cross considers Hurricane Isabel to have been the death knell of the ODU motorsports enterprise. At the evacuated Full-Scale Tunnel, the flooding, damage, and aftermath were devastating. Flooding within the structure rose to depths greater than 4 feet in the offices, workshops, model-preparation areas, and tunnel test section. The facility was shut down for over 4 months for repairs and cleanup activities after the hurricane, during the year's highest-demand time for testing by the NASCAR community. Most of the NASCAR teams went elsewhere because, as will be discussed, new tunnel testing services for motorsports had become available, and the loss of revenue at the Full-Scale Tunnel was significant. In addition, ODU had to absorb the costs of refurbishing the facility after Isabel.[16]

Nature continued to plague the ODU operation following Isabel, especially in 2006 when a severe nor'easter caused yet another shutdown for repairs, during which additional revenues were lost. Despite a $400,000 recovery gift from the Air Force, the tunnel enterprise lost significant income.

Passed by the Competition: Other Tunnel Test Facilities

As enthusiasm and test activities with race teams at the Full-Scale Tunnel continued to flourish, it became known that a new competitor for motorsports wind tunnel testing was being organized in North Carolina, the hotbed of NASCAR country. A new organization known as AeroDyn was being formed in Mooresville, in close proximity to NASCAR teams. Led by Gary W. Eaker, a former engineer at the General Motors Aerodynamic Laboratory and Hendrick Motor Sports, AeroDyn aggressively pursued a position providing wind tunnel testing capability within the NASCAR testing community. The wind tunnel designed by AeroDyn was optimized for full-scale stock racecars and included an optimally slotted–wall test section with an open return. The closed test section was 11.9 feet high, 19 feet wide, and 56 feet long, with test section speeds up to about 130 mph. Jim Cross and his staff were aware of the emerging competition, but they were not immediately concerned because they felt the planned facility was too complex for AeroDyn to facilitate. They were wrong.[17]

AeroDyn provided NASCAR customers with three major items of interest: the convenience of conducting brief wind tunnel testing in close proximity to the racing teams lowered travel costs and reduced the deployment of personnel; the direct wind tunnel fees in the AeroDyn tunnel were extremely competitive (less than $1,400 per hour); and simulation of under-car aerodynamics was provided by a boundary-layer control system.[18]

The interest in motorsports and ground-vehicle testing in wind tunnels has grown exponentially in the years since the Full-Scale Tunnel was closed. For example, in September 2008, Windshear Incorporated of Concord, NC, began operation of a single-return, 180-mph rolling-road wind tunnel with an open-throat test section.[19] Clemson University also explored the possibility of a new $40 million rolling-road tunnel in Greenville, SC.[20]

In 2010, ODU briefly considered returning to the wind tunnel business with a new facility. Colin Britcher, then Chairman of the University's Aerospace Engineering Department, led an ODU effort to determine the feasibility of building a wind tunnel in Danville, VA, with the aid of funding from the state's economic development funds for tobacco-dependent communities.[21] Analysis of the project's building and operational costs quickly revealed that the concept was unpractical.

Mission Restored: Return of Aircraft Testing

The Eagle's Beak: The F-15

The first aerospace test under the ODU arrangement had been conducted in 1997 as a follow-on experiment to subscale wind tunnel tests that had attempted to determine the contribution of the F-15 Eagle's pointed fuselage forebody to asymmetric yawing moments at high angles of attack. As discussed in previous chapters, fighter aircraft configurations had exhibited extreme sensitivity to geometric conditions on the forebody during subscale and full-scale flight tests. In many cases, the magnitude of the asymmetric aerodynamic yawing moment exceeded that provided by conventional rudders. Although the subscale model tests had provided information on the potential problem, questions remained regarding the direct application of tunnel data to full-scale conditions. In particular, what might be the impact of Reynolds number for typical nose shapes and perturbations in geometry for service aircraft?

In 1995, a high-angle-of-attack uncontrollable departure (i.e., "nose slice") occurred for an F-15 during a typical combat maneuver at Nellis Air Force Base. Inspection of the aircraft after the incident revealed irregular surface conditions on the forebody radome, raising suspicions that the aerodynamics of the irregular nose might have been the cause of the departure. When the radome was replaced, the aircraft demonstrated acceptable behavior. As a result of the incident, the Air Force requested tests in the Full-Scale Tunnel to assess the impact of the forebody imperfections for representative service conditions and thereby provide information on radome/forebody maintenance requirements to minimize future incidents.[22]

Test setup for a full-scale F-15 forebody for Bihrle Applied Research and the Air Force in 1996. Tests were conducted to identify the effects of the physical conditions of the nose tip and forebody on large asymmetric yawing moments at high angles of attack. The forebody was sting-mounted, with an internal strain-gage balance and pressure instrumentation. (BAR)

The F-15 radome test was a combined effort, with participation by ODU, Bihrle Applied Research, and the Air Force. Production radomes were used in the tests, which included extensive pressure, force, and moment measurements for angles of attack up to about 60°. All F-15 radomes include a metal nose cap at the apex of the nose, over the end of a rain erosion boot to minimize deterioration during flight. Several nose caps were tested in a very detailed examination of the impact of nose-tip and rain-erosion-boot configurations on yawing moments at high angles of attack.

The results of the study clearly demonstrated that significant yawing moments were generated by very small geometric imperfections near the apex of the radome. Minute nose-cap geometric properties such as trailing-edge overlap, flat spots, extruded sealants, and uneven paint erosion were found to trigger significant yawing moments at moderate and high angles of attack. Similar effects were found for the rain-erosion boot. The results also showed that radome repair patches and surface damage located further aft on the radome had minimal effects on high-angle- of-attack yawing moments.

Formation Flight

In the late 1990s, BAR was awarded a Small Business Innovative Research (SBIR) contract by the Air Force Research Laboratory to explore the aerodynamic interference effects of a lead aircraft on a trailing aircraft in close formation. The objectives of a study in the Full-Scale Tunnel would be to acquire force and moment data that would be used as inputs to a 6-degrees-of-freedom simulation to study the effects of aerodynamic interference phenomena on performance, stability, and control. A 1⁄10-scale model of the F/A-18E was used in the trail-aircraft configuration.[23] Results of the investigation provided quantitative measurements that were used to assess performance enhancements of the trailing aircraft in formation flight and the magnitude of asymmetric moments generated by the flight condition.

This early exploratory study proved the feasibility of obtaining aerodynamic data in simulated formation flight and led to additional SBIR efforts to acquire aerodynamic characteristics for several other aircraft configurations. The goal of the testing was to determine formation positions and aircraft conditions that would maximize performance benefits (e.g., lift-to-drag ratios) while minimizing forces and moments that would degrade aircraft handling qualities. During testing, static and dynamic forces and moments, surface-pressure data, and wake-survey data were measured. In addition, studies were made to determine the control power required to enter, exit, and remain in beneficial formation positions.[24]

The scope of testing focused on the characteristics of two different trail configurations involving the F/A-18E and a delta-wing configuration representative of future uncrewed aerial vehicles (UAVs). Data were generated for two F/A-18E models as a paired-aircraft formation and for the UAV configuration as both a paired configuration and as a trail aircraft to study issues related to the challenge of air-to-air refueling of uninhabited aircraft. Several UAV configurations were tested behind a model of the KC-135R tanker aircraft, which included a refueling boom and electric fans within the engine nacelles to simulate the exhaust flow from the engines. The results of the studies demonstrated that tools and test techniques for measuring formation-flight effects in the wind tunnel were valuable in the analysis of interference effects. The data included force and moment measurements on both lead and trail test subjects; control required to trim; surface pressures; and wake-flow angularity.

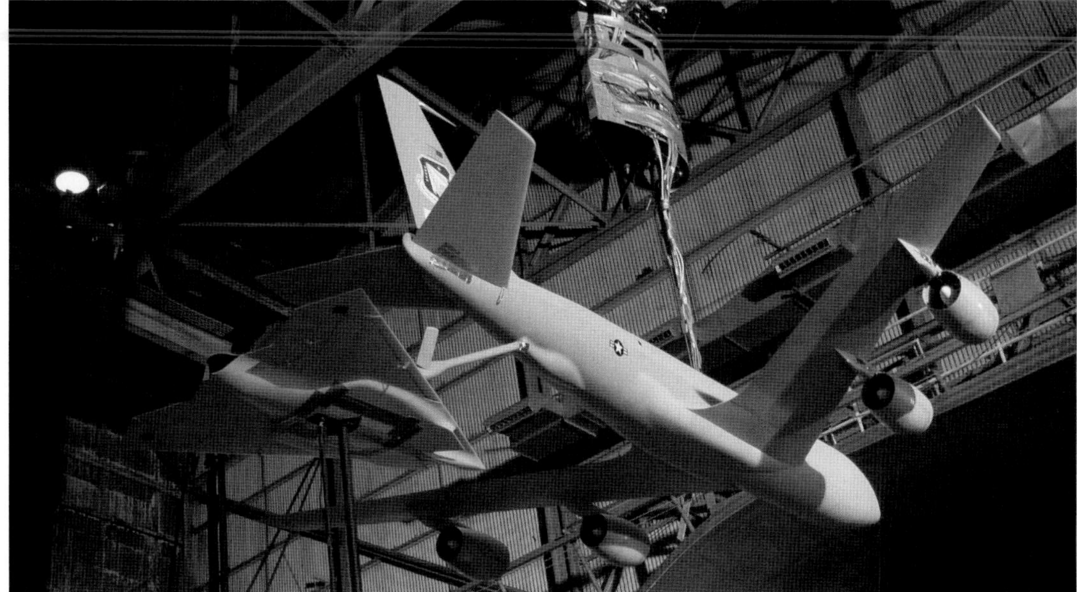

A model of a delta-wing uncrewed aerial vehicle is positioned to acquire aerodynamic data during a simulated air-to-air refueling mission behind a model of a KC-135R tanker. (BAR)

Return of the Wright Spirit

One of the most publicized test activities at the Full-Scale Tunnel during the early 2000s triggered a wave of interest in the engineering prowess of the Wright brothers and their experiences in unlocking the doors to practical controlled flight. The Full-Scale Tunnel became an integral component of the quest to document technical data on the characteristics of the Wright aircraft and the approaches used in the design of their vehicles.

Tongue-in-Cheek Humor Proves Prophetic

The early NACA organization at Langley in the 1930s and 1940s was well known for its social cohesiveness, parties, and fraternity-like atmosphere. Gossip and humor were mainstays of the internal Langley newspaper, which began in the 1940s and became known as *The Air Scoop*. In its February 2, 1945, issue, the newspaper contained a tongue-in-cheek cartoon directed at the Full-Scale Tunnel and its involvement in the historic early days of flight. The cartoon depicted an old biplane flying in the test section of the Full-Scale Tunnel with the caption, "Oh, that flies through here every year on Wilbur Wright's birthday!" The artist could not have imagined that 58 years later a similar event would actually happen!

This cartoon appeared in Langley's internal newspaper in 1945, poking fun at the old Full-Scale Tunnel. The cartoonist could have never known that such a test would occur 58 years later. (NASA)

The Discovery of Flight Foundation

The Discovery of Flight Foundation was established in 1999 to attract the resources necessary to preserve and promote the legacy of early aeronautical invention. The nonprofit foundation sought to rediscover the methods of experimentation and discovery used by the Wright brothers, provide for authentic recreation of their original aircraft, and create a living classroom for school children and people of all ages. To achieve this ambitious plan, the Foundation assembled the world's leading experts. The production team, headed by Ken Hyde, would research, design, build, and test authentic, full-scale reproductions of the developmental aircraft and engines used by the Wrights.

Full Circle: The Wright Experience

Ken Hyde, president of the Warrenton, VA, organization known as The Wright Experience, was commissioned by the Experimental Aircraft Association of Oshkosh, WI, to build a duplicate of the original Wright 1903 Flyer using identical materials and the Wrights' original design.[25] Hyde's passion for aviation came at an early age, and he earned both his pilot and mechanic licenses while still in high school. An American Airlines pilot for 33 years, he subsequently founded Virginia Aviation, an antique aircraft restoration company that has gained national attention. He formed The Wright Experience in 1992 with the objective of having the replica fly at Kitty Hawk, NC, on or near the December 17, 2003, anniversary date of the Wright Flyer's first flight. The project was designed with a deeply scientific focus and was definitely not planned as a media stunt. Despite the brilliant achievements of the Wright brothers, documentation of their technical data and achievements (e.g., powerplant, propellers, airfoils, controls, etc.) was virtually nonexistent. By conducting carefully controlled tests in the Langley Full-Scale Tunnel, Hyde envisioned an opportunity to learn the secrets of the Wrights that had been hidden for so many years.[26]

Previous testing of the 1903 Wright Flyer configuration had taken place in other wind tunnels, most notably the GALCIT 10-Foot Tunnel and the NASA Ames Research Center's 40- by 80-Foot Tunnel in 1999.[27] However, none of the previous efforts had approached the authenticity and detailed accuracy proposed by Hyde's project.

Hyde contacted A.G. "Gary" Price of the Langley Public Affairs Office to explore the possibility of testing in the Full-Scale Tunnel. After Jim Cross indicated his interest in supporting the project from an educational perspective, Bob Ash was assigned the responsibility of raising funds to cover the estimated $250,000 required for the test program. Although unsuccessful in raising the full funding, ODU conducted the tests anyway. Rather than leaping ahead with tests of a complete Flyer airplane, Hyde rightfully proposed that it was first necessary to evaluate the performance of replicas of the Flyer's propellers, and of the gliders with which the Wrights learned many secrets of the science of flight.

Propellers

One of the critical steps in Hyde's project was the authentic remanufacture, testing, and evaluation of the Wright propellers.[28] The Wright brothers had been among the first to recognize propellers as rotating-wing sections moving air that had been accelerated by the propeller disc before meeting the blade section. However, there is little documentation on their methodology (other than their use of a wind tunnel for testing) and how they were able to provide adequate thrust for flight. The first propeller to be fabricated (handcrafted by a woodworking expert on the Wright Experience team) and tested in the Full-Scale Tunnel was from the 1911 Model "B" Wright Flyer because it is the best-documented early Wright airplane. NASA Langley supported the propeller test program under a cooperative educational agreement.

Testing of the Wright propellers was conducted using a NASA-developed thrust-torque balance, shown on the left, and the propeller-test-stand setup on the right. (ODU)

In October 1999, Bob Ash and Stan Miley led a team during tests of two propellers that were authentically crafted by Hyde's Wright Experience team. During the propeller tests, the test articles were mounted on a test tower and a 25-horsepower variable-speed motor was used for power.[29] A calibrated thrust-torque balance that had been conceived by H. Clyde McLemore of the NASA Full-Scale Tunnel staff for propeller testing in the 1970s was provided by NASA for the projects. The balance was connected between the propeller and the driveshaft for measurements of thrust and torque.[30] Before the tests in the Full-Scale Tunnel it had been believed that the 1910–1911 Wright propellers were 70-percent efficient, but data from the tunnel tests demonstrated a much higher nominal peak efficiency of 81.5 percent. The performance of the remanufactured Wright propeller was amazing, considering that today's wooden propellers are only 85 percent efficient. The gain of only 5 percent in 90 years clearly demonstrated the Wright brothers' ability as engineers.

The 1911 propeller has become known as the "bent-end" propeller design because of its unique shape. Additional testing of the 1903 Flyer propeller and the 1904 Flyer propeller was completed in December 2000. The static thrust performance of the 1903 and 1904 propellers showed very good to excellent agreement with the Wright's own tests.

The First Step: Gliders

In 1900, the research conducted by the Wrights had proceeded with encouraging kite tests and the establishment of North Carolina's Outer Banks as a field laboratory. That year, they attempted to fly a crewed glider at Kitty Hawk, but the glider proved incapable of carrying the weight of the pilot. They were, however, able to accumulate a few minutes of flight time to assess controllability before the 1900 glider suffered a major crash. The Wrights returned to the Outer Banks in 1901 with the largest glider ever built (with a wing span of 22 feet) for more assessments of their design approach and controls. After conducting uncrewed kite tests of the glider, they conducted piloted tests where they learned that the glider produced insufficient lift and exhibited severe control problems—especially lateral directional control. Upon returning to Dayton, they built a small-scale wind tunnel to evaluate nearly 200 airfoil and wing models.

The test of the 1901 glider was conducted a century later in 2001 by a team led by Bob Ash of the ODU staff and Dr. Kevin Kochersberger, a professor on sabbatical from the Rochester Institute of Technology who was working with the Wright Experience. The test program included a range of tunnel speeds that encompassed the stall and maximum gliding conditions for angles of attack up to 20° and sideslip angles up to 15°.[31] Results of the investigation provided measurements of lift, drag, and moments including control power from deflecting the canard and warping the wing for roll control.

As received for testing in the Full-Scale Tunnel, the glider was found to have a lift-to-drag ratio of only 3.9. The result was in marked disagreement with flight results, as the Wrights had meticulously measured a much higher ratio of about 6.0 during their flights. The test team considered several potential effects that might have increased lift-to-drag ratio in flight, including ground effect, dynamic soaring, and wind factors. One item of particular interest was the potential effects of the wing covering material's physical characteristics. Ken Hyde's team was certain that they had reproduced the geometry, structure, and material of the covering down to the correct thread count per inch of fabric; however, the effects of porosity of the wing fabric became an issue. Hyde had discovered an entry in one of the Wright notebooks that suggested that the glider wings had been coated with something. After deciding that the coating was probably wallpaper paste (flour and water), the team prepared a mixture of wallpaper paste and brushed it on the wings. When the modified glider was tested in the Full-Scale Tunnel, the new measurements indicated a maximum lift-to-drag ratio of about 6.0.[32] The results also confirmed the large adverse yawing moments that had been experienced by the Wrights and was produced by using wing-warping for roll control.

Overhead front view of the 1901 Wright glider test setup in the Full-Scale Tunnel. (ODU)

After returning from Kitty Hawk with many lessons learned from the 1901 glider flights, the Wrights conducted intense studies of airfoils that might provide sufficient lift for their flying machine. Systematic investigations of wing camber, thickness, and aspect ratio provided guidance for the design of a new glider to be flown in the fall of 1902. The aspect ratio of the wing of the new 1902 glider was twice that of the 1901 glider (6.8 compared to 3.4), and it had a wingspan of 32 feet. Initially, two vertical-tail surfaces were included in an attempt to alleviate the lateral directional control of the earlier glider. However, the problem persisted, and Orville suggested that a single-surface rudder interconnected to the wing-warping mechanism might mitigate the issue. The resulting breakthrough in controllability greatly enhanced the flying characteristics of the glider, and extended flights of over 300 feet became commonplace. In October, they logged 250 flights and set a distance record of 622 feet for 26 seconds. In their 9-week test program, both brothers learned to fly the glider and completed about 1,000 flights. When they returned to Kitty Hawk in 1903 with a powered aircraft, the 1902 glider was flown again for pilot proficiency. The single vertical tail was replaced with a two-surface rudder, and the Wrights achieved a flight duration record of 1 minute and 12 seconds.[33]

In June 2002, wind tunnel tests of a replica of the 1902 glider were carried out in the Full-Scale Tunnel by a team of ODU, The Wright Experience, and BAR. The test setup was unique because the glider was mounted to a three-strut support system using inboard–wing attach points at the bottom wing's leading-edge spar and the center of the rear spar. Bearings were used on the rear-spar attach point to permit wing-warping inputs.

The test setup for the 1902 glider in the Full-Scale Tunnel in 2002 shows the increased wing aspect ratio compared to the earlier 1901 glider. (ODU)

Results of the tests yielded quantitative information regarding the performance, stability, and control of the vehicle. The maximum lift-to-drag ratio for the glider was about 7—a remarkable engineering achievement by the Wrights. Data gathered in the entry were also used as inputs to a simulator of the 1902 glider developed by BAR and The Wright Experience. Dynamic aerodynamic inputs (e.g., pitch-damping and roll-damping) were determined by correlating simulator results with flight-test experience.

The 1902 glider provided solutions to many of the problems faced by the Wrights by demonstrating an effective three-axis control system, verifying the airfoil and wing performance derived from the small Wright wind tunnel, and providing pilot proficiency before the first powered flight of the Flyer in 1903.[34]

Grand Finale: The 1903 Flyer

After almost 10 years of painstaking research and detailed, faithful reproductions of propellers, engine, and airframes, Ken Hyde's 1903 Wright Flyer replica was ready for testing in the Full-Scale Tunnel in February 2003. The stage was set to move on to the First Flight Centennial Celebration at Kitty Hawk on December 17, 2003, following the wind tunnel tests and intense pilot training via simulators and preliminary flights.

To even casual observers, the tests of the 1903 Flyer in the tunnel evoked a flood of memories and notable relationships to the legacy of the Full-Scale Tunnel, most of which are documented in this book. The role of Orville Wright as a key participant in the NACA committees and his many trips to the Langley Memorial Aeronautical Laboratory in the 1930s coincided with the construction and initial operations of the Full-Scale Tunnel, and his presence in the group photograph taken during the 1934 Engineering Conference led to reflections on the historic significance of the 2003 tunnel test. Orville could never have anticipated in 1934 that a replica of his airplane would be mounted in the test section of the Full-Scale Tunnel 69 years after he was seated in the bleachers just a few feet away.

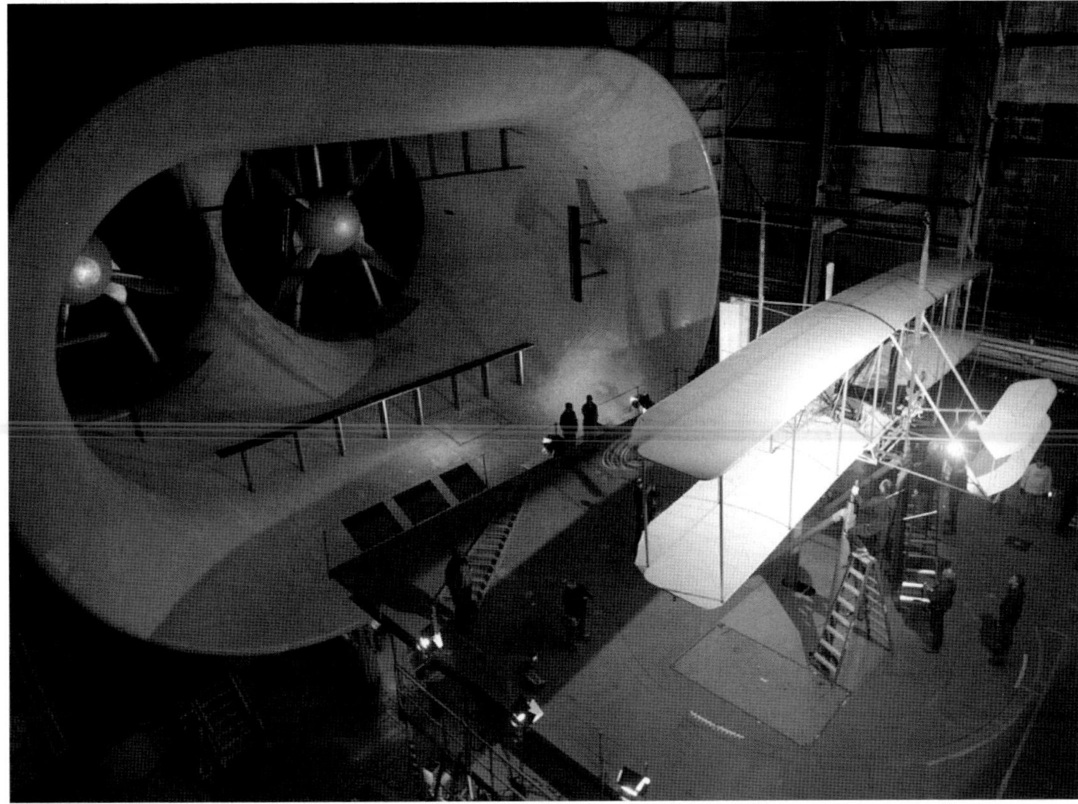

The Wright Experience test series concluded with tests of the 1903 Wright Flyer in 2003. (ODU)

Three years of construction and fabrication by craftsmen had come to fruition as the Flyer was mounted and the test program began. Ken Hyde was taking a leap forward in uncovering and documenting how the Wright brothers, neither of whom finished high school, managed to conquer the principles of controlled, powered flight in 5 short years.

Built with help from the Ford Motor Company and the Experimental Aircraft Association, the Wright Flyer used a 20-horsepower AC electric motor that could be controlled precisely during wind tunnel testing.

The excitement and anticipation over the tunnel tests were captured by a NASA press release:

> Hyde said, "Our journey will continue through December 17th this year with the flight of this 1903 Wright flyer reproduction at Kitty Hawk. These wind-tunnel tests will help us recreate the Wrights' historic accomplishment and help us reduce the risk involved in the flight.

> "We can't predict what the weather will be on December 17th 2003, when the Wright Experience plans to fly the EAA Flyer reproduction," said Professor Robert Ash, Wright test program manager for ODU. "We only know that the original Flyer could be flown on a cold day into a 27 mph wind. The wind-tunnel test results will give us the necessary knowledge to guide and train pilots for virtually all eventualities."
>
> Four Wright Experience pilots trained for the upcoming flight. They've gotten expert guidance from a simulator, created using the Langley Full-Scale Tunnel data, and a former NASA test pilot, Scott Crossfield. Crossfield was the first American to fly at twice the speed of sound, 50 years ago.
>
> One of those aviators is Kevin Kochersberger, an associate professor from the Rochester Institute of Technology in Rochester, N.Y., who helped oversee most of the wind tunnel tests. During training in North Carolina last month, he successfully got the Flyer reproduction off the ground.
>
> "Being a scientist and engineer are important qualifications for flying this aircraft," said Kochersberger. "I've been looking at the characteristics of the Wright Flyer for four years. Being in the wind tunnel with it really made a difference."[35]

Team members analyzed the aerodynamic data that was obtained in the Full-Scale Tunnel tests, and Kevin Kochersberger and other pilots trained on a BAR-developed flight simulator with data based on the results of the tunnel tests.[36] Experience with the simulator was especially valuable because the Flyer was extremely unstable in pitch (about 17 percent unstable). Kochersberger subsequently made three flights with the Flyer at an encampment at the Wright Memorial in Kill Devil Hills, NC, from November 3 to December 7. The flights were made with an onboard digital data flight recorder provided by ViGYAN, Inc.

On December 17, the fourth flight was attempted on the anniversary of the Wrights' flight a century before. A lack of wind and an engine that progressively misfired during the ground roll resulted in a 6-inch lift from the rail-launching dolly and a stall that settled the plane back to the track. The lack of engine power, partly due to unfavorable atmospheric conditions, precluded a successful flight. No more flights were made on this aircraft, and after the Centennial celebration, the aircraft was shipped to Dearborn, MI, for permanent display at the Henry Ford Museum.[37]

Hollywood Arrives: *The Box*

The Full-Scale Tunnel proved to be the mystic cathedral of wind tunnels that George Lewis had envisioned in 1929. Photographers and the public were deeply impressed by its sheer size and the powerful sounds of its drive system. The first-time visitor who climbed the stairs to the ground board and observed the huge test section was awed by the sight. Any media discussion of the NACA always contained a photo of the gigantic open test section and the unique twin-drive motors. The nickname "Cave of the Winds" became popular and was used in widespread literature in the 1930s and 1940s. Through the years, numerous photographs appeared in a wide range of coverage, including such prestigious publications as *National Geographic* magazine and many aerospace periodicals such as *Aviation Week & Space Technology* magazine. During the late 1950s, Jules Bergman, the famous science editor of ABC News, featured a nationally televised segment on the tunnel with spectacular smoke-flow visualization of test subjects.

The test section of the Full-Scale Tunnel was transformed into a mystical headquarters office of one of the central characters in the film *The Box*. View is looking forward into the entrance cone of the tunnel. (NASA)

Following on the heels of the extensive media exposure afforded to The Wright Experience activities, the Full-Scale Tunnel would return to the public's eye. In January 2008, Hollywood came to Langley for the filming of the psychological horror movie *The Box*, in which a couple finds a wooden box on their doorstep and are told that they will become rich if they push a button on the box—but if they do so, someone will die. Starring actress Cameron Diaz and actors James Marsden and Frank Langella, the movie was shot at several locations at the Langley Research Center, including the aircraft hangar and the gantry formerly known as the Lunar Landing Training Facility. The father of the movie's director, Richard Kelly, had worked at Langley in the 1970s and 1980s. Scenes of Diaz, Marsden, and Langella were shot in the test section and the west return passage of the Full-Scale Tunnel on January 28–30. *The Box* premiered in Virginia in November 2009 and received mixed reviews after its worldwide release.[38]

The Centerpiece: The Blended Wing Body

The last 5 years of operations at the Full-Scale Tunnel included a major emphasis on the radical aircraft configuration concept known as the Blended Wing Body (BWB). The BWB was stimulated by Dennis Bushnell, senior scientist of the NASA Langley Research Center in the late 1980s, when he challenged the aerospace community to explore novel approaches for new aircraft designs that might provide breakthroughs in aircraft lift-to-drag ratios—a performance index for subsonic transports that was rapidly becoming stagnant because of diminishing advances.[39]

Bushnell's colleague, Robert H. Liebeck of the Long Beach, CA, division of the McDonnell Douglas Corporation (now the Boeing Company), accepted his challenge and pursued several revolutionary configurations, including a radical BWB configuration with the potential to revolutionize aircraft efficiency standards. Aided by Langley funding in April 1993, Liebeck and his associates began to refine their initial BWB concept, and interest in the concept soon accelerated with numerous tests in Langley facilities, including static and dynamic force tests, free-spinning tests, and tests in the Langley National Transonic Facility.

By the turn of the century, a vast number of research efforts on the BWB had been conducted by Boeing and NASA. In the late 1990s, interest in the concept resulted in a proposal to NASA Administrator Daniel S. Goldin from industry and Langley BWB enthusiasts for a piloted X-plane program with a projected cost of about $130 million. There were two technology areas identified early in the BWB development process that required further research: flight dynamics and control, and noncylindrical pressure vessels. These were the focus of the cooperative NASA-Boeing research, and a piloted X-plane would address both areas. Unfortunately, the proposed X-plane program was turned down by the NASA Administrator. After a piloted plane was dismissed by NASA management, the cooperative research team turned their attention to the potential uses of unpiloted subscale vehicles to provide technical data and further progress in the development of the radical concept.

In 1997, the highly successful NASA space mission to land a rover named Sojourner on Mars created a huge spike of interest in space exploration from the scientific community, the media, and the public. Darrel R. Tenney, Langley's director of the Airframe Systems Program Office, met with Joe Chambers, then chief of the Aeronautics Systems Analysis Division, to plan an aeronautics effort that might bring similar excitement and support to the stagnating aeronautics elements of the NASA program. The resulting plan consisted of identifying revolutionary aircraft concepts that had been developed by NASA, industry, and DOD, and submitting them to flight evaluations using unpiloted subscale models at NASA Dryden. The program, known as the Revolutionary Concepts for Aeronautics Program, included plans for a 35-foot-wingspan BWB flying model designated the X-48A. Construction of the X-48A model was started in Langley's shops, but the program was terminated by NASA because of higher priority commitments. After yet another disappointing cancellation within NASA, the Langley BWB team turned to less expensive, traditional free-flight model tests in the Full-Scale Tunnel.

By the early 2000s, over 10 years of research and development had been invested in the BWB, and from the public and media perspectives, it had become the centerpiece of advanced concepts and excitement within the NASA aeronautics program. The time had come to conduct some very critical free-flight tests in the Langley Full-Scale Tunnel.

Return Home: BWB Free-Flight Tests

After conducting extensive tests of many different BWB variants in the Langley 14- by 22-Foot Tunnel, the 12-Foot Low-Speed Tunnel, and the Langley Spin Tunnel, the NASA Flight Dynamics Branch pursued NASA and industry interests in conducting a free-flight study of the dynamic stability and control of the BWB configuration.[40] The unique data obtained from free-flight tests could not be obtained from other types of test techniques, and the risk reduction provided by flight tests—especially regarding controllability and specific use of control surfaces—was highly desirable.

The Pains of Homecoming

As previously discussed, when the Vehicle Dynamics Branch (later changed back to Flight Dynamics Branch in 2004) moved from the Full-Scale Tunnel and installed its subscale model–testing techniques in the 14- by 22-Foot Tunnel, virtually all of the test equipment had been removed from the installations at the old tunnel. Two exploratory free-flight investigations had been conducted within the test section of the 14- by 22-Foot Tunnel using models of the X-29 and F/A-18E.

Unfortunately, the procedure for arriving at the design parameters for dynamically scaled free-flight models requires closure for physical parameters involving weight, geometry, and moments of inertia. For some models, it is difficult or even impossible to arrive at physical characteristics that are compatible with capabilities of the test facility. In the case of the BWB, the NASA researchers found that achieving a free-flight model for the large, low-density BWB configuration would be impossible to test in the 14- by 22-Foot Tunnel, but a BWB dynamic model could be designed for testing in the Full-Scale Tunnel. With great urgency, an approach to re-installing the free-flight test technique in the Full-Scale Tunnel was mapped out.

Sue Grafton led the team responsible for reviving free-flight tests at the Full-Scale Tunnel. As might be expected, numerous efforts were required to re-install and update the free-flight capability, including the acquisition of a new flight computer, the installation of compressed-air power for model thrust, and the installation of electrical circuits for onboard controls and instrumentation. After Grafton's team completed its considerable tasks involving hardware and software, the NASA staff faced yet another challenge—this time involving human factors.

Over the 50 years that free-flight models had been tested in the Full-Scale Tunnel, it had been demonstrated time and again that providing pilot proficiency and training in flying dynamic models was absolutely critical to the success of such projects. Piloting the models was a unique challenge requiring special skills and experience. In fact, on many occasions, visiting industry and DOD pilots associated with aircraft development programs had been given the opportunity to fly the free-flight model, only to find that they quickly created pilot-induced oscillations that terminated the flights. This experience was not like flying an actual airplane. In preparing for the BWB-model flight tests, it became evident that many of the researchers who had served as experienced pilots for free-flight models in the past had not participated in a free-flight test for over 5 years and were extremely lacking in piloting skills and proficiency. The existing X-29 free-flight model was brought back to the Full-Scale Tunnel for free-flight training, and it was crashed on several occasions while bringing the pilots back up to a level of proficiency for the BWB program. Another impact of not conducting free-flight testing at the 14- by 22-Foot Tunnel during the previous years was that experienced key personnel had moved on to other assignments and were not readily available for the flight testing.

The Last Free-Flight Test

The BWB free-flight model was a three-engine, 0.05-scale model of a configuration that had been extensively tested at Langley. With a 12.4-foot wingspan, the 92-pound model was the largest free-flight model ever tested in the Full-Scale Tunnel. The dynamic-scaling requirements for the model were very challenging, particularly in complying with moments of inertia in roll. Because of the scaling sensitivity in roll inertia, the wings of the model were left unpainted and the scale of the model was chosen to be as large as practical for free-flight testing in the tunnel.[41]

The design and evaluation of control laws for the model's flight-control system was a primary focus of the free-flight investigation that was conducted in September 2005. With over 20 individual wing-trailing-edge control surfaces available, plus a fixed interchangeable leading edge, the optimal application of individual surfaces over the desired range of angles of attack was of importance.[42] Landman and NASA's Dan Vicroy led an experiment

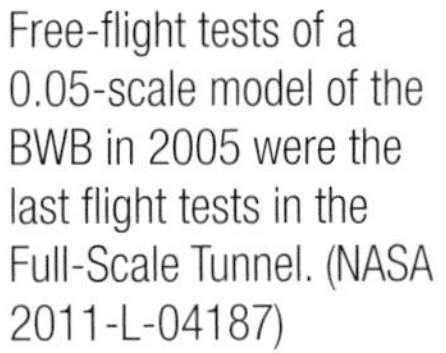

Free-flight tests of a 0.05-scale model of the BWB in 2005 were the last flight tests in the Full-Scale Tunnel. (NASA 2011-L-04187)

designed to evaluate the interactive effects of the numerous control surfaces.[43] The overall objectives of the flight test were to characterize the 1-g high-angle-of-attack characteristics of the BWB, including a definition of the minimum-control speed with asymmetric thrust and the effectiveness of using thrust vectoring for the center engine for additional control in yaw. Configuration variables such as center-of-gravity location, deflected wing leading-edge slats, and asymmetric thrust were included within the scope of testing, and preliminary force and moment testing of the powered model were conducted before the flight tests.

Some of the more critical results of the free-flight investigation included the fact that the configuration had limited directional control authority. Use of center-engine thrust vectoring provided significant augmentation of directional control in an outboard engine–out condition, resulting in a significant lowering of the minimum control speed. It was also demonstrated that control-interference effects could be significant with multiple trailing-edge control deflections and that such effects must be accounted for as aerodynamic inputs for simulation models.

The Boeing X-48B

When NASA management derailed enthusiasm for potential crewed or uncrewed X-48 subscale vehicles, Bob Liebeck and the Boeing Phantom Works organization forged ahead in 2002 with arrangements for the fabrication and flight testing of two turbojet-powered uncrewed vehicles designated X-48B. The 0.085-scale models were built by Cranfield Aerospace, Ltd., in England. Made primarily of advanced lightweight composite materials, the X-48B models had wingspans of 20.4 feet, weighed 523 pounds, and were designed for speeds up to about 140 mph and altitudes as high as 10,000 feet, with flight durations of about 30 minutes. Testing of the first model (Ship 1) began in the Full-Scale Tunnel in April 2006 and ended in May.[44] The testing was a cooperative venture between NASA,

Ship 1 of the X-48B flight vehicles was tested in the Full-Scale Tunnel in 2006. (NASA)

Boeing, and the Air Force Research Laboratory. The Air Force was interested in exploring the potential of the BWB configuration as a multirole, long-range, high-capacity military transport or tanker.

Ship 1 completed 250 hours of tunnel testing in the Full-Scale Tunnel and was shipped to NASA Dryden as a backup for the flight testing of Ship 2. The second X-48B flew for the first time on July 20, 2007, in the first phase of a multiyear flight investigation. The last flight of the first-phase testing took place in March 2010 with flight Number 80. Flights resumed later that same year.

Preliminary results of the X-48B flight testing indicated that the vehicle was very responsive in roll and that the flight-control system was very robust.[45] Stalls were docile, and attempts to defeat a departure-limiter feature of the flight-control system were unsuccessful. At the same time, results indicated significant differences in pitch-trim predictions from several wind tunnels when compared to the flight results. The discrepancy was caused by model support-strut interference effects in certain test setups. However, pitch-trim data from the earlier X-48B Ship 1 tests in the Full-Scale Tunnel in 2006 showed good agreement with flight results.

Bob Liebeck praised NASA for its continued multi-Center support for the BWB, saying, "Without them this would have dried up a long time ago."[46]

The X-48C

The last wind tunnel project in the legendary 78-year operational history of the Full-Scale Tunnel was the Boeing X-48C, a modification of the Ship 1 X-48B model. Based on

The test program for the X-48C has the distinction of being the last test project in the Langley Full-Scale Tunnel. Note the revised engine and vertical tail configurations. Testing ended on September 4, 2009. (NASA)

cooperative Boeing and NASA ground-based acoustic research, the X-48B configuration was modified to further enhance the reduced-noise potential of the BWB configuration. The modifications included removal of the winglets and the addition of noise-shielding twin vertical tails with rudders, replacement of the original three turbojet engines with two larger fanjet engines, and an extended deck area.

Testing of the X-48C began in the Full-Scale Tunnel in June 2009 and ended on Friday, September 4, 2009.[47] After completion of the Full-Scale Tunnel test program, the X-48C model was sent to Dryden in October 2009 to be prepared for flight testing that was scheduled for 2012.

End of the Line: Final Closure

The Decision

Starting in the mid-1980s and continuing into the 1990s and 2000s, many studies were made by several national committees of the status and future requirements for national research test facilities. Over a dozen individual studies focusing on the future needs of NASA, industry, and DOD were conducted to plan the future national testing infrastructure that would support and advance anticipated U.S. aerospace needs.[48] In 2008, NASA completed an Agency Facilities Study, a mission-driven assessment of 300 major technical facilities for program utilization requirements through 2028. From the study, a NASA list of disposable facilities was developed, and NASA determined that the Full-Scale Tunnel was not a part of the core capability outlook for subsonic aeronautics research and development. The decision was accepted by a national assessment team and documented in a National Facilities Study. Subsequent national facility review committees did not reverse this decision.

The retirement and disposal of major Government wind tunnels had begun in earnest in the 1990s. In 1993, NASA and DOD had operated a total of 60 major wind tunnels, of which 39 were NASA tunnels. In 1997, the total had been reduced to 39 tunnels, with NASA operating 25 facilities. By 2006, the total major Government wind tunnels had been reduced to 30, of which 16 were NASA tunnels.

In arriving at its decision to dispose of the Full-Scale Tunnel, NASA solicited indications of interest in owning the facility from non-NASA organizations. Although an entity was interested in operating the tunnel, that interest did not extend to owning the facility because of its poor physical condition. The Agency also explored the possibility of transferring ownership of the Full-Scale Tunnel to other organizations, but this was constrained by the fact that the Full-Scale Tunnel was part of a land grant from the Air Force to NASA, and transferring ownership to a non-Federal entity would be very difficult.

NASA had transferred wind tunnels to other organizations in the past, including the transfer of the Langley Stability Tunnel to Virginia Tech in 1958, the transfer of the Langley 11-Inch Hypersonic Tunnel to Virginia Tech in 1973, and the transfer of the Langley Hypersonic Pulse Facility to General Applied Science Laboratory (GASL) in 1989. Leasing the Full-Scale Tunnel was also explored; the use of leasing is not a transfer of ownership but does allow the sharing of cost and the preservation of technical capability. Examples of leasing of NASA research facilities include the lease of the Ames Research Center National Full-Scale Aerodynamics Complex to the Air Force and the lease of the Ames Research Center's 7- by 10-Foot Subsonic Tunnel to the Army.

The Agency also considered other uses for the Full-Scale Tunnel, including conversion into an aerospace museum. The aftermath of the events of September 11, 2001, resulted in an intense increase in security at Langley Air Force Base, and public access to the secure military base would make such an application impossible. The official visitor's center for the Langley Research Center is the off-Center Virginia Air and Space Center in Hampton, VA, which provides access to the public. As will be discussed, NASA conducted extensive meetings with national museums including the Smithsonian for potential displays of artifacts from the Full-Scale Tunnel.

In a major briefing to the House Science and Technology Committee on September 1, 2009, NASA provided details on the status and condition of the Langley Full-Scale Tunnel, the results of national facility-review committees, and the overall process for demolition. The Agency also responded to congressional inquiries, highlighting the cost of returning the facility to a level considered safe and sustainable for continued occupancy, the lack of criticality for national needs as determined by many reviewers, and the adequacy of the Langley 14- by 22-Foot Tunnel and the Ames National Full-Scale Aerodynamics Complex (NFAC) for future national subsonic testing requirements. This briefing led to the decision to demolish the facility.

Preserving History

The decision to demolish the Full-Scale Tunnel had been made by NASA in the mid-2000s with the acknowledgment that the preservation of the legacy of the facility was mandatory, particularly in view of its National Historic Landmark status. Langley worked with state and Federal historic offices for more than 5 years to ensure that the facility was properly documented and that appropriate artifacts were preserved. Langley has made an exceptional effort to retain documents, photographs, motion-picture records, and documentaries that detail the 80-year history of the facility.

Firestorm of Protests

In its briefing to Congress, NASA emphasized that the renewal of technical capability and facilities is an ongoing part of the history and culture of the NACA and NASA that is critical to advancing and validating technology. In the introduction of his noted history of the NACA at Langley, *Engineer in Charge*, historian James R. Hansen recalled early observations by NACA managers of the intimate, almost religious, permanent bond that exists between researchers and their facilities:

> In a famous paper on wing section theory published by the NACA in 1931, Langley physicist Theodore Theodorsen suggested that the laboratory staff sometimes tied the progress of their work so completely to the use of test equipment

> that the equipment started to use them. For example, while possession of the world's first full-scale propeller research tunnel presented Langley in 1926 with a unique opportunity to explore systematically the potential of dozens of different cowling shapes and arrangements, having this large and costly research plant also obligated the lab's researchers to make full and routine use of the facility.
>
> The symbiosis between engineer and wind tunnel would grow so strong over the years that it was often almost impossible for management to put a machine out of business. The closing of some tunnels that had reached the point of diminishing returns—like the Propeller Research Tunnel in the 1940s—was accomplished only by overpowering stubborn defenders. Sometimes even after equipment was formally abandoned, old operators tried surreptitiously to run tests with it. Demolition proved the only sure way to end a tunnel's life.[49]

The fallout of the decommissioning of the Full-Scale Tunnel in 1995 was a prime example of the situation, but the tunnel's NASA civil-service staff at the time publicly accepted management's decision, albeit with great disappointment and concern. Many felt that the angst over the tunnel closure was tremendously aggravated by immediately reopening the operation under ODU management.

In 2005, news was spreading through the media regarding potential NASA plans to demolish the Full-Scale Tunnel and other old wind tunnels.[50] When NASA announced its intention to demolish the Full-Scale Tunnel after the X-48C studies in 2009, an immediate, very public protest and questioning of the impending closure arose in various forms, including letters to the editor in local newspapers and national publications such as *The Wall Street Journal* and *Smithsonian Air & Space*.[51] The closure was even mentioned on national television newscasts.[52] Most of the debate was waged by former customers who used the tunnel during the ODU era.

Planning for Demolition

After the conclusion of X-48C testing and the departure of the ODU staff in October 2009, the Full-Scale Tunnel joined several of its sibling NACA wind tunnels—the 8-Foot Transonic Tunnel (previously the 8-Foot High-Speed Tunnel), the 8-Foot Transonic Pressure Tunnel,

When vacated in October 2009, the Full-Scale Tunnel was showing the effects of corrosion and a lack of maintenance. The photo on the left shows large areas of corrosion on the side of the building facing the Back River, and the photo on the right shows the deterioration of the downspouts on the front of the building. (NASA)

the Low Turbulence Pressure Tunnel, and the 16-Foot Transonic Tunnel—as an abandoned hulk awaiting its turn for demolition. The demolition activities would begin a year later, in the fall of 2010.

Historic Artifacts

Mary Gainer, Langley's historic preservation officer, led the collection of historical artifacts from the Full-Scale Tunnel and served as a point of contact for museums interested in displaying artifacts from the tunnel. In 2010, representatives from the Smithsonian's National Air and Space Museum visited the tunnel to inspect the hardware for potential exhibits. After viewing the facility, the visitors indicated an interest in the propeller blades, propeller hubs, and scales. Subsequently, one complete fan assembly and a set of Toledo scales were requested by the Smithsonian and were transported to storage for a planned display at the museum in the near future. Several of the wooden fan blades that were installed in 1939 remained at Langley for display in the Center's newest building, for which construction is scheduled to begin in 2012. Other items discovered by Gainer and her crew, such as ornate metal firehouse reels, were found in hard-to-access areas. The designers of Langley's revised West Area buildings, known as "New Town," plan to incorporate several of these articles in the design of the new building.

Interns under Gainer's direction (with the assistance of the author) uploaded documents and videos of testing in the tunnel to a special Langley Web site at *http://crgis.ndc.nasa.gov/historic/30_X_60_Full_Scale_Tunnel*, where the material is available to the public. As the author noted in an interview at the time, although many of the tunnel's testing contributions were well known, tunnel personnel had also collected more than 3,000 photos and films that had never been seen before. Making such information available to the public on the Web was an effective tribute to the hundreds of personnel who had worked there. This effort allowed NASA to capitalize on an unprecedented opportunity to preserve its heritage for NASA's stakeholders: aerospace enthusiasts, historians, academia, and especially the public.[53]

The Celebration

On October 14, 2009, special guests and tunnel alumni gathered at Langley for a final "goodbye" reception and tour of the tunnel. The 300 attendees were treated to a multimedia presentation at Langley's Reid Conference Center, followed by guided tours of the Full-Scale Tunnel. Some of the historic free-flight models used in NASA tests were displayed, and a continuous slide show depicted many of the test activities conducted at the facility. Joe Chambers presented a multimedia review of the history of the tunnel, including videos of drag cleanup tests of early fighter aircraft during World War II and flight tests of models demonstrating the effectiveness of thrust vectoring for high-angle-of-attack maneuverability. During the tunnel tours, retirees such as H. Clyde McLemore and others shared their memories of the culture and projects in which they had participated.

The event was highlighted by interview sessions with personnel who had worked at the facility; a souvenir booklet titled "Full-Scale Tunnel Memories" and a DVD that contained the day's events were provided to attendees.

Summary of the ODU Years

The last 13 years of operations of the Full-Scale Tunnel were characterized by challenges and contributions to aeronautical and ground-vehicle technologies. The small ODU staff weathered an extremely difficult startup process to become the mangers of the world's largest university-run wind tunnel. Along the way, modifications to the tunnel were made to permit testing that had never been attempted in the past for a clientele that was totally unfamiliar with the tunnel and its capabilities. Attracting and growing a customer base was a major accomplishment, as was the integration of academic onsite courses.

Endnotes

1. The Langley 7- by 10-Foot High-Speed Tunnel was decommissioned in 1994 and demolished in 2009.
2. Ernest J. Cross, interview by author, January 24, 2011.
3. The NACA and NASA had built and operated the Full-Scale Tunnel under permits from the Air Force and its legacy organizations. When the Space Act Agreement was signed in 1997, the permit (NASA Permit, contract no. DA-44-110-ENG-4299) had been issued by the 1st Fighter Wing of the Air Combat Command.
4. Subsequent modifications to the SAA were made to allow ODU to borrow components such as internal strain-gage balances and other equipment from NASA for tests.
5. James Schultz, "Tunnel Soars Into New Era," *Quest* 4, no. 1 (June 2001).
6. NASA Langley Research Center and the Old Dominion University/Old Dominion University Research Foundation, Space Act Agreement (SAA No. 409) Concerning Operation of the 30- by 60-Foot Full-Scale Wind Tunnel, August 18, 1997. Signees were Jeremiah F. Creedon (director of the NASA Langley Research Center), James V. Koch (president of ODU), and Bob E. Wolfson (executive director of the ODU Research Foundation).
7. Ibid., p. 3.
8. E. James Cross, "Aerospace Engineering at Old Dominion University," *Aerospace Engineering Education During the First Century of Flight*, edited by Barnes W. McCormick, Conrad F. Newberry, and Eric Jumper (AIAA: 2004), pp. 712–719.
9. Shultz, "Tunnel Soars Into New Era."
10. Drew Landman and Colin Britcher, "Development of Race Car Testing at the Langley Full-Scale Tunnel," Society of Automotive Engineers Paper 98MSV-21 (1998).
11. Drew Landman, Colin Britcher, and Preston Martin, "A Study of Grounds Simulation for Wind Tunnel Testing of Full-Scale NASCAR's," AIAA Paper 2000-0153 (2000).
12. Schultz, "Tunnel Soars Into New Era."
13. Richard Stradling, "NASA Tunnel Puts the Wind at Their Backs," *Newport News Daily Press* (April 13, 1997).
14. Ibid., p. 12.
15. Robin Imber, Ernest Rogers, and Jane Abramson, "Initial Experimental Evaluation of a Circulation Controlled Sail on a Submersible Vehicle for Enhanced Maneuverability," Naval Air Warfare Center Aircraft Division Technical Report No. NAWCADPAX/TR-2007/12 (2007).
16. Ernest J. Cross, interview by author, August 4, 2011.
17. Jim Cross, interview by author, January 24, 2011.
18. Gary W. Eaker, interview by author, November 29, 2011.
19. Wind Shear, Inc., homepage, *http://www.windshearinc.com/index.htm*, accessed November 25, 2011.
20. John Warner, "Ultimate Wind Tunnel Intrigues NASCAR Racers," *http://www.swampfox.ws/ultimate-wind-tunnel-intrigues-nascar-racers*, accessed June 4, 2011.

21. Cory Nealon, "ODU Considers Building Wind Tunnel in Danville," *Newport News Daily Press* (March 30, 2010).
22. Jacob Kay and John Ralston, "Forebody Aerodynamic Asymmetry on a Full-Scale F-15 Radome," AIAA Paper 2000-4104, AIAA Atmospheric Flight Mechanics Conference, Denver, CO, August 14–17, 2000.
23. David R. Gingras, "Experimental Investigation of a Multi-Aircraft Formation," AIAA Paper 99-3143, 17th Applied Aerodynamics Conference, Norfolk, VA, June 28–July 1, 1999.
24. David R. Gingras and William B. Blake, "Static and Dynamic Wind Tunnel Testing of Air Vehicles in Close Proximity," AIAA Paper 2001-4137, AIAA Atmospheric Flight Mechanics Conference and Exhibit, Montréal, Canada, August 6–9, 2001.
25. James Schultz, "Taking Flight: Tunnel Testing of Duplicate Wright Flyer Begins," *http://www.odu.edu/ao/instadv/quest/WrightFlyer.pdf*, accessed November 2, 2011.
26. An outstanding review of activities of The Wright Experience is available at The Wright Experience, "Education Presentations and Resources," *http://www.wrightexperience.com/edu/index.htm*, accessed November 1, 2011.
27. Jack Cherne, Fred E.C. Culik, and Pete Zell, "The AIAA 1903 Wright Flyer Project Prior to Full-Scale Tests at NASA Ames Research Center," AIAA Paper 2000-0511 (2000); Henry R. Jex, Richard Grimm, John Latz, and Craig Hange, "Full-Scale 1903 Wright Flyer Wind Tunnel Test Results from the NASA Ames Research Center," AIAA Paper 2000-0512 (2000).
28. Robert L. Ash, Stanley J. Miley, Drew Landman, and Kenneth W. Hyde, "Evolution of Wright Flyer Propellers between 1903 and 1912," AIAA Paper 2001-0309 (2001).
29. R.L. Ash, S.J. Miley, D. Landman, and K.W. Hyde, "Propeller Performance of Wright Brothers' 'Bent End' Propellers," AIAA Paper No. 2000-3152, 36th AIAA/ASME/SAE/ASEE Joint Propulsion Conference, Huntsville, AL, July 2000.
30. Robert L. Ash, Colin P. Britcher, and Kenneth W. Hyde, "Prop-Wrights," *Mechanical Engineering Magazine: 100 Years of Flight* (special publication, 2003).
31. Kevin Kochersberger, Robert Ash, Drew Landman, Colin Britcher, Robert Sandusky, and Ken Hyde, "An Evaluation of the Wright 1901 Glider Using Full Scale Wind Tunnel Data," AIAA Paper 2002-1134 (2002).
32. Robert L. Ash, personal communication with author, November 30, 2011.
33. Kevin Kochersberger, Drew Landman, Robert Ash, Jenn Player, and Ken Hyde, "An Evaluation of the Wright 1902 Glider Using Full Scale Wind Tunnel Data," AIAA Paper 2003-0096 (2003).
34. Ibid.
35. Bob Allen, NASA Press Release, February 18, 2003.
36. C.P. Britcher, D. Landman, R. Ash, K. Kochersberger, and K. Hyde, "An Analysis of the Flight Performance of the 1903 'Flyer' Based on Full-Scale Wind Tunnel Data," AIAA Paper 2004-0104 (2004). See also K. Kochersberger, N. Crabill, J. Player, C. Britcher, K. Dominguez, and K. Hyde, "Flying Qualities of the Wright 1903 Flyer: From Simulation to Flight Test," AIAA Paper 2004-0105 (2004).

37. Ibid.
38. Jim Hodges, "Some Extras in 'The Box' Learn About Cutting Room Floor," *Langley Researcher News* (November 7, 2009).
39. Chambers, *Innovation in Flight.*
40. Dan D. Vicroy, "Blended-Wing-Body Low-Speed Flight Dynamics: Summary of Ground Tests and Sample Results (Invited)," AIAA Paper 2009-0933, 47th AIAA Aerospace Sciences Meeting and Exhibit, Orlando, FL, January 5–8, 2009.
41. Ibid.
42. E. Bruce Jackson and C.W. Buttrill, "Control Laws for a Wind Tunnel Free-Flight Study of a Blended-Wing-Body Aircraft," NASA TM-2006-214501 (2006).
43. D. Landman, J. Simpson, D. Vicroy, and P. Parker, "Response Surface Methods for Efficient Complex Aircraft Configuration Aerodynamic Characterization," *AIAA Journal of Aircraft 44*, no. 4 (July–August 2007): pp. 1,189–1,195.
44. Ship 2 was under construction in England as tests were under way on Ship 1 at Langley.
45. Dan Vicroy, "X-48B Blended Wing Body Ground to Flight Correlation Update," AIAA Aero Sciences Meeting, Orlando, FL, January 4–7, 2011.
46. Guy Norris, "Boeing's X-48B Pushes Boundaries of UAVs, Future Transport Designs," *Aviation Week & Space Technology* 167, No. 6 (August 6, 2007), p. 23.
47. The NASA-ODU SAA of 1997 had been amended four times prior to the X-48C test and called for the agreement and X-48C testing to end on August 18, 2009. Because of technical problems encountered in the testing, a fifth modification was signed calling for the X-48C testing to be extended but to end no later than September 30, 2009.
48. Some of the studies included the National Facility Study (1992), National Wind Tunnel Complex (1994), NASA-DOD Aeronautics and Astronautics Coordinating Board (1995), National Wind Tunnel Alliance and Air Breathing Propulsion Test Facilities Alliance (1998), National Aeronautical Testing Alliance (2000), National Aeronautics Research and Development Policy (2006), National Partnership for Aeronautical Testing (2007), DOD Study Team Report on NASA Aeronautics Facilities Critical to DOD (2007), U.S. Industry Wind Tunnel User's Group Study (2008), updated RAND Study (2008), and NASA Facilities Study (2008). Many of the studies included substudies of specific facilities, including subsonic wind tunnels.
49. Hansen, *Engineer in Charge*, p. xxxi.
50. Jeremiah McWilliams, "Future Uncertain for NASA's Historic Wind Tunnel," *The Virginian-Pilot* (July 24, 2006).
51. Cory Nealon, "Winds of Change at NASA," *The Daily Press* (August 26, 2009); Barry Newman, "Shutting this Wind Tunnel Should be a Breeze, but its Fans Won't be Silent," *Wall Street Journal* (August 26, 2009); Michael Klesius, "Last Breath," *Smithsonian Air & Space* (September 10, 2009), *http://www.airspacemag.com/history-of-flight/Last-Breath.html?c=y&story=fullstory*, accessed November 20, 2013.

52. "Out of Breath," Fox News, April 28, 2011, *http://video.foxnews.com/v/3939600/out-of-breath*, accessed January 13, 2012.
53. Denise Lineberry, "Langley's Full-Scale Tunnel Lives On," *Langley Researcher News* (July 30, 2010).

The demolition of the Full-Scale Tunnel was completed in early 2011. This view of the vacant space previously filled by the tunnel was taken in June 2011. The cylindrical building in the center is the Langley 20-ft Spin Tunnel and the tunnel circuit for the Langley 8-ft Transonic Pressure Tunnel is seen at the right. The 8-ft TPT was also demolished in 2011. (Lee Pollard, NASA)

CHAPTER 10

Demolition of the Cave of the Winds

2010–2011

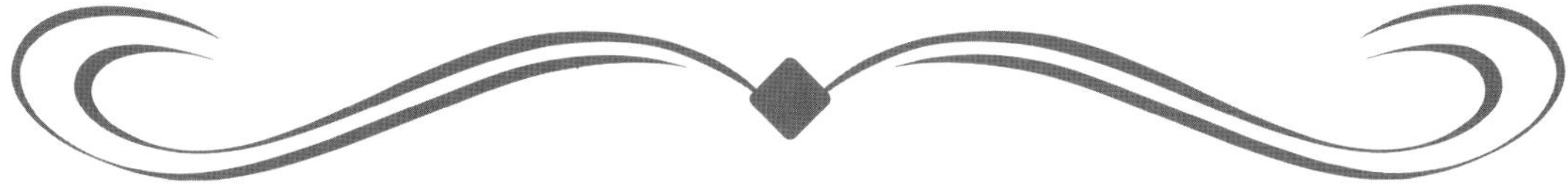

The Task at Hand

By 2009, the NASA Langley Research Center faced the issue of disposition of several historic wind tunnels that had been decommissioned and removed from service. The facilities included the Low-Turbulence Pressure Tunnel, the 8-Foot High-Speed Tunnel, the 8-Foot Transonic Pressure Tunnel, and the Full-Scale Tunnel in the East Area, as well as the 16-Foot Transonic Tunnel in the West Area. The demolition task would be massive, with special concerns over cost, schedule, safety, environmental issues, impact on other daily operations, and historic preservation. Extensive in-house meetings were held between various groups and management to arrive at an appropriate plan and procedures. Activities to retrieve historic artifacts and documents from the deserted buildings had begun 5 years earlier under the direction of Mary Gainer. Historically valuable documents and photographs were digitally scanned, and old 16-millimeter (mm) movie film was collected and converted to digital video. Old wind tunnel models were collected and transferred to educational institutions, and discussions were held with the Smithsonian Institution regarding potential displays of hardware from the Langley tunnels.

Langley and the U.S. Army Corps of Engineers

As part of its approach to this huge job, Langley officials approached the Army Corps of Engineers for assistance. The primary mission of the Corps is construction of new facilities and facility maintenance—not demolition. However, the Corps initiated a Facilities Reduction Program (FRP) in 2004 to assist Government agencies in eliminating excess facilities and structures.[1] The FRP had been involved in other DOD and Federal agency demolition tasks, including an activity with the NASA Marshall Space Flight Center in 2008. The organization is managed by the U.S. Army Engineering and Support Center (USAESC) in Huntsville, AL, with the mission to plan, estimate, contract, and provide management for efforts to remove excess capability for the Federal Government. The FRP business model is elegantly simple and effective: expertise in the commercial-demolition industry is used to reduce excess inventory in the Federal Government, and competition between professional demolition contractors is used to obtain the lowest facility-removal costs. The program repeatedly demonstrated significant cost savings (typically at least 50 percent) in projects from 2005 to 2009.[2]

Langley Construction of Facilities Program Manager Cheryl L. Allen was attending a NASA facilities meeting when she heard Thad Stripling, the Corps' FRP manager, make a presentation on his program's capabilities.[3] "At the time, we were working with a small local contractor who really didn't have the level of capability that Thad was presenting," Allen said. "Thad's numbers were too good. So after the meeting, I cautiously enquired about removing what we had at Langley. Thad had talked about removing traditional structures such as housing, office buildings, etc., but he was excited about the challenge of the wind tunnels. Removing the wind tunnels was not a capability NASA had. But being part of the same Government family, through an MOA we were able to tap into the FRP. It's been a great find."[4]

The initial FRP activity at Langley would include the demolition of the two 8-Foot Tunnels, the Full-Scale Tunnel, and the 16-Foot Transonic Tunnel; work on the Low-Turbulence Pressure Tunnel was delayed as a follow-on task. The initial cost estimate for demolition of only the Full-Scale Tunnel and the 16-Foot Transonic Tunnel was $8.4 million. The final demolition contract award cost was $3.65 million for all four wind tunnels—a remarkable savings of 43 percent over the initial estimate for only two tunnels. The projected schedule for the four demolition tasks was 11 months.

Project Management

The multiple award task order contract (MATOC) for demolition of the four Langley tunnels was awarded to Charter Environmental (prime) and Neuber Environmental (subcontractor) in January 2010. Partners on the project included NASA Langley, the Army Engineering and Support Center, and the Norfolk District of the Corps of Engineers. NASA's project manager for the demolition of the Full-Scale Tunnel was Kim F. "Skip" Schroeder, and Mindy Shelton was the project manager for the USAESC. Charter Environmental and Neuber Environmental reported directly to Shelton.

The demolition team faced many challenges during the deconstruction of the tunnel, including the critical issue of removal and disposal of the Careystone asbestos-impregnated sheets used in the construction of the facility. The extensive expertise of the contractor team defined a safe, efficient[5] approach to the potentially hazardous situation. The transite asbestos in the building was removed intact and sent to an appropriately permitted landfill. Although NASA explored methods to recycle the wood contained in the facility, recycling was not feasible because of the way the building was to be demolished. The wood removed from the facility went to a commercial waste landfill. All steel, metals, and concrete were sent to recyclers for processing and reuse (the Corps contract required the contractors to recycle a minimum of 51 percent of the materials from the demolition by weight). All foundations and structures were to be removed to 4 feet below grade level and the site covered with 8 inches of crushed concrete.

Deconstruction work began at the Full-Scale Tunnel in late October 2010, with a planned duration of about 5 months. The removal of Careystone panels was performed by two-man teams in a boom lift. After removal of the fasteners, the panels were placed on scissor lifts and lowered to containers on the floor. After abatement of asbestos materials, the demolition crews proceeded to demolish all interior structures, leaving only the steel skeleton of the building. The concrete floor was broken up with hydraulic hammers, processed, crushed, and spread over the building's footprint. NASA and Langley Air Force Base environmental units conducted extensive inspections and testing at the site during and after the demolition.

The final remnants of the Full-Scale Tunnel's steel structure came down on schedule in May 2011, and any structural evidence of the presence of the tunnel was removed by May 18, 2011—almost 80 years to the day after the facility was dedicated in 1931.[6]

Photographs of the Demolition Process

The following photographs provide a chronological display of progress during the deconstruction process for the Full-Scale Tunnel.

In August 2010, representatives from the Smithsonian Air and Space Museum visited the vacant Full-Scale Tunnel and inspected possible artifacts for display at the museum. The new Toledo scales and the west propeller hub were of interest. The Smithsonian also requested a set of the tunnel's wooden blades. (Mary Gainer)

The artifacts were loaded on a flatbed truck for transport to the museum. (Mary Gainer) Demolition activities began at the Full-Scale Tunnel at the end of 2010. (NASA 2011-L-00019)

After completing work on the office area at the south end of the building, the crew began to remove the balance house and ground plane on January 10, 2011. The orange circular structure was the final turntable. (NASA 2011-L-00019) Overhead view of the initial demolition of the balance-house area. (NASA 2011-L-00069)

The return passages of the Full-Scale Tunnel were deconstructed to ground level. The photo on the left shows the west return passage with debris following a nor'easter in January 2011, and the photo on the right shows the beginning of deconstruction 2 weeks later. (NASA 2011-L-00019 and NASA 2011-L-00161)

The balance-house area (left) was almost cleared by January 19, 2011. (NASA 2011-L-00163) The crew began removal of the original offices of the head mechanic and tunnel operator (right). (NASA 2011-L-00019)

Workmen cutting up exterior structures removed from the Full-Scale Tunnel in January 2011. (NASA 2011-L-00152)

Comparison of a photograph taken during the construction of the Full-Scale Tunnel annex hangar (left) with a photograph showing the hangar during deconstruction in 2011 (right) shows the structure of the facility after 80 years. (G. Lee Pollard)

Photograph of the east side of the entrance cone shows details of the modifications to the corners installed following the unacceptable first tests in 1931. (NASA 2011-L-00037)

Demolition of the west return-passage ramp at the end of January 2011. (NASA 2011-L-00131) Careystone panels have been removed from the interior wall of the east return passage. Note the yellow-gold acoustic panels on the east test section wall. (NASA 2011-L-00141)

View of the test section looking upstream at the entrance cone in January 2011 (left). (NASA 2011-L-00144) View looking downstream in the east return passage (right) on February 1, 2011. The test section and acoustic panels are on the immediate left. The turning vanes have been removed from the downstream corner, and the entrance cone is on the left. (NASA 2011-L-00202)

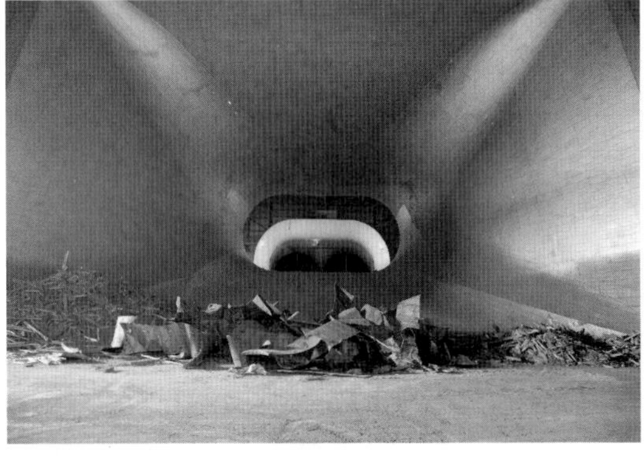

View from the entrance cone looking downstream as demolition begins in that area (left). (NASA 2011-L-00204) The underside of a drive motor and exit-cone structure (right) shows the area below the flow circuit on February 14, 2011. (NASA 2011-L-00323)

By February 17, most of the drive motor and exit-cone areas had been demolished and removed. (NASA 2011-L-00335)

Poignant photo of the removal of exit-cone material from the Full-Scale Tunnel. (NASA 2011-L-00344)

By April 22, the Careystone panels covering the inner and outer walls had been removed. This view is looking north. (NASA 2011-L-01270)

One of the most challenging aspects of the process was the safe removal of the asbestos-impregnated Careystone panels. On the left, the two-man crew is removing a panel that will be bagged for disposal. (NASA 2011-L-01275) On the right, a Careystone panel is lowered to the ground for disposal. (NASA 2011-L-01284)

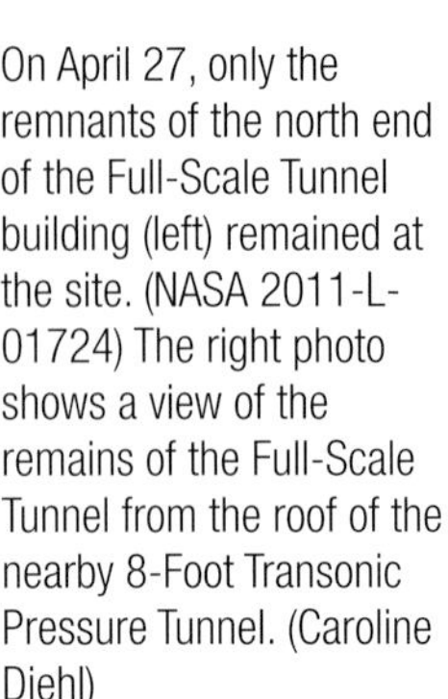

On April 27, only the remnants of the north end of the Full-Scale Tunnel building (left) remained at the site. (NASA 2011-L-01724) The right photo shows a view of the remains of the Full-Scale Tunnel from the roof of the nearby 8-Foot Transonic Pressure Tunnel. (Caroline Diehl)

Demolition of the Cave of the Winds

The last structure of the Full-Scale Tunnel was demolished on May 2, 2011—almost exactly 80 years after the tunnel became operational. (G. Lee Pollard)

On June 1, the Full-Scale Tunnel had been removed. (NASA L-02160)

In August 2011, the oldest wind tunnel in the NASA inventory had disappeared. At the time this picture was taken, the remaining NASA facilities near the site were (from top of photo, clockwise) the Langley 16-Foot Transonic Dynamics Tunnel; the old NACA complex consisting of the 15-Foot Free-Spinning Tunnel building, the 12-Foot Low-Speed Tunnel (housed in the sphere), and the Langley 20-Foot Spin Tunnel; and the 8-foot tunnels complex (at the lower left) consisting of the 8-Foot High-Speed Tunnel and the 8-Foot Transonic Pressure Tunnel. The 8-foot tunnels were demolished in 2011 and 2012. The 12-Foot Low-Speed Tunnel is now the oldest continuously operating NASA wind tunnel. (NASA/Langley)

In the Presence of History

I visited the former site of the old Cave of the Winds in February 2012 to view its status and reflect on the history that happened there and the legendary people that had walked the grounds. On the day of my visit, its historic next-door neighbors—the Langley 8-Foot Transonic Tunnel (formerly the 8-Foot High-Speed Tunnel) and the 8-Foot Transonic Pressure Tunnel—had also been demolished, leaving a city-block-sized area of gravel to replace the famous facilities. As I stood there looking out over the open space to the nearby Back River, amidst soaring and squawking sea gulls, I considered the events that had happened there:

- The very beginnings of Full-Scale Tunnel operations, helping to establish the NACA as a leader in aeronautical research.
- Distinguished visitors during normal research activities and annual NACA inspections—Wilbur Wright, Howard Hughes, Charles Lindbergh, Hap Arnold, Jimmy Doolittle, Walter Diehl, Leroy Grumman, and many others.
- The first wind tunnel studies of complete, full-scale, powered aircraft.
- Critical, performance-enhancing tests that boosted the capabilities of U.S. military aircraft in World War II.

- The identification and resolution of unexpected problems for advanced civil and military aircraft.
- Over 1,000 tests of different aerospace configurations.

The history of wind tunnels and their applications by the NACA and NASA is a special story of personal dedication by thousands of workers, brilliant leaders, world-class leading-edge technology, and outstanding service to the Nation. Hundreds of wind tunnels were ultimately developed and utilized with remarkable effectiveness by the agencies. Most of the tunnels were capable of higher speeds and more unique flow properties than the Full-Scale Tunnel, but I submit that no other facility provided contributions for more critical studies of complete aerospace vehicles. The tunnel truly lived the mission declared during the formation of the NACA in 1915 to focus on "the scientific study of the problems of flight, with a view to their practical solution."[7]

As I walked back to my car, an F-22 Raptor thundered by, only a few blocks away from the now-empty site. As the demonstration pilot practiced his air-show routine, he crisply piloted the big fighter in impressive maneuvers that vividly illustrated the effectiveness of the airplane's thrust-vectoring system. As I left, I wondered whether the young pilot knew that his world-dominant vehicle had resulted in part from studies conducted over 25 years earlier in an ancient wind tunnel that had been first used to test biplanes in the 1930s.

Endnotes

1. Debra Valine, "Wind Tunnels at NASA Langley Being Removed by Facilities Reduction Program," U.S. Army homepage, June 7, 2011, *http://www.army.mil/article/59097/Wind_tunnels_at_NASA_Langley_being_removed_by_Facilities_Reduction_Program/*, accessed September 5, 2011.
2. Thad Stripling, "Facilities Reduction Program," briefing to NASA Headquarters, Huntsville, AL, May 11, 2011, *http://www.hq.nasa.gov/office/codej/codejx/Assets/Docs/ConferenceNashville2011/Wednesday/Thad_Stripling_USACE-FacilityReductionProgram.pdf*, accessed December 10, 2011.
3. Valine, "Wind Tunnels at NASA Langley Being Removed by Facilities Reduction Program."
4. Ibid.
5. The task was finished well ahead of schedule.
6. A poignant video tribute to the history of the Full-Scale Tunnel is available at *http://www.youtube.com/watch?v=wMpYvXjvARc*, accessed January 5, 2012.
7. George W. Gray, *Frontiers of Flight: The Story of NACA Research* (New York, NY: Alfred A. Knopf, Inc., 1948), p. 13.

APPENDIX

Personalities at the Full-Scale Tunnel

Attempting to single out individuals in historical essays is always potentially dangerous, especially when attempting to cover 80 years of critical contributions to the Nation by thousands of engineers, technicians, administrative personnel, fabrication specialists, and many others at Langley and around the country who supported research operations in the Full-Scale Tunnel. However, the author is compelled to provide additional information on some selected individuals of the NACA and NASA who have been mentioned in various chapters of this book. The contributions of the following people have already been discussed herein—it is hoped that brief thumbnail biographies of their personalities and careers might be of interest to the reader.

Heads of the Full-Scale Tunnel

NACA	**Years**
Smith J. DeFrance (1931–1935)	4
Clinton H. Dearborn (1935–1941)	6
Abe Silverstein (1942–1943)	1
Herbert A. Wilson, Jr. (1943–1948)	5
Gerald W. Brewer (1948–1958)	10
NASA	
John P. Campbell (1959–1962)	3
Marion O. McKinney, Jr. (1962–1974)	12
Joseph R. Chambers (1974–1981)	7
Joseph L. Johnson, Jr. (1981–1990)	9
Luat T. Nguyen (1990–1992)	2
Dana L. Dunham (1992–1995)	3
Old Dominion University	
Ernest J. Cross (1997–2002)	5
Robert Ash (2002–2009)	7

Elliott G. Reid

Elliott G. Reid (1900–1968) first suggested the concept of a "full-scale" wind tunnel to management of the Langley Memorial Aeronautical Laboratory in 1925. A native of Ohio, he obtained bachelor's and master's degrees from the University of Michigan. He entered duty at Langley in July 1922—coincidentally, on the day that Smith DeFrance arrived at Langley. Reid and DeFrance worked together on projects including a study using free-falling spheres to determine the critical Reynolds number for turbulence studies.

Reid became dissatisfied with Langley management and left in 1927. Based on his demonstrated expertise in aerodynamics at Langley, he was accepted at Stanford University as a full professor at the age of 27, and he continued in his positions at Stanford until he retired in 1965.

At Stanford, Reid worked with the legendary William F. Durand, who had instituted some of the first major wind tunnel studies of propeller performance. Reid demanded that students have a thorough understanding of principles and subject matter, and he was regarded as one of the most effective professors in the university community.

Elliott Reid appears in this photograph taken during a visit of Theodore von Kármán to Langley in 1926. Reid is at the far right in the second row. Other notables include Henry Reid, Langley's engineer in charge (left in second row on steps); Max Munk (third from left in front row); von Kármán (on step in middle of second row); Fred Weick (first from left in rear row); Tom Carroll (third from left in rear row); and George Lewis (far right in first row). (NASA EL-2003-00332)

Dr. George W. Lewis

Dr. George W. Lewis (1882–1948) graduated from Cornell University in 1910 with a master's degree in mechanical engineering and taught mechanical engineering at Swarthmore College from 1910 to 1917. Lewis became the executive officer of the NACA in 1919. In 1924, he was given the title of director of aeronautical research, which he kept until his resignation from the NACA in 1947.

George Lewis, NACA Director of Research. (NASA EL-1997-00143)

Lewis was the liaison between the NACA committee and the Langley laboratory. In 1974, Russ Robinson commented that "Lewis was a man for his time. He ran a one-man show."[1] When Robinson started at Langley, Lewis used to visit the lab on a weekly basis to see "his boys"—not necessarily the Engineer in Charge, Henry J.E. Reid. Robinson said, "In a sense he directed the research at Langley. He knew every single employee at Langley."[2]

When Langley engineer Elliott Reid unsuccessfully suggested that a full-scale wind tunnel was needed at Langley instead of the smaller Propeller Research Tunnel, Lewis personally became involved in promoting the concept and obtaining the approval of the NACA committee. During his career, Lewis was the leading advocate for the design, construction, and use of Langley's Variable Density Tunnel, Full-Scale Tunnel, Icing Tunnel (later the Low-Turbulence Tunnel), Free-Flight Tunnel, Gust Tunnel, and High-Speed Tunnel.

Lewis was self-driven beyond the limits of human endurance. He never took a single vacation during the 5 years of World War II, and in 1945 his heart began to fail. He had three heart attacks and resigned from the NACA in 1947 because of failing health. He died at his home near Scranton, PA, in July 1948 at the age of 66. When his death was announced at Langley, a moment of silence was formally observed across the entire laboratory for this special man who so loved the Langley personnel and facilities. The body of George Lewis was cremated and his ashes were scattered over Langley Field.[3]

The NACA laboratory at Cleveland, OH, was initially named the Aircraft Engine Research Laboratory in June 1940, and was renamed the Flight Propulsion Research Laboratory in 1947. In honor of George Lewis's many contributions to aviation and the Nation, the name was changed to the Lewis Flight Propulsion Laboratory in 1948. When NASA was created in 1958, the laboratory became the NASA Lewis Research Center. The Center was renamed again in 1999 as the Glenn Research Center, in tribute to astronaut John Glenn.

Smith J. DeFrance

Smith J. DeFrance (1896–1985) was a military aviator with the Army's 139th Aero Squadron during World War I. After the war, he earned a bachelor's degree in aeronautical engineering from the University of Michigan in 1922 before beginning a career with the NACA and NASA. He started his career at Langley as its 63rd employee in July 1922, joining the team that put the Variable Density Tunnel into operation. DeFrance worked on the development of a balance system for the tunnel. In 1924, he transferred to the Langley flight section and developed pressure instrumentation and cameras to define the flightpaths of aircraft. DeFrance also flew as a test pilot in flight studies.

Smith J. DeFrance, the designer and first head of the Full-Scale Tunnel. (NASA L-54565)

After returning its first three borrowed Curtiss JN-series planes, or "Jennies," back to the Air Service in 1923, Langley acquired a later-model Jenny known as the JN-6 in 1923. On the afternoon of August 20, 1924, DeFrance was flying this aircraft with a junior engineer named Stevens Bromley in the rear seat as an observer. While flying near the Back River, DeFrance lost control and the plane crashed, killing Bromley and severely injuring DeFrance; he spent 10 months in the Walter Reed hospital and lost an eye. Stevens Bromley is now honored at a memorial in the Langley West Area as one of the NACA/NASA individuals who gave their lives in Government service. After recovering, DeFrance promised his wife he would never fly on an airplane again—and he did not. Even when he became Director of Ames and had to travel frequently to NACA Headquarters in Washington, DC, DeFrance never flew in an airplane despite facing a 4-day train trip instead.[4]

After his airplane accident, DeFrance continued to work in the flight group until he was assigned by the chief of aerodynamics, Elton Miller, to the design of the Full-Scale Tunnel. His actions in overseeing the construction of the tunnel, keeping it under budget, and supervising its early research operations as the first head of the facility earn him, in the opinion of the author, the title of "Father of the Full-Scale Tunnel." DeFrance designed other Langley tunnels and supervised their operations before becoming director of the new Ames

This photo of the first staff of the Ames laboratory in August 1940 shows some of the ex-Langley employees that transferred to the new NACA installation, including: Smith DeFrance (1), Ed Sharp (2), Manley Hood (3), Donald Wood (4), George Bulifant (5), Harry Goett (6), Jack Parsons (7), and Ferril Nickle (8). All had participated in the design and/or early operations of the Langley Full-Scale Tunnel.[5] Several other ex-Langley personnel are also in the picture. (Ames M-704-1)

Aeronautical Research Laboratory in 1940. He quickly recruited many Langley employees during Ames's startup, including several former members of the Full-Scale Tunnel staff. He remained director of Ames until his retirement in 1965.

When asked in a 1974 interview what he considered to be the NACA's greatest contribution, DeFrance unhesitatingly replied, "The information that came from the Full-Scale Tunnel at Langley and the 40- by 80 Wind Tunnel at Ames. I wouldn't call it research—it was strictly engineering."[6]

Edward R. Sharp

Edward R. "Ray" Sharp (1894–1961) was born in Elizabeth City, VA, and joined Langley in 1922. He then earned a law degree from the College of William and Mary in 1924 through a correspondence course. At Langley, he quickly rose from hangar boss to construction administrator. During his 36-year NACA/NASA career, Sharp held positions of importance at each of three principal research establishments. From 1925 to 1940, he was administrative officer at Langley (in the 1930s he was called chief clerk). During this time, he was active in the construction of the Full-Scale Tunnel. In 1941, he served as administrator of the building program for Smith DeFrance at the Ames Aeronautical Research Laboratory in California, overseeing the establishment of research facilities. In 1942, he was recalled to Langley to begin the planning process for the establishment of the NACA Aircraft Engine Research Laboratory in Cleveland, OH. Upon its completion, he became manager and was appointed director of the laboratory in 1947. He remained director of Lewis until his retirement from NASA in 1960.

Edward R. Sharp was chief clerk and participated in the construction and early operations of the Full-Scale Tunnel. (H.J.E. Reid retirement album, LHA)

Clinton H. Dearborn

Clinton H. "Clint" Dearborn (1897–1965) served as Smith DeFrance's protégé during the design, construction, and early days of operations at the Full-Scale Tunnel. He received his bachelor's degree in mechanical engineering from the University of Michigan in 1922 and joined the Langley staff on August 13, 1927, as an assistant aeronautical engineer. During the design of the Full-Scale Tunnel, he led the design, construction, and data acquisition for the ¹⁄₁₅-scale model tunnel in addition to his duties in the design of the Full-Scale Tunnel. When Smith DeFrance was promoted to higher responsibilities in 1935, Dearborn was appointed the second head of the Full-Scale Tunnel

Clinton H. Dearborn succeeded DeFrance as head of the Full-Scale Tunnel. (NASA L-47523)

Section, where he led research projects and managed the facility through the early days of World War II. He was widely recognized as the laboratory's primary point of contact on engine-cooling issues. Within that responsibility, he was in charge of the so-called "Cooling College," which included members of industry onsite at Langley. In 1950, he transferred to NACA Headquarters in Washington, where he served as assistant to the director for research management. He retired in 1954 and died at his retirement home in Florida in 1965.

Abe Silverstein

Abe Silverstein (1908–2001) earned a bachelor's degree in mechanical engineering in 1929 and a master's degree in engineering from Rose Polytechnic Institute. He joined the NACA at the Langley Memorial Aeronautical Laboratory in June 1929, and despite his total lack of training in aerodynamics, Silverstein was assigned to the aerodynamics group to assist Smith DeFrance's team in the design of the Full-Scale Wind Tunnel. He also participated in other wind tunnel development activities at Langley, including the design of the Langley Icing Tunnel (the precursor to the Low-Turbulence Tunnel). In August 1940, Silverstein was promoted to become the third head of the tunnel following DeFrance's transfer to the NACA's new Ames Aeronautical Research Laboratory. During the early years of World War II, Silverstein undertook engine-cooling tests that were a harbinger of his future propulsion research. He demonstrated that the use of internal baffles to direct airflow over hot cylinders resulted in improved engine cooling. The engine-cooling studies melded his mechanical engineering background with the aerodynamics work at the Full-Scale Tunnel.

Abe Silverstein was the third head of the Full-Scale Tunnel. (NASA C-1998-217)

George Lewis considered Silverstein to be one of the NACA's bright young stars. When the NACA began creating a new Aircraft Engine Research Laboratory in Cleveland, OH, Lewis personally selected Silverstein to manage its marquee new facility, the Altitude Wind Tunnel (AWT). In October 1943, after 14 years at Langley, Silverstein was transferred to the AERL just as the AWT was being completed. He played a key role in directing the AWT, the Nation's first tunnel capable of simulating altitude conditions. Silverstein also helped design the 8- by 6-Foot Supersonic Wind Tunnel, one of the first supersonic tunnels in the Nation, and the 10- by 10-Foot Supersonic Wind Tunnel. In 1949, he was appointed chief of research at the Lewis laboratory, and in 1953 he became associate director there.

Silverstein transferred to NACA Headquarters in 1958 to assist in the organization of the new NASA. Later that year, Silverstein was named chief of space flight programs. Seventeen days after the official founding of NASA, a group led by Silverstein presented plans for the Mercury Program to new NASA Administrator T. Keith Glennan. The Space Task Group was created in October 1958 to oversee the Mercury Program, and it later oversaw the Apollo

Director of the Lewis Flight Propulsion Laboratory Edward Ray Sharp (left), and Chief of Research Abe Silverstein (right) discuss a model of a ramjet aircraft at Lewis in October 1951. (NASA GPN-2000-001823)

Abe Silverstein (center) enjoys the festivities at Smith DeFrance's 30-years-of-service party on July 24, 1952. H.J.E. Reid of Langley is shown at the right. (Smith DeFrance retirement album, Langley Technical Library)

Program as well. Although it was based at Langley, the group reported to Silverstein at NASA Headquarters. In 1960, Silverstein worked with the Space Task Group (STG) to outline the Apollo Program. He is credited with having named the Mercury and Apollo Programs. Silverstein also helped plan the Apollo, Ranger, Mariner, Surveyor, and Voyager missions.

In 1961, Silverstein returned to Lewis as director of the Center, and he retired after the Apollo 11 lunar landing in 1969. In June 1979, he attended an honorary session at the Full-Scale Tunnel, where he reunited with many old Langley friends and gave a presentation on his days at the tunnel.

In 1994, the NASA Lewis 10- by 10-Foot Supersonic Wind Tunnel was renamed the Abe Silverstein 10- by 10-Foot Supersonic Wind Tunnel. Silverstein died on June 1, 2001, at the age of 92.

Russell Robinson

Russell Robinson (1908–2003) earned a degree from Stanford in 1930. Ironically, one of his professors was Elliott Reid, who had first suggested while a researcher at Langley that the NACA construct a full-scale wind tunnel. Upon graduation, Robinson joined the NACA at the Langley Memorial Aeronautical Laboratory, which had a staff of about 200 and was like "a graduate school with pay."[7] Upon reporting for work he was assigned to Smith DeFrance's group designing the Full-Scale Tunnel, and he was especially proud of the fact that Langley returned money to the Treasury after the tunnel was completed. Robinson designed several key components of the Full-Scale Tunnel, including the balance-scale system and the fairings for the motor support structures.[8]

Russell Robinson participated in the design and early operations of the Full-Scale Tunnel. (NASA A-14783)

Later, in 1933, he was assigned (along with Manley Hood of the Full-Scale Tunnel staff) to design the new 8-Foot High-Speed Tunnel conceived by Eastman Jacobs. He then became the first head of the new tunnel. Later, Robinson was assigned to be the NACA's liaison with West Coast aircraft manufacturers, during which time he supervised the first construction at the Ames Aeronautical Research Laboratory.

In 1939, George Lewis requested that Robinson go to NACA Headquarters in Washington and become his assistant, which he did in 1940. Lewis then sent him on a special intelligence-gathering mission to Europe to uncover German aeronautical progress from 1944 to 1945. Upon his return to NACA Headquarters, he became active in advocacy for high-speed flight and participated in a joint NACA–Department of Defense (DOD) activity for a Unitary Plan Wind Tunnel. He returned to Ames in 1950 and remained there until his retirement in May 1970.

Russell Robinson (right) supervises the first groundbreaking construction at the new NACA Ames Aeronautical Research Laboratory in 1939. (NASA G-325 (0-82))

Dale McConnaha

The early group photograph of the first staff of the Full-Scale Tunnel includes Dale McConnaha, who served as the first chief of mechanics at the facility. McConnaha had served as a chauffeur in World War I and entered the NACA at Langley in 1923 as an engine mechanic. He was assigned to the flight division at Langley and worked on many of the laboratory's research aircraft. He then joined DeFrance's group at the Full-Scale Tunnel, where he worked until serving in World War II; after the war, he transferred to the NACA Aircraft Engine Research Laboratory in Cleveland.

After serving the NACA for 30 years, McConnaha was honored for his service by Dr. Ray Sharp, director of the Lewis Flight Propulsion Laboratory, in June 1953. In a poignant ceremony at a local Veterans Administration Hospital, Sharp and several other dignitaries visited McConnaha, who had undergone several surgeries for cancer. A certificate presented to him commended McConnaha for his work as chief of mechanics at Langley's Full-Scale Wind Tunnel and for his service at Lewis during the transition from reciprocating to jet engines.

Manley Hood

Manley Hood was hired by the Langley laboratory in 1929 and assigned to the physics laboratory to develop instrumentation. In an interview with Walter Bonney in 1974, Hood described how he transferred to DeFrance's group during the construction of the Full-Scale Tunnel, where his job was to "tap rivets, look over paint and steel structure, and climb all over the building to see if things were right."[9] After a few years at the Full-Scale Tunnel, Hood was assigned to help Russ Robinson design the 8-Foot High-Speed Tunnel. Hood thought the name first selected for the new tunnel— The Langley Full-Speed Tunnel—was absurd, considering that the maximum design speed was only about 500 mph.

When Russ Robinson left for Headquarters, Hood conducted extensive tests in the new facility on the effects of wing roughness and rivet heads on drag. After World War II, Hood transferred to Ames to design and manage new wind tunnels along with 30 other Langley personnel. He became head of the 7- by 10-Foot Tunnels at Ames, followed by an assignment as head of the Ames 16-Foot Tunnel. Hood later became Don Wood's assistant.

John F. Parsons

John F. "Jack" Parsons (1908–1969) earned both a bachelor's degree (1928) and an advanced degree (1930) in aeronautical engineering from Stanford University. In 1931, he joined the staff of the Langley Memorial Aeronautical Laboratory as a junior aeronautical engineer in Smith DeFrance's group at the Full-Scale Tunnel. He later was given responsibility for the design of the Langley 19-Foot Pressure Tunnel. He transferred to Ames in 1939 when the lab was established and worked on planning, design, and construction of the new Center.

Jack Parsons, Assistant Director of Ames in 1958. (NASA A-23886)

DeFrance used him as his top assistant during the construction of the new laboratory. From 1943 to 1947, he served as the chief of the Full-Scale and Flight Research Divisions and also chief of the Construction Division, and he then served as assistant to the director of the Center until 1949. From 1949 to 1956, he supervised the wind tunnel construction program, among other duties. He was named associate director of Ames in 1952, a position he held until his death in 1969.

Jack Parsons (left) and Ferril Nickle (right) were the first two employees of the Ames laboratory. They were both former members of the staff at the Langley Full-Scale Tunnel. This photo was taken on January 29, 1940, in front of the construction shack. (NASA Ames M-253)

Jack Parsons (right) presents a plaque to Smith DeFrance in DeFrance's office at Ames on the occasion of his 35 years of service in 1957. (NASA GPN-2000-001525)

Personalities of the Full-Scale Tunnel in the 1930s and 1940s

Harry J. Goett

Harry Goett speaks at a reception for test pilot Bill McAvoy at Ames in 1957. (NASA GPN-2000-001525)

Harry J. Goett (1910–2000), a native of the Bronx, NY, earned a degree in physics and mathematics from Holy Cross College in 1931 and a degree in aeronautical engineering from NYU in 1933. He attended the Fordham University Law School from 1933 to 1935. After holding a number of engineering posts with private firms, he became a project engineer at the Langley Full-Scale Tunnel under Clint Dearborn in 1936. He later transferred to the Ames Aeronautical Research Laboratory with DeFrance and took over the management of the 7- by 10-Foot Tunnels when Hood was promoted. Goett was chief of the Full-Scale and Flight Research Division at Ames from 1949 to 1959. He became director of the Goddard Space Flight Center from 1959 to 1965 and then became a special assistant to NASA Administrator James E. Webb. Later he was director for plans and programs at Philco's Western Development Labs in California. He retired from Ford Aerospace and Communications.

Dr. Samuel Katzoff

Dr. Samuel Katzoff led an analytical studies group at the Full-Scale Tunnel. (NASA)

Dr. Samuel Katzoff (1909–2010), a native of Baltimore, MD, received his bachelor's degree and doctorate in chemistry from Johns Hopkins University in 1929 and 1934, respectively. His early work experiences at college included a diverse range of topics, including x-ray technology, colloid chemistry, antifreeze mixtures, development of commercial oil and wax polishes, and antiknock compounds for motor fuels. He was hired by the NACA and entered duty at the Langley Memorial Aeronautical Laboratory on March 23, 1936, as an associate physicist at the Full-Scale Tunnel. Being the only physicist among aeronautical engineers, he stressed theory in his work assignments, particularly the correlation of experiments and theory. He was appointed head of the Full-Scale Analysis Section and was later chosen as the assistant chief of the Full-Scale Research Division in 1946. In 1959, after the formation of NASA, Katzoff served as the assistant chief of the Applied Materials and Physics Division and became deeply involved in space-related research, including the Lunar Orbiter Project.

At the Full-Scale Tunnel, Katzoff's work focused on the correlation of experimental data and theory, but "Doctor Sam" was best known for his extraordinary skills at technical writing. He later authored a guide for technical writers entitled "Clarity in Technical Writing," which became the standard for all NASA organizations. His guide was always included in the orientation material given to new engineers at the Center. Following retirement in 1974 as Langley's chief scientist, he tutored children and wrote teaching manuals. Katzoff died on September 25, 2010, at a retirement community in Pikesville, MD, at the age of 101.[10]

John P. Reeder

John P. "Jack" Reeder (1916–1999) graduated from the University of Michigan in 1938 with a bachelor's degree in aeronautical engineering. Coincidently, one of his graduate-student instructors was Ken Pierpont, who would later join the NACA and work at the Full-Scale Tunnel. Jack was hired by the NACA in 1938 and was assigned to the Full-Scale Tunnel despite his forcefully stated interest in flight testing. His request to be a test pilot had been turned down because the laboratory already had a full complement of pilots. Reeder worked as an engineer at the Full-Scale Tunnel for 4.5 years, conducting drag reduction and general assessments of aircraft such as the Curtiss XP-40, the Bell XP-39B, the Curtiss XSOC-1, the Grumman XTBF-1, the Douglas A-20A, and several unorthodox airplanes, including Charlie Zimmerman's Chance Vought V-173 "Flying Flapjack."

John P. "Jack" Reeder began his career as a researcher at the Full-Scale Tunnel. (Reeder Collection, LHA)

Melvin Gough, the first NACA engineering test pilot and later head of NACA Flight Operations at Langley, received a directive from his superiors at NACA Headquarters in the early 1940s to choose and groom qualified volunteers for flight-research duty among the engineers at Langley. Reeder quickly applied for the opportunity, was selected, and started a new career as a flight-research pilot in October 1942.

During World War II, Reeder flew and evaluated most of the emerging Army and Navy aircraft, and in 1944 he became the NACA's first helicopter test pilot. After the war, he became an internationally known expert on the handling qualities of vertical takeoff and landing aircraft and helicopters, and he later initiated a terminal-area research program for jet-transport aircraft using a specially modified Boeing 737 aircraft. In 1962, he became involved in the flight evaluations of the British P.1127 V/STOL aircraft at the request of the British government.

During his extraordinary career with the NACA and NASA, Reeder flew over 230 different types of fixed-wing, rotary, and V/STOL aircraft. He retired from NASA in 1980 after a distinguished 42-year career. In 1990, he suffered a severe head injury in an automobile accident and shortly thereafter was diagnosed with Alzheimer's Disease. He died in May 1999. Reeder was inducted posthumously into the Virginia Aviation Hall of Fame in November 2005. The first NASA astronaut to step on the Moon, Neil Armstrong, called Reeder the best test pilot he had ever known.

Joseph Walker

Joseph Walker (1898–1976) was one of the most beloved and respected mechanics to ever work at the Full-Scale Tunnel. Walker began working at the local Newport News Shipbuilding and Dry Dock Company when he was 12 years old. After graduating from the Apprentice School as a machinist, he served in the Navy during World War I and returned to the shipyard after the war. Walker was hired by the Langley Memorial Aeronautical Laboratory in 1928 and was one of the first persons to walk in the area that would become the site of the Full-Scale Tunnel.

Joe Walker enjoys a crab claw at the retirement party for H.J.E. Reid. (H.J.E. Reid retirement album, Langley Technical Library)

Joe Walker became the chief of mechanics after Dale McConnaha left during the war. He was quick-witted, full of colorful humor, and a perfectionist at practical jokes. One of the rituals of initiation for new engineers at the tunnel was climbing the long, spindly "balloon" ladders that were used to access aircraft test subjects mounted high in the tunnel while Walker spewed verbal abuse and shook the ladder. No one knew the facility as well as he, and operational issues always ended up on his desk for solutions.

Walker was always participating in visitor briefings and was befriended by all who met him, including Langley's senior management—especially engineer in charge H.J.E. Reid. Reid and Walker became very close friends, and the story is told that, on V-J Day, Joe Walker had strolled out into the celebrating crowd in the street next to the Full-Scale Tunnel. Reid's office was just a few blocks away at the NACA Headquarters building, and he too joined the crowd. When Walker joined Reid in the crowd, he pulled out a flask of whiskey and asked Reid if he wanted a celebratory drink. Reid accepted with a long draw on the flask and handed it back to Walker, who put it back in his pocket. Reid asked, "Joe, aren't you going to have a drink?" and Walker replied to Langley's top man, "Hell no, you can get fired for drinking on the job!"[11]

Walker was frequently invited by Reid to formal receptions and parties, including Reid's own retirement party, where he celebrated with Center directors and dignitaries. Joe Walker retired from Langley in 1966 and died in 1976.

Herbert A. Wilson, Jr.

Herbert A. "Hack" Wilson, Jr. (1914–1992), was a native of Inverness, MI. While still in high school in 1930, he won the state's Thomas A. Edison Scholarship contest and then competed at the national level at Edison's lab in New Jersey, where he met Edison, Henry Ford, and Harvey Firestone. He graduated with a bachelor's degree in aeronautical engineering from Georgia Tech in 1934 and was hired by Langley in 1937 and assigned to the Full-Scale Tunnel. During WWII, he led or participated in tests of many military aircraft, and he became the fourth head of the Full-Scale Tunnel in 1943 when Abe Silverstein transferred to the NACA Aircraft Engine Research Laboratory at Cleveland. In 1948, Hack Wilson was named by Langley management to head up a Supersonic Facilities Unit for planning Langley's supersonic facilities under the new Unitary Plan Act, and he was relieved of his position as head of the Full-Scale Tunnel Section. After the Unitary Supersonic Tunnel was constructed and put into operation, he was named chief of the Unitary Plan Wind Tunnel Division at Langley in 1955, and he remained in that position until the end of the laboratory's NACA years.

Herbert A. "Hack" Wilson, Jr., was the fourth head of the Full-Scale Tunnel. (NASA L-03265)

In 1960, Wilson was appointed to head a panel that was responsible for the problems surrounding the reentry of spacecraft into Earth orbit. He later headed Project FIRE (Flight Investigation Reentry Environment), which researched those problems and led to the development of heat shields for the Apollo spacecraft. He was appointed chief of the Applied Materials and Physics Division in 1964, overseeing research programs in space stations, launch vehicles, and rocket research. Wilson was appointed Langley's assistant director for space in 1970. He retired from NASA in 1972 but remained active in space and engineering work. Hack Wilson was also an expert ocean sailor, and many in the local community remembered him as an outstanding, confident sailor. During the 1970s, he served as executive secretary at the National Academy of Engineering in Washington, DC, and as a consultant to the Department of Energy. He died in Newport News, VA, at the age of 78 in 1992.

Personalities of the Full-Scale Tunnel in the 1950s and 1960s

Gerald W. Brewer

Gerald W. "Jerry" Brewer served as the fifth head of the Full-Scale Tunnel from 1948 to 1958. Despite his long tenure as manager of the facility (10 years), documentation of Brewer's career is very sketchy and few details are available. His projects at the Full-Scale Tunnel in the 1940s included testing of the Northrop MX-334 glider and America's first jet, the Bell YP-59A. Brewer left the Full-Scale Tunnel in 1959 and joined the Space Task Group, becoming the head of the Flight Control Branch during Project Mercury. He later was a member of the Lunar Orbiter Project Office.

Gerald W. Brewer rose through the ranks of researchers to become head of the Full-Scale Tunnel. (NASA L-60448)

John P. Campbell, Sr.

John P. Campbell, Sr. (1917–2007), was a native of Scottsboro, AL. He attended the U.S. Naval Academy from 1935 to 1937 and received his bachelor's degree in aeronautical engineering from Alabama Polytechnic Institute (now Auburn University) in 1939. He joined the NACA at Langley as a junior aeronautical engineer on November 15, 1939, at the Langley 12-Foot Free-Flight Tunnel. After 5 years of leading-edge research on dynamic stability and control using free-flight models in the facility, Campbell was named head of the Free-Flight Tunnel on October 2, 1944. At the age of 27, he became the youngest Langley employee to become a facility head. He personally led exploratory free-flight testing in the Full-Scale Tunnel in the early 1950s and successfully advocated for his group to take over the facility in 1958. Campbell became the sixth head of the Full-Scale Tunnel in 1959.

John P. Campbell, Sr., was an internationally recognized expert in V/STOL technology. (NASA L-07490)

John Campbell was an internationally recognized expert in the field of V/STOL aircraft technology and held a patent on the externally blown jet-flap concept, which is used on today's military C-17 transport. He was an exceptional technical writer and was widely respected for his writing clarity and leadership in adhering to high-quality technical reports.

After retiring as a Langley division chief in 1974, he spent the next 14 years as a gifted professor for George Washington University teaching graduate studies in aeronautics onsite at Langley. His son, Richard L. Campbell, became a noted NASA engineer at Langley in the field of computational fluid dynamics. Campbell passed away in February 2007 after a long illness.

Marion O. McKinney, Jr.

Marion O. "Mac" McKinney, Jr. (1921–1999), was a native of Chattanooga, TN. He received a bachelor's degree in aeronautical engineering from Georgia Tech before joining the Langley staff in 1942 at the Free-Flight Tunnel. He was a specialist in dynamic stability and control, working closely with Charles Zimmerman and John Campbell during studies using free-flight models. McKinney became Campbell's protégé, and after the group moved to the Full-Scale Tunnel he became head of a section conducting research on V/STOL aircraft concepts. He was awarded the coveted Wright Brothers Medal in 1964 for his work on the aerodynamics of V/STOL aircraft. After Campbell was promoted to assistant division chief in 1962, McKinney was selected as head of the Full-Scale Tunnel. His career as head of the tunnel was the longest (12 years) in the tunnel's history.

Marion O. "Mac" McKinney, Jr., had the longest tenure as head of the Full-Scale Tunnel. (NASA L-07490)

McKinney's personality was as different from that of John Campbell as night is from day. Whereas Campbell was a soft-spoken technical manager and sensitive supervisor of his personnel, McKinney was a blunt, demanding individual who struck fear into the hearts of young engineers. He was a brilliant technical leader, quick to grasp the potential of new aircraft concepts and knowledgeable of the "real world" constraints that had to be resolved before new technology could be accepted. At the same time, his dictatorial style of personnel management intimidated most of the staff.

Mac McKinney demanded a full and productive workday from every member of his organization. He always left work with a clean desktop and expected others to be as aggressive toward their work load as he was. Every day at lunch time, McKinney could be found engaged in a serious game of bridge in a conference room directly below the return passage of the Full-Scale Tunnel. During his brief lunch break he listened carefully for the sounds of the tunnel running and always wanted to know, "Why isn't the tunnel running?"

McKinney retired in 1980 as the assistant chief of the Subsonic-Transonic Aerodynamics Division, and he died in 1999 in Hampton, VA, after a short illness.

Personalities of the Full-Scale Tunnel from the 1970s to the end of NASA Operations

Joseph R. Chambers

Joseph R. Chambers (1940–) is a native of Houma, LA. He graduated from Georgia Tech in 1962 with a bachelor's degree in aeronautical engineering and from Virginia Tech with a master's degree in aerospace engineering in 1964. He was hired by the NASA Langley Research Center in 1962 and was assigned to the Full-Scale Tunnel for studies of aircraft dynamic stability and control. Chambers became head of a research section at the tunnel using piloted simulators and analytical studies for evaluations of V/STOL aircraft, parawing vehicles, and reentry lifting bodies. In 1969, he began a major program on high-angle-of-attack behavior of fighter aircraft that continued for over 40 years until the tunnel was closed. He became the eighth head of the tunnel in 1974 and remained in that position until 1981, when he was promoted to an assistant division chief assignment. Chambers headed the Langley Flight Applications Division from 1989 until 1994, when he became chief of Langley's Aeronautics Systems Analysis Division. He retired in 1998 after a 36-year career at Langley.

Joseph R. Chambers in 1993. (NASA L-04332)

Joseph L. Johnson, Jr.

Joseph L. Johnson, Jr. (1922–2014), was born in Atlanta, GA. He earned a bachelor's degree in aeronautical engineering from Georgia Tech in 1944 and was hired by the NACA for research in the Langley Free-Flight Tunnel under John P. Campbell. After conducting 14 years of research on the dynamic stability and control of aircraft, he moved to the Full-Scale Tunnel in 1959 with Campbell's group. Johnson worked primarily for John Paulson, Sr., on non-V/STOL aircraft and reentry vehicles. He became the assistant branch head when Joe Chambers became head of the branch in 1974; and when Chambers was reassigned to assistant division chief in 1981, Johnson was selected to become the ninth head of the Full-Scale Tunnel—and the fourth head of the Full-Scale Tunnel to graduate from Georgia Tech. During his entire 46-year career he remained a member of the same organization and only worked in two buildings: the Free-Flight Tunnel and the Full-Scale Tunnel.

Joseph L. Johnson in 1963. (NASA L-02043)

He was nicknamed "J.J." by his associates and was highly respected at Langley for his technical expertise and outgoing personality. Usually seen with a beat-up cigar in his mouth, he always inspired his coworkers with an attitude of "Why not?" Johnson never hesitated to promote first-ever research on radical new configurations and concepts, and he was a master at conducting "jack-leg" research in which unauthorized models suddenly appeared for quick-look testing by two-person crews in the 12-Foot Low-Speed Tunnel. He would often assign co-ops and a mechanic to such tests, with a philosophy that, "If the idea doesn't work, we'll just send the co-op back to school and never mention the test to management!"

Johnson thoroughly enjoyed one-on-one sessions with the lead engineers from industry and they reciprocated with long-term friendship. For example, in the 1970s, Chambers and Johnson established a close working relationship with the famous designer Elbert "Burt" Rutan in which Johnson shared Rutan's excitement for the unconventional. Johnson passed away at home in February 2014 after a long illness.

Luat T. Nguyen

Luat T. Nguyen (1947–) received undergraduate and graduate degrees in aeronautics and astronautics at the Massachusetts Institute of Technology, and became a member of Joe Chambers's Simulation and Analysis Section at the Full-Scale Tunnel in 1970. He never conducted a research test on a full-scale aircraft in the Full-Scale Tunnel, but he participated in free-flight model tests and used aerodynamic data derived from the wind tunnel for inputs to his computer-based studies.

Luat T. Nguyen in 2007. (NASA L-01570)

Nguyen spent his first 15 years at Langley as a research engineer in the area of flight dynamics and control. Using aerodynamic data and the results of free-flight tests in the Full-Scale Tunnel, he developed high-angle-of-attack control technology that was applied to front-line U.S. fighter aircraft such as the F-14 and F-16. He worked diligently to maintain the tunnel's traditional role in supporting the development of critical Navy and Air Force aircraft, including the F-14, F-16, F-18, F-22, and B-2. Based on his leadership and technical expertise, he was selected as assistant head under Joe Johnson and then became the 10th head of the Full-Scale Tunnel in 1990. Nguyen led various projects within the NASA High-Angle-of-Attack Technology Program, including as the chairman of the Intercenter Steering Committee. In 1992, he began a series of management assignments of increasing importance. From 2002 to 2004, he served as the Return to Flight Manager for the NASA Hyper-X/X-43A hypersonic research vehicle after it suffered a catastrophic accident in an earlier flight. As a result of his management, highly successful flights at Mach 7 and Mach 10 were subsequently made, resulting in two world speed records for air-breathing vehicles.

In 2007, Nguyen was assigned to the position of director of the Flight Projects Directorate at Langley. In that position, he led the Ares I-X Systems Engineering and Integration and Crew Module/Launch Abort System projects that were key elements of the very successful Ares I-X flight in November 2009. He also made critical contributions to the Orion Launch Abort System and Flight Test Article projects, culminating in the highly successful Pad Abort-1 flight in May 2010. He retired from NASA in 2013.

Dana J. Dunham

Dana J. Dunham was the last NASA manager of the Full-Scale Tunnel. (Dana Dunham)

Dana J. Dunham (1946–) graduated from Huntingdon College with a bachelor's degree in mathematics and French in 1967, and from Auburn University with a master's degree in mathematics in 1970. She began her career at Langley in 1970 conducting analytical and computational fluid dynamics studies in a variety of areas, including spacecraft trajectories and shock-interference patterns and aircraft wake-vortex minimization. She participated in studies of the effect of heavy rain on aircraft aerodynamics and experimental tunnel studies of advanced turboprop configurations. She also conducted research on predicting the wakes behind aircraft for assessing aerial dispersant patterns. Her various projects included conducting experiments in the Langley 14- by 22-Foot Tunnel to validate prediction methodology.

In 1990, Dunham was selected to be the assistant branch head to Luat Nguyen for the Flight Dynamics Branch at the Full-Scale Tunnel, and she later became the 11th head of the Full-Scale Tunnel in 1992. She was the only female head of the organization. If Joe Johnson is considered an "insider" who came up through the field of dynamic stability and control to head the organization, then Dunham was a true "outsider" because she came from another organization and, in fact, had not been educated in the world of dynamic stability. In the author's opinion, Dunham was much like Abe Silverstein when he joined the Full-Scale Tunnel without formal training in aerodynamics—and like Silverstein, she persevered and adapted to the expertise of the research organization. She was a superior supervisor, and her former staff members consider her to have been one of their most respected managers.

She was best known for her steady leadership of Branch activities during the upheavals and chaos of the 1990s leading up to the move from the Full-Scale Tunnel and the tunnel's decommissioning by NASA. Dunham retired from Langley in 2002.

Sue B. Grafton

Sue B. Grafton (1937–) graduated from Wake Forest College in 1958 with a degree in mathematics and entered duty at the NASA Langley Research Center immediately thereafter. Her first assignment was at the Langley Spin Tunnel as a participant in computer-generated predictions of aircraft spinning behavior. Although limited in any formal engineering background, she was personally driven to understand the testing methods and procedures in use at the Spin Tunnel and the neighboring Full-Scale Tunnel. As her interest in participating in "hands-on" experimental activities increased, she made her interests clearly known to Mac McKinney, who was the Branch Head in charge of both tunnels at the time.

Sue B. Grafton maintained an exceptional knowledge of the Full-Scale Tunnel and its operations for over 45 years. (NASA L-94-01649)

In 1964, Joe Chambers returned from graduate school to lead a new section at the Full-Scale Tunnel focused on the analysis of dynamic stability and control of aerospace vehicles using data derived from model tests, including the use of dynamic force-test methods. Grafton was transferred to the new section with instructions from McKinney to "watch and learn" under Chambers's tutelage. Her intense interests expanded to learning about all facets of the projects conducted in the tunnel, from fabrication of models to operating the unique hardware and software required for static and dynamic force tests. In her continuing on-the-job training, she became adept at conducting all aspects of free-flight model tests. By the 1970s, Grafton had led several projects, written technical reports, and established a rapport with shop personnel and tunnel technicians that resulted in team efforts and successful studies in the Full-Scale Tunnel. She became the main point of contact and leader in subscale testing within the branch in both the Full-Scale Tunnel and the 12-Foot Low-Speed Tunnel (the old Free-Flight Tunnel).

She coauthored many technical reports with others and was photographed with most of the free-flight models she directed in fabrication and testing. Grafton appeared in so many photographs that she became the public face of the Full-Scale Tunnel.

If, as suggested herein, Smith J. DeFrance is regarded as the "Father of the Full-Scale Tunnel," then Sue B. Grafton is the "Mother of the Full-Scale Tunnel." She has maintained an exceptional knowledge of the tunnel, its equipment, and the entire test program conducted in the facility, even after the branch was transferred to the Langley West Area. She remained in close contact with day-to-day operations in the tunnel for over 45 years—much longer than any other person in Langley's history—and she also had a common career fact with Joe Johnson in having worked in only one Langley organization. After retiring from NASA, she returned as a contractor and continued to play a key role in supporting the activities of her old organization.

Endnotes

1. Russ Robinson, interview by Walt Bonney, September 24, 1974, LHA, p. 15.
2. Ibid.
3. James R. Hansen, *Engineer in Charge: A History of the Langley Aeronautical Laboratory, 1917–1958* (Washington, DC: NASA SP-4305, 1987), p. 27.
4. Glenn E. Bugos, *Atmosphere of Freedom: Sixty Years at the NASA Ames Research Center* (Washington, DC: NASA SP-4314, 2000), pp. 25–26.
5. Ibid., p. 24
6. Smith DeFrance, interview by Walter Bonney, September 23, 1974, LHA, p. 12.
7. Russell G. Robinson, "Memoir for Three Sons, Toryn and Kra" (unpublished manuscript, 1997), Archives Reference Collection, NASA Ames History Office, p. 20.
8. Ibid.
9. Manley Hood, interview by Walter Bonney, September 24, 1974, LHA.
10. Jim Hodges, "Technical Writing Guru Katzoff Dies at 101," *Langley Researcher News* (September 28, 2010).
11. Phil Walker, interview by author, June 29, 2010. Phil Walker is Joseph Walker's son.

About the Author

Joseph R. Chambers is an aviation consultant who lives in Yorktown, VA. He retired from the NASA Langley Research Center in 1998 after a 36-year career as a researcher and manager of military and civil aeronautics research activities. He began his career in 1962 as a member of the research staff of the Langley Full-Scale Tunnel, where he specialized in flight dynamics research on a variety of aerospace vehicles, including V/STOL configurations, parawing vehicles, re-entry vehicles, and fighter-aircraft configurations. In 1974, he became the head of the Full-Scale Tunnel, the Langley 20-Foot Spin Tunnel, and outdoor free-flight and drop-model testing. In 1989, he also became head of aircraft flight research at Langley in addition to his other responsibilities. In 1994, he was assigned to organize and manage a new group responsible for conducting systems-level analysis of the potential payoffs of advanced aircraft concepts and technology to help guide NASA research investments.

Mr. Chambers is the author of over 50 NASA technical reports and publications, including NASA Special Publications SP-514 on airflow condensation patterns for aircraft, SP-2000-4519 on contributions of the Langley Research Center to U.S. military aircraft of the 1990s, SP-2003-4529 on contributions of Langley to U.S. civil aircraft of the 1990s, and SP-2005-4539 on Langley research on advanced concepts for aeronautics.

He has written several books for the NASA Aeronautics Research Mission Directorate, including SP-2009-575 on the development and application of dynamic free-flight models by the NACA and NASA (2010), and was also a contributor to SP-2009-570 on the contributions of NASA to aviation (2010).

He has made presentations on research and development programs to audiences as diverse as the von Kármán Institute in Belgium and the annual Experimental Aircraft Association (EAA) Fly-In at Oshkosh, WI, and has consistently shown the ability to address both a technical audience and the general public.

Mr. Chambers has served as a representative of the United States on international committees and has given lectures on NASA's aeronautics programs in Japan, China, Australia, the United Kingdom, Canada, Italy, France, Germany, and Sweden.

Mr. Chambers received several of NASA's highest awards, including the Exceptional Service Medal, the Outstanding Leadership Medal, and the Public Service Medal. He also received the Arthur Flemming Award in 1975 as one of the 10 most outstanding civil servants for his management of NASA's stall/spin research for military and civil aircraft. He has a bachelor of science degree from the Georgia Institute of Technology and a master of science degree from the Virginia Polytechnic Institute and State University (Virginia Tech).

Index

Numbers in **bold** indicate pages with illustrations.

Index

C

D

E

F

K

L

N

Q

R

S

T

U

W

X